D0938989

ANTARCTIC
RESEARCH
SERIES

American Geophysical Union

ANTARCTIC
RESEARCH
SERIES

American Geophysical Union

Volume 1 BIOLOGY OF THE ANTARCTIC SEAS
Milton O. Lee, *Editor*

Volume 2 ANTARCTIC SNOW AND ICE STUDIES
Malcom Mellor, *Editor*

Volume 3 POLYCHAETA ERRANTIA OF ANTARCTICA
Olga Hartman

Volume 4 GEOMAGNETISM AND AERONOMY
A. H. Waynick, *Editor*

Volume 5 BIOLOGY OF THE ANTARCTIC SEAS II
George A. Llano, *Editor*

Volume 6 GEOLOGY AND PALEONTOLOGY OF THE ANTARCTIC
Jarvis B. Hadley, *Editor*

Volume 7 POLYCHAETA MYZOSTOMIDAE AND SEDENTARIA OF ANTARCTICA
Olga Hartman

Volume 8 ANTARCTIC SOILS AND SOIL FORMING PROCESSES
J. C. F. Tedrow, *Editor*

Volume 9 STUDIES IN ANTARCTIC METEOROLOGY
Morton J. Rubin, *Editor*

Volume 10 ENTOMOLOGY OF ANTARCTICA
J. Linsley Gressitt, *Editor*

Volume 11 BIOLOGY OF THE ANTARCTIC SEAS III
Waldo L. Schmitt and George A. Llano, *Editors*

Volume 12 ANTARCTIC BIRD STUDIES
Oliver L. Austin, Jr., *Editor*

Volume 13 ANTARCTIC ASCIDIACEA
Patricia Kott

Volume 14 ANTARCTIC CIRRIPEDIA
William A. Newman and Arnold Ross

Volume 15 ANTARCTIC OCEANOLOGY
Joseph L. Reid, *Editor*

Volume 16 ANTARCTIC SNOW AND ICE STUDIES II
A. P. Crary, *Editor*

Volume 17 BIOLOGY OF THE ANTARCTIC SEAS IV
George A. Llano and I. Eugene Wallen, *Editors*

Volume 18 ANTARCTIC PINNIPEDIA
William Henry Burt, *Editor*

Volume 19 ANTARCTIC OCEANOLOGY II: THE AUSTRALIAN–NEW ZEALAND SECTOR
Dennis E. Hayes, *Editor*

Volume 20 ANTARCTIC TERRESTRIAL BIOLOGY
George A. Llano, *Editor*

Antarctic Terrestrial Biology

FRONTISPIECE

Lake Bonney, which lies in a deep trough left by the receding Taylor Glacier, is one of a half-dozen salty, partially ice-covered lakes found in the dry valley desert region fronting McMurdo Sound, Ross Sea. This southwest view looking up the valley shows the Kukri Hills reflected in the ice-free waters. The high point on the left is Sentinel Peak, adjacent to the Hughes Glacier. The upper end of Lake Bonney is hidden by the Bonney Riegel; the area that is visible is partially covered with permanent ice. Dark patches in the far background are cinder cones of volcanic origin. The large glacier midway between the cinder cones and Sentinel Peak is the Calkin Glacier, which backs up toward Mt. Coates. Almost no snow falls on the valley floor, which is barren except for a meager soil microbiota. Lichens and mosses may occur along the upper slopes of the valley walls where snow provides melt water. This zone of vegetation is marked on the Kukri Hills by a light mantle of freshly fallen snow. Photograph by G. A. Llano.

Volume 20

ANTARCTIC
RESEARCH
SERIES

Antarctic Terrestrial Biology

George A. Llano, *Editor*

Published with the aid of a grant from the National Science Foundation

PUBLISHER

AMERICAN GEOPHYSICAL UNION

OF THE

National Academy of Sciences—National Research Council

December 15, 1972

Volume 20 | ANTARCTIC RESEARCH SERIES

ANTARCTIC TERRESTRIAL BIOLOGY

GEORGE A. LLANO, *Editor*

1707 L Street, N.W.
Washington, D. C. 20036

Library of Congress Catalog Card No. 72-92709
International Standard Book No. 87590-120-4

List Price $30.00

Printed by
THE HORN-SHAFER COMPANY
DIVISION OF
Geo. W. King Printing Co.
Baltimore, Maryland

THE ANTARCTIC RESEARCH SERIES

The Antarctic Research Series is designed to provide a medium for presenting authoritative reports on the extensive and detailed scientific research work being carried out in Antarctica. The series has been successful in eliciting contributions from leading research scientists engaged in antarctic investigations; it seeks to maintain high scientific and publication standards. The scientific editor for each volume is chosen from among recognized authorities in the discipline or theme it represents, as are the reviewers on whom the editor relies for advice.

Beginning with the scientific investigations carried out during the International Geophysical Year, reports of research results appearing in this series represent original contributions too lengthy or otherwise inappropriate for publication in the standard journals. In some cases an entire volume is devoted to a monograph. The material published is directed not only to specialists actively engaged in the work but to graduate students, to scientists in closely related fields, and to interested laymen versed in the biological and the physical sciences. Many of the earlier volumes are cohesive collections of papers grouped around a central theme. Future volumes may concern themselves with regional as well as disciplinary aspects, or with a comparison of antarctic phenomena with those of other regions of the globe. But the central theme of Antarctica will dominate.

In a sense, the series continues the tradition dating from the earliest days of geographic exploration and scientific expeditions—the tradition of the expeditionary volumes which set forth in detail everything that was seen and studied. This tradition is not necessarily outmoded, but in much of the present scientific work one expedition blends into the next, and it is no longer scientifically meaningful to separate them arbitrarily. Antarctic research has a large degree of coherence; it deserves the modern counterpart of the expeditionary volumes of past decades and centuries which the Antarctic Research Series provides.

With the aid of a grant from the National Science Foundation in 1962, the American Geophysical Union initiated the Antarctic Research Series and appointed a Board of Associate Editors to implement it. A supplemental grant received in 1966, the income from the sale of volumes in the series, and income from reprints and other sources have enabled the AGU to continue this series. The response of the scientific community and the favorable comments of reviewers cause the board to look forward with optimism to the continued success of this endeavor.

To represent the broad scientific nature of the series, the members of the Board were chosen from all fields of antarctic research. At the present time they include: Avery A. Drake, Jr., representing geology and solid earth geophysics; A. P. Crary, seismology and glaciology; George A. Llano, botany and zoology; Martin A. Pomerantz, aeronomy and geomagnetism; Morton J. Rubin, meteorology and oceanography; David L. Pawson, biology; Waldo L. Schmitt, member emeritus; and Laurence M. Gould, honorary chairman. Fred G. Alberts, secretary to the U. S. Advisory Committee on Antarctic Names, gives valuable assistance in verifying place names, locations, and maps.

Morton J. Rubin
Chairman, Board of Associate Editors
Antarctic Research Series

PREFACE

These 13 original papers on terrestrial biological research initiate a companion to *Biology of the Antarctic Seas*, which now comprises four volumes of the Antarctic Research Series. The articles in this volume range in subject matter from the limnology, physiology, and ecology of aquatic systems to the taxonomy of fresh-water algae, lichens, mosses, fungi, protozoa, and land arthropods of Antarctica. It concludes appropriately, and for reasons stated below, with a paper on subantarctic rain forests.

This terrestrial volume brings together research papers that are less suitable for publication under existing Antarctic Research Series volumes and, as a consequence, reveals another dimension of the U.S. effort in antarctic biological research. Although the volume includes systematic and ecologic papers, the longer reports point to changes in the design and completion of field biological work. These changes are exemplified by greater use of increasingly sophisticated instrumentation and by emphasis on in situ experimental studies. This transition began as early as 1961–1962 when Goldman, Mason, and Wood, and also Koob and Leister, began their field work. To some extent, the rationale and methodology of these two groups stimulated the 1970 studies by Parker and his associates. Much of biological research involves graduate research assistants, and seven of these papers represent, in part or in whole, their contributions presented in partial fulfillment of requirements for higher degrees.

Three of the larger papers concern fresh-water lakes, and they reveal an unsuspected variety in antarctic aquatic systems. Goldman, Mason, and Wood pioneered in antarctic limnologic studies, giving their attention to the high latitude, coastal zone ponds and lakes of Ross Island. The work of Koob and Leister in the dry valleys brings out some of the unique characteristics of the perennially ice-covered lakes of the cold desert climate found in the dry valleys, one of the earth's natural wonders. Parker, Samsel, and Prescott report on lakes north of the Antarctic Circle and point to the need for more intensive investigations of the terrestrial and aquatic ecosystems of Antarctica.

Lichens and mosses, two major elements of antarctic vegetation, are discussed in the papers by Lange and Kappen, Robinson, Rastorfer, and Schofield and Ahmadjian. Except for Robinson's contribution, the primary emphasis is on photosynthesis, metabolism, physiology, and ecology. This emphasis reveals the strong interest in experimental biology for information on adaptation to polar conditions. Singer's contribution on a Basidiomycete new to Antarctica is of interest because the larger fungi are rare to the Antarctic. This paper is an extension of Singer's earlier studies on antarctic fungi.

Young's report on the subantarctic vegetation of the Magellanic region represents one research activity supported by RV *Hero* during the austral winter. Other studies on the cryptogamic vegetation of the Falkland Islands, Staten Island, Islas Juan Fernández, Campbell Island, Îles Kerguelen, and the Auckland Islands are basically complete. These subantarctic studies furnish a basis essential for the preservation and conservation of subantarctic island ecosystems and contribute to a better understanding of the vegetational relationships of Antarctica to southern land masses.

George A. Llano

CONTENTS

COMPARATIVE STUDY OF THE LIMNOLOGY OF TWO SMALL LAKES ON ROSS ISLAND, ANTARCTICA

Charles R. Goldman, David T. Mason, and Brian J. B. Wood[1]

Environmental Studies and Institute of Ecology, University of California, Davis, California 95616

Abstract. Algal and Skua lakes on Cape Evans illustrate the extremes in the variety of conditions found among the shallow bodies of water of the McMurdo Sound region. Skua Lake is linked to the productivity of the sea by avian fertilization and responds to this organic and inorganic enrichment by a seasonal phytoplanktonic bloom comparable in productivity to that of eutrophic temperate waters. The water becomes highly colored and turbid and is rich in both sestonic and dissolved organic matter. Its phytoplanktonic production is almost 10 times as great per unit area as that of Algal Lake. A fine black sediment is deposited in the deeper parts of Skua Lake, and a felt of filamentous algae covers the shallower regions. This felt is dense and remains attached to the gravels from one year to the next. The water of Algal Lake is clear throughout the season and contains relatively few planktonic forms. The benthos grows thicker than it does in Skua Lake, and large portions of it may be removed from the water and piled up to the lee of the lake by wind action each season. The chemical characteristics of these small lakes also reflect both their proximity to the ocean and the kenyte lava surrounding them. Salts are introduced into the lakes by winds and by surface and subsurface waters, Skua Lake receiving an additional contribution from birds. During the year the concentrations and the distribution of ionic substances are also altered by evaporation, sublimation, leaching, melting and freezing, and microbial mineralization. The snows were virtually free of detectable trace element contamination. Water temperatures, which are closely dependent on insolation, fluctuate greatly. A maximum water temperature of 9°C was observed during the year. During periods of bright light, water temperatures are usually highest at the bottom, where much of the radiant energy is absorbed and converted to sensible heat. The predominant biological factor in the environment of the two lakes is the skua. Its activities remove nothing from the cape and bring to it some of the organic production and trace element concentration of the fertile Ross Sea. The rates of production and the resulting efficiencies for the biological productivity of each lake are remarkably similar. The benthic production is more important than the planktonic production in these shallow lakes. The benthic community thrives probably because of the comparative warmth of the felt, where temperatures may be 10°C higher than those in the overlying water. The solar heating of the benthos early in the season also opens large productive areas before a truly planktonic community can exist. In addition the benthos has developed an upper protective layer from the adverse effects of inhibitory light intensities. The over-all planktonic production and chlorophyll estimates indicate that Skua Lake is about 8 times more productive than Algal Lake per unit volume, whereas other sestonic crop estimates indicate that Skua Lake is only 5 times more productive. This relationship suggests that a greater portion of the total seston of turbid Skua Lake than of that of Algal Lake is actively producing. This phenomenon is perhaps largely attributable to the greater light inhibition in the clear water of Algal Lake.

Lakes are particularly attractive ecologic units for study, since they provide a biological integration of an entire drainage system. A large part of the solar energy they receive is involved in abiotic thermal exchanges, whereas a very small part is channeled as biochemical energy through the living aquatic organisms. The metabolic activities of the living organisms are to varying degrees dependent on the quantity and the quality of incoming radiation and the resulting thermal conditions. Chemical and physical factors peculiar to each lake provide unique conditions that must be studied individually and that add much to the fascination of limnologic research.

At the environmental extremes resulting from high latitude or altitude these processes may display a remarkable clarity. This paper reports on the investigation of two small antarctic lakes on Cape Evans (Figure 1). The field work was accomplished during

[1] Now at Department of Applied Microbiology, Strathclyde University, Glasgow Cl, Scotland.

Fig. 1. The study area, as seen in an oblique aerial photograph of Cape Evans and adjacent McMurdo Sound. The camera was facing southwest, and Scott's hut is visible on the beach at the right (arrow); Skua Lake is in the center, and Algal Lake is on the left. The photograph was taken in late January 1962 after the winter sea ice had broken up and the lakes had refrozen. Official U.S. Navy photograph.

the austral summers of 1961–1962 and 1962–1963. The work reported here was directed mainly toward understanding the flows of energy through the unique physical and biological conditions found in antarctic inland waters, but it includes many elements of survey and experimental work often extending beyond the water's edge for a more thorough understanding of the basic fertility and ecology of the phytoplankton and the periphyton found in these unusual lakes.

Most of Antarctica is a biological desert, but in regions near the fertile sea, where the rigorous climate is moderated, aquatic life develops rapidly during the brief antarctic summer. Between November and February some of the shallow fresh-water lakes of the McMurdo Sound area are completely or partially melted and support a surprising diversity of biological activity. The two lakes that we studied are located near the site of R. F. Scott's 1913 base camp on Cape Evans (77°38′S, 166°24′E), Ross Island, Antarctica. One was named Skua Lake by E. W. Nelson, a biologist on Scott's expedition, for the resident bird population (*Stercorarius skua maccormicki*); the other was named Algal Lake by the authors because of the striking mat of blue-green

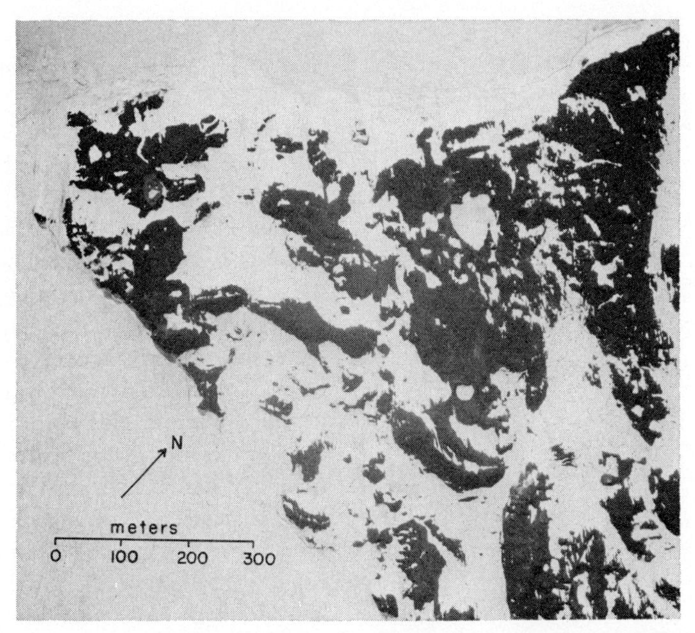

Fig. 2. Vertical aerial photograph of Cape Evans, Ross Island, Antarctica, from 915 meters (3000 feet) early in December 1961. Skua and Algal lakes were just beginning to melt (dark areas on western margins of ice). Other lakes and landmarks on the cape may be identified with reference to the 'sketch map' (Figure 3). Note that the configuration of snowdrifts indicates that the prevailing winds are from the south-southeast. Official U.S. Navy photograph.

algal remains around its leeward edge. Features of the area have been described in the published results of the 1910 British Antarctic Terra Nova Expedition, by Scott [1913], Taylor [1916], Debenham [1921], and Cherry-Gerrard [1930]. During the early expeditions certain aspects of the biology of the lakes were recorded [Taylor, 1916]. References are made to 'Flagellata' in the melt waters, and Taylor wrote, 'In the lakes a reddish plant akin to seaweed coats the bottom, and dries to a leathery wrinkled mass.' He also reported that the skuas used the lakes for bathing. However, the Scott party made no observations of the fresh-water life comparable to the excellent studies of Murray [1909] in Shackleton's expedition. The area investigated by the Shackleton group, Cape Royds, is <12 km from Cape Evans and is of similar topography. Many of the organisms collected on Cape Royds and at Hut Point (24 km away) by Murray [1909, 1910], West and West [1911], and Fritsch [1912] are found on Cape Evans. More recently Armitage and House [1962] made some preliminary limnologic investigations of antarctic lakes, including Skua Lake and two ponds (ponds 2 and 3). Aspects of phytoplanktonic radiation ecology have been treated by Goldman et al. [1963]. Survey work on invertebrates from the lakes of the area has been published by Dougherty and Harris [1963], and antarctic fresh-water ecology has been reviewed by Goldman [1970].

Cape Evans is a low triangular ice-free area composed almost entirely of black basaltic lava (originally described as kenyte by Jensen [1916] but apparently incorrectly so [Smith, 1954; Treves, 1962]) with an area of about 0.5 km^2 and a maximum elevation of 23 meters above the adjacent sea (Figure 2). Outcroppings of lava are interspersed with morainal material, and there are extensive aeolian gravel deposits in the lee of every hill. The prevailing winds have extremely variable velocity and are from the south-southeast [Simpson, 1923]. The scanty snow is distributed into patterns similar to those of the gravels.

The general glacial history of the McMurdo Sound area has been discussed by Péwé [1960]. He postulates that at present the ice is receding and that four separate advances have previously covered Cape Evans. It is probable that the present recession has opened virgin areas to biotic colonization. Llano [1959] estimates from studies of lichen that the cur-

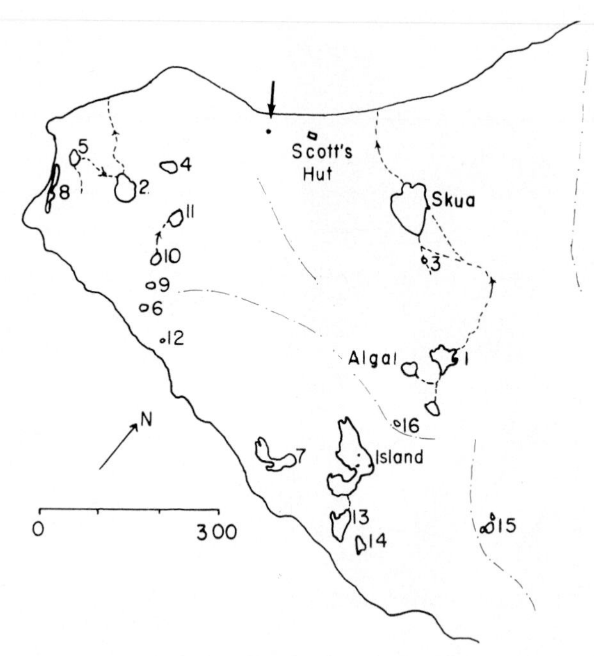

Fig. 3. Sketch map of Cape Evans. Skua, Algal, and Island lakes are indicated, and other lakes are numbered to correspond to the designations in Appendix 1. The dashed lines indicate drainage courses, and the dashed and dotted lines the approximate boundaries of drainage regions; the arrow indicates the field laboratory trailer used during this study.

rent exposure of Cape Evans may date back no more than 500 years.

Algal and Skua lakes are in the northern drainage (Figure 3) and comprise about one-third of the area of Cape Evans, which receives melt water from local snowdrifts and from the glacial ice to the east. During periods of melting and runoff several well-defined drainage patterns appear. These stream courses show up on aerial photographs (Figure 2) because they harbor a deposit of fine light colored clay and organic material. Precipitation and melt conditions vary greatly from year to year on the cape but in general are low. Precipitation at McMurdo Station to the south is reported to be 11.9 cm yr^{-1} [Rudolph, 1967], whereas the mean annual temperature at that station (1956–1961) was −17.8°C. Ball and Nichols [1960] have estimated 5 cm of water equivalent of precipitation per year at Marble Point across McMurdo Sound. They suggest that only about 2.5 cm yr^{-1} could be available for runoff because of high sublimation losses. Little new snow fell at the cape during the summer of 1961–1962, and no surface runoff was observed, although Scott [1913] and Ponting [1923] photographed and reported 'quite a rushing stream' emerging from Skua Lake in January 1911. Debenham [1921] also described a steady flow of water into even the smallest lakes of the cape in 1911. During December 1962 a storm left a few centimeters of snow on the cape, and a small amount of surface runoff fed into the lakes.

During our first visit to the area in November 1961 we noted certain striking differences between the two lakes. Skua Lake is the lowest and largest lake of the cape's northern drainage. It lies exposed in a broad flat depression and during wet summers drains by a short steep channel directly to the sea. Algal Lake is small and roughly circular and has an ice mound (hydrolacolith) in the center. Lying as it does at the foot of the central ridge of the cape, it is relatively protected from the wind. It receives runoff from a small area of snowdrifts nearby and, at high water, overflows into a second lake that in turn empties into Skua Lake. A thin orange cyanophycean 'felt' develops on the benthic gravels of Skua Lake to a depth of about 0.5 meter. Large areas of the winter ice are yellow-brown, whereas the ice of Algal Lake is clear to greenish and the bottom growths are thick and tend to be concentrated at the northwestern edge of the lake.

During the 1961–1962 season we concentrated our efforts on chemistry and general biology, whereas during the second field season we considered the heat budget and some of the physiological adaptations of the primary producers to high light intensities [Goldman et al., 1963]. Although our studies have centered around the general problem of biological productivity under the rigorous conditions of Antarctica, we have also considered a number of environmental variables, such as the study of the contribution of the skuas to the system, which, although peripheral to productivity itself, leads to a better understanding of the unusual ecosystems.

Brief descriptions and data from other lakes of the area will be found in Appendix 1. They are included to provide a basis for the comparisons and generalizations made from the data presented in the main body of this report.

METHODS

Morphometric measurements. Morphometric measurements were made during both seasons because the levels of the lakes changed from the one year to the next. The first season's data [Mason, 1963] are less accurate but were used for converting other measurements made that year because of the volume change. Estimates of the total melted volume were made from

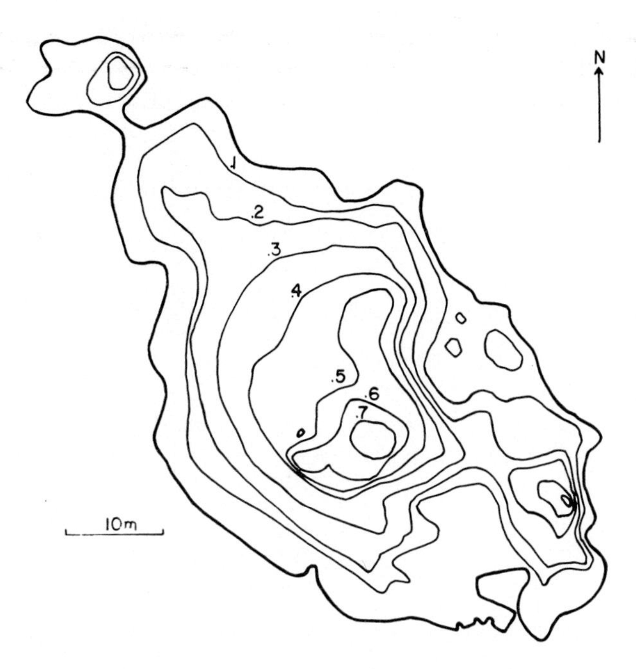

Fig. 4. Morphometric map of Skua Lake, Cape Evans. The lake was surveyed on February 17, 1963. The contour interval is 0.1 meter. The lake level had dropped several centimeters since the preceding season [Mason, 1963].

aerial photographs in combination with our morphometric maps. The second season's data were obtained by a transit stadia survey (Skua Lake, 312 soundings) and the transit-metered line method (Algal Lake, 231 soundings). The resultant maps (Figures 4 and 5) were treated by standard planimetric procedures to give the data in Table 1.

Incident light. A recording pyrheliometer (Belfort Instruments Co., Baltimore, Md.) was operated continuously throughout each season at Cape Evans to obtain values for total incident radiation. The manufacturer claimed 5% precision and ±0.1-l min^{-1} accuracy. Checks made during the 1962–1963 season with a 50-junction Eppley pyranometer at Cape Evans confirmed this degree of accuracy. The instrument was mounted on the top of the laboratory trailer and was unshaded by nearby objects. The incident light therefore closely resembled the light at Skua Lake; Algal Lake, however, lay at the northern foot of a steep hill and was partially shaded when the sun was low in the southern sky. This statement is supported by field records indicating the time of shading of Algal Lake and by quantitative data from simultaneous light recordings at both lakes on January 17–18, 1963, when Algal Lake in shadow received 29 fewer langleys than Skua Lake. This amount represents about 4% less light at Algal Lake than at Skua Lake. All recordings and experiments were set to standard time at McMurdo Station (Greenwich mean time plus 12 hours); solar noon consequently fell 54 min before noon local time.

Net radiation and meterology. A Fransilla-Suomi ventilated net radiometer powered by a 110-volt generator was mounted on a tripod 1.5–1.8 meters above the water or ice. It was calibrated with the Eppley pyranometer and used in conjunction with it for several short-term heat budget studies on Skua Lake. The meteorologic station occupied by the Scott expedition at Cape Evans was used to house a Belfort recording thermometer and a hair hygrometer. The thermograph was calibrated against a standard mercury thermometer, whereas the hygrograph was not altered from the manufacturer's specifications but conformed to daily wet–dry bulb measurements within ±5% relative humidity. The wind measurements at Cape Evans were made with a recently calibrated hand-held (2-meter) anemometer.

Light penetration. Light penetration in the lakes was measured with a Whitney underwater light meter having a Weston selenium photronic cell (Whitney Underwater Instruments, San Luis Obispo, Calif.).

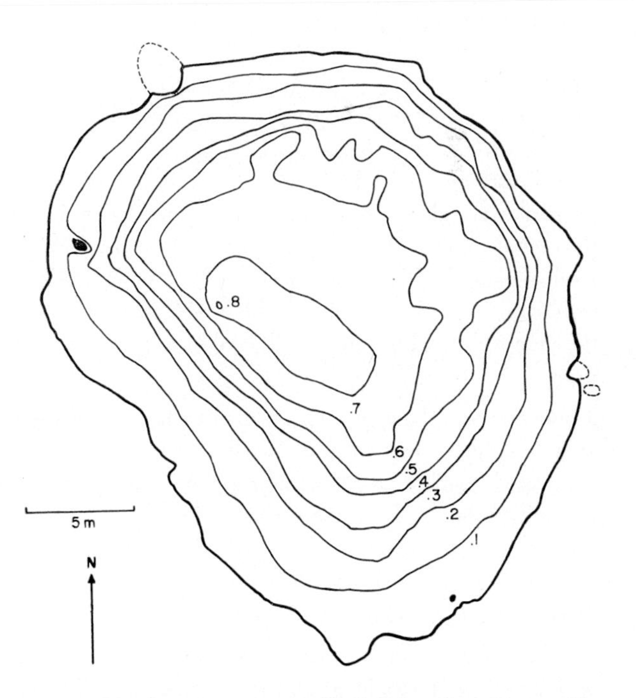

Fig. 5. Morphometric map of Algal Lake, Cape Evans. The lake was surveyed on January 29, 1963. The contour interval is 0.1 meter. The solid black areas and those outlined with dashes are large rocks.

TABLE 1. Comparative Morphometry of Algal and Skua Lakes, Cape Evans, McMurdo Sound, Antarctica, for the 1961–1962 and 1962–1963 Seasons

Parameter	1961–1962 Algal Lake	1961–1962 Skua Lake	1962–1963 Algal Lake	1962–1963 Skua Lake
Greatest length l, meters	36.5	99	26.8	78.0
Greatest breadth b max, meters	30.2	65.7	20.7	37.5
Mean breadth $\bar{b} = A/l$, meters	22	46	16.6	21.5
Maximum depth z max, meters	0.9*	1	0.8	0.75
Area A, m^2	810	4,600	446	1,680
Volume V, m^3	290	1,500	154	370
Shore line length L, meters	111	335	82	232
Average depth $\bar{z} = V/A$, meters	0.4	0.3	0.35	0.22
Development of shore line $D_L = \frac{1}{2}\,[L(A)^{-2}]$	1.1	1.4	1.1	1.6
Development of volume $D_V = 3\bar{z}/z$ max	1.2	1.0	1.3	0.88
Integrated melt area over season, m^2 days	34,700	156,900	31,950	128,300
Integrated melt volume over season, m^3 days	7,790	52,300	8,450	34,950
Average melt depth over season, meters	0.22	0.33	0.26	0.27
Maximum melt water volume, m^3	226	1,420	154	370
Maximum melt water area, m^2	722	4,370	446	1,680
Average depth at maximum melt, meters	0.3	0.3	0.35	0.22

* The maximum depth of Algal Lake may have been greater than this value since ice was still grounded in the center at the time of this measurement.

A 'season' was taken to be from November 26 to the end of February. The ice melt was relatively uniform and continuous in 1961–1962, and data on melting conditions for this season are therefore somewhat more accurate. Both lakes lost considerable amounts of water to evaporation between the two surveys. Maximum melt occurred in mid-January when the first year's soundings were made [Mason, 1963]. The lakes were surveyed on January 29 (Algal), and February 17 (Skua), 1963, with considerably more precision.

The equipment was calibrated for linearity at field temperatures and light intensities and was used with Schott colored glass filters 2 mm thick (BG 12, VG 9, and RG 1), which were placed under a diffusing glass. The data are expressed as per cents of the surface readings when all exterior surfaces have been wetted. The absorbency of lake water at a number of wave lengths was also measured with 10-cm cells in a Beckman model DU spectrophotometer in which freshly glass-distilled water was used as a blank. The cells were placed in contact with a diffusing plate at the phototube end to improve the collection of scattered light.

Temperature. Temperatures were regularly measured to 0.1°C accuracy with a Whitney thermistor unit calibrated against a standard mercury thermometer. A multiprobe Yellowsprings instrument with similar accuracy was infrequently used, whereas thermocouples buried in the lake ice were used for measurements early in the summer.

Conductivity. Conductivity was measured in situ with a Whitney bright platinum electrode probe combined with the thermistor unit. This instrument was calibrated against a standard KCl solution, and field measurements were corrected to 0°C by using the customary thermal coefficient of conductivity of 0.025. The coefficients determined for the two lakes (0.021 for Skua and 0.023 for Algal) were sufficiently close to the conventional value to justify its use. Laboratory measurements were made with an Industrial Instruments RC 16 B2 bridge also calibrated against KCl standards. A set of salinity hydrometers calibrated in parts per thousand (sea salts) was used for a few specific gravity determinations to corroborate the conductivity measurements.

Dissolved inorganic chemical analyses. A Beckman model N *p*H meter was used for both *p*H determinations and alkalinity titrations; in the alkalinity titrations, *p*H was monitored while 0.0100 *N* sulfuric acid was added to exhaust the natural buffering of the freshly collected water sample.

Oxygen and chloride methods followed the unmodified Winkler and Mohr procedures [American Public Health Association, 1960]. The procedure for divalent metals expressed as calcium also followed the EDTA (ethylenediaminetetraacetic acid) method of the American Public Health Association [1960] but substituted calmagite [Lindstrom and Diehl, 1960] as indicator. Dissolved nitrate was measured by the method of Mullin and Riley [1955]. The dissolved ammonia determination employed the Wattenberg method [Barnes, 1959] but used a 7.0-ml sample, a 0.5-ml Rochelle salt solution, 1.0 ml of the KOH solution, and 1.5 ml of Nessler reagent; readings made at 430 nm (nanometers) by the Beckman model B spectrophotometer were compared to a standard curve. Dissolved phosphates were estimated by the method of Greenfield and Kalber [1954].

Sestonic and dissolved organic compounds. Sestonic and dissolved organic compounds were determined by parallel methods. The seston was concentrated by filtration onto a magnesium carbonate prefilter pad, and the dissolved matter was concentrated when necessary by boiling. Sestonic organic carbon was determined by digesting samples with acid dichromate solution [Strickland and Parsons, 1960] and relating the decrease in extinctions to those given by a glucose standard. Organic nitrogen samples were digested in a boiling water bath with 0.4 ml of the sulfuric acid/mercuric chloride reagent of Harvey [1951] until the samples became colorless. One to three drops of 30% H_2O_2 were routinely added after 10 min of digestion. The bulk of the water needed for a final 10 ml of volume was then added followed by 0.5 ml of 30% weight/vol. Rochelle salt solution. Next 1.0 ml of 6.6 *N* KOH was added for dissolved organic nitrogen (2.0 ml for seston), and finally 1.5 ml of Nessler reagent [Umbreit et al., 1957]. After 5 min the optical density at 520 nm was read and compared to a standard curve prepared by using analytical grade ammonium sulfate.

Pigments. Seston was collected on weighed prewashed filters (Whatman GF/C), dried in the dark, weighed, and extracted with 90% redistilled acetone for 24 hours in the dark at 4°C. Little pigment remained in the seston after this procedure. Spectrophotometric measurements (Beckman model DU or B) of the absorbency of these extracts permitted evaluation of the chlorophyll *a* and carotenoid contents of the seston in standard pigment units. Disks of periphyton from the lake bottoms were cut, dried, weighed, and similarly extracted. The extracted disks retained some color and thus indicated incomplete removal of the pigment, but errors introduced here are believed to be no larger than those arising in sampling the heterogeneous substrate. The equations of Parsons and Strickland [1963] were used for computer calculation of the pigment values.

Plankton enumeration. Plankton samples from the lakes were preserved during the first season with 3–4% formaldehyde and during the second with acid Lugol's solution. The plankton were allowed to settle for 24 hours in small (3- to 5-ml) chambers and were then counted with an inverted microscope. Fifteen fields from each sample were counted, the magnification being varied so that >50 individuals of the common organisms were seen. Usually >100 individuals of each category were enumerated per sample.

Primary productivity. Measurements of carbon uptake by the sensitive ^{14}C method were employed both in the field and in the laboratory. The methods followed those outlined by Goldman [1963] for plankton samples. Calibration of ^{14}C and counting efficiency was done in the gaseous phase [Goldman, 1968]. Neutral density screens and thermostatically controlled water baths were used for physiological studies in the field. In these studies ^{14}C was always added to the whole water sample, mixed, and dispersed to the several bottles. A 'standard' carbon uptake assessment independent of field lighting and temperature conditions was performed routinely in the first year by returning water to the laboratory and incubating the water sample with $^{14}CO_2$ at 3°C and 300 ft-c (cool white fluorescence) for 4 hours. The carbon uptake of periphyton samples was determined in a similar fashion by using disks 2.14 cm^2 in area cut from the benthos with a plastic vial. The disks were incubated in situ or in the laboratory in clear and opaque 125-ml bottles containing lake water to which $^{14}CO_2$ had been added. After incubation the disks were drained and rinsed repeatedly with lake water and finally preserved with an extended rinse of 3% formalin followed by rapid desiccation. Portions of the dry disks were then combusted with van Slyke reagent, and the evolved $^{14}CO_2$ was measured in the gaseous phase [Goldman, 1963]. Calculations are based on the amount of carbon fixed per unit area of periphyton. For estimates of total carbon fixation the benthos was assumed to be active in open water to a depth of 0.5 meter in Skua Lake and 1.0 meter in Algal Lake. This assumption was based on our observation of the development of growths of periphyton in the two lakes.

RESULTS AND DISCUSSION

Lake morphometry and meteorology: morphometry. The morphometric maps made in 1963 appear as Figures 4 and 5. The volume and area of melted water were plotted against time (Figures 6–9). The morphometric calculations are summarized in Table 1. The 1962 values are larger, since the water level was

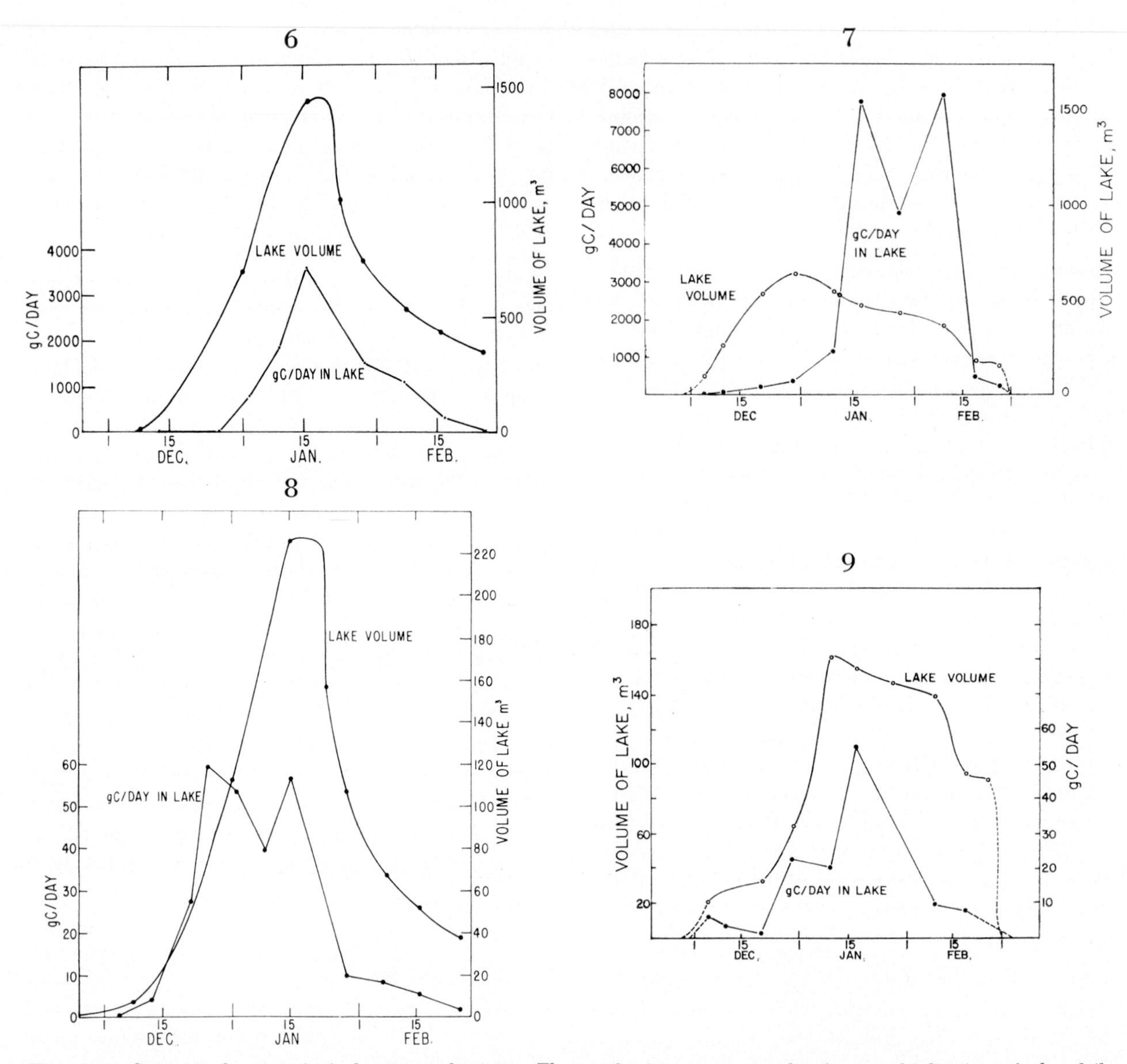

Figs. 6–9. Seasonal changes of planktonic productivity. The productivity curve results from multiplication of the daily rates by the corresponding approximate volume of the lake. Fig. 6, Skua Lake, 1961–1962 (daily rate is shown in Figure 10). Fig. 7, Skua Lake, 1962–1963 (daily rate is shown in Figure 11). Fig. 8, Algal Lake, 1961–1962 (daily rate is shown in Figure 10). Fig. 9, Algal Lake, 1962–1963 (daily rate is shown in Figure 11).

somewhat higher during that year; however, because of their relative inaccuracy they have been used only for calculations involving the 1961–1962 data. The hypsographic curve and the map of Skua Lake for the 1962 season show two stages of the melt pattern that have been discussed by Mason [1963].

The basic differences in the size and the basin shape of the two lakes are apparent from Figures 4–9 and Table 1. The area and the volume of Skua Lake are several times those of Algal Lake. Melting in Algal Lake left an ice 'island' roughly concentric with the contours, whereas melting in Skua Lake was elongate in the direction of the main axis of the lake and the prevailing wind.

Wind. Wind speeds are moderate in summer and except for February fall >15% below the yearly average speed of 22.4 km hr^{-1} at McMurdo Station [U.S. Naval Weather Research Facility, 1961]. The winds flowing across the cape are predominantly from the south-southeast, and their direction is clearly indicated by the resulting snowdrift patterns seen on the aerial photographs (Figure 1). Algal Lake lies di-

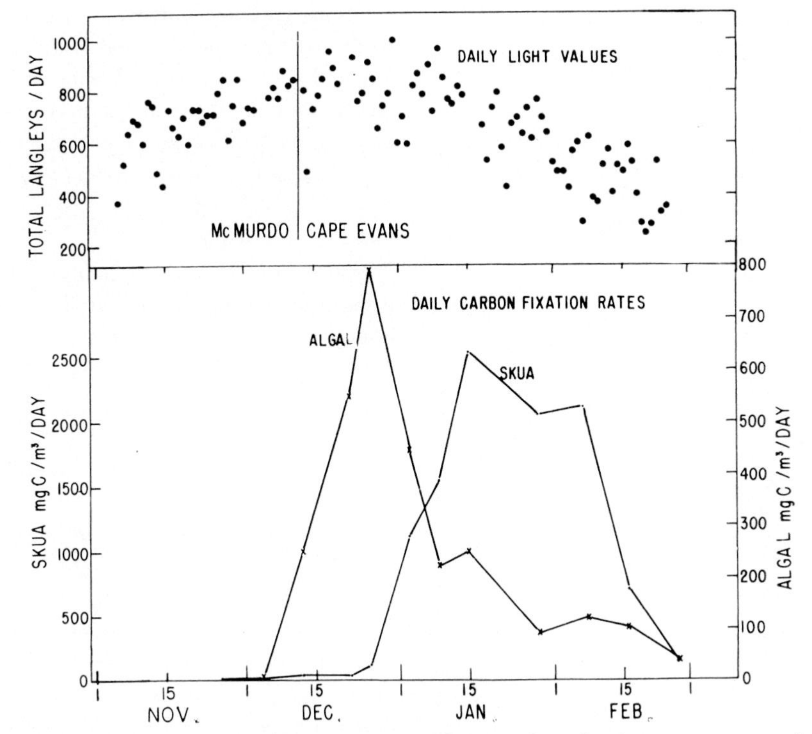

Fig. 10. Total daily insolation and planktonic carbon fixation rates measured during the 1961–1962 season. The recording pyrheliometer was moved from McMurdo Station to Cape Evans in early December. All carbon fixation rates were determined by the ^{14}C method in situ and corrected to whole-day values by means of independent diel experiments [Goldman, 1963].

rectly in the lee of the tallest ridge on the cape (24 meters above sea level) and is therefore exposed to less severe wind action than Skua Lake. Several parallel anemometer readings at Skua Lake, Algal Lake, and the field laboratory indicate that Skua Lake had velocities about 13% higher than those of the other two localities. On the basis of 81 paired observations through the summer of 1962–1963 the wind at the Cape Evans laboratory averaged 0.9 km hr^{-1} faster than that at McMurdo Station. The average wind velocity observed at Cape Evans was 17.5 km hr^{-1} during December, 16.7 km hr^{-1} during January, and 21.6 km hr^{-1} during February.

Solar energy. Solar radiation was continuous from early November to late February. Integrations of the daily pyrheliometer tracings are plotted for both seasons (Figures 10 and 11) and show an apparent seasonal maximum in late December. The monthly totals were (1961–1962) 17,340 ly in November 1961, 24,160 ly in December 1961, 22,720 ly in January 1962, and 12,840 ly in February 1962 and (1962–1963) 25,883 ly in December 1962, 23,583 ly in January 1963, and 10,555 ly in February 1963. This total short-wave radiation of 60,000 ly in the 1962–1963 season at Cape Evans compares reasonably with a value of 57,160 ly at Scott Base for the same 3-month period in 1957–1958 [Thompson and MacDonald, 1959]. Insolation in the 4 summer months provides >80% of the yearly total of incoming radiation and must be considered largely responsible for melting the lake ice, since the average air temperatures at Cape Evans during December, January, and February of 1962–1963 were −4.6°, −2.8°, and −10.3°C, respectively (Figure 11).

Limnomicroclimatology. Melting due to (sublacustral) solar heating began beneath the relatively shallow transparent ice and produced extensive pockets of water insulated from atmospheric contact by several centimeters of ice [Goldman et al., 1967]. The subsurface capillary movement of water through terrestrial gravels was possible even early in the season, owing to solar warming of the dark substrate [Llano, 1959]. Kelly and Zumberge [1961] have reported rock surface temperatures as much as 17°C higher than air temperatures.

An example of this lacustral microclimatic hetero-

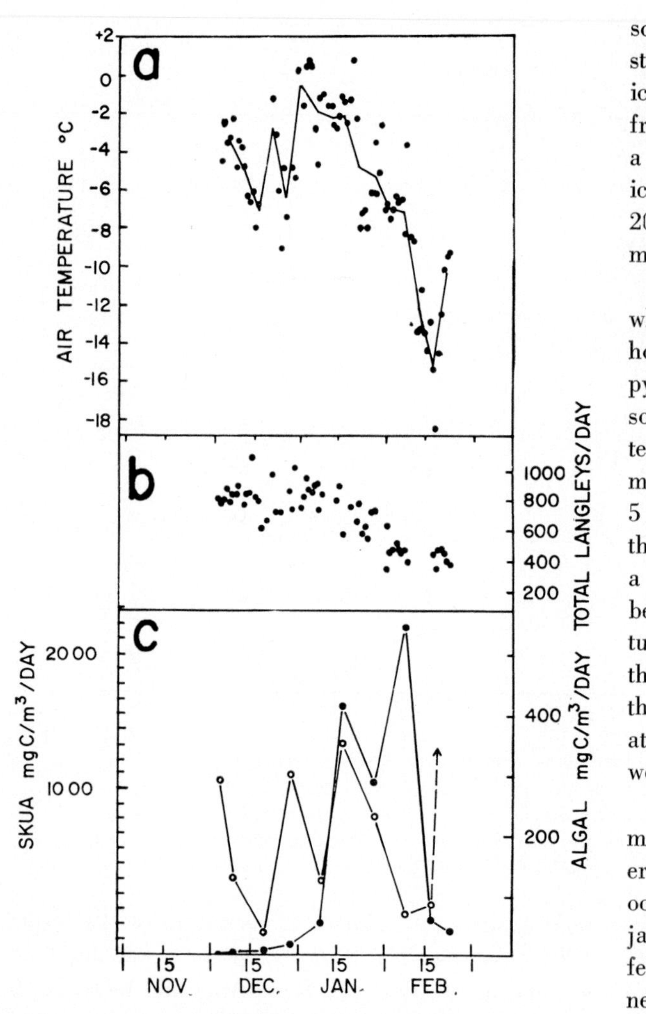

Fig. 11. Total daily insolation and planktonic carbon fixation rates measured at Cape Evans during the 1962–1963 season. (*a*) The air temperature, which was continuously monitored at 1.5 meters and planimetrically integrated to give the daily average values plotted; 5-day averages are connected with a solid line. (*b*) Daily light values; the pyrheliometer was maintained continuously at Cape Evans. (*c*) Daily carbon fixation rates, all of which were determined by the ^{14}C method in situ and corrected to whole-day values by means of independent diel experiments [Goldman, 1963] (open circles are Algal Lake values and solid circles Skua Lake values). Note the changes of scale from Figure 10 for carbon fixation.

geneity is seen in Figures 12 and 13, which show the results of an experiment in which thermistor probes were placed on the substrate and in open water at several depths. Heat sources of the system were recorded by simultaneous measurements of air temperature, insolation, and net radiation. On the first occasion of study the lake was covered with the previous winter's ice, and on the second occasion with new ice that had frozen to a depth of about 6 cm during a snowstorm a few days previously. Only a small island of winter ice remained in the center of the lake on December 20; a few centimeters of water were exposed at the margins both times.

December 10 was virtually cloudless until noon, when a thin overcast appeared (Figure 12). By 2100 hours the sky was nine-tenths occluded, but the pyrheliometer still showed a smooth tracing. South to southeast winds varied from 4 to 7 m sec^{-1}. Water temperatures rose sharply (0.6°C hr^{-1}) through midmorning and fell again in the afternoon. The water 5 cm below the surface was often 2°–3°C cooler than the adjacent dark benthos, whereas the open water at a depth of 15 cm was generally <1°C cooler than the benthos at the same depth. The maximum temperature measured in the open water was 4.6°C, whereas that on the benthos rose to 7.5°C. At the same time the air temperature remained below −1.8°C. Temperatures on moist ground 25 cm from the lake shore were as high as 6.7°C.

The weather on December 20 (Figure 13) was marked by considerable cloudiness (average sky cover being 60%), calm or only very light breezes, and occasional light snow. The water temperatures adjacent to the substrate rose by >11°C at midday and fell sharply as incoming and net radiation dropped near 1800 hours. The temperature changes at greater depths were less pronounced. The air temperatures at 1.5 meters in the Scott Base weather station closely paralleled the changes of insolation after perhaps an hour's lag. The open water slowly gained slightly less than ½°C during the 24-hour period.

December 21 was much less cloudy except for a short period in midmorning, and the winds were moderate from the east until 1400 hours, when they began blowing from the north. This shift, which brought the full length of the melted portion of the lake into effective wind exposure, apparently caused a breakdown of stratification both in the open water and in the substrate microzone. However, the open water temperature changes are roughly symmetrical about 4°C, and consequently little or no mixing energy may have been involved in the recorded tempera-

Figs. 12–13. Diel changes in open water and benthos ('felt') temperatures together with air temperature, insolation, and net radiation values. Note the considerably higher temperatures adjacent to the dark benthos; this area provides a significantly warmer microzone for benthic plants and animals. Fig. 12, December 10, 1962. Fig. 13, December 20 and 21, 1962.

12

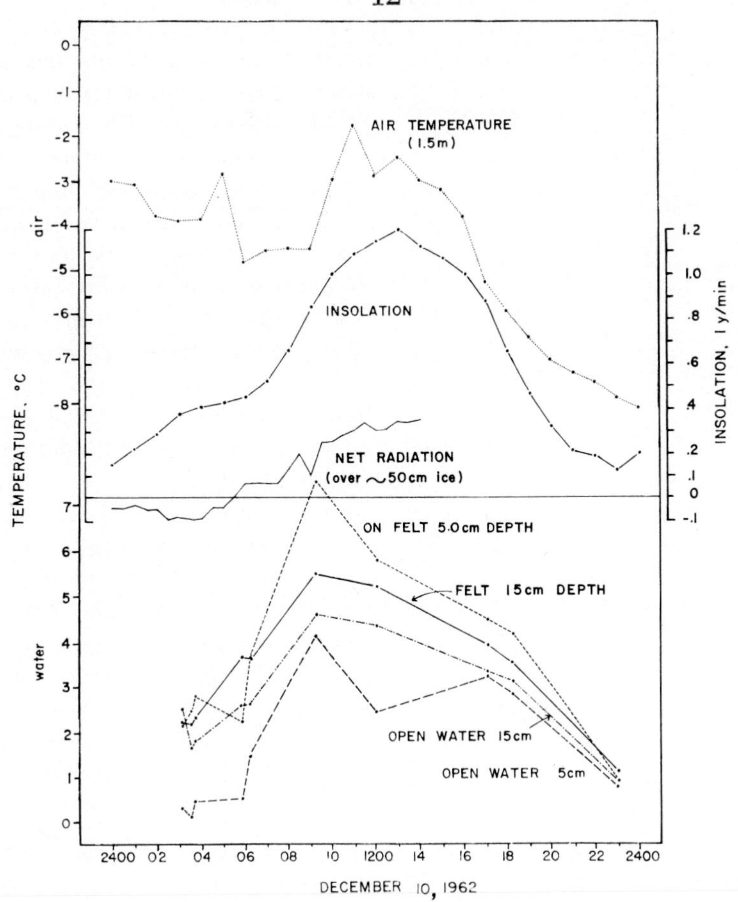

AIR TEMPERATURE (1.5m)
INSOLATION
NET RADIATION (over ~50cm ice)
ON FELT 5.0cm DEPTH
FELT 15cm DEPTH
OPEN WATER 15cm
OPEN WATER 5cm
TEMPERATURE, °C
air
water
INSOLATION, ly/min
2400 02 04 06 08 10 1200 14 16 18 20 22 2400
DECEMBER 10, 1962

13

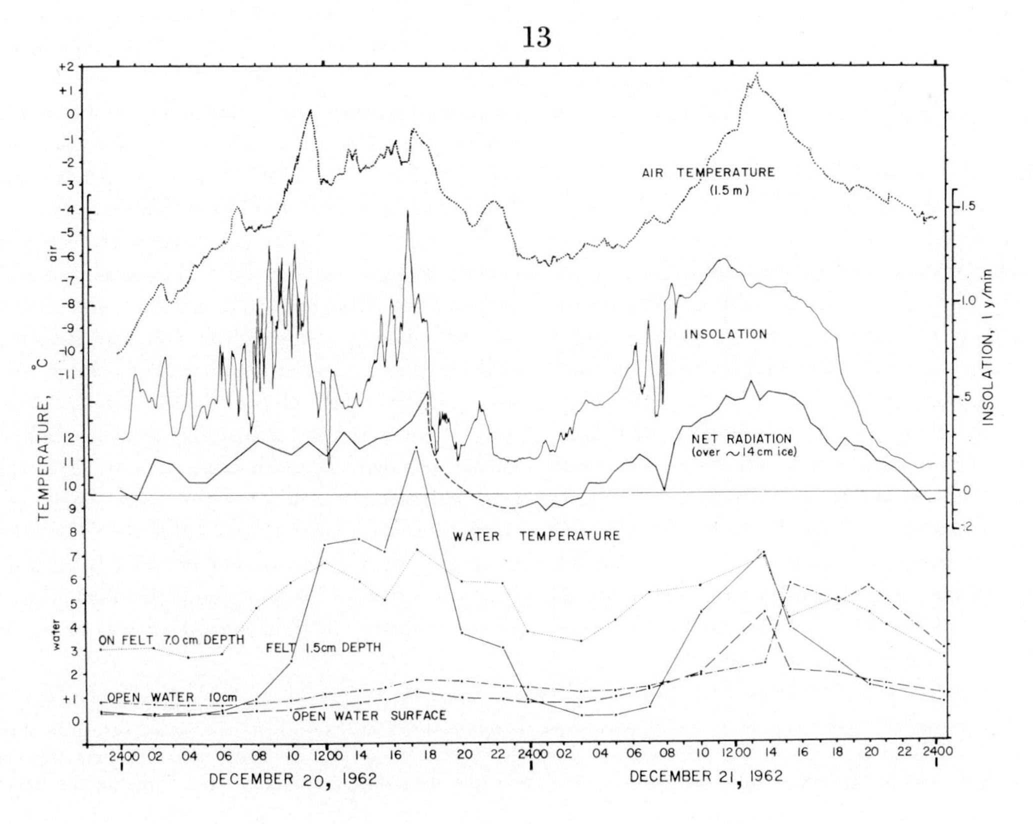

AIR TEMPERATURE (1.5m)
INSOLATION
NET RADIATION (over ~14cm ice)
WATER TEMPERATURE
ON FELT 7.0cm DEPTH
FELT 1.5cm DEPTH
OPEN WATER 10cm
OPEN WATER SURFACE
TEMPERATURE, °C
air
water
INSOLATION, ly/min
DECEMBER 20, 1962
DECEMBER 21, 1962

TABLE 2. Ice Albedo at Skua and Algal Lakes

Date	Time, hours	Description of Substrate	Average Albedo, %
		Skua Lake	
Dec. 9, 1962	1300	8 cm water at lake edge	17
		white bubbly ice, 2 years old	56
		rough yellow ice	37
		smooth ice	46
		bubbly ice	44
Jan. 30, 1963	1240	recently frozen ice, dark	39
Feb. 18, 1963	1645	refrozen ice, dark with some bubbles	22
		refrozen ice, dark with few bubbles	14
		Algal Lake	
Dec. 9, 1962	1300	white bubbly ice, 2 years old	52
		typical ice, rough	40
Feb. 18, 1963	1655	refrozen ice, few bubbles	14
		refrozen ice, no bubbles, transparent	13

The albedo was measured at a height of 1 meter with an Eppley pyranometer.

ture shifts. An over-all temperature rise occurred again the second day, whereas the total heat storage for the entire period was very close to the integrated 2-day net radiation surplus (53 ly). This result suggests that exchanges due to evaporation, heat of fusion, and conduction played only minor roles in the heat budget of the lake during this period.

Patterns of net radiation. The insolation and net radiation data from seven separate observations at Skua Lake are presented in Figure 14, the net radiation data being transformed to show total back radiation from the lake. A significant shift in the patterns of back radiation occurred after the previous season's ice and the intervening slush ice had melted. The midday back radiation on December 21 from 0600 to 1800 hours was 62% of the incoming radiation, whereas that on January 19 from 0600 to 1800 hours was only 41% of the incoming radiation. This shift reflects albedo changes (Table 2) from the old and turbid ice to the water of clear new ice. The radiation budget through the summer season was thus modified by reflectance changes of the lake itself. The peak heating effect in mid-January is probably due more to this decrease of reflection than to the actual accumulation of heat in these small bodies of water. Additional evidence for this conclusion comes from the erratic nature of the melting–freezing process, which was noted especially the second year. As a consequence the thermal climate of these shallow lakes appears to be buffered little by the thermal capacity of the water and the adjacent basin materials but to depend highly, as it does in terrestrial environments, on the effects of incoming radiation.

Algal Lake differed from Skua Lake in the flux of heat needed for melting. Of the 26,000 ly incident on the cape in December 1962 we calculate that about 10,000 ly were needed to melt and to sublime the water of Skua Lake. Since net radiation over the ice of this lake for the month was estimated from our data to be only 7300 ly, we suspect a significant fraction of the water loss may have occurred by lateral percolation into the gravels at the lake margin rather than by direct evaporation from the water surface. Epilittoral efflorescence around the receding lake shore supports this hypothesis. Algal Lake, which has a less well developed shore line, required only 5000 ly for the melting and evaporation process in December.

Light penetration. Visible light striking the lake is subject to reflectance from the water or any ice surface. Albedo measurements of the ice were made with the Eppley pyranometer (Table 2). The water had a characteristically low albedo, although not as low as is common for deep open waters where the bottom cannot be seen. Old ice with many bubbles had the highest reflectance (52–58%), whereas clear newer ice had values approaching those of water.

The amount of light penetrating the ice varied of course with ice quality and thickness: 10 cm of 1-year-old ice absorbed 30% of the light entering the ice, whereas 10 cm of clear new ice absorbed only slightly more than a comparable column of water would. The solar energy finally entering the water was attenuated by absorption and scattering that varied markedly between lakes (Figure 15). The vertical extinction coefficients for Skua Lake were 3–5 times those for Algal Lake (Table 3). The increasing extinction during the season in Skua Lake was paralleled by changes in the plankton population. Little seasonal change in light extinction was noted in Algal Lake.

Fig. 14. Incoming (dashed lines) and back (solid lines) radiation measured at Skua Lake on seven occasions during the 1962–1963 season. The back radiation was derived by algebraic subtraction of the incoming radiation from the net radiation measured with a Fransilla-Suomi net radiometer. Note the depression of midday back radiation in January.

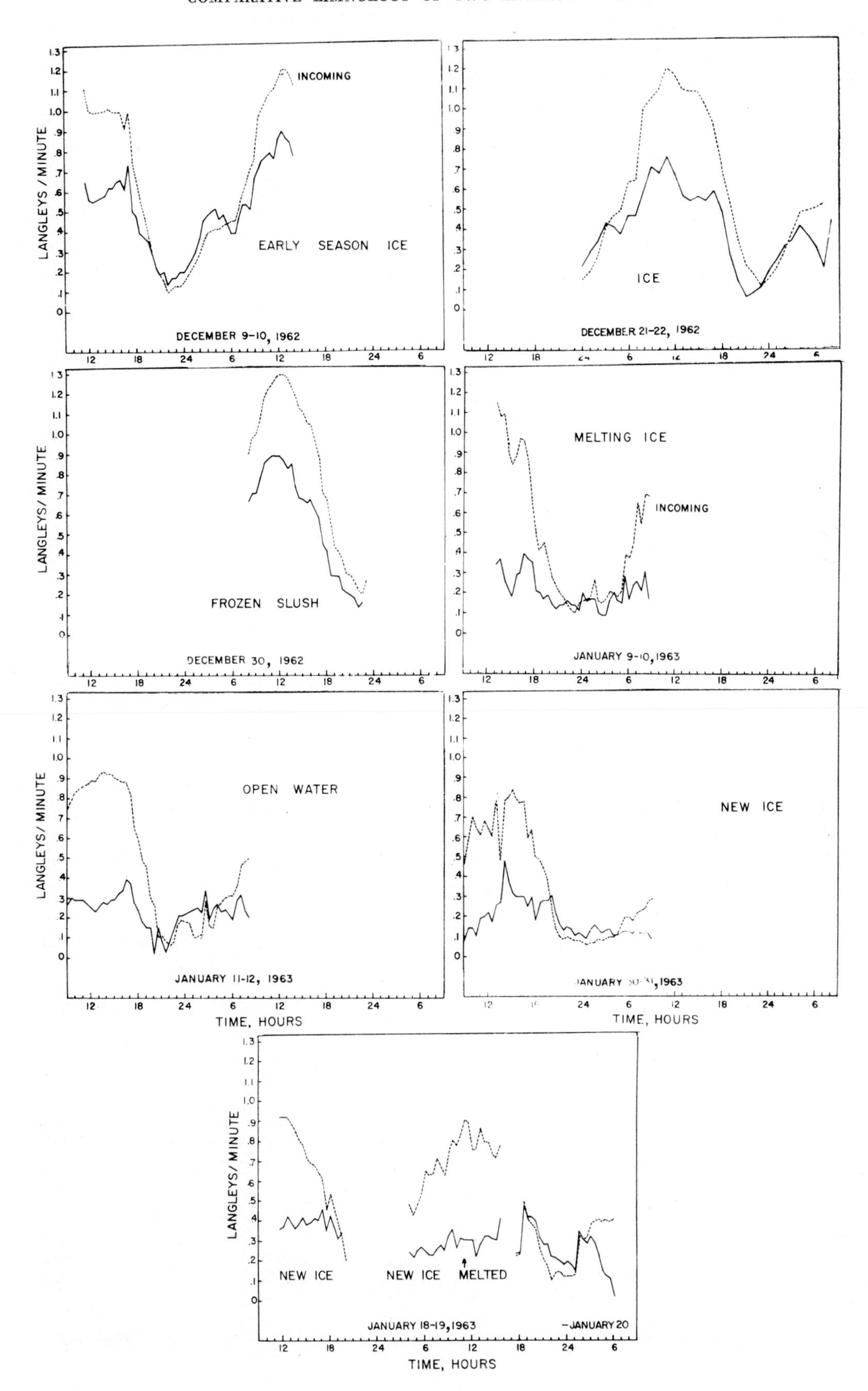
LANGLEYS / MINUTE
INCOMING
EARLY SEASON ICE
DECEMBER 9-10, 1962
ICE
DECEMBER 21-22, 1962
FROZEN SLUSH
DECEMBER 30, 1962
MELTING ICE
INCOMING
JANUARY 9-10, 1963
OPEN WATER
JANUARY 11-12, 1963
NEW ICE
TIME, HOURS
NEW ICE
NEW ICE
MELTED
JANUARY 18-19, 1963
JANUARY 20
TIME, HOURS

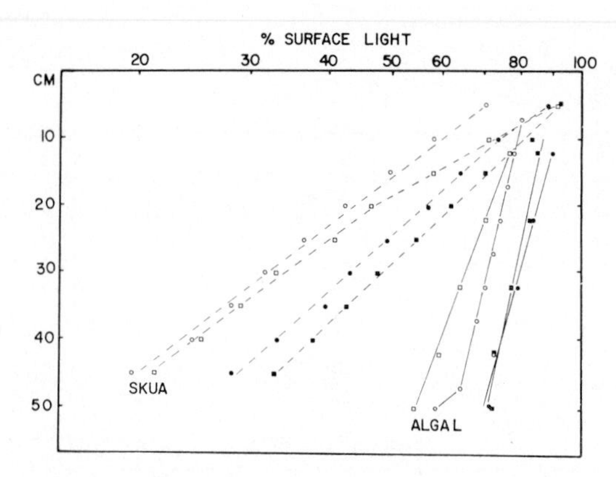

Fig. 15. Light penetration to depths in Skua and Algal lakes, January 17, 1963 (open circles represent no filter, open squares blue filter, solid circles red filter, and solid squares green filter). The abscissa is a logarithmic scale.

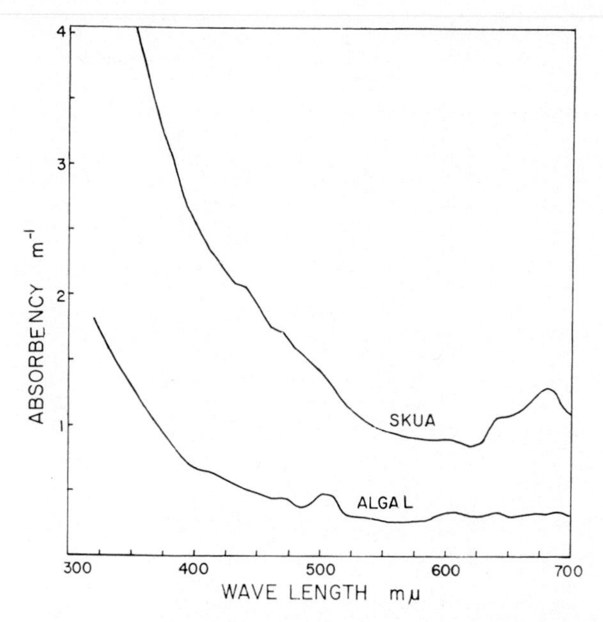

Fig. 16. Absorbency of unfiltered lake waters measured on December 30, 1962, in 10-cm cells with a diffusing glass in a Beckman DU spectrophotometer. The blank was water twice-distilled in glass, and measurements were made at 10-nm intervals.

Unfiltered water collected on December 30, 1962, showed comparable differences in light absorption for Skua and Algal lakes (Figure 16) when absorbency was measured in the laboratory.

CHEMISTRY

Melting patterns. The seasonal progression of melting and freezing greatly influenced the water chemistry of the two lakes. In late November 1961 both lakes were just beginning to melt at their edges. The melting proceeded from the lake margins toward the centers from early December to mid-January, when the first lasting freeze occurred (January 23). Small islands of old ice remained the first year, their surfaces lying a few inches above the surface of the new ice. The ice grew downward throughout February and concentrated the dissolved material below it through a freezing out process.

The melt pattern in the second season (1962–1963) was complicated by several cold periods that refroze substantial portions of the water. Several centimeters of new ice formed on the developing moats between December 4 and 9; this ice had remelted by December 13, but 6 cm of new ice was again present on December 19, although substantial melting had occurred between December 13 and 19. By December 24 most of Algal Lake and all of Skua Lake had melted except the 'roots' of the previous year's ice islands; however, a storm on December 26 added considerable snow to the lakes, and the floating slush refroze to 5 cm and trapped much plankton within it. By Janu-

TABLE 3. Extinction Coefficients for Skua and Algal Lakes

	Skua Lake					Algal Lake				
Date	Time, hours	Unfiltered	Green	Red	Blue	Time, hours	Unfiltered	Green	Red	Blue
Dec. 10, 1962	noon	2.56								
Dec. 21, 1962	1815	2.88								
Dec. 29, 1962	1700	2.32	1.92	1.77	4.14					
Jan. 9, 1963	2150	2.93				2330	0.92	0.42		0.55
Jan. 11, 1963	1640	2.70	2.25	2.65	3.30	1715	0.68	0.42	0.70	0.75
Jan. 17, 1963	1815	3.17	2.53	2.76	3.14	1430	0.54	0.47	0.68	0.94
Jan. 30, 1963	noon	3.08	3.47	3.00	5.76	1815	0.54	0.60	0.51	0.93
Feb. 18, 1963	1600	5.32	4.86	4.50	7.74					

Values for coefficients are derived from filtered (green, red, and blue) and unfiltered photocell readings and are expressed in reciprocal meters.

TABLE 4. Percentage Composition of Predominant Lava at Cape Evans Compared to World and Antarctic Averages for Basalts

Compound	Kenyte	Vitrophyric Kenyte	World Average	Antarctic Average
SiO_2	55.62	54.84	54.14	49.65
TiO_2	1.35	1.34	0.88	1.41
Al_2O_3	19.07	18.57	16.97	16.13
Fe_2O_3	6.06	6.26	4.05	5.47
FeO	nil	0.47	5.39	6.45
MnO	0.20	0.21	0.03	0.30
MgO	1.20	1.06	4.71	6.14
CaO	2.72	3.00	7.66	9.07
Na_2O	7.56	6.96	3.58	3.24
K_2O	4.59	4.51	2.32	1.66
P_2O_5			0.17	0.48
Others			0.01	

Values for Cape Evans lava are from Jensen [1916], and averages for world and antarctic basalt are from Daly [1914] and Clarke and Washington [1924], respectively. The Cape Evans minerals were called kenyte in the original description; more recent work has challenged this identification [Smith, 1954].

ary 9 both lakes were open, and by January 19 permanent refreezing had begun. During and between both seasons the lake levels (as measured from a permanent station) dropped, their volumes consequently decreasing.

Our chemical analyses are presented in the perspective of this seasonal progression of events, emphasis being placed first on the major ionic components of the ice and water and second on the trace metal constituents. Finally we treat the chemical parameters closely associated with the biological factors.

Major elements. As Mawson [1916] suggested, it is necessary to look to both the sea and the rock material of the cape for the sources of the ionic constituents of the inland waters of Cape Evans. Decomposition of the lavas of the cape may be a significant source of some ionic materials (Table 4). These lavas are richer in the monovalent cations and less rich in the divalent forms than the average for antarctic igneous rocks. A white crystalline efflorescent material is common among the gravels of the cape and testifies to the importance of terrestrial areas as at least secondary sources of ionic material for the lake waters. An analysis of this efflorescent material by the irrigation laboratory at the University of California at Davis showed the following percentage composition by weight: sodium, 27.2%; potassium, 1.05%; mag-

TABLE 5. Major Ionic Constituents of Skua and Algal Lakes on Several Dates

	HCO_3^-	Cl^-	$SO_4^=$	Na^+	K^+	Ca^+	Mg^+
Nov. 25, 1961							
Skua Lake ice							
Top	37.8	304		150	12.1	20.0	29.4
Middle	25.6	212		103	9.0	9.2	21.0
Bottom	25.6	172		80.0	5.9	7.6	16.4
Algal Lake ice							
Top	25.6	276		141	7.4	13.0	26.4
Middle	25.6	328		160	12.1	15.0	31.4
Bottom	254	6810	1070	3500	267	160	624
Edge melt	31.7	1964	256	976	55.9	60.1	182
Dec. 4, 1961							
Skua Lake, east	50.6	551	59.6	274	23.8	30.1	48.6
Skua Lake, west	111	1570	317	784	55.9	80.1	156
Dec. 22, 1961							
Skua Lake	50.6	519		256	24.6	20.0	47.7
Algal Lake	25.6	903	136	466	30.1	28.1	85.1
Jan. 16, 1962							
Skua Lake							
Surface	37.8	523	76.4	274	23.8	22.0	52.6
35 cm	63.4	579	82.6	300	28.2	22.0	52.6
Algal Lake							
Surface	37.8	915	148	466	30.1	28.1	85.1
35 cm	37.8	1030	161	516	30.1	44.1	91.2
Feb. 16, 1962							
Skua Lake subice	102	1095	144	548	48.1	44.1	97.3
Algal Lake subice	47.6	1850	301	924	64.1	60.1	172

Values are in mg l^{-1}. The analyses were performed by the University of California Irrigation Laboratory.

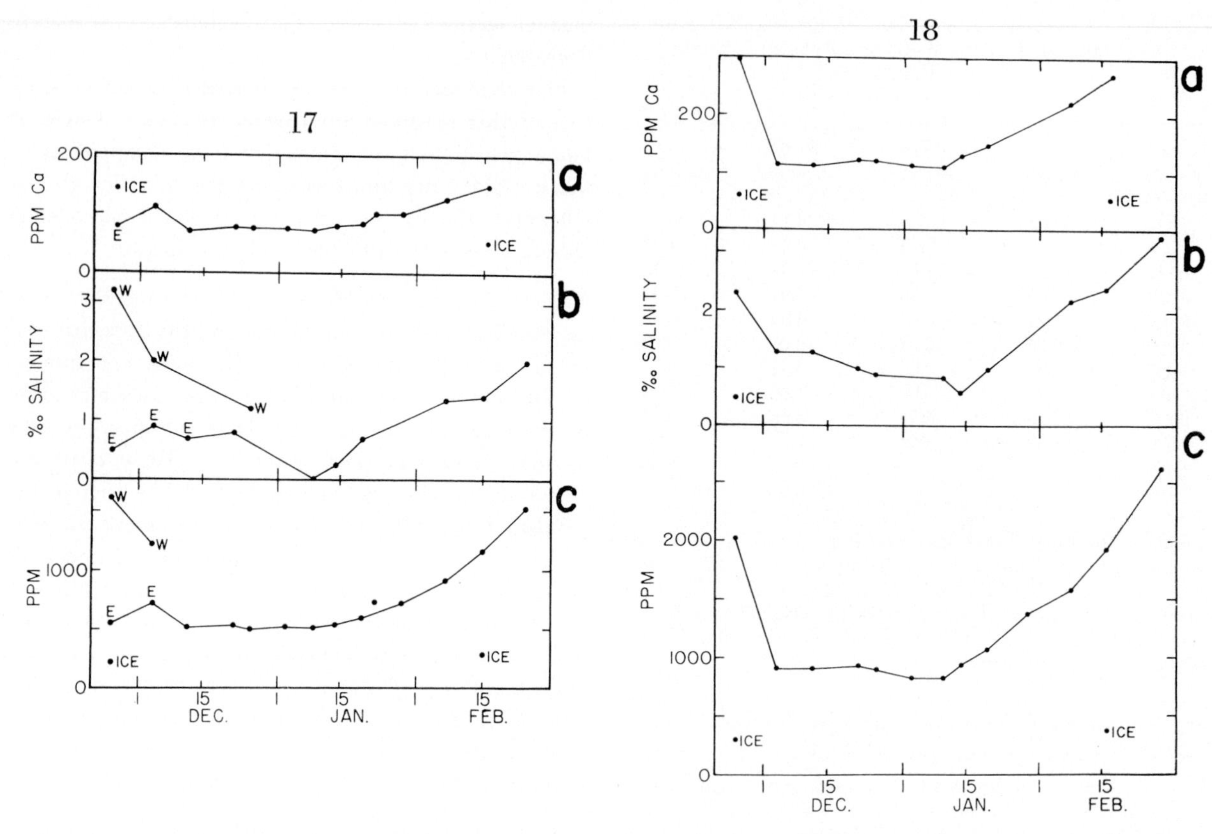

Figs. 17–18. Seasonal (1961–1962) changes in (*a*) divalent ions (EDTA titration with calmagite indicator), (*b*) specific gravity (hydrometer), and (*c*) chloride ion (Mohr titration). Values for the midlake surface ice are also indicated. Fig. 17, Skua Lake (where the western and eastern sections were isolated, the values are plotted separately). Fig. 18, Algal Lake.

nesium, 0.13%; and calcium, 0.04%. Ball and Nichols [1960] found a similar efflorescence from Marble Point to contain a comparable abundance of sodium but significantly less magnesium. If the source of these efflorescences is typical sea salt, wind removal of this salt from the cape would leave a residuum of potassium and divalent ions and deplete the cape in sodium (relative to sea water). The formation of sea spray itself may involve a discrimination against sodium [Sugawara, 1961].

The waters of the cape lakes indeed show a depletion of sodium and an enrichment of potassium, calcium, and magnesium in relation to the sea water proportions. The loss of sea-derived sodium through the ablation of sodium rich efflorescences is suspected, and the parent rock material may supply part of the potassium enrichment. Part of the magnesium enrichment in Skua Lake may derive from the bathing activities of the resident bird population (see section below on bird contribution).

Early in the first season Algal Lake showed both the kind of ice structure and the distribution of ionic constituents that has been found in other brackish lakes that freeze to the bottom (cf. Mawson's [1916] description of Green Lake at Cape Royds). The top 30 cm of bubbly ice contained 276 ppm Cl^- (Table 5), a massive layer 60 cm thick below it contained 328 ppm Cl^-, and the gray briny ice of the bottom 10 cm contained 6810 ppm Cl^-. These chloride concentrations indicate the trends in all the major ions and undoubtedly result from freezing out processes during the ice formation the previous season.

The ice at Skua Lake presented an anomalous distribution of ionic constituents (Table 5), the 15 cm of bubble-filled surface ice being nearly twice as concentrated as the crystalline bottom ice. Several mechanisms may have contributed to this phenomenon: the top ice may have been formed as a matrix of snow and water, salts being trapped thereby in the upper layers; sublimation during the winter may have concentrated salts in the upper layers of ice; wind-borne sea salts,

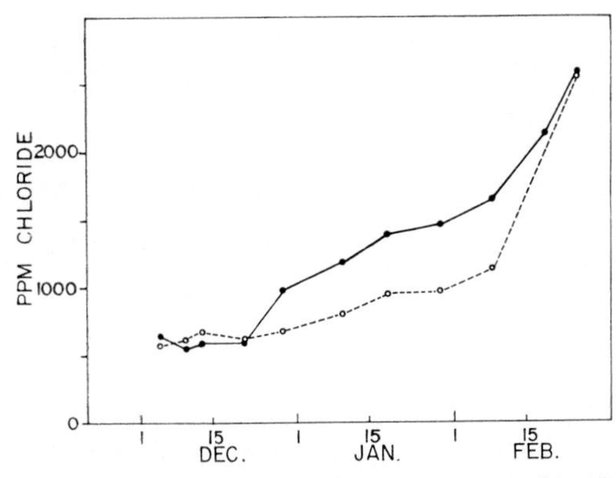

Fig. 19. Seasonal (1962–1963) changes in chloride ion concentration for waters of Skua (open circles) and Algal (solid circles) lakes. Multiplication of these values by the corresponding estimated lake volumes (Figures 6 and 8) gave relatively constant values for total chloride in each lake in midseason: Skua and Algal lakes contained approximately 430 and 220 kg of chloride, respectively.

skua feces, and efflorescent crystals may have lodged in the surface ice. Note also that the Skua ice profile was not made at the deepest part of the lake; consequently the effects of freezing out processes would be less pronounced in the Skua Lake samples than in the Algal Lake samples.

The first melt water around the edges of the lakes contained more solutes than the surface ice from which it was formed (Table 5 and Figures 17 and 18). We believe that the margins of the lake concentrate salts through an outward leaching of water and the subsequent evaporation from the gravels; this process would be particularly effective early in the season, when there is extensive melting under the ice at the margins and considerable heating of the surrounding exposed gravels.

As the melting progressed inward in both lakes during the first season, the salt content of the waters

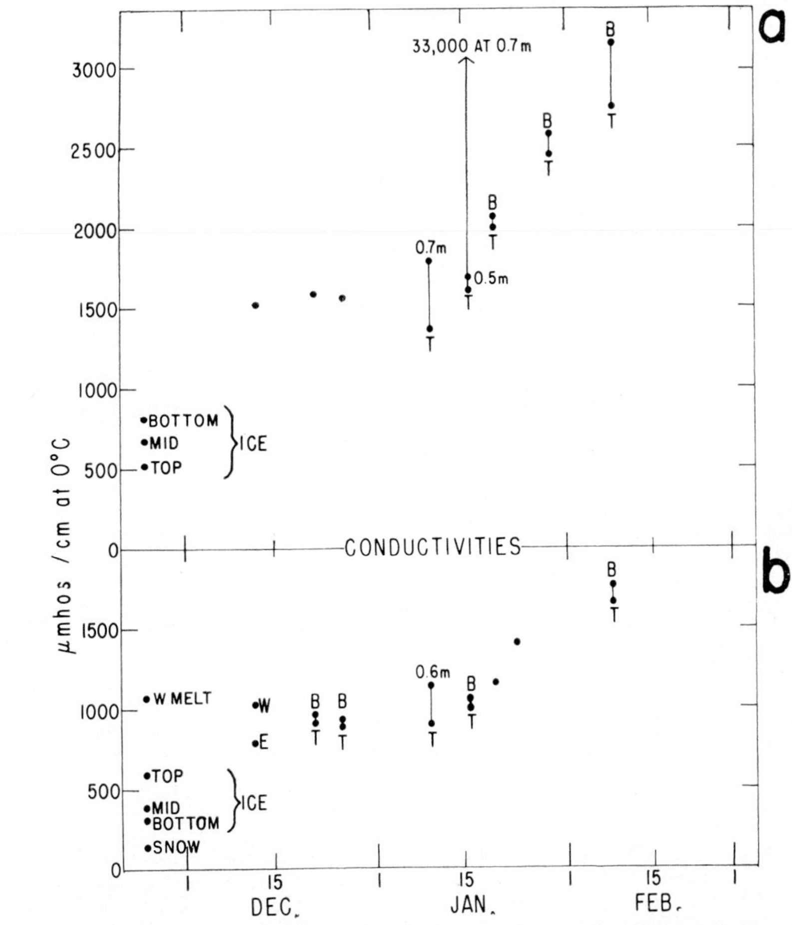

Fig. 20. Changes of conductivity during the season (1961–1962) for both (*a*) Algal and (*b*) Skua lakes. Letters denote the lake top T or bottom B, and numerals denote specific depths in meters. Points with no designation are average values from open water, where little vertical variation was noted.

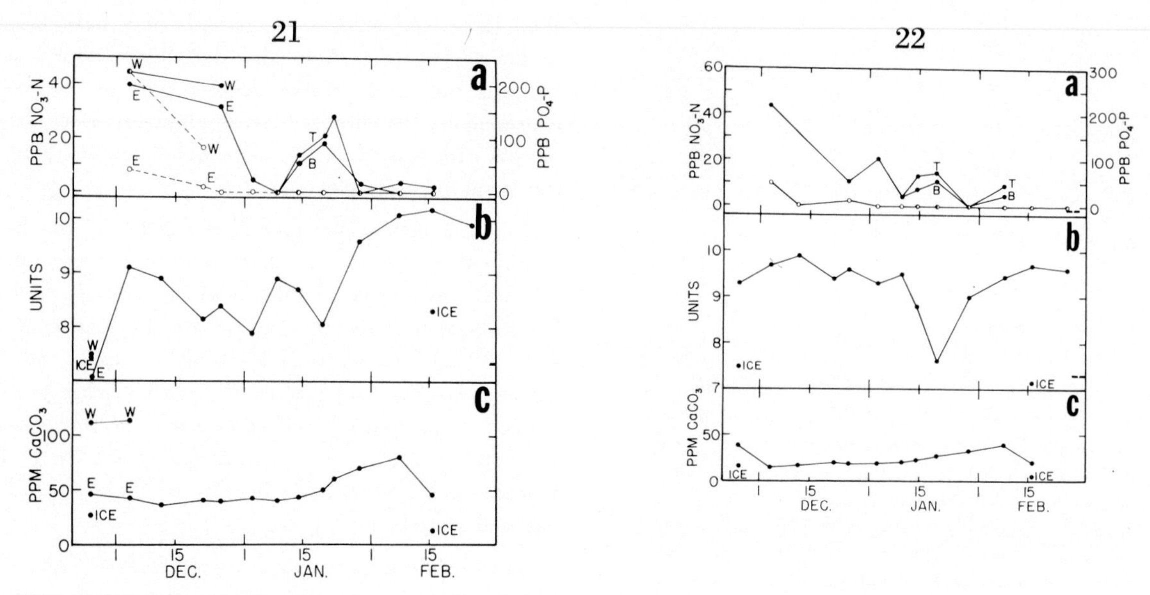

Figs. 21–22. Seasonal changes in (*a*) nitrate nitrogen (solid circles) and phosphate phosphorus (open circles), (*b*) *p*H, and (*c*) alkalinity (as $CaCO_3$) during 1961–1962. Values for the top T and bottom B of the accessible water column are also indicated in some cases, as well as the concentrations of the various parameters in the ice at the beginning and end of the season. Fig. 21, Skua Lake (where ice separated the eastern E and western W regions, they are separately designated). Fig. 22, Algal Lake.

dropped, presumably as the result of dilution by the melting of the upper ice layers (Figures 17 and 18). This process continued into January, but then another factor began to affect the salinity. The melting process had reached into the zones of salt concentration in the deep ices, and the interstitial brines began to be released into the water. On January 15, 1962, we observed such a brine deep in Algal Lake, having the characteristic greenish tint of phytoplanktonic growth.

Freezing began in both lakes in the latter part of January, and there was a consequent concentration of ions in the underlying water (Figure 19). Brine release may still have been occurring, since the bottom ice would be melting at a temperature lower than that at which the top ice had formed. In addition the ice cover insulated the lakes from atmospheric contact, while it permitted insolative heating and the subsequent turbulence. Conductometric analyses showed an

TABLE 6. Spectrographic Analyses of the Major Trace Metals in Skua and Algal Lakes in Snow and in Skua Feces

Element	Skua Lake Water	Skua Lake Ice	Algal Lake Water	Algal Lake Ice	Cape Evans Snow	McMurdo Station Snow	Sea Water	Skua Feces, $\mu g\ kg^{-1}$ dry weight
Aluminum	316	238	270	60	<2.5*	28	8.8	2.9×10^6
Copper	?	<2.5*	<1.2*	<5.0*	<2.5*	7.2	<10	32,000
Iron	?	200	55	34	<2.5*	12	16	>20,000
Manganese	28	80	22	<5.0*	<1.2*	<0.83*	<10	125,000
Molybdenum	1.5	0.75	1.3	<1.0*	<0.50*	<0.17*	<2.0	420
Nickel	2.8	9.0	6.8	1.9	2.0	0.93	4.0	25,000
Titanium	2.8	5.5	2.6	1.8	<1.0*	3.2	<4.0*	2,200
Vanadium	<0.25*	<0.50*	<0.25*	<1.0*	<0.50*	<0.17*	1.4	640

* Subanalytical quantities.

Sample sizes for Skua Lake water, Skua Lake ice, Algal Lake water, Algal Lake ice, Cape Evans snow, McMurdo Station snow, sea water, and skua feces were 4, 2, 4, 1, 2, 6, and 0.5 liters and 5.09 grams, respectively.

All values except those for skua feces are in micrograms per liter of original water. A question mark indicates that analytical values have been discounted because of a wide discrepancy in their values.

Water samples are averages of two samples taken and concentrated on January 17, 1962, and February 16, 1962. Ice samples were taken on February 16, 1962. Snow was collected near the McMurdo Station laboratories on November 26, 1961, and at Cape Evans on January 1, 1962. Sea water was collected in a plastic sampler from a depth of 40 meters in water 240 meters deep near McMurdo Station on November 13, 1962.

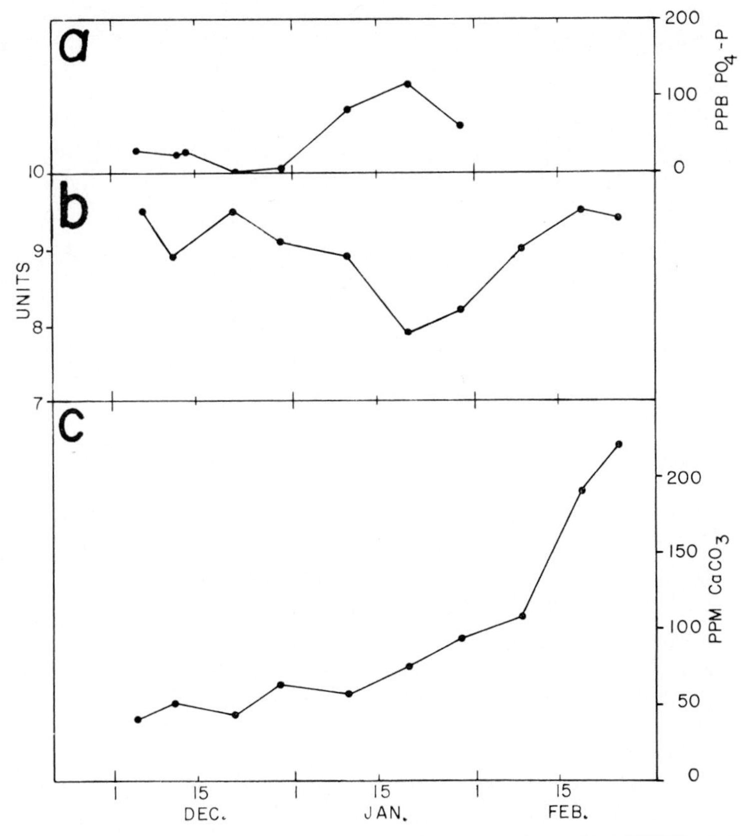

Fig. 23. Seasonal changes in (*a*) phosphate phosphorus, (*b*) *p*H, and (*c*) alkalinity (as $CaCO_3$) for Skua Lake during 1962–1963.

increasing ionic content in both the top and the bottom waters from mid-January to the end of the season (Figure 20). Therefore the rise in the ion concentration in the water after late January may not have been entirely the result of the freezing out process.

Minor elements. Through the courtesy of Mr. William Silvey of the U.S. Geological Survey, Sacramento, California, we obtained trace metal analyses (aluminum, beryllium, bismuth, cadmium, chromium, copper, iron, gallium, germanium, manganese, molybdenum, nickel, lead, titanium, vanadium, and zinc) of several waters and possible ion sources. Snow (collected on January 3, 1962, at Cape Evans) appears to be a poor source of these elements, detectable amounts (2 ppb) of only iron and nickel being present. These small impurities may have come from contact with stainless steel during preanalytical concentration procedures. In contrast, snow in the McMurdo Station vicinity is high in a variety of trace elements. This finding undoubtedly reflects the contribution of dust from McMurdo Station. Sea water, however, contained significant quantities of iron, aluminum, nickel, and vanadium. Analyses of skua feces collected in December 1961 show that a significant amount of the aluminum, manganese, iron, copper, nickel, titanium, and

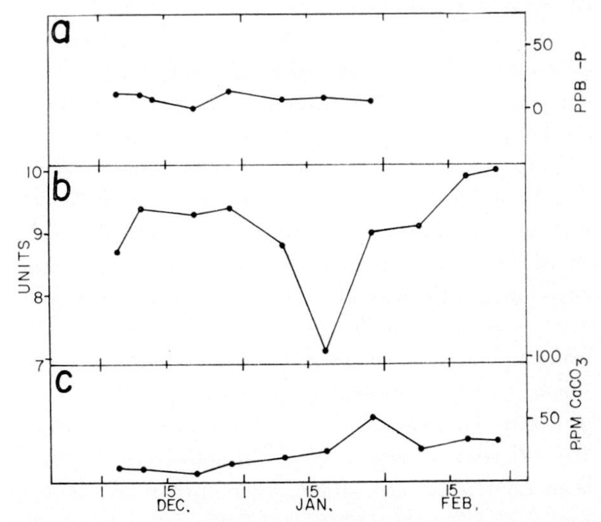

Fig. 24. Seasonal changes in (*a*) phosphate phosphorus, (*b*) *p*H, and (*c*) alkalinity (as $CaCO_3$) for Algal Lake during 1962–1963.

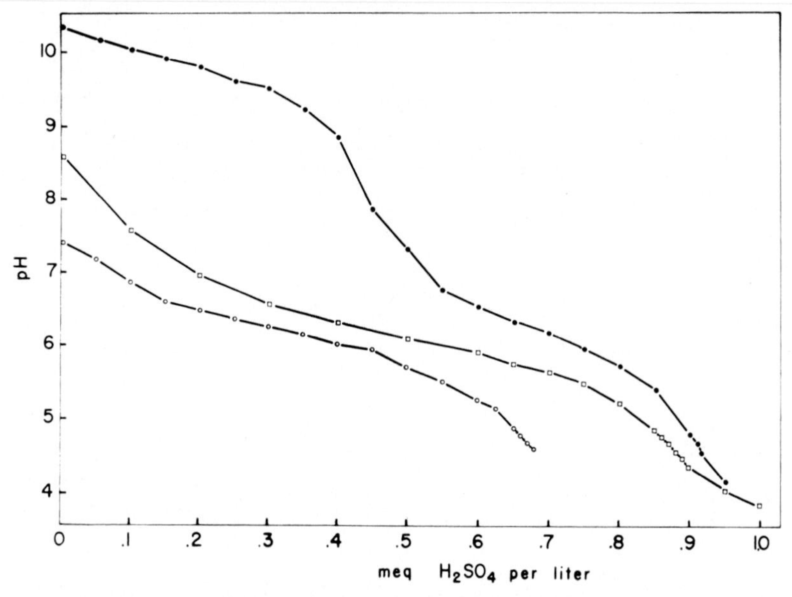

Fig. 25. Buffer curve for titration of Skua Lake water with standard acid on three representative dates in the 1961–1962 season (open circles represent December 12, 1961; open squares, January 16, 1962; and solid circles, February 16, 1962). The curves for 1962–1963 were substantially the same.

vanadium may come to Skua Lake from this source (Table 6). Analyses of the lake waters reveal significant concentrations of aluminum, iron, manganese, nickel, titanium, and molybdenum in both lakes. Of these, all except nickel and molybdenum have been reported in the cape minerals [Jensen, 1916].

Biologically important parameters. The buffer capacity and the *p*H, as well as the dissolved nitrate, phosphate, and organic carbon and nitrogen concentrations, in the waters are especially important to the biology of Skua and Algal lakes. For comparison we give our analyses of the lakes of Cape Royds, where the penguin population exerts an important influence on the lakes.

During both seasons the hydrogen ion concentration in the two lakes at Cape Evans was variable in the first part of the season and was undoubtedly influenced by a combination of biological and melting phenomena (Figures 21–24). A downward *p*H trend was evident in both lakes in mid-January, when freeze-up began, and an upward *p*H trend thereafter. These trends were less pronounced in Skua Lake, probably because of its greater buffering capacity. The late season increase of *p*H in both lakes may have been due to the use of inorganic carbon by the still actively photosynthesizing organisms remaining in the water beneath the ice, a continuing net fixation of carbon dioxide thus being implied. And thus, even when the lake was shaded by the ice layer, the photosynthetic uptake of carbon dioxide still exceeded its production by respiratory processes. Since both respiration and photosynthesis are largely halted when the lakes are completely frozen, the preceding results suggest that the lakes represent a substantial net contribution of organic matter to the soil of the McMurdo Sound area. Support for this hypothesis was found in our examinations of several dried-up ponds near the shore, where excavation revealed layers of black algal peat several inches deep. These layers gave off a strong aroma of organic sulphur compounds, which are indicative of active bacterial decay processes. Changes in salt concentrations (Table 5) caused by freezing would also contribute to the changes in *p*H.

The shapes of the buffer curves (Figures 25 and 26) may be described in terms of the behavior of the $CO_3^{=}/HCO_3^{-}/H_2CO_3/CO_2$ system. The much lower buffer capacity of Algal Lake despite its higher concentration of dissolved salts suggests a significant qualitative difference in the minerals present in the two bodies of water. This difference is reflected in cation analyses (Table 5). The pronounced changes of *p*H noted at the onset of freezing are reflected in the Skua Lake buffer curves made immediately before and after the beginning of the winter freeze (Figure 27) and probably result from a combination of the freezing out of salts and the isolation of the water

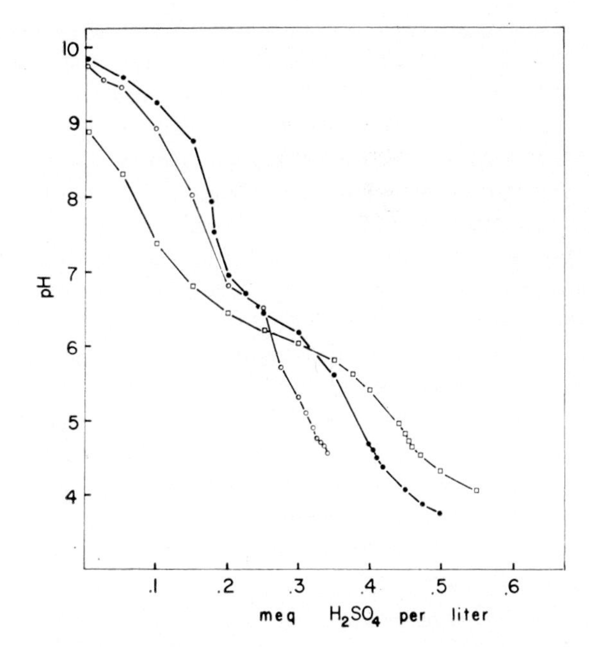

Fig. 26. Buffer curve for titration of Algal Lake water with standard acid on three representative dates in the 1961–1962 season (open circles represent December 12, 1961; open squares, January 16, 1962; and solid circles, February 16, 1962). The curves for 1962–1963 were substantially the same.

from contact with air. Substantially the same curves were obtained the second season.

A buffer curve from Pony Lake (Home Lake), Cape Royds (Figure 27), is included for comparison. This water was much richer in phosphates and organic nitrogen than that of Skua or Algal lake (Appendix 1). Despite these levels of nitrogen and phosphorus and a higher *p*H, Pony Lake showed a lower buffering capacity than the Cape Evans lakes. The buffer system in the highly polluted Pony Lake was no doubt complicated by organic acids.

Oxygen was measured by means of the Alsterberg azide modification of the Winkler technique in midmorning on December 12, 1961, and showed supersaturation values for both Cape Evans lakes. This condition presumably prevailed throughout the active season; on calm days bubbles could be seen emerging from the water in Skua Lake, and the algal felt of both lakes produced much intestrial gas. Reducing conditions, which were indicated by the odor of hydrogen sulfide, were present beneath the surface felt layers of both lakes.

At the beginning of the first season the melted pools of Skua Lake showed appreciable amounts of phosphate. The west pool, which the skuas frequented, consistently had much higher concentrations than the east pool. Algal Lake also showed a measurable level of phosphate early in the season. By January 3, 1962, however, the phosphate in both lakes had dropped to levels too low to be measured and stayed low for the rest of the season. It is interesting to compare these lakes with those near the penguin rookery at Cape Royds (Appendix 1), where the concentration of phosphate phosphorus was 10 times the greatest concentration observed in Skua Lake. In a small lake

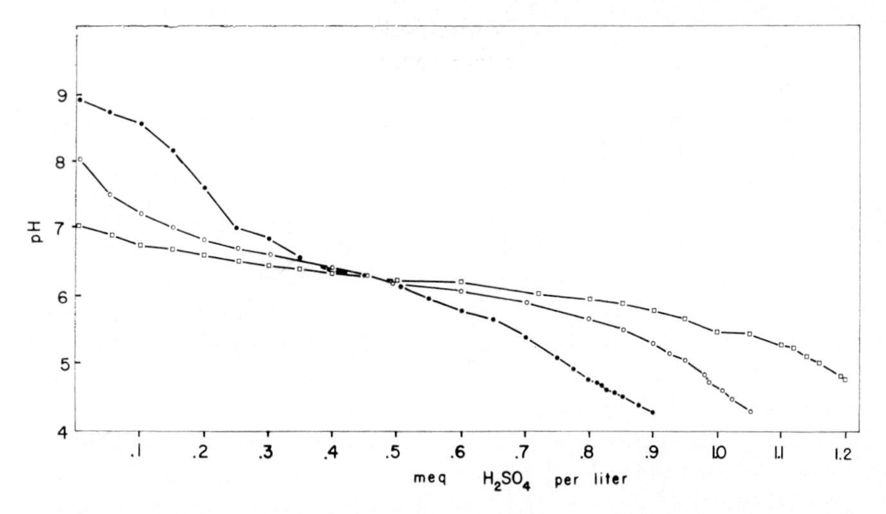

Fig. 27. Buffer curves determined just before (January 22) and just after (January 24) the onset of freezing of Skua Lake in 1962 (open circles represent Skua Lake on January 22, 1962; open squares, Skua Lake on January 24, 1962; and solid circles, Pony Lake on January 24, 1962). The curve for Pony Lake on Cape Royds is included for comparison.

just north of the present Cape Royds rookery the phosphate reached a level of 7.7 ppm, 30 times the highest level observed at Skua Lake. These Cape Royds figures were obtained from samples collected on January 24, 1962, 3 weeks after the phosphate in Skua and Algal lakes had fallen below detectable levels.

The pattern of phosphate concentration during the second season at Cape Evans was quite different from that during the first. High initial levels were not observed, although phosphate again dropped throughout December. In contrast to the pattern of the first year a second phosphate peak occurred in mid-January. Note, however, that a comparably high nitrate value was observed in January of the first year. In association with the development of these nutrient levels the population of planktonic cyanophyceans declined. We may postulate that the nutrient release from this early population and the melting of the deepest ice provided the essential requirements for the succeeding planktonic forms. It is also possible that the nitrogen/phosphorus ratio was unfavorably high for the blue-green algae.

Throughout most of the first season the nitrate level in both lakes was low but measurable and generally higher at the surface than at the bottom (Figures 21 and 22). Like the phosphate level, the nitrate level at the beginning of the season in the west melt pool in Skua Lake exceeded that in the east melt pool. A similar set of seasonal variations was observed in both lakes. From an early high level of almost 45 ppb nitrate nitrogen in both lakes, the concentration dropped rapidly, increased at midseason, and fell off again after freeze-up. It is possible that the cyanophycean benthic algae fixed and released important amounts of atmospheric nitrogen.

The nitrate levels recorded at Pony Lake, Cape Royds, on January 24 were between 5 and 10 times those at Skua Lake on the same date. The small pond north of Pony Lake was so rich in nitrate that not even a twentyfold dilution with distilled water brought the concentration down to a measurable level. We estimate that it contained >3 mg l^{-1} of nitrate nitrogen.

The decrease in the nitrate and phosphate levels in both Skua and Algal lakes during the season undoubtedly reflects the use of the compounds by the active phytoplanktonic populations. The slight nitrate rise (phosphate rise the second season) in the second part of January may be attributed in part to regeneration from planktonic or sedimentary sources or to release from trapped brine layers.

In January and February the dissolved organic nitrogen varied slightly about an average of 2.0 mg l^{-1} for Skua Lake and 1.6 mg l^{-1} for Algal Lake. No seasonal trend was observed in the eight analyses. In Skua Lake the dissolved organic carbon increased from 2.0 (on November 25, 1961) to 5.7 (on January 15, 1962) to 14.2 mg l^{-1} (on February 11, 1962). This increase was more rapid than that of chloride and probably represents a real accumulation of dissolved organic matter during the first season. This carbon may be available to facultative heterotrophs when light becomes limiting in late fall. On the same sampling dates the dissolved organic carbon concentrations in Algal Lake rose from 2.8 to 5.4 to 5.8 mg l^{-1}. This increase, however, was less than the chloride ion change and may therefore represent an actual decrease in the dissolved organic carbon content of the lake.

Dissolved organic nitrogen. During the latter part of the 1961–1962 season the dissolved organic nitrogen content was routinely determined. Values ranged between 1.0 and 3.0 ppm of nitrogen in the ponds at Cape Evans and between 13.0 and 17.0 ppm of nitrogen in water taken from the lakes at the Cape Royds penguin rookery on January 24, 1962. On the assumption that the chief source of the organic nitrogen was a mixture of peptides and amino acids, the figures for organic nitrogenous compounds were calculated by means of the standard factor 6.25 to convert Kjeldahl nitrogen to crude protein. This quantity of material must clearly represent an important source of both carbon and nitrogen for bacteria and for the algae and protozoa able to use it. Furthermore these compounds can be expected to act as chelating agents and thus to help keep important heavy and alkaline earth metals in solution and available to the algae.

It is interesting to speculate on the origin of this material. Obviously the material in Pony Lake must originate in large part from the penguin rookery; however, the organic nitrogen concentration in this lake is only about twice the ammonia nitrogen concentration, whereas the organic nitrogen concentration in both Skua and Algal lakes is nearer 5 times the ammonia nitrogen concentration. At first it was assumed that the skuas contribute significantly to the organic nitrogen in Skua Lake, but in fact the levels recorded in Skua and Algal lakes were fairly similar. A more probable origin is the algae themselves. It is well known that nitrogen-fixing cyanophycean algae release considerable amounts of peptides into the surrounding water. That leaching from old algal felts also contributes to the dissolved organic nitrogen is demonstrated in the following experiment.

A portion of dry algal felt from the margin of Algal Lake was dried in the oven to constant weight and was then ashed. The following values are the means of the results of two experiments in per cent:

Moisture	15.6
Ash	31.7
Organic matter (by difference)	52.7

Native felt (20 grams) was homogenized with water (100 ml) in the Waring blender, filtered through cheesecloth, and then centrifuged.

Portions of this extract were applied to Whatman 3MM chromatography paper. Then the chromatograms were developed overnight by using butanol, acetic acid, and water in the ratio 125:30:125, respectively. For the detection of carbohydrates, chromatograms were treated with 2% aniline hydrogen phthalate dissolved in a mixture of acetone and butanol. Only one faint spot appeared, which corresponded in position with authentic glucose.

Amino compounds were detected by spraying chromatographs with a 1.0% solution of ninhydrin in acetone. Nine bright clearly defined spots appeared. By comparison with authentic samples it was possible to assign very tentative identifications to the compounds present in the extract. From the origin there were (1) cystine, (2) lysine, (3) arginine and/or histidine and/or aspartic acid, (4) glycine and/or serine, (5) threonine and/or glutamic acid, (6) alanine, (7) tyrosine, (8) phenylalanine, and (9) leucine and/or isoleucine.

In addition a considerable amount of ninhydrin positive material remained at the origin. Thus it is clear that the old felt represents a source of a variety of organic nitrogen compounds.

BIOLOGY

Bird contribution. One of the important differences between the two lakes that led to their selection for study was the habitual and exclusive bathing of the skuas in the relatively sheltered western area of Skua Lake. As soon as any melt water was accessible at the margin of the lake, the birds congregated from their nesting sites scattered around the cape and spent considerable time bathing and resting at the western edge. An average of 50 birds was noted at the lake in 31 counts made throughout the study. The eastern littoral regions consequently received a significant amount of fecal pollution, and birds taking off into the prevailing wind frequently defecated onto the lake. A grid was set up, and the fecal droppings on the lake ice were counted and removed during a period of several days. From these data and measurements of the weight and the soluble phosphate and chloride of the droppings, values of 42 mg dry weight of skua excrement m^{-2} day^{-1}, 110 μg of phosphate phosphorus m^{-2} day^{-1}, and 3.7 mg of chloride m^{-2} day^{-1} were determined. On the basis of surface area these values mean a contribution to Skua Lake of about 90 grams of excrement, 230 mg of soluble phosphorus, and 7.9 grams of chloride ion per day. An additional enrichment of inorganic ions may be expected from the birds' nasal excretions, which presumably are rich in sodium and

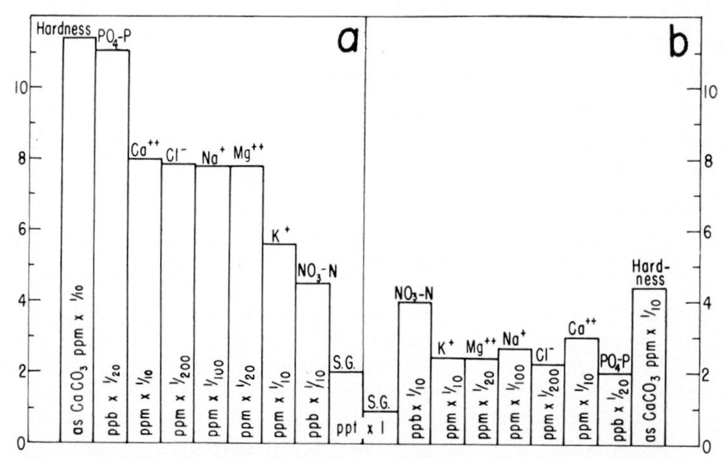

Fig. 28. Several chemical parameters illustrating the differences between the isolated (*a*) western (skua bathing area) and (*b*) eastern parts of Skua Lake on November 25, 1962. Factors relating each parameter to the general scale are given within the block; S.G. is specific gravity.

chloride. Before beginning to bathe its body, a newly arrived bird spent many minutes standing in the water, dipping and shaking its head.

The suspected contribution of allochthonous materials to the lake by the birds is supported by the findings of Allen et al. [1967] and Kear [1963] and by chemical measurements obtained when the lake melted in two separate regions (Figure 28). For every parameter measured the concentration was greater in the western melt pool. This pattern was particularly true of phosphate, an ion probably important in maintaining the high productivity measured in the lake. Skua Lake consistently had higher phosphate concentrations than the more saline Algal Lake, and the experiments on phytoplanktonic nutrient-limiting factors discussed below revealed no demonstrable phosphate deficiency in Skua Lake, in contrast to a short-lived deficiency in Algal Lake.

Benthos composition. Organisms in the lakes were sampled periodically and studied in both a living and a preserved state. The benthos communities of the two lakes were quite similar and will thus be treated as a unit. Figure 29 illustrates some of the dominant benthic forms found in the lakes. The benthic felt was primarily composed of filaments of *Oscillatoria* of several species (*tenuis, agardhii,* and others), *Phormidium* sp., and *Nostoc* sp. The surface filaments frequently contained 'gas vacuoles,' were poor in chlorophylls, were high in carotenoids, and presumably protected a more active photosynthetic layer beneath from the high light intensities at the surface [Goldman et al., 1963]. A variety of diatoms and grazing protozoans and metazoans were within the meshwork of the filaments. Since the lakes lacked truly planktonic herbivores other than ciliates, this felt community was the only significant site of secondary productivity. The animals presumably benefit from the localized warming produced within the felt by solar radiation. Rhizopods, holotrichous and hypotrichous ciliates, colorless and colored flagellates, acoelic flatworms, nematodes, rotifers, and tardigrades were common in the living samples, and the metazoans were frequently seen devouring their cyanophycean substrate.

Planktonic composition. The plankton population of Algal Lake was largely made up of flagellated forms, mostly colored. A cryptomonad, *Chlamydomonas intermedia,* and *C. subcaudata* (Figure 30), *Brachiomonas* sp., and *Ochromonas* sp. were recognized, along with several small unidentified flagellates. Skua Lake plankton contained these forms, but the bloom conditions during the late season were dominated by *Ankistrodesmus falcatus* var. *acicularis* (an alga not previously reported from Antarctica) and a small (about 5-μ) yellow-green unicell tentatively assigned to the xanthophycean genus *Chloridella* (Figure 30). Some blue-green algae were present in the plankton of both lakes, notably a *Spirulina* sp. Holozoic protists and diatoms were also among the plankton common to both lakes.

Diatom flora. In January 1962 phytoplankton samples were collected from Algal Lake and Skua Lake on Ross Island, Antarctica. From observations of these samples it was concluded that the composition of the diatom community was dominated by two endemic antarctic species: *Navicula muticopsis* and *Pinnularia cymatopleura.* Two other endemics were noted in small numbers, and the remaining four types were cosmopolitan. Each form and its distribution, status (frequency of occurrence in our samples), and locality are listed in Table 7. The forms are described in Appendix 2.

Both Algal and Skua lakes have the *Navicula muticopsis–Pinnularia cymatopleura* association reported by Fukushima [1967] in Cape Royds, Ross Island. *Navicula gausii,* which appears to be very similar to *Navicula muticopsis* except for the distinctly convex sides of *N. gausii,* was seen only in the Algal Lake samples. It has been reported from a Lake Miers (Victoria Land) bottom sample by Baker [1967]. The average sizes of our *N. gausii* were smaller than those he described. The fourth endemic diatom (*Gomphonema kamtschatica* var. *antarctica*) was observed in

Figs. 29–30. Microphotographs of organisms from Algal and Skua lakes. The photographs were taken of formalin-preserved materials in water mounts by means of an inverted microscope. Fig. 29, benthic organisms tentatively identified as (*a*) rhizopod test and *Oscillatoria* sp., (*b*) variety of diatoms and tardigrade test with three embryos, (*c*) three tardigrade eggs and two species of *Oscillatoria* (*tenuis* and *agardhii*), (*d*) rhizopod, (*e*) tardigrade *Macrobiotus* sp. and a contracted rotifer, (*f*) nematode (*Plectus* sp.), (*g–i*) diatoms, and (*j*) *Oscillatoria tenuis.* Fig. 30, planktonic and pseudoplanktonic organisms tentatively identified as (*a*) *Chlamydomonas subcaudata,* (*b*) *Chloridella* sp., (*c*) *Chlamydomonas subcaudata* and *Ankistrodesmus falcatus* (var. *acicularis*), (*d*) two ciliates, possibly *Urotricha* and *Halteria,* (*e*) small diatom, (*f*) *Euplotes* sp., (*g*) *Ankistrodesmus* and *Chloridella,* (*h*) variety of diatoms and cyanophycean fragments including a *Synedra* (*ulna*), (*i*) *Nostoc* sp., and (*j*) *Urotricha* sp.

29

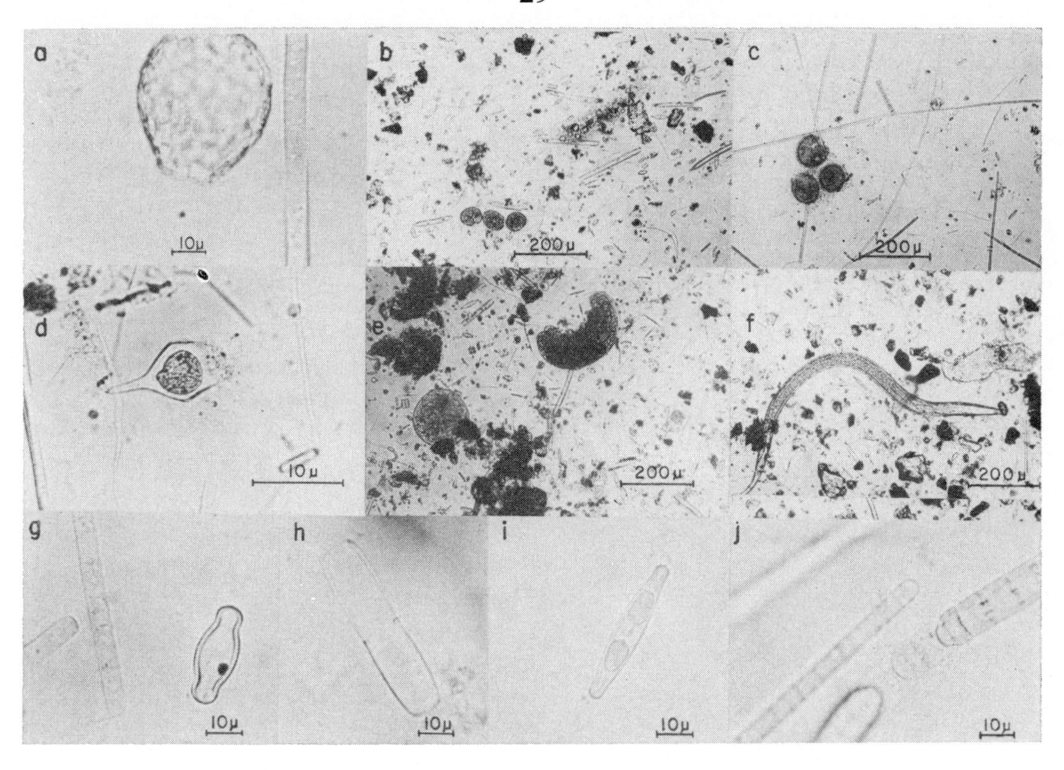
a
10μ
b
200μ
c
200μ
d
10μ
e
200μ
f
200μ
g
10μ
h
10μ
i
10μ
j
10μ

30

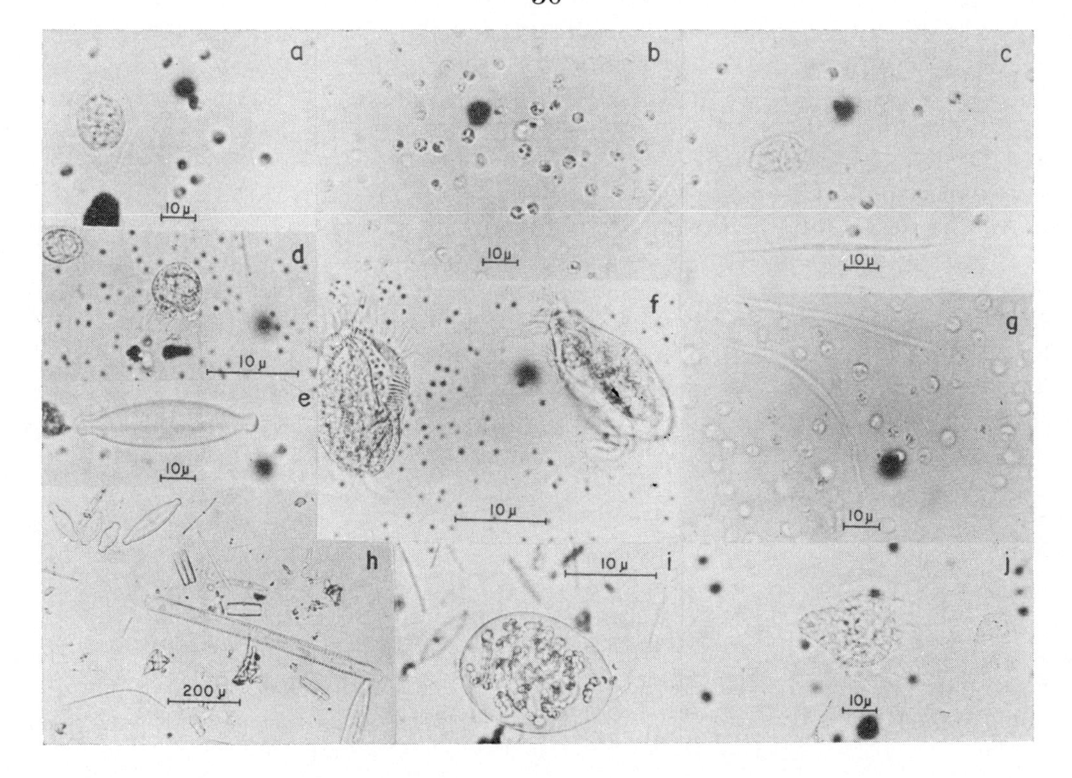
a
10μ
b
10μ
c
10μ
d
10μ
e
f
10μ
g
10μ
10μ
h
200μ
10μ
i
j
10μ

TABLE 7. Distribution and Status of the Diatoms of Skua and Algal Lakes

	Status	Distribution
Achnanthes brevipes var. *intermedia*		cosmopolitan
Algal Lake	rare	
Skua Lake	rare	
Navicula gausii		antarctic
Algal Lake	subdom.	
Navicula muticopsis		antarctic, subantarctic islands
Algal Lake	dom.	
Skua Lake	dom.	
Pinnularia cymatopleura		antarctic
Algal Lake	subdom.	
Skua Lake	subdom.	
Stauroneis anceps		cosmopolitan
Algal Lake	occ.	
Skua Lake	occ.	
Gomphonema kamtschatica var. *antarctica*		antarctic
Algal Lake	rare	
Skua Lake	rare	
Hantzschia amphioxys		cosmopolitan
Algal Lake	occ.	
Skua Lake	occ.	
Hantzschia amphioxys var. *maior*		cosmopolitan
Algal Lake	rare	
Skua Lake	occ.	

dom., dominant form; subdom., subdominant form; occ., occasional form.

both lakes. It is evidently a rare form and was first reported from Graham Land in 1908 by Peragallo [1921]. Fukushima [1962] records its occurrence in the Kasumi Rock ice-free area. This species is normally considered to be a marine or brackish water form, but its presence in Skua Lake may be explained by our findings concerning the contribution of allochthonous materials by birds (see preceding discussion). Algal Lake was found to be more saline than Skua Lake; therefore the presence of *G. kamtschatica* var. *antarctica* was more predictable in this instance.

Achnanthes brevipes var. *intermedia* is reported from North America, Europe, and Antarctica [Baker, 1967] and has been observed to form a natural association with the antarctic endemic *Navicula muticopsis* in the Kasumi Rock region [Fukushima, 1967] as well as in our lake samples. Another communal association between *N. muticopsis* and *Stauroneis anceps* was observed in our samples and in the Cape Evans area [Fukushima, 1967]. *Stauroneis anceps, Hantzschia amphioxys,* and *H. a.* var. *maior* are common diatoms in fresh or slightly brackish water all over the world; their presence in Algal and Skua lakes is another example of the adaptive ability of many diatoms.

STANDING CROP

The standing crop of planktonic forms was estimated at various times by seven methods: (1) dry weight, (2) nitrogen, (3) carbon, (4) chlorophyll, and (5) carotenoid contents of the seston, (6) direct counts of preserved plankton samples, and (7) photosynthetic carbon uptake by the plankton under standard conditions of light and temperature. These parameters of the standing crops of each lake are illustrated in Figures 31–37 and are discussed separately below.

Sestonic weight. The relatively large error in determining sestonic weight does not permit more than semiquantitative results. However, a trend toward increasing seston may be observed during both seasons in both lakes, the sestonic weight of Skua Lake being about 5 times that of Algal Lake (Figures 31–34). Seston trapped in the ice when the lakes froze was of the same order of weight as that in the water from which the ice formed. Considerably more seston developed in Skua Lake toward the end of the second season (Figure 33) than had been measured during the first.

Sestonic nitrogen and carbon. Measurements of sestonic nitrogen and carbon, which were made only in the second half of the first season, showed a peak after

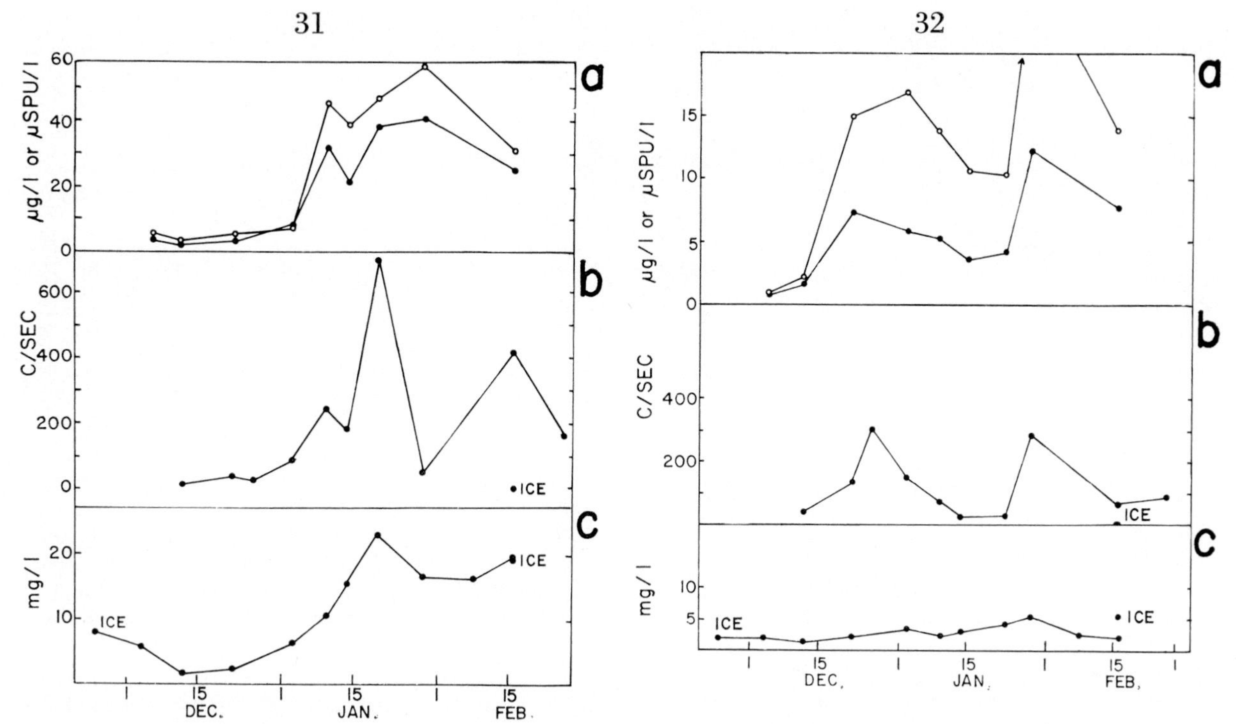

Figs. 31–32. Seasonal changes of various measures of planktonic standing crop during 1961–1962. (*a*) Pigments (chlorophyll *a* is represented by solid circles, and total carotenoids by open circles). (*b*) Carbon uptake under standard conditions of light and temperature. (*c*) Sestonic weight. Fig. 31, Skua Lake (all values are for the eastern, birdless part of the lake in surface water except where ice is indicated). Fig. 32, Algal Lake (all values are for surface water except where ice is indicated).

the middle of January followed by a second rise in February in Skua Lake (Figure 35). After the January peak in Algal Lake a modest decline occurred throughout February. Again the ratio of Skua Lake to Algal Lake for these sestonic parameters was just below 5. The increasing carbon/nitrogen ratio after the rapid freeze of January 22–24, 1962, may reflect a metabolic shift toward carbohydrate storage by the late season plankton. Additional evidence for altered metabolic patterns comes at this time from the change in dissolved organic nitrogen accompanying the freeze: although all the other measured dissolved components of Skua Lake water increased as expected as the water froze, the dissolved organic nitrogen decreased from 2.60 to 2.47 ppm.

Quantitative results for pigments. The development of chlorophyll and carotenoid maximums in late December in Algal Lake and the succeeding peak in February suggest a midseason nutrient exhaustion and replenishment from deep ice sources in this lake. The ratio of pigment extract absorbencies (430:665 nm was used by Margalef [1968] as an index of community diversity) increased in Algal Lake plankton from an early season value near 3 to a mid-January high of nearly 5. After the second pulse of growth the ratio again dropped below 3. Chlorophyll in Skua Lake showed gradual development during late December and January and leveled off at freeze-up.

Planktonic counts. The planktonic counts are presented in Figures 36 and 37. The dominant influence in Skua Lake seems to be the late season blooms of the newly discovered *Ankistrodesmus* and *Chloridella.* Numerous members of these genera were found frozen in the ice from the previous season and had colored it a dull orange-yellow. The significantly larger numbers of phytomonads observed in Skua Lake during the second year may represent a long-term successional change in this lake. The phytomonads in Algal Lake, which were subjected to the same conditions of climate and preservation, showed no such dramatic change from the first year to the second. It is also noteworthy that the filamentous cyanophycean population in both lakes tended to decrease in mid-January; this decline, coupled with the melting of the deepest ices, may release nutrients at midseason (Figures 21–24) and trigger or enhance the late season increases of species other than cyanophyceans.

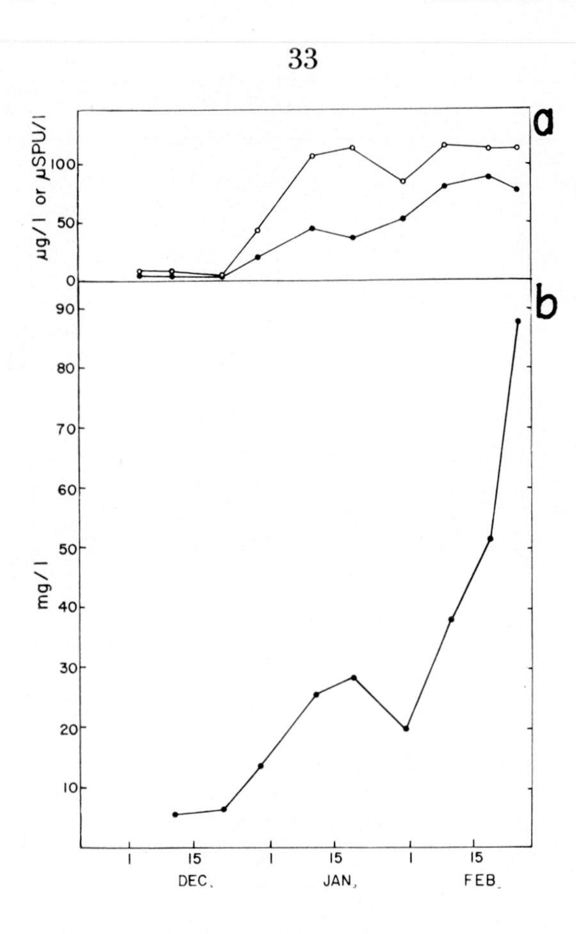

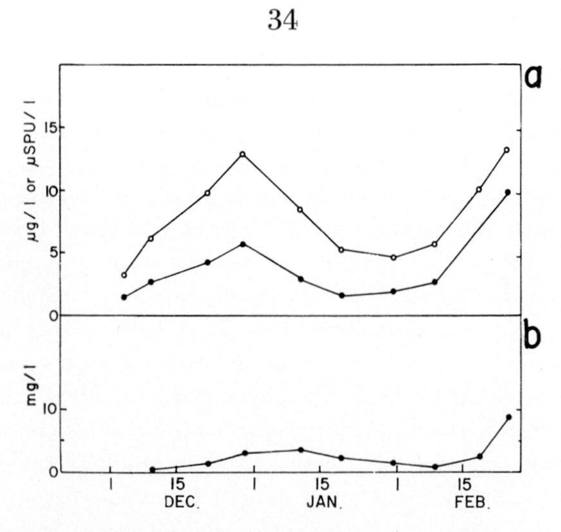

Figs. 33–34. Seasonal changes of various measures of planktonic standing crop during 1962–1963. (*a*) Pigments (chlorophyll *a* is represented by solid circles, and total carotenoids by open circles). (*b*) Sestonic weight. Fig. 33, Skua Lake (all values are for the eastern, birdless part of the lake in surface water). Fig. 34, Algal Lake (all values are for surface water).

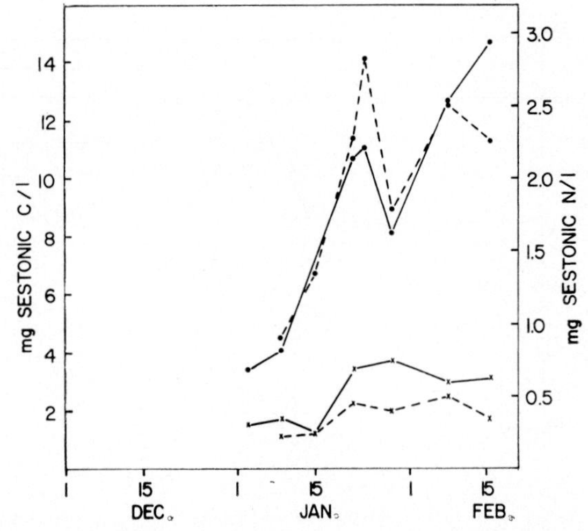

Fig. 35. Seasonal changes in sestonic nitrogen (dashed line) and carbon (solid line) in Skua (dots) and Algal (crosses) lakes, 1961–1962. These measurements of standing crop were begun only in the latter part of the season.

Carbon uptake. Carbon uptake under standard conditions (Figures 31 and 32), although the seasonal changes in available carbon dioxide (Figures 21 and 22) were ignored, showed a sensitive correspondence to the other measures of standing crop.

Qualitative results for pigments. Pigment extracts of the planktonic and benthic communities reflect the species composition noted previously. The absorption spectra of the acetone extracts of both the seston and the benthos of each lake have certain distinctive features. Typical curves are shown in Figure 38 for pigment extracted from equal volumes of water and equal areas of benthos.

The seston of Skua Lake showed a qualitatively consistent absorption spectrum throughout the season and had a distinct blue peak at about 430 nm, a shoulder at 470–480 nm, small broad peaks at 575 and 615 nm, and a sharp chlorophyll *a* peak at about 660 nm. A quite highly colored (rusty) ice from the previous season was noted in small patches, and both a ferruginous precipitate (possibly associated with iron bacteria) and planktonic cells (*Chloridella* and *Ankistrodesmus*) were found in this colored ice. The sestonic pigment from new ice formed late in the season compared to that from the adjacent water showed one-tenth of the chlorophyll *a* peak and only one-fifth to one-third of the absorption in the blue. These relationships appear to indicate a selective destruction of chlorophyll in the fraction of seston trapped in the ice during the freezing process and may partially account

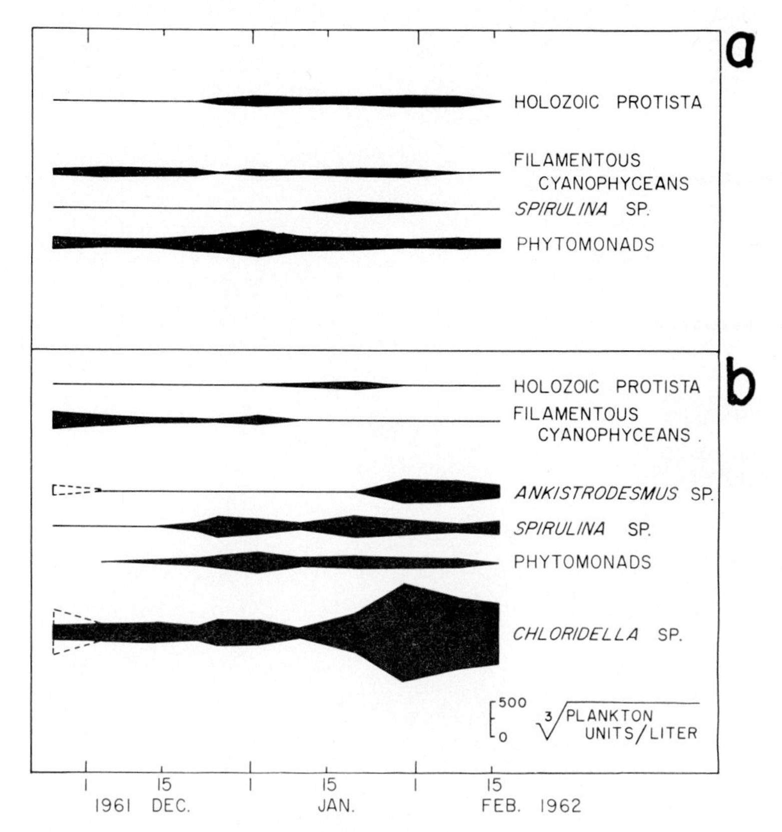

Fig. 36. Seasonal change of planktonic populations in (*a*) Algal and (*b*) Skua lakes, 1961–1962 season. The widths of the black bars are proportional to the cube root of the number of plankton per liter, according to limnologic convention. A 'unit' of the filamentous forms was considered to be a strand 100 μ long, or one spiral in the coiled forms. The dashed lines indicate numbers occurring in the ice melts.

for the yellow ice of the previous season. The exposure of Skua Lake to the wind would favor the incorporation of plankton into developing ice.

In mid-December an Algal Lake sestonic extract had a high absorption at 430 nm and a shoulder at 470–480 nm and was comparable to Skua Lake seston in the red portion of the spectrum. By December 22 a strong peak had developed at 450 nm, and the 470- to 480-nm shoulder had become more pronounced. These peaks declined again in February. No pigment was extractable from the ice of the previous season; nor was pigment detectable in the newly formed ice. The highly saline layer at the bottom of the ice column, however, was very rich in planktonic remains and thus suggested that the freezing out of the salts had been accompanied by a similar downward movement of the sestonic element.

The algal felt at the margin of Skua Lake was bright orange in the shallow regions sampled, and there was no change of spectrum quality during the season. The fact that extracts from the algal mat exposed to the air by receding water showed a drop, particularly in the 430- and 665-nm peaks, again suggests a selective decomposition of the chlorophyll *a* component.

The felt from Algal Lake was yellow-orange on the upper surface and blue-green beneath. Extracts from the total felt samples had a characteristic peak just below 400 nm and shoulders at 425, 470, 510, 575, and 615 nm. Drying in the laboratory did not appear to affect the relative absorption at these wave lengths. As we expected, the division of the felt into its orange and green portions before extraction indicated that the pigments absorbing at 425 and 475 nm were more predominant in the upper layers.

Identification of any of the components of the carotenoid pigment system is virtually impossible without further analysis. However, the composition of the algal flora sampled allows some reasonable speculation. The benthic material, which is almost exclusively

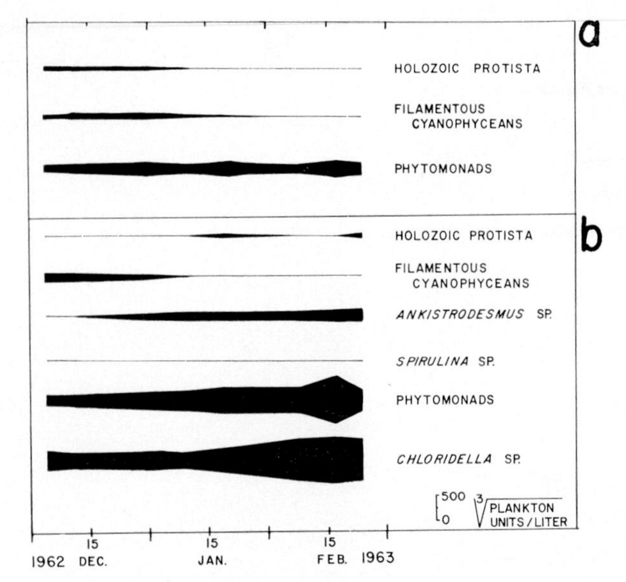

Fig. 37. Seasonal change of planktonic populations in (*a*) Algal and (*b*) Skua lakes, 1962–1963 season. The widths of the black bars are proportional to the cube root of the number of plankton per liter, according to limnologic convention. A 'unit' of the filamentous forms was considered to be a strand 100 μ long, or one spiral in the coiled forms.

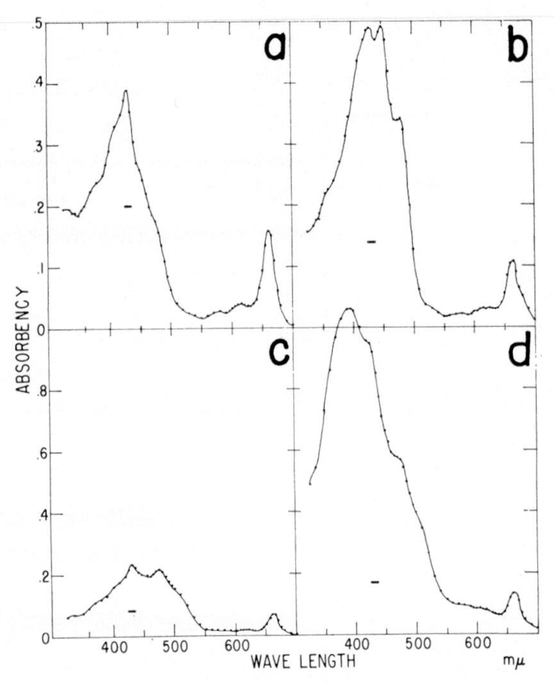

Fig. 38. Spectral absorption of acetone extracts of seston and benthos from Skua and Algal lakes. (*a*) Skua Lake seston (dry weight, 12.6 mg). (*b*) Algal Lake seston (dry weight, 7.0 mg). (*c*) Skua Lake benthos (wet weight, 108 mg). (*d*) Algal Lake benthos (wet weight, 206 mg). The seston from a 2-liter sample of each lake was extracted with 10 ml of 90% acetone. A benthos disk 0.54 cm^2 in area was extracted. A 1-cm cell was used for all samples. The bars at $\lambda \cong$ 431.5 represent the expected chlorophyll *a* absorption at this wave length as calculated from the values at $\lambda \cong 665$ and the constants given by Strickland [1960].

filamentous cyanophyceans, would be expected to contain large proportions of beta carotene, echineone (= myxoxanthin), myxoxanthophyll, and possibly antherxanthin [Goodwin, 1957; Parsons, 1961]. The expected contribution of the blue absorption of chlorophyll *a* (at 431.5 nm) is indicated by the dash on each graph; it was calculated by using the constants given by Strickland [1960]. Additional absorption at 430 and 450 nm may be attributed to myxoxanthophyll [Goodwin, 1957]. It is likely that beta carotene would produce strong absorption at about 450 and 480 nm; such an effect could account for the other obvious absorption peaks. Absorption at 615 and 570 nm may be due to phycocyanins.

The Xanthophyceae and Cryptophyceae, which are predominant in the plankton, probably contain alpha and beta carotenes, lutein, zeaxanthin or diatoxanthin, and neoxanthin as major carotenoids [Goodwin, 1957; Haxo and Fork, 1959]. Again, peaks near 450 and 480 nm are probably assignable to the carotenes and lutein. The significant numbers of the photosynthetic bacterium *Chromatium minutissimum* present in Skua Lake plankton (W. L. Boyd, personal communication, 1963) may also contribute to the sestonic pigment complement. The strong absorption at 395 nm in Algal Lake seston is not assignable but may be a pigment of the common cryptophyceans.

Benthos. The only available measures of benthic standing crop were photosynthetic pigment contents and dry weights. Because of benthic heterogeneity and occasional contamination by gravels, both of these estimates lack precision. The benthic felt at Skua Lake, which extends to a depth of only about 0.5 meter, appears to synthesize chlorophyll *a* during the second half of January (Table 8), and an unusually high value was found for Algal Lake at this time as well. This growth came at a time when deep, presumably nutrient rich ice was melting and the planktonic populations were also experiencing a rejuvenation. A more meaningful interpretation may be placed on the average values, however: Skua Lake benthic felt contained an average of 150 grams of chlorophyll *a* per square meter, 380 spu (standard pigment units) of carotenoids, and an average A_{430}/A_{665} ratio of 3.6; Algal Lake felt contained an average of 100 grams of chlorophyll *a* per square meter, 500 spu of carotenoids,

TABLE 8. Pigment Concentrations in the Benthic Algal Felt of Skua and Algal Lakes and the Ratio of Absorbencies at 430 and 665 nm, 1961–1962

	Skua Lake			Algal Lake		
Date	Chlorophyll *a*, mg cm^{-2}	Carotenoids, mspu cm^{-2}	A_{430}/A_{665} Ratio	Chlorophyll *a*, mg cm^{-2}	Carotenoids, mspu cm^{-2}	A_{430}/A_{665} Ratio
Nov. 4, 1961, 1-year-old dry felt				0.2	1.3	50.0
Dec. 6, 1961						
West	14.0	34.3	3.3	13.8	45.4	5.5
East	5.2	13.8	3.1			
Dec. 13, 1961	7.5	19.7	3.2	7.8	44.3	8.6
Dec. 22, 1961	7.6	21.8	3.4	7.8	38.5	9.9
Jan. 3, 1962	5.0	18.5	4.7	6.4	49.4	7.2
Jan. 10, 1962	3.9	13.3	3.8	11.9	41.0	7.7
Upper layer				8.8	47.6	9.4
Lower layer				7.3	28.8	8.5
Jan. 16, 1962	24.1	67.5		7.9	33.0	
Jan. 22, 1962	25.8	65.9		20.7	108.0	
Jan. 29, 1962	35.0	53.2	3.3	8.0	54.1	7.8
Feb. 16, 1962						
Wet	25.7	68.6	3.6	6.5	33.9	7.7
Recently dry	15.2	69.5	5.3			
Upper layer				4.0	21.9	8.2
Lower layer				4.6	14.2	7.4

and an average ratio of 7.8. From average production figures to be derived later, Algal Lake felt is observed to be approximately twice as efficient in its use of chlorophyll *a* for carbon fixation as Skua Lake felt. This greater efficiency may be due in part to the light penetration differences between the two lakes, although the felt, like the plankton, was apparently photosynthesizing under inhibiting light conditions [Goldman et al., 1963]. More probably the high carotenoid content of Algal Lake felt indicates an adaptive shading of the photosynthetic mechanism either intracellularly or in bulk by the development of an orange protective upper layer in the Algal Lake felt. Characteristically the Algal Lake felt was about 10 mm thick and had a definite color difference from top to bottom (Table 8), whereas the Skua Lake felt was only half as thick and had no obvious zonation.

Despite these differences in felt thickness the average dry weight of Skua Lake felt (for 48 samples) was 0.110 g cm^{-2}, whereas that of Algal Lake felt (for 41 samples) was only 0.072 g cm^{-2}.

PRODUCTIVITY

Plankton. Photosynthetic carbon fixation in both lakes proceeded for the 24-hour diel period during the summer. The intensity and the pattern of photosynthesis varied during the seasons for both the standing crop and the light regime (Figures 39–43) [Goldman et al., 1963, Figure 1]. In general, as light increased in the early morning hours, carbon fixation also increased until it became light saturated; saturation occurred at between 0.1 and 0.3 ly min^{-1}. At the beginning and the end of the season, when ambient light levels were low or when an ice cover reduced the energy reaching the plankton, photosynthesis con-

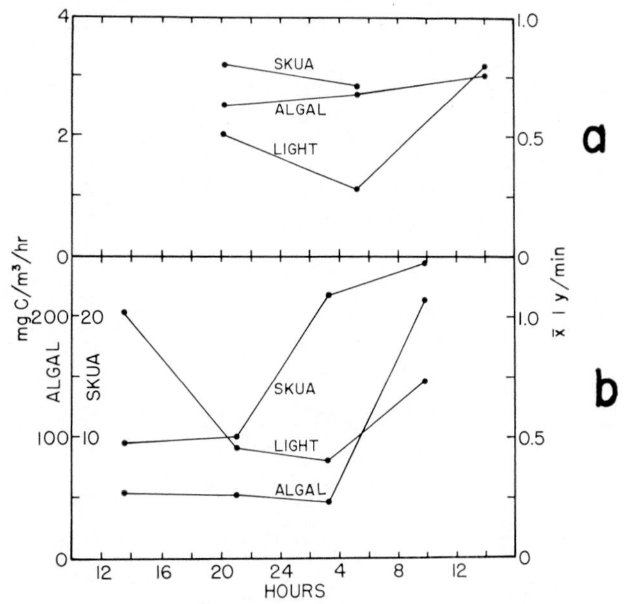

Fig. 39. Diel changes in productivity of Skua and Algal lakes measured in situ on (*a*) December 4–5 and (*b*) December 12–13, 1961. The average light intensity (total short-wave) for each period of incubation is also plotted.

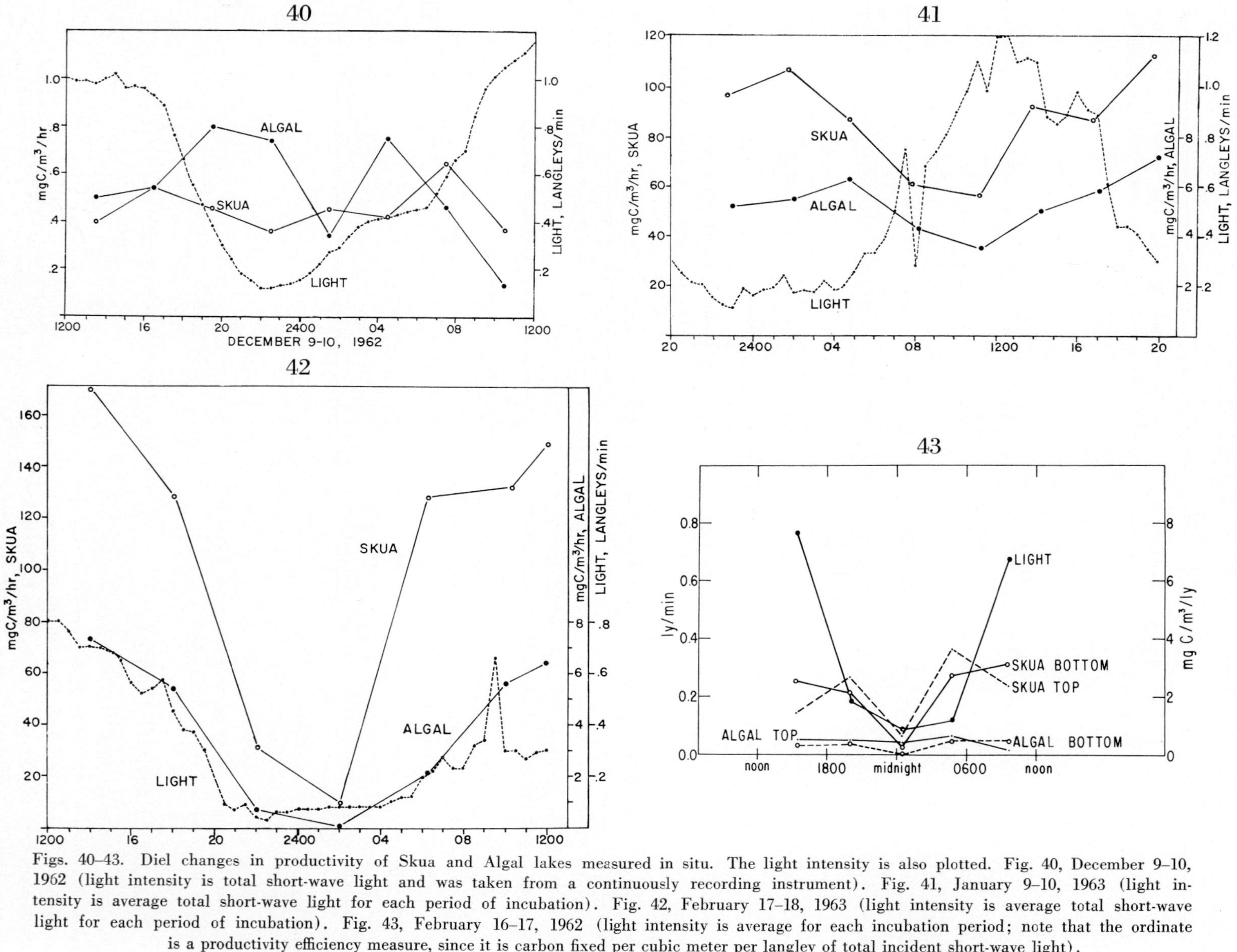

Figs. 40–43. Diel changes in productivity of Skua and Algal lakes measured in situ. The light intensity is also plotted. Fig. 40, December 9–10, 1962 (light intensity is total short-wave light and was taken from a continuously recording instrument). Fig. 41, January 9–10, 1963 (light intensity is average total short-wave light for each period of incubation). Fig. 42, February 17–18, 1963 (light intensity is average total short-wave light for each period of incubation). Fig. 43, February 16–17, 1962 (light intensity is average for each incubation period; note that the ordinate is a productivity efficiency measure, since it is carbon fixed per cubic meter per langley of total incident short-wave light).

TABLE 9. Summary Values for Productive Rates and Total Production in Skua and Algal Lakes for the Summer Seasons, 1961–1963

	1961–1962	1962–1963
Average seasonal planktonic photosynthetic rate, g C m^{-3} day $^{-1}$		
Skua Lake	1.55	6.72
Algal Lake	0.26	0.19
Total carbon fixed, plankton, kg		
Skua Lake	81	235
Algal Lake	2.0	1.6
Average seasonal benthic photosynthetic rate, g C m^{-2} day $^{-1}$		
Skua Lake	1.56	2.56
Algal Lake	1.91	3.63
Total carbon fixed, benthos, kg		
Skua Lake	244	328
Algal Lake	66.4	116
Ratio of planktonic total and benthic total		
Skua Lake	0.33	0.72
Algal Lake	0.03	0.01
Total carbon production, kg		
Skua Lake	325	563
Algal Lake	68.4	117.6
Incident langleys in 90-day season (Dec., Jan., and Feb.)	59,720	60,021
Efficiency, %		
Skua Lake	0.6	1.5
Algal Lake	0.2	1.3

Methods of summation are discussed in the text. The caloric efficiency values are calculated by assuming that each gram of fixed carbon represents 11 kcal of energy and that half of the incoming radiation was available for photosynthesis. The second part of this assumption, which neglects both changing albedo and ice absorption, leads to minimal values for efficiency.

tinued to increase with light until noon and then dropped again as the solar altitude decreased (Figures 39 and 42). In midseason the diel pattern of photosynthesis was complicated by the inhibition of the phytoplankton at high light levels [Goldman et al., 1963]; carbon fixation decreased from a morning high to a midday low and then increased again as the inhibitory light intensities decreased. Maximum photosynthetic efficiency might occur in the morning or afternoon (Figure 43) or even near midnight on bright days (Figure 41). We return to a discussion of the photosynthetic response to light intensity in a later section.

Throughout the season, as standing crop increased (Figures 31–34), daily carbon fixation rates did also (Figures 10 and 11). Algal Lake showed maximum productivity in late December, whereas Skua Lake showed maximum productivity in late January. The erratic nature of the carbon fixation rates in Algal Lake during the second season was probably due to the repeated partial freezing of this small body of water. Plankton from the last sampling day (February 24, 1963) had been nearly doubled in concentration by downward freezing of the ice and exhibited a high rate of photosynthesis (1300 mg C · m^{-3} day^{-1}). At that time nutrients had apparently been fully released from the previous season's ice and had also been concentrated along with the plankton.

Planktonic photosynthesis in Skua Lake was considerably greater in the second season. This increase may reflect an increasing nutrient supply provided by the avian fertilization and the decrease of volume between the two seasons. This elevated productivity apparently came primarily from the cryptomonad population that flourished throughout the latter part of the second season in Skua Lake (Figure 37). The unit volume primary productivity rates in Skua Lake during the first season were roughly 4 times those in Algal Lake, whereas the rates in Skua Lake during the second season were approximately 40 times those in Algal Lake (Table 9). Extrapolation of this trend indicates a remarkable eutrophication of Skua Lake that may continue until the nutrient input from the birds

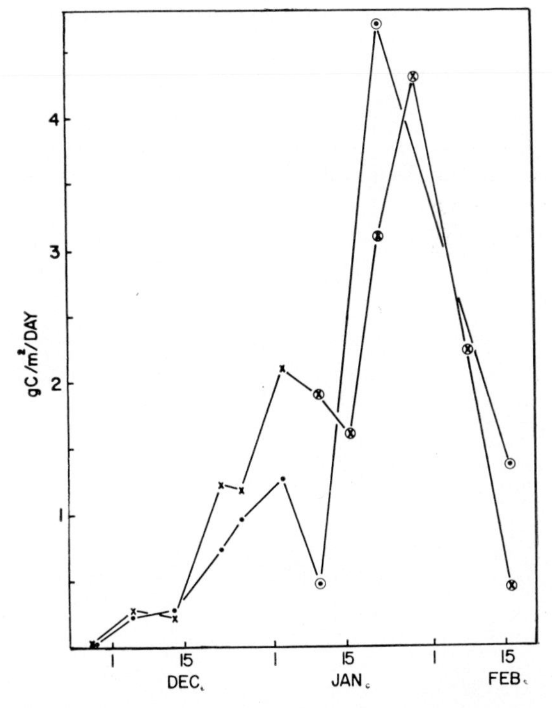

Fig. 44. Seasonal course of benthic productivity in Skua (dots) and Algal (crosses) lakes, 1961–1962. The 'lab run' (circled) values are derived from laboratory measurements multiplied by an empirically determined conversion factor of about 8. Other points were derived from in situ measurements.

ceases, the lake is flushed during a wet year, or the physiological constraints of the high salt concentration become limiting.

When the total chloride in the lake melt was computed for both seasons, a close correspondence of the two lakes in the rate of increase of the total chloride in the melted water was noted. For Skua Lake this rate was 18 kg day^{-1}, and for Algal Lake it was 8 kg day^{-1}. In both cases the increase proceeded linearly to January 15 and then leveled off abruptly except for Skua Lake in the second season, when leveling off occurred about January 1. Both lakes approached their final chloride totals (100%) at a rate of about 4% day^{-1}. The maximum planktonic production rates for each of the lakes likewise developed quasilinearly in the first year (1961–1962), Skua Lake at 5% day^{-1} and Algal Lake at 4.5% day^{-1}. However, in the second season, production increased in a more logarithmic fashion, $k = 0.12$ day^{-1} for both lakes, or $k = \ln$ (population$_{\text{time 1}}$/population$_{\text{time 2}}$). Benthic production rates in the first season climbed more slowly than planktonic rates (Skua Lake at 2.6% day^{-1} and Algal Lake at 2.2% day^{-1}). Observations for the second season were not timed properly to permit calculation of this rate of benthic increase.

The preceding analysis suggests that primary planktonic production in the first year could have been controlled by a process related to melting: the release of nutrients, the trapped viable algal cells, or both. Similarly, total benthic production in the lake showed features suggesting the presence of nutrient limitation or some other environmental constraint on growth. In the second year, however, total planktonic production grew increasingly rapidly with time and produced, at least in Skua Lake, a significantly higher standing crop than it had in the first year.

When the daily productivity rates were adjusted to the volume of the available melted water, the planktonic photosynthetic capacity of the whole lake could be estimated. Figures 6–9 show the open water volume and the planktonic productivity of the whole lake throughout the season. On this basis Algal Lake performed similarly during both seasons despite the diminution of volume the second year. Skua Lake also had less volume the second year and showed considerably more planktonic productivity, presumably owing to the increase in nutrient concentrations in the second year.

The integration of Figures 6–9 yields the total amount of carbon fixed by the plankton of each lake, and the division of this total by the amount of melted water integrated over the season gives a mean rate of photosynthesis for the season. These summary figures appear in Table 9. Skua Lake had an average rate of planktonic primary production during the second season >4 times that during the first season and produced nearly 3 times the amount of fixed carbon. The rate of production in Algal Lake, however, dropped during the second season to about 70% of the first

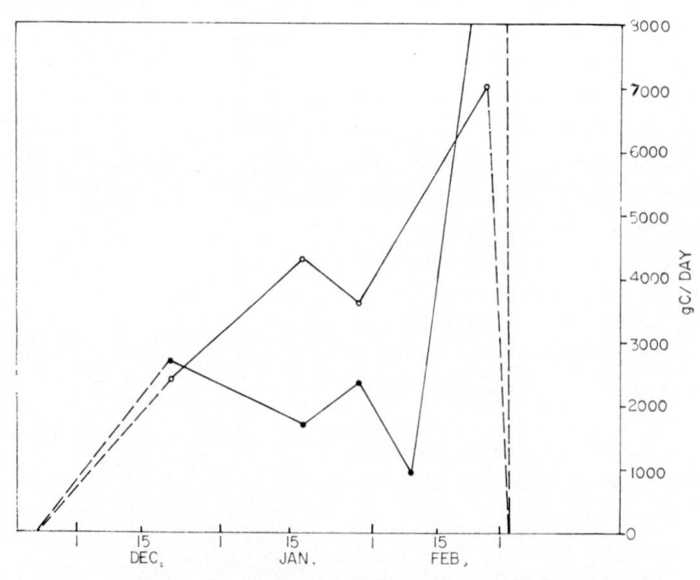

Fig. 45. Seasonal course of benthic productivity in Skua (solid circles) and Algal (open circles) lakes, 1962–1963. All values were derived from in situ measurements.

TABLE 10. Percentages of Total Photosynthetic Carbon Fixation for Particles Larger Than 5–10 μ in Surface Water on January 15–16, 1962, during a Diel Productivity Measurement

	Skua Lake Carbon Fixation			Algal Lake Carbon Fixation	
Time, hours	%	Total, mg C m^{-3} hr^{-1}	Average Total Light Level, ly min^{-1}	%	Total, mg C m^{-3} hr^{-1}
1200–1600	60	114	0.81	30	10
1631–2030	55	134	0.59	42	13
2040–2440		163	0.12		16
0100–0500		115	0.20		12
0515–0915	50	70	0.46	60	1.8
0930–1330	30	81	0.98	69	8.5

season's rate, and the total fixed carbon was correspondingly less. Without enrichment from the sea one might expect small lakes of this type to decrease in planktonic production because of the low temperatures and the slow regeneration of nutrients from sedimented materials.

Benthos. The benthic primary production estimates, though fewer in number, were augmented by laboratory extrapolations and were treated in a similar fashion. Figures 44 and 45 show the daily rates of benthic carbon fixation for both lakes throughout the season. Both lakes showed remarkably similar rates and seasonal trends. Carbon fixation rose throughout the open water period, apparently without a discernible nutrient limitation in the second season but with a possible nutrient limitation in the first. Skua Lake showed a very active carbon fixation in the benthos as well as in the plankton toward the end of the second season.

Comparisons of lakes and seasons. When these data were placed on an areal basis and converted to values for the whole lake (Figures 46–49), the general similarity of the 2 years is apparent. The results of the integration to give the total benthic and planktonic production appear in Table 9 and show similar trends. Skua Lake benthos had a markedly higher average benthic production rate in the second season than in the first. Algal Lake benthos (unlike its plankton) fixed carbon in the second season at nearly twice the rate that it did in the first season. In these lakes the nutrients available to the plankton during the first year may have been available by way of sedimentary regeneration only to the benthos the second year.

When the total planktonic production is compared with the total benthic production (Table 9), we see that the algal felt of both lakes fixes considerably more carbon than the phytoplankton in the overlying water. This effect is especially true in Algal Lake, which was in fact named for the excesses of felt at its margins. The total of both modes of production permits the most general comparison of the two lakes. During the first season Skua Lake was 4.8 times as productive as Algal Lake and had 5.7 times the area, whereas during the second season Skua Lake was again 4.8 times as productive and had only 3.8 times the area.

A measure of the total productive efficiency (Table 9) based on the assumptions in Goldman et al. [1963] shows that both lakes increased their use of the available photosynthetic light energy from the first season to the second, Algal Lake showing the more dramatic increase.

ECOLOGIC PHYSIOLOGY

To understand the physiology of the planktonic and benthic primary producers more completely, a number of laboratory and field experiments were performed to examine the dark fixation of carbon and the size fractionation of the primary producers as well as the effects of snowfall and temporary slush formation on primary productivity. Then interactions of the benthic and planktonic communities and the effects of nutrients, light, and temperature on photosynthesis were also considered.

Particle size. An estimate of the particle size of the organisms fixing carbon was sought by using filters of two porosities. Each sample was filtered through both a 5- to 10-μ membrane filter (Gelman, Sleicher, and Schull) and a 0.45-μ filter (Millipore HA). In general, dark fixation showed somewhat lower percentages of activity in the large size fraction than light fixation did, presumably because much of the dark fixation could be attributed to bacterial activity or adsorptive phenomena (discussed below). Four duplicate filtrations during a diel experiment in an unusually windfree period (Table 10) showed the 5- to 10-μ fraction to be of diminishing importance in Skua Lake (from noon one day to 1330 hours the next) and of increasing importance in Algal Lake; this variation showed no apparent relation to light or total productivity but may represent the migration or the settling of weakly motile phytoplankters in Skua Lake during the calm period. Initially (at 1200–1600 hours) the large size fraction was low (33%) in the bottom waters of Skua Lake and high (79%) in those of Algal Lake. Another observation (on January 29, 1962) again revealed that just about half of the carbon fixation occurred in the large size fraction (59 and 40% in Skua

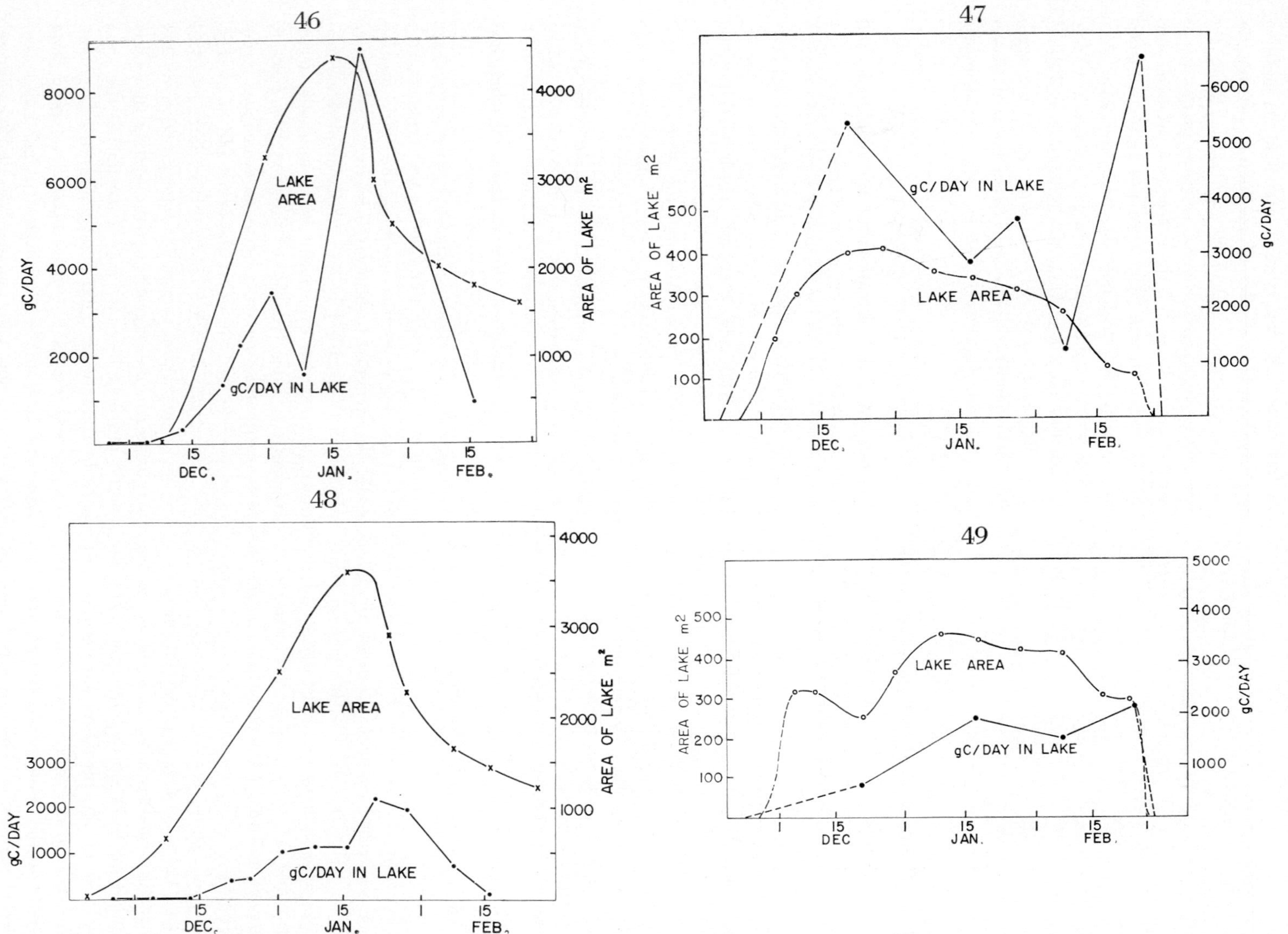

Figs. 46–49. Seasonal changes in benthic productivity computed for the whole lake and for the area of melt available to the benthos. The productivity curve results from multiplying the daily rates by the corresponding approximate area. Fig. 46, Skua Lake, 1961–1962 (benthos limited to the area above 0.5 meter in depth; daily rates are shown in Figure 44). Fig. 47, Skua Lake, 1962–1963 (benthos limited to the area above 0.5 meter in depth; daily rates are shown in Figure 45). Fig. 48, Algal Lake, 1961–1962 (benthos extended to the maximum depth; daily rates are shown in Figure 44). Fig. 49, Algal Lake, 1962–1963 (benthos extended to the maximum depth; daily rates are shown in Figure 45).

Lake for light and dark fixation, respectively, and 53 and 47% in Algal Lake for light and dark fixation, respectively).

Slush entrapment. Stormy weather on December 26 and 28, 1962, brought several centimeters of snow to the cape and produced an ice cover on Skua Lake composed of 9 cm of dense frothy ice on the surface with 6 cm of slushy water beneath. The water trapped within the slush was rich yellow and contained an abundance of plankton. Samples of both the slush water and the lake water underlying the ice cover were inoculated with radioactive carbon and incubated both in surface light and under the ice cover (9.8% of the red component, 11.5% of the green component, and 4.8% of the blue component of the surface light penetrated this slush). Table 11 shows the results of this experiment. In contrast to the open water sample, the slush water sample contained forms somewhat better adapted to the high light conditions of shallow water. The slush water showed a high dark uptake and required longer filtration times than the underlying water, a larger population of organisms in the slush thus being indicated. The planktonic counts of the predominant forms showed a ratio of slush water to underlying water of 3.1:1.0 for cryptomonads, 1.4:1.0 for *Chloridella*, and 1.2:1.0 for small green flagellates. These data suggest that cryptomonads are the important light tolerant forms trapped by the slush. Under the same conditions of incubation the phytoplankton from Algal Lake also showed a marked inhibition in full incident light (Table 11), though much less severe than that shown by the plankton from the Skua Lake slush. This difference may reflect the greater transparency of Algal Lake.

On January 19, 1963, a polyethylene pan of Skua Lake water was placed near the lake and allowed to freeze slowly to a depth of 2 cm. The underlying water and the melted ice were compared with respect to their carbon-fixing ability. The photosynthesis of the ice melt was <4% of that of the unfrozen water. We concluded from this result that active photosynthetic cells are largely absent in developing permanent ice.

Adsorption of ^{14}C. Since the dark bottle in all ^{14}C measurements represents both a 'reagent blank' and a control variable, an effort was made to determine the extent to which the particulate matter and the filter itself localized radioactive carbon in the absence of metabolic activities. Experience at Lake Vanda had shown that an average of 0.07 count sec^{-1} was retained by the Millipore HA filter when 50 ml of prefiltered water containing ^{14}C was run through it. This value was well below the background value (0.25 count sec^{-1}) and was ignored. Formaldehyde was added to the ^{14}C experimental bottles (to 1.6% vol./vol.), and the radioactivity of the poisoned plankton was determined. The values ranged between 2 and 53% of the dark values (both sets were corrected for the filter adsorption mentioned above), and the largest percentages came from Skua Lake when the sestonic values were also high. Felt experiments showed carbon uptake after poisoning to be between 2 and 22% of the normal dark values, the poisoned Algal Lake felt giving consistently higher uptake than the Skua Lake felt. In the very unproductive surface waters of Lake Bonney and Lake Vanda formaldehyde poisoning led to values insignificantly different from those of the dark (or light) samples [Goldman, 1964].

TABLE 11. Carbon Uptake on December 29, 1962, by Skua Lake Phytoplankton Interstitial in the Surface Slush and the Underlying Water

Incubation Site	Skua Lake Water: Slush	Skua Lake Water: Underlying	Algal Lake Water
Light			
Shallow water	39.0	17.4	42.5
Under ice	163.5	151.3	71.0
Dark	27.3	11.7	3.8

Carbon uptake values are in counts per second corrected for background. Algal Lake water was run concurrently at the same location and incubated in two locations at 1230–1630 hours; the light intensity averaged 0.93 ly min^{-1}.

The addition of antibiotics (10 mg of penicillin and 5 mg of streptomycin sulfate per liter) to the plankton showed no consistent results in a number of experiments.

Felt–plankton interactions. To determine if the presence of felt influenced the planktonic carbon uptake, on six occasions the photosynthesis of plankton was measured in duplicate bottles, one of which contained a disk (2.14 cm^2) of benthos. In both lakes there were as many instances of 'stimulation' as of 'inhibition' of the plankton. These equivocal results do not rule out the possibility of interaction but fail to demonstrate it above the experimental noise encountered.

Nutrient-limiting factors. A program to determine mineral limitations to primary productivity was instituted in December 1961. Water and felt collected from Algal and Skua lakes on December 22 were assayed in the laboratory for nutrient deficiencies by the separate additions of 100 ppb phosphate phosphorus, 1000 ppb nitrate nitrogen, 100 ppb chelated

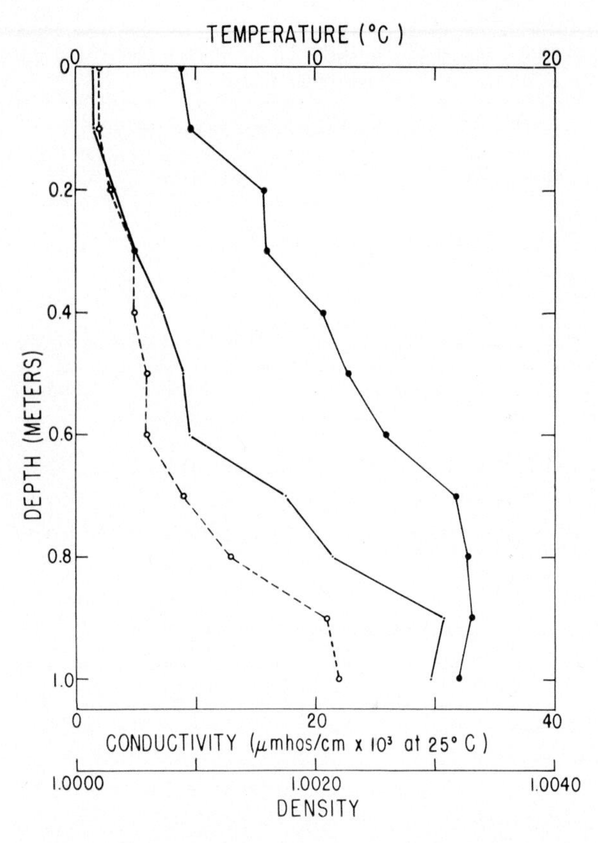

Fig. 50. Thermal and conductimetric profiles of pond 14 made on January 16, 1962 (large dots represent temperature, small dots conductivity, and open circles density). The density stratification of salts presumably arose through previous freezing out processes preserved in this pond after thawing, because the pond, surrounded on all sides by rock walls, was completely sheltered from the strong prevailing winds; the resultant meromictic condition may therefore be termed 'cryogenic.' The density figures were calculated from conductivity, NaCl being assumed the only salt present.

iron to both filtered and unfiltered water [Goldman and Mason, 1962], and a trace metal mixture. These levels of nutrients were in some cases significantly above the initial environmental levels and thus may have caused inhibition or lack of stimulation. However, significant stimulation of planktonic carbon uptake (72% over the control in 4 hours) was found for Algal Lake plankton to which phosphate had been added. The next sampling (on December 27) of Algal Lake yielded plankton that responded only weakly to a range of phosphate additions, the maximum being 13% over the control in 4 hours at the 10-ppb level. These results are consistent with our analytical data, which show no detectable phosphate in the water on December 22 but 7.45 ppb on December 27.

On January 2, 1962, the plankton and the benthos of neither lake responded photosynthetically to a small addition of a water extract of freshly collected skua dung (such that the phosphate phosphorus content was 60 ppb); inhibition was indeed produced by this extract and thus suggests that the 'raw' dung may require significant modification or further dilution before it will serve as a stimulating nutrient source for photosynthetic organisms. It is not uncommon for domestic sewage effluent that has had secondary treatment to be a better photosynthetic stimulant than sewage that has had only primary treatment. Another experiment on the same date involved the addition of 100 mg of dextrose per liter. This additive also failed to produce an increase of either light or dark uptake in either felt or planktonic cultures. By January 9, 1962, the planktonic carbon fixation was strongly inhibited (to 41% of control) by the 100-ppb level of the phosphorus, and there was no stimulating effect on the plankton in either lake by nitrogen, calcium, sulfate, or a trace element mixture. At that time the Skua Lake algal felt may have responded with some stimulation to phosphorus and calcium, although the effect, if any, was minimal and was obscured by the usual variability inherent in benthic sampling. The disappearance of phosphate stimulation during early January suggests that some other factor or condition was limiting the carbon uptake system in Algal Lake at that time despite the analytically undetectable phosphate levels and the apparent sufficiency of nitrogen in the water. Renewed growth in mid-January 1962 was coincident with the observed melting and the release of the deepest ice (brine) layer in the lake. For a short time a portion of this layer remained at the bottom, its own isolated population of plankton being visibly in bloom (on January 15). This bottom water was 40% richer in calcium (with respect to chloride) than the surface water (Table 5). When mixing into the overlying waters began, an abrupt rise in the specific gravity, the chloride, and the divalent ions ensued; this natural addition of both nutrients and an independently developed standing crop (principally cyanophyceans) gave a second stimulus to the primary production of Algal Lake. A repeat of the phosphate addition experiments in the second season (on January 10, 1963) showed no planktonic stimulation by 500 ppb of phosphorus at that time.

A final broad spectrum nutrient-limiting factor analysis of Skua Lake in February 1962 showed no stimulations. These results confirmed the importance of the organic and mineral additions to Skua Lake

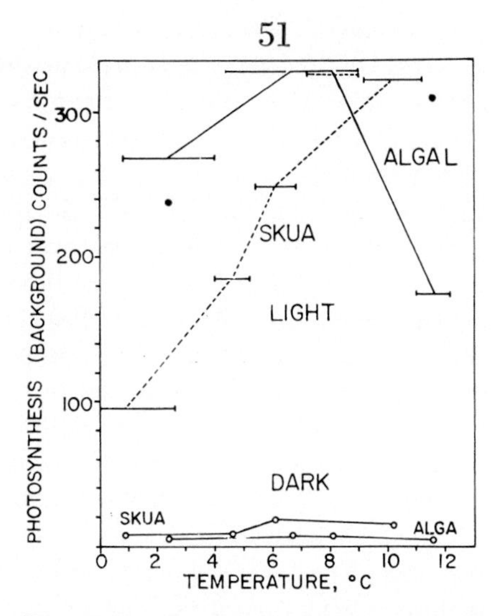

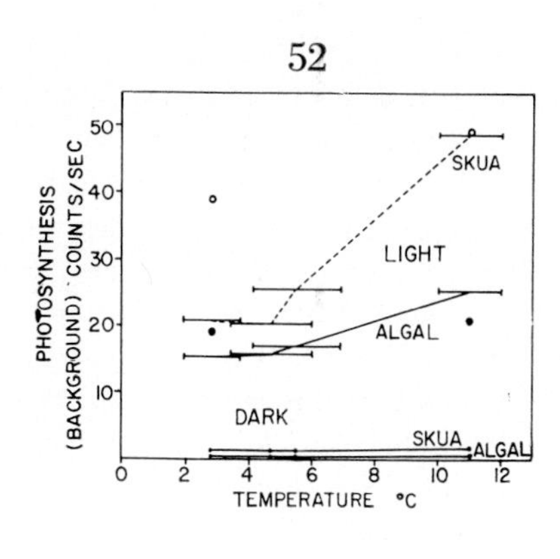

Figs. 51–52. Response of members of the Skua and Algal lakes photosynthetic communities to temperature. Incubation was done at lakeside in water-filled thermostatically heated white pans that, owing to their high reflectivity, elevated the ambient light by 10%. The bars indicate the temperature range measured in the pans during incubation. Fig. 51, Skua (December 20, 1962) and Algal (December 21, 1962) lakes. During incubation (1430–2010 hours on December 20 and 1440–2100 hours on December 21) the phytoplankton received an average of 0.74 ly min^{-1} (range of 0.25–1.05) on December 20 and 0.81 ly min^{-1} (range of 0.22–1.08) on December 21. The circled values are from shielded bottles of Algal Lake water receiving an average of 0.08 ly min^{-1}. Fig. 52, Skua and Algal lakes on December 30, 1962. During incubation (Skua Lake, 1545–2045 hours; Algal Lake, 1645–2145 hours) the phytoplankton received an average of 0.75 ly min^{-1} (range of 0.28–1.04) for Skua Lake samples and 0.55 ly min^{-1} (range of 0.26–0.90) for Algal Lake samples. The circled values are from shielded bottles of water from each lake, the upper points being Skua Lake samples receiving an average of 0.08 ly min^{-1} and the lower points being Algal Lake samples receiving 0.06 ly min^{-1}.

associated with the skua bathing activities previously discussed. The general lack of demonstrable nutrient-limiting factors suggests that other parameters of the environment may be limiting productivity. It has previously been shown [Goldman et al., 1963] that both continuous high light intensities and low temperatures impose limitations on photosynthesis in these shallow ponds. In virgin environments of this sort, closely associated with fertile ocean waters, nutrient limitation would be expected to be of secondary importance.

Temperature and light. In the absence of clearly defined nutrient limitations the production of the phytoplankton and the benthos of the Cape Evans lakes is probably most limited by light, temperature, or some combination of the two. The natural environment presents important spatial and temporal gradients of temperature (Figures 12 and 13), values varying by well over 10°C within the distance of a single meter (Figure 50) and by similar amounts within a few hours of time at the same spot. Photosynthetic organisms must be prepared not only to take advantage of this diverse and changing milieu but also to be able to withstand intermittent periods of supercooling, freezing, and perhaps desiccation. Some aspects of the response of the photoplankton of Cape Evans to light have been discussed in earlier papers [Goldman et al., 1963, 1969].

The responses of photosynthetic organisms to temperature were measured in the field by using four plastic water baths warmed with conventional thermostatic electric aquarium heaters. Equilibration of the baths with the ambient ice water temperatures was rapid, and the bath temperature usually varied within 1°C about the nominal mean. When bottles containing water at 3°C were introduced into a 9°C bath, the water reached 92% of the required 6°C change in 23 min.

When a series of four temperature measurements was run (on December 20, 21, and 30, 1962), both the light and the dark uptakes of carbon were stimulated by increasing temperatures (Figures 51 and 52).

In one case (on December 21, 1962, in Algal Lake) an apparent optimum near 7°C was found, although this effect did not reappear at the later date. The thermal effect on carbon uptake in the light was twice as great for Skua Lake phytoplankton as for Algal Lake phytoplankton in both experiments, and both lakes showed larger temperature responses on the later date. Both of these experiments were performed in inhibiting light intensities, the average total radiation during incubation being 0.74 ly min^{-1} for December 20, 0.81 ly min^{-1} for December 21, and 0.65 ly min^{-1} for December 30.

The response of noninhibited photosynthesis to temperature would appear less sensitive than that of inhibited photosynthesis. Shielded bottles included in the preceding experiments at extreme temperatures permitted the evaluation of Q_{10} values for several light levels: On December 21 bottles receiving full sunlight (81 ly min^{-1}) had photosynthetic Q_{10} values of 4.4 for the first 10°C above freezing for Skua Lake and of 1.4 for that of Algal Lake (the 11.6°C datum being disregarded), bottles receiving 10% sunlight (0.08 ly min^{-1}) had Q_{10} values of 1.2 for Skua Lake and 0.9 for Algal Lake, and bottles receiving 1% sunlight had a Q_{10} value of 0.9 for Skua Lake. Thus it appears that the photosynthetic thermal response may depend on light intensity such that, as light increases, higher temperatures become more and more favorable than lower temperatures for photosynthesis.

This conclusion is supported by data for December 30, when surface light intensities produced Q_{10} (1°–10°C) values of 4.2 for Skua Lake and 2.1 for Algal Lake and 10% of ambient light produced Q_{10} values of 1.3 for Skua Lake and 1.2 for Algal Lake. Again on January 18, 28, and 30, 1963, the plankton showed photosynthetic Q_{10} values to decrease with a decreasing light level (Table 12). By that time the differences in the temperature responses of the plankton of Skua and Algal lakes had essentially vanished. These data are comparable to those collected the previous season by Goldman et al. [1963] and confirm the speculation advanced by them concerning the relative importance of temperature to phytoplankton inhibited by light. Viewing the relationship another way, one could say that higher temperatures ease the problems posed by inhibiting light intensities. The physiological basis for these effects is unknown, but one may speculate that the local biochemical repair of the damage caused by the intense light proceeds more rapidly at the higher temperatures. Furthermore the products of photosynthesis may be metabolized more rapidly at the higher temperatures and thus may prevent the accumulation of possible inhibitory quantities of intermediates.

In summary the temperature responses of the phytoplankton in the few hours of experimental exposure indicate that the measured photosynthetic process is not narrowly stenothermal and that elevated temperatures not only augment photosynthesis in light-limited

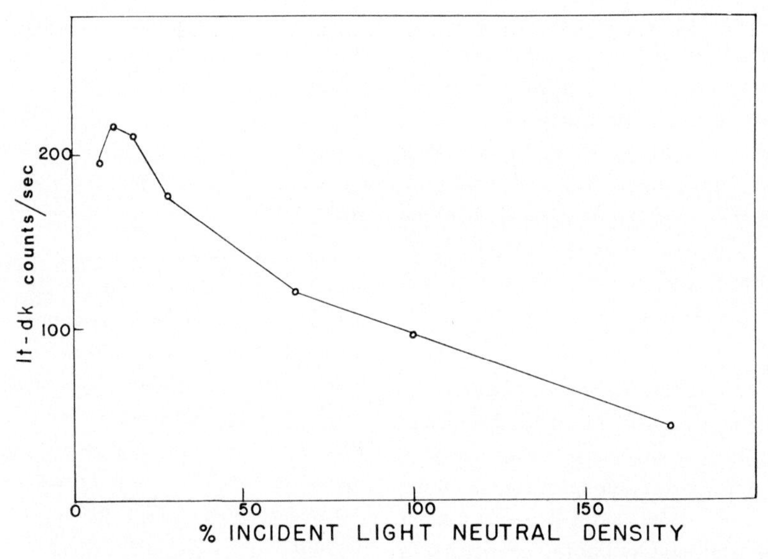

Fig. 53. Relationship of photosynthesis to light intensity for Algal Lake, December 9, 1962. The experiment was performed in situ by using neutral density screens at 1200–1500 hours. The average total short-wave light intensity was 0.99 ly min^{-1} (range of 0.97–1.02). The ambient temperature was near 4°C.

Table 12. Response of Photosynthetic Carbon Uptake to Light and Temperature

Relative Light, %	Average Energy Level, ly min^{-1}	Response at Lower Temperature, counts sec^{-1}	Response at Higher Temperature, counts sec^{-1}	Q_{10}
Skua Lake water, Jan. 18, 1963		*3.5° ± 1.5°C*	*14.0° ± 1.9°C*	
100	0.92	67.8	102.7	1.6
56	0.51	74.8	116.3	1.6
32	0.29	76.2	108.0	1.4
10	0.09	66.1	83.1	1.2
0	dark	4.8	7.8	1.6
Algal Lake water, Jan. 28, 1963		*0.5° ± 0.5°C*	*10.0° ± 2.0°C*	
135	1.13	5.66	20.88	4.9
68	0.57	9.99	23.34	2.7
43	0.36	13.11	27.32	2.3
14	0.12	13.76	18.21	1.3
0	dark	0.97	1.34	1.4
Skua Lake water, Jan. 30, 1963		*2.1° ± 0.1°C*	*15.0° ± 2.0°C*	
135	0.93	15.48	60.12	6.0
68	0.47	24.24	54.92	2.3
43	0.30	28.87	51.17	1.6
14	0.10	24.03	14.94	0.9
0	dark	1.44	1.82	1.2

Samples were incubated for 3 hours near midday at lakeside, during which the average light intensity was well above the photosynthetic optimum. The values for photosynthetic carbon uptake response do not include background.

phytoplankton but also tend to offset the effects of inhibiting light levels. Furthermore these results suggest that the brackish water algae of the Antarctic are not narrowly psychrophylic.

Some of the effects of light in producing photosynthetic inhibition and injury in Cape Evans phytoplankton have been documented elsewhere [Goldman et al., 1963]. This work stemmed primarily from the first field season, and further experiments covering the second field season are reported below.

In early December 1962 the first melt waters of Algal Lake were relatively productive (Figure 11), and bright ambient light averaging 0.99 ly min^{-1} of total short-wave radiation produced the photosynthetic response shown in Figure 53. Note that there was an optimum at about 14% of surface light (which was 0.14 ly min^{-1} that day) and that it was succeeded by an initially rapid decrease followed by a slower decline. This two-stage nature of the inhibition curve suggests that two important processes may be controlling the inhibiting part of the photosynthetic response. The first and steepest stage extrapolates upward to a

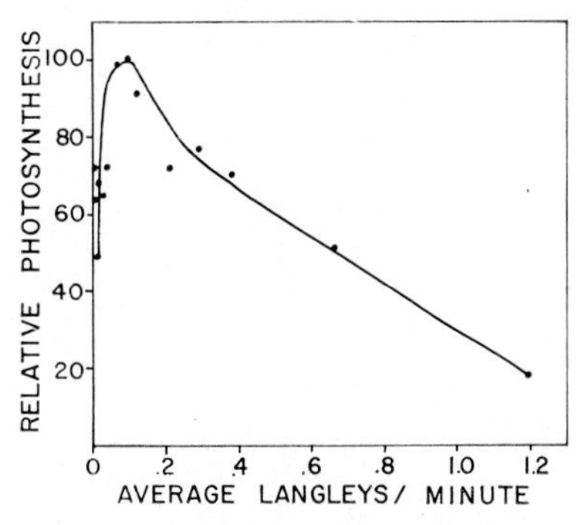

Fig. 54. Relationship of photosynthesis to light intensity for Skua Lake, December 30, 1962. The experiment was performed in situ by using neutral density screens at 1200–2400 hours in three successive 4-hour periods. The average light intensity for each of these successive periods was 1.19 (range of 1.04–1.32), 0.66 (range of 0.37–1.04), and 0.29 (range of 0.18–0.38) ly min^{-1}. The temperature in the ambient water was just above freezing.

'total inhibition' value of about 100% of incident light, whereas the second extrapolates to about twice this value and appears to begin affecting photosynthesis at about 0.5 ly min^{-1}. An attractive hypothesis would be to assign to the first, steeply sloping section of the curve the designation 'inhibiting,' which would connote short-lived easily reversible damage, and to the second, less steeply sloping section the designation 'injurious,' which would connote more permanent damage. The possible countereffect of increased intracellular heat at high light intensities [Goldman et al., 1963] could also be important in this region.

A comparable curve (Figure 54) fits the data for Skua Lake (on December 30, 1962). Again inhibition began at low light intensities (0.1–0.3 ly min^{-1}) and declined fairly rapidly; at >0.3 ly min^{-1} this rate decreased to a value, $-62\%(\text{ly min}^{-1})^{-1}$, close to that found in the December 9 Algal Lake experiment, $-71\%(\text{ly min}^{-1})^{-1}$. The optimum light intensity in this series was very close to 0.10 ly min^{-1}.

In addition to the clear and opaque bottles in the January 9–10, 1963, diel run, bottles with neutral density screens passing 10% of the light, bottles with aluminum foil reflectors increasing the light to about 175% of ambient, and bottles suspended at 35 cm in each lake (at the level of 73% of full sunlight in Algal Lake and 36% in Skua Lake) were included for each period. The course of the photosynthesis and the *p*H in these bottles throughout the day is plotted in

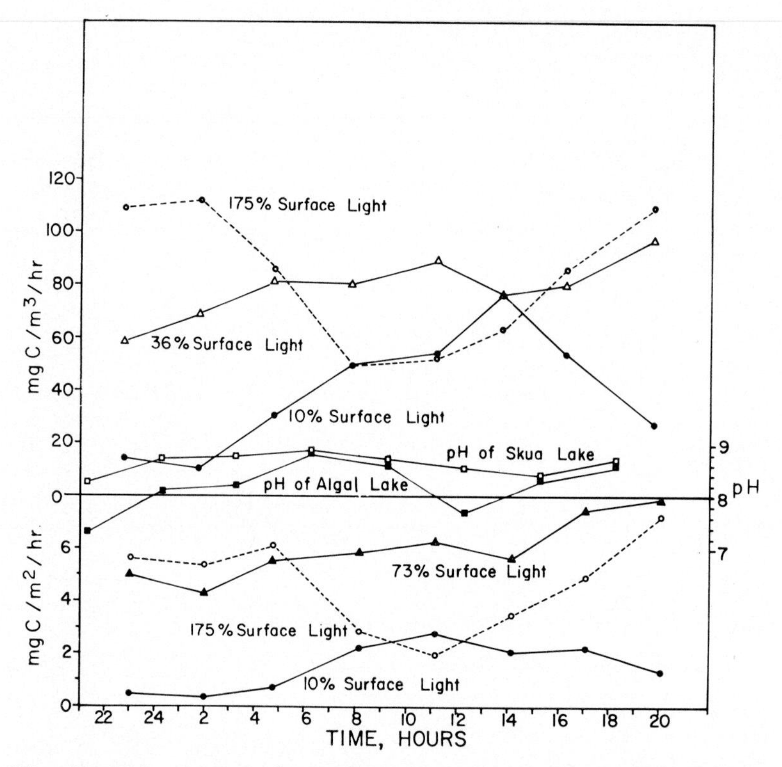

Fig. 55. Diel course of photosynthesis in bottles equipped with reflectors (175% of ambient light), screens (10%), and incubated at a depth of 35 cm (36% in Skua Lake, 73% in Algal Lake) on January 9–10, 1963. The ambient temperature varied from 6° to 9°C. Compare this figure to Figure 41, in which the data from unaltered surface bottles and light measurements are plotted.

Figure 55, which should be compared with Figure 41. The carbon uptake in the reflector bottles and the *p*H of the lakes closely paralleled the 'standard' measurement of productivity. The morning photosynthetic maximum occurred at about 0.3 ly min^{-1}. (Here the assumption is made that, when lines cross, there is a bracketing of the intensity of maximum photosynthesis and the curve of photosynthesis in response to light may be considered symmetrical in this region.) The six crossings of the lines between 0400 and 0600 hours from the plain bottles, bottles with reflectors, and bottles suspended at 35 cm in both lakes point to this value. Since 0.3 ly min^{-1} is considerably higher than any of the photosynthetic optimums determined previously near midday, the shape or the position of the photosynthetic curve may change as the light intensity rises or the phytoplanktonic population at this time may have become adapted to the higher intensities.

The bottles receiving 10% of the surface light closely followed the insolation curve, as one might expect from phytoplankton functioning at suboptimal light intensities. The bottles suspended at 35 cm, which were shielded by water rather than by a truly neutral density screen, showed relatively little variation and fixed more carbon than any other regime. Whether this result was due to some specific adaptation of the organisms to this light environment (35 cm was close to the mean depth of these waters) or to an artifact arising from the differences of 'neutrality' of the light-attenuating devices is not known.

When the same experimental data were plotted to show the dependency of photosynthesis on light, curves not unlike those in Figures 53 and 54 fitted them. In addition, when the rate of change of photosynthesis at each experimental light intensity was plotted against the average light intensity for the period over which the change had occurred, the data again showed maximum positive values at light levels of <0.2 ly min^{-1}, dropped to maximum negative values of 0.2–0.7 ly min^{-1}, and returned to somewhat smaller negative values at higher intensities. This manipulation again strengthened the hypothesis of a two-part inhibitory portion of the photosynthetic curve.

TABLE 13. Control of Photosynthesis by Light, Assessed by Using Both Neutral Density Screens and Water Columns To Filter the Light

	Skua Lake		Algal Lake	
Incubation	Relative Light, % surface	Photosynthesis (Light–Dark), counts sec^{-1}	Relative Light, % surface	Photosynthesis (Light–Dark), counts sec^{-1}
Unscreened bottles, depth				
5 cm	74	83.3	85	107
25 cm	42	116	75	121
45 cm	24	104	65	134
65 cm	14	27.3	57	125
Screened bottles				
1 screen	41	104	47	107
3 screens	13	93.4	15	90.5
4 screens	7.4	74.3	8.5	28.7
5 screens	4.1	43.5	4.8	7.3

The screened bottles were incubated at a depth of 5 cm. The average light intensity during the incubation (1530–1930 hours, January 11, 1963) was 0.65 ly min^{-1}.

The results of an experiment (on January 11, 1963) in which bottles were incubated at several depths, including a depth just beneath the surface, with screens of a neutral density filter suggested that the plankton had a higher optimum light intensity (possibly near 0.4 ly min^{-1}) on that date than they had had previously (Table 13). Differences between the type of experimental light alternation showed little consistency.

Again on January 18 (Figure 56) the optimum intensity for Skua Lake plankton appeared in the area of 0.4 ly min^{-1}, the higher incubation temperature (14°C) apparently having a higher light optimum than the lower temperature (3.5°C). The data from January 28 and 30 (Table 12) also supported this conclusion; inhibition at low temperatures occurred at lower light levels than it did at high temperatures.

The final experiment of the season (on February 8, 1963, Figure 57) showed comparable behavior of Skua and Algal lake phytoplankton, the optimum photosynthetic intensity being once again near 0.15 ly min^{-1}. By that time the lakes had been frozen for >2 weeks, and the plankton were presumably adapted to somewhat reduced light and stable hydrodynamic conditions. The inhibitory portion of these curves failed to show the two-part nature discussed earlier and exhibited near linearity, which would extrapolate to total inhibition at about 1.6 ly min^{-1}.

These experiments indicated a seasonal trend of optimal photosynthetic light intensity for phytoplank-

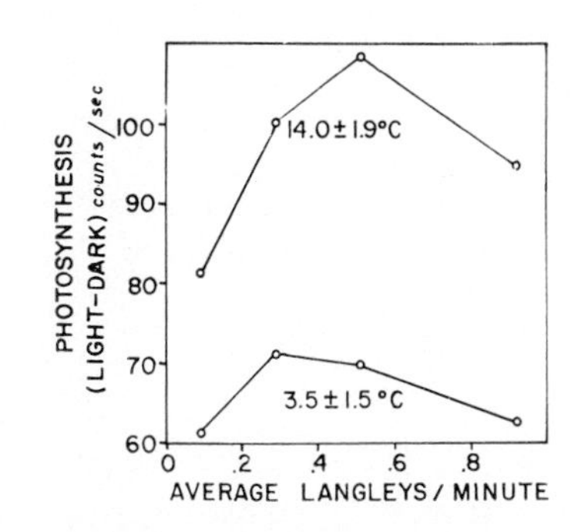

Fig. 56. The effect of light on the photosynthesis of Skua Lake plankton at two temperatures. Incubation was in water-filled white pans at 1345–1645 hours on January 18, 1963.

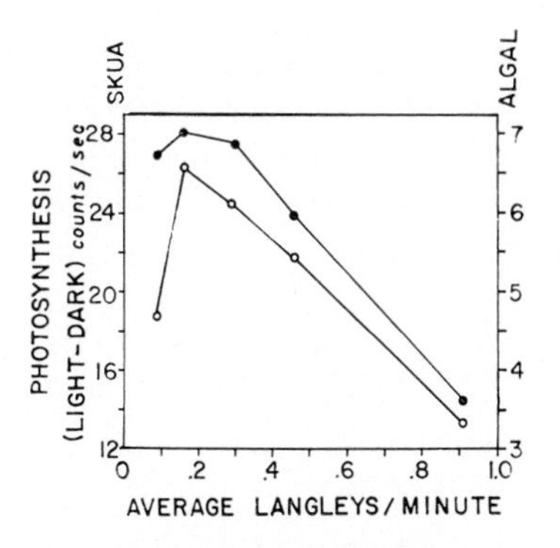

Fig. 57. The effect of light on the photosynthesis of Skua (open circles) and Algal (solid circles) lake plankton at two temperatures. Incubation was in water-filled white pans at 1315–1620 hours on February 8, 1963.

tonic photosynthesis that rose from near 0.1 ly min^{-1} in December to 0.4 ly min^{-1} in mid-January, when the water was ice free and insolation was high, and that dropped once again in February, when the onset of ice cover began. This trend occurred along with a shifting phytoplanktonic composition (Figures 36 and 37) and may therefore not be an 'adaptation' of the existing cells to summer light conditions. However, the community as a whole appears to be adaptively responsive to the changing light climate of the short antarctic growing season. The January 9–10, 1963, diel experiment (Figure 41) also testifies to this fitness of the phytoplanktonic community and shows maximum photosynthetic carbon fixation in bottles held near the mean depth of the lakes.

Studies of the effect of the interaction of temperature and light on algal photosynthesis once more suggest a fitness for the extreme environment. The thermal properties of these shallow lakes have been shown to be highly dependent on the radiation budget: a period of bright sunshine may warm the waters to temperatures near 15°C even at subfreezing air temperatures. The phytoplankton therefore characteristically face simultaneously high temperature and light regimes, and the experimental evidence suggests that the light optimum for photosynthesis increases with temperature. Conversely the highest temperature response (Q_{10}) was found at the highest light intensities.

CONCLUSION

Lakes like Algal and Skua are sufficiently abundant in the Antarctic to provide environmental laboratories that are relatively unaffected by man and able to yield a greater understanding of the physiology of a biota remarkably well adapted to environmental extremes. As such, these lakes may be expected to continue to lure ecologists from the lower, warmer, and more settled latitudes.

APPENDIX 1: OTHER LAKES IN THE McMURDO SOUND AREA

Cape Evans

See Figure 3 for the locations of the lakes on Cape Evans.

Pond 1. Pond 1 is a late melting pond having extensive deposits of partially buried algal peat on its western shore. It receives drainage from Algal Lake and drains in turn into Skua Lake. The conductivity on December 12, 1961, was 1050 μmhos cm^{-1} at 0°C. The temperature was 3.6°C. The algal felt collected on February 16, 1962, had an absorbency ratio of D430/D665 of 3.2, a chlorophyll *a* content of 3.8 mg cm^{-2}, and a carotenoid content of 2.8 mspu cm^{-2}.

TABLE 14. Conductivity and Temperature of Cape Evans Pond 2 on Three Dates

Date	Conductivity, μmhos cm^{-1} at 0°C	Temperature, °C
Dec. 27, 1961	12,500	4.6
Jan. 10, 1962	10,400	3.6
Jan. 16, 1962	>25,000	10.6

Pond 2. Pond 2 is a large flat pan west of the hut having an abundance of pigmented felt that smells of H_2S when it is disturbed. Skuas nest nearby but do not bathe in the pond, although it melts early in the season. Patches of green terrestrial algae are common below snowbanks in this drainage.

The planktonic count on December 22, 1961, was as follows (in millions per liter):

Chlamydomonas subcaudata	74
C. intermedia	10
Pennate diatom (61 μ)	2
Phormidium sp. (100-μ strands)	0.7
Ciliates	1.4
Choanoflagellates (3–4 μ)	21
Bacterial clumps (100 $μ^2$)	0.4

The conductivity and the temperature of pond 2 on three dates are given in Table 14. The following measurements were made from a water sample taken from pond 2 on December 27, 1961:

Specific gravity	14.1‰
Chloride	8670 mg l^{-1}
pH	9.5
CO_2	93 mg l^{-1}
Divalent cations (as Ca)	1110 mg l^{-1}
Dissolved PO_4-P	103 μg l^{-1}
NO_3-N, NH_3-N	undetectable

The following measurements were made from a water sample taken from pond 2 on January 3, 1962:

pH	9.5
CO_2	93 mg l^{-1}

Pond 2a. Pond 2a is a relict of receding pond 2. The conductivity on December 27, 1961, was 24,200 μmhos cm^{-2} at 0°C. The temperature was 3.6°C.

Pond 3. Pond 3 is up the southern inflow to Skua Lake and is a small (3-meter long) pond. Intense bacterial development clouded the water white on December 4, 1962. The high carbon uptake early in December is noteworthy.

The planktonic count on December 4, 1962, was as follows (in millions per liter):

Chlamydomonas subcaudata	5
C. intermedia	6
Pennate diatom (20 μ)	1
Phormidium sp. (100-μ strands)	3
'Sulfur granules'	401

The productivity (measured from 1535 to 1955 hours, December 4, 1962) was 47 mg C m^{-3} hr^{-1} (6.2 g C m^{-3} day^{-1}). The available inorganic carbon was 35.4 mg l^{-1}. The alkalinity ($CaCO_3$) was 154 mg l^{-1}. The conductivity (in μmhos cm^{-1} at 0°C) was 4720 on January 10, 1962, and 13,000 on December 4, 1962.

Pond 4. Pond 4 had melted only on the northwestern edge on December 27, 1961; the felt growth was scanty, and the water clear at that time. On January 16, 1962, a planktonic bloom of about 75 million cells l^{-1} was noted. Between depths of about 0.2 and 0.4 meter there was a shelf protruding from the unmelted ice island. This shelf coincided in depth with a sharp chemocline and thermocline and probably represented a region of little turbulence and horizontal heat transfer from the bottom. Cells deriving heat from contact with the dark bottom were presumably circulating above and below it.

The temperature and the conductivity measured at varying depths in pond 4 are given in Table 15.

TABLE 15. Temperature and Conductivity Measured at Varying Depths in Cape Evans Pond 4

Depth, meters	Temperature, °C	Conductivity, μmhos cm^{-1} at 0°C
0	3.2	1,090
0.3	2.0	1,250
0.4	0.7	6,680
0.9 (bottom)	0.0	10,320

Pond 5. Only about 50 m^2 of pond 5 had melted on December 27, 1961. The conductivity was 2210 μmhos cm^{-2} at 0°C. The temperature was 2.9°C. On January 16, 1962, pond 5 was draining into pond 2 and displayed a rich pink felt and patches of green *Ulothrix.* Nowhere was it >0.3 meter deep, and a planktonic bloom was noted only in the northern end near the outlet. The conductivity (in μmhos cm^{-1} at 0°C) was 2520 in the south end surface, 3350 in the north end surface, and 6390 in the north end bottom (0.2 meter). The temperature was 6.4°C in the south end surface, 5.5°C in the north end surface, and 10.1°C in the north end bottom.

Pond 6. Pond 6 is called 'Red Rotifer' pond; much snow accumulates in its limited sheltered drainage; there is skua guano on an adjacent bluff; penguin molt feathers are abundant; in wet seasons it may drain directly to the sea over the cliff just south of it.

On January 2, 1962, living material revealed numbers of *Chlamydomonas* sp. 10–15 μ in length that were probably *C. intermedia;* the 'chloroplast [was] a yellow green parietal band, but sometimes a cup or lateral glob.' *Philodina roseola* was very abundant. Pink sulfur bacteria were seen in visible concentrations in mud on January 16, 1962. There were 37 million *Chlamydomonas intermedia* per liter. The conductivity of pond 6 measured on three dates was as follows (in μmhos cm^{-1} at 0°C):

Dec. 27, 1961	
Surface	2,570
Bottom	8,900
Jan. 10, 1962	
Surface	1,645
0.5 meter	1,700
0.75 meter on ice 'root'	8,280
Jan. 16, 1962	
Surface	1,500
0.9 meter	1,700
1.2 meters (bottom)	10,900

The following measurements were made from a water sample taken on December 27, 1961:

Chloride	1290 mg l^{-1}
pH	10.5
CO_2	30.6 mg l^{-1}
Divalent cations (as Ca)	176 mg l^{-1}
Dissolved PO_4-P	345 μg l^{-1}
NO_3-N, NH_4-N	undetectable

Pond 7. Pond 7 is a coastal pond behind the strand line of the beach. It was mostly frozen on December 27, 1961, and poor in algal felt. Beach lines were 1.5 meters above the present level. The northwest arm conductivity was 667 μmhos cm^{-2} at 0°C. The temperature was 2.8°C. The southeast area conductivities were 725 (water) to 747 (at sediments) μmhos cm^{-2} at 0°C. The temperatures were 5.0°C (water) and 5.4°C (at sediments). On January 16, 1962, the northwest arm had uniform thermal and conductimetric conditions surface to bottom; the conductivity was 1130 μmhos cm^{-1} at 0°C. In the deeper southeastern area, however, a saline layer 10 cm thick lay at the bottom (1 meter). The conductivities (in μmhos cm^{-1} at 0°C) were 1360 at the surface and >29,200 on the bottom. The temperatures were 5.7°C at the surface and 3.9°C on the bottom.

Pond 8. Pond 8 is an elongate multibasined pond apparently dammed by sea ice thrust on the shore by tidal action. When it was observed on January 16, 1962, the upper, eastern basin had been rapidly dropping in level (0.2 meter in 24 hours) as it flowed through a recently breeched ice dam into the lower western basin and thence out over snow-covered ice. The eastern basin had a melt pattern like that of pond 4, and the outflowing water was quite saline. The western section had a surface conductivity of 3160 μmhos cm^{-1} at 0°C and a bottom water conductivity of 3920 μmhos cm^{-1} at 0°C. The temperature of both sections was 2.4°C. A small area of gas-containing felt was observed below 0.2 meter. The maximum depth was not more than 0.5 meter.

Pond 9. Pond 9 is a small pool whose main attraction was a dead seal partly submerged at the western edge. A rich growth of *Ulothrix* was noted on January 16, 1962, probably fertilized by decomposition of the seal. The conductivities (in μmhos cm^{-1} at 0°C) were 1400 at the surface and 13,200 on the bottom. The temperatures were 9.0°C at the 'pelagic' surface, 9.8°C at the 'littoral' surface, and 6.7°C on the bottom.

Pond 10. The northern half of pond 10 melted and flowed into pond 11 on January 16, 1962. A visible yellow-green planktonic bloom was noted, along with bits of *Ulothrix* and red rotifer concentrations. The felt growth was moderate. An arcuate underwater ridge cuts across the northern third of the pond. The conductivities (in μmhos cm^{-1} at 0°C) were 1870 at the surface and 23,600 on the bottom (0.6 meter). The temperatures were 4.9°C at the surface and 0.6°C on the bottom.

Pond 11. Pond 11 receives water from pond 10 but had no obvious outflow on January 16, 1962. The maximum depth was 0.9 meter, and there were orange gas-containing felt and associated red rotifers. The conductivities (in μmhos cm^{-1} at 0°C) were 189 at the surface and 2720 on the bottom (0.8 meter). The temperatures were 4.9°C at the surface and 0.6°C on the bottom.

Pond 12. Pond 12 is a small pool (about 7 meters long) on the opposite side of the bluff from pond 6 and has a rich white-orange benthic felt. The northern half was ice covered on January 16, 1962. The conductivities (in μmhos cm^{-1} at 0°C) were 1750 at the surface and 29,200 on the bottom (0.35 meter). The temperatures were 5.8°C at the surface and 0.5°C on the bottom.

Island Lake. Island Lake appears with this name on the sketch map of Cape Evans in British Museum [1921–1964]. Shore lines corresponding to the early map were noted 1.5 meters above the water level on January 16, 1962. There was much felt in the intervening area and also in the melted marginal waters. White bacterial growths and red rotifer masses were observed. The conductivities at the surface varied from 360 to 790 μmhos cm^{-1} at 0°C. The temperatures were 3.6°–5.5°C.

Pond 13. Pond 13 was formerly connected to Island Lake, the high beach line being continuous. A strongly developed green algal bloom was noted in the water on January 16, 1962. The lake was partly sheltered by a snowdrift 4 meters high and immediately south. The maximum depth was 1.1 meters. The conductivities (in μmhos cm^{-1} at 0°C) were 1250 at the surface and 1750 on the bottom (0.9 meter). The temperatures were 4.0°C at the surface and 3.6°C on the bottom.

Pond 14. Pond 14 was behind a sheltering snowdrift 3 meters high and to the south. The bottom had a flocculent white felt (*Oscillatoria, Phormidium, Pleurococcus*), clear water overlying it. The unusual thermal and conductimetric profiles have been discussed by Goldman et al. [1967] and are shown in Figure 50 along with a calculated density profile based on the assumption that NaCl was the only salt present. Like most of the other ponds, pond 14 displays a water column stratified by dissolved salts whose distribution can probably be attributed to the previous year's freezing out process. This type of meromixes may therefore be termed 'cryogenic.'

Pond 15. Pond 15 was four small pits in the moraine of the ramp. Some planktonic bloom was noted in all the pits on January 16, 1962, but benthic growth was sparse. The conductivities in four sections, southwest to northeast (in μmhos cm^{-1} at 0°C), were (1) surface, 1930; bottom (0.25 meter), 13,800; (2) surface, 3700; bottom (0.2 meter), 13,300; (3) surface, 2290; bottom (0.3 meter), 13,000; and (4) surface, 5660; bottom (0.4 meter), 12,900. The temperatures were (1) surface, 10.0°C; bottom, 13.5°C; (2) surface, 8.0°C; bottom, 5.8°C; (3) surface, 6.7°C; bottom, 2.4°C; and (4) surface, 8.8°C; bottom, 9.6°C.

Pond 16. Pond 16 was in a rift on the highest ridge of the cape just above Algal Lake. It was completely melted on January 16, 1962, and had an abundant but thin pale felt colored tan to gray-green. It was shallow, nowhere deeper than 0.2 meter. The conductivity was 6860 μmhos cm^{-1} at 0°C. The temperature was 6.6°C. On December 4, 1962, the pond was partially melted and had a conductivity of 4600 μmhos cm^{-1} at 0°C. The temperature was 1.5°C.

Cape Royds

Pony Lake (Home Lake). When Pony Lake was sampled on January 24, 1962, there was an intense green bloom and abundant evidence of penguin pollution. The planktonic count on January 24 was as follows (in millions per liter):

Chlamydomonas subcaudata	34
Ch. sp. (rugose, possibly intermedia, 10 μ)	140
Phormidium (100-μ strands)	2
Rhizomastigophoran, *Tetramitus* (12–18 μ)	18
Small (3- to 5-μ) colorless flagellates	51
Coccoid bacteria (100-μ^2 clumps)	48
Filamentous bacteria (100-μ strands)	53

(A *Chlamydomonas subcaudata* microgamete was observed apparently undergoing sexual fusion with a typical cell.) The following measurements were made from a water sample taken on January 24, 1962:

Sestonic dry weight	64 mg l^{-1}
Specific gravity	38‰
Conductivity	3620 μmhos cm^{-1} at 0°C
Chloride	1740 mg l^{-1}
*p*H	9.2
CO_2 (Figure 27)	31.2 mg l^{-1}
Divalent ions (as Ca)	203 mg l^{-1}
Dissolved PO_4-P	2300 μg l^{-1}
Dissolved NO_3-N	261 μg l^{-1}
Dissolved NH_4-N	6300 μg l^{-1}
Dissolved organic N	13 mg l^{-1}
Sestonic N	4.7 mg l^{-1}
Sestonic C	2.5 mg l^{-1}
Chlorophyll *a*	347 μg l^{-1}
Carotenoids	311 μspu l^{-1}

Cape Royds upper. Cape Royds upper is a small pond just north of Pony Lake that may drain a former penguin rookery. The planktonic count on January 24, 1962, was as follows (in millions per liter):

Chlamydomonas subcaudata	61
C. intermedia	30
Rhizomastigophoran, *Tetramitus*	14
Vorticellid telotrochs	0.2

The following measurements were made from a water sample taken on January 24, 1962:

Sestonic dry weight	63 mg l^{-1}
Specific gravity	2.6‰
Conductivity	4300 μmhos cm^{-1} at 0°C
Chloride	1730 mg l^{-1}
*p*H	8.4
CO_2	34.8 mg l^{-1}
Divalent cations (as Ca)	238 mg l^{-1}
Dissolved PO_4-P	7700 μg l^{-1}
Dissolved NO_4-N	>300 μg l^{-1}
Dissolved NH_4-N	9300 μg l^{-1}
Dissolved organic N	17 mg l^{-1}
Sestonic N	2.6 mg l^{-1}
Sestonic C	1.1 mg l^{-1}
Chlorophyll *a*	134 μg l^{-1}
Carotenoids	119 μspu l^{-1}

TABLE 16. Concentrations of Various Constituents of Two Ponds at McMurdo Station

	Upper Pond	Lower Pond
Chloride, mg l^{-1}	14.2	79.4
*p*H	8.7	9.3
CO_2, mg l^{-1}	22.1	15.1
Divalent cations (as Ca), mg l^{-1}	11.6	21.4
Dissolved NO_3-N, μg l^{-1}	...	2.6
Dissolved PO_4-P	0	0

Two ponds at McMurdo Station. Two ponds at McMurdo Station were located just southeast and above the biology laboratory. They were sampled on December 29, 1961, and the results are shown in Table 16.

Marble Point ponds. On December 15, 1962, we visited five ponds north and west of the camp at Marble Point. The temperatures were all <2°C, and the conductivities varied between 74 and 210 μmhos cm^{-1} at 0°C. Benthic growth was less than that at Algal Lake on Cape Evans but was noted in all cases. *Nostoc* sp. was observed in abundance in several lakes and streams. All ponds had melted and refrozen to a depth of a few centimeters. Since the Cape Evans lakes were still partly frozen from the preceding year, it was concluded that the lacustrine climate at Marble Point was somewhat warmer than that at Cape Evans.

Don Juan pond, Wright valley. A water sample was collected on January 7, 1962, from the main lake; living benthic material from the inflowing stream was also examined. *Pleurococcus* sp., *Chlorococcum* sp., *Phormidium* sp., *Oscillatoria* sp., a naviculoid diatom (15 μ), a few ciliates and flagellates, numerous rod-shaped and spiral, actively mobile bacteria, and a lobopodate amoeba that had ingested a *Phormidium* strand were observed in this stream sample. This unusual pond has been discussed by Meyer et al. [1962] and Tedrow et al. [1963]. The specific gravity (weighed volume) was 1.35‰. The salinity (dried to 115°C for 6 hours) was 840 g l^{-1}.

Since a small sample was used, the only elements detectable by emission spectrographic methods (W. Silvey, personal communication, 1962) were (in mg l^{-1}):

Cobalt	4
Manganese	58
Molybdenum	0.4
Nickel	0.4

(Note that no carbon fixation would be detected by the ^{14}C method in water freshly collected on January 8, 1962.)

APPENDIX 2: DESCRIPTIONS OF DIATOMS OF ALGAL AND SKUA LAKES

Achnanthes brevipes Ag. var. *intermedia* (Kutz.) Cleve.

Ko-Bayashi [1963, pp. 5–7, Plates 1–3].

Valve: linear elliptical with broadly rounded ends and slightly inflated sides. Striae radiate, 14 in 10 μ. Length 30 μ, breadth 8 μ.

Navicula gausii (Heiden)

Hustedt [1930b, p. 617, Figure 1615].

Valve: elliptical lanceolate with capitate ends and distinctly convex sides. Striae radiate, 20 in 10 μ. Length 18 μ, breadth 8 μ.

Navicula muticopsis van Heurck

Hustedt [1930b, p. 614, Figure 1614].

Valve: broadly lanceolate with capitate ends and parallel to slightly convex sides. Striae radiate, 15 in 10 μ. Length 20 μ, breadth 7 μ.

Pinnularia cymatopleura W. and G. S. West (1911)

West and West [1911, p. 285, Plate XXVI, Figures 33–34].

Valve: linear and triundulate margins and subcapitate ends. Striae radiate and shortened near the central area, 23 in 10 μ. Length 23 μ, breadth 5 μ.

Stauroneis anceps Ehrenb.

Patrick and Reimer [1966, p. 361, Plate 30, Figure 1].

Valve: lanceolate with protracted subcapitate ends. Striae radiate, 20 in 10 μ. Length 45 μ, breadth 9 μ.

Gomphonema kamtschatica Grun. var. *antarctica* Per.

van Heurck [1880, Plate 25, Figures 28–29].

Valve: clavate with broad rounded apex and base; narrow linear axial area; central area rounded, stigma lacking. Striae punctate, radiate, 13 in 10 μ. Length 23 μ, breadth 6 μ.

Hantzschia amphioxys (Ehrenb.) Grun. var. *amphioxys*

Hustedt [1930a, p. 394, Figure 747].

Valve: arcuate, linear with produced capitate ends; concave on keel margin. Keel puncta, eight in 10 μ. Striae, 18 in 10 μ. Length 90 μ, breadth 10 μ.

Hantzschia amphioxys (Ehrenb.) Grun. var. *maior* Grun.

Hustedt [1930a, p. 394, Figure 749].

Valve: more elongate than nominate variety. Keel puncta, eight in 10 μ. Striae, 18 in 10 μ. Length 140 μ, breadth 10 μ.

Acknowledgments. This study was supported by grants from the National Science Foundation, U.S. Antarctic Research Program. Special thanks are due to Dr. George Llano, the foundation program director, and to Dr. John Hobbie, who assisted in the research during the second field season. Mrs. Anne Sands gave valuable assistance in identifying the diatom flora, and Mrs. Denne E. Bertrand gave valuable assistance in assembling the figures and the tables. Excellent logistic support was provided by the U.S. Navy at McMurdo Sound, Antarctica.

REFERENCES

Allen, S. E., H. M. Grimshaw, and M. W. Holdgate
1967 Factors affecting the availability of plant nutrients on an antarctic island. J. Ecol., *55:* 381–396.

American Public Health Association
1960 Standard methods for the examination of water and wastewater. 11th ed., 626 pp. New York.

Armitage, K. B., and H. B. House
1962 A limnological reconnaissance in the area of McMurdo Sound, Antarctica. Limnol. Oceanogr., *7:* 36–41.

Baker, A. N.
1967 Algae from Lake Miers, a solar-heated antarctic lake. N. Z. J. Bot., *5*(4) : 453–468.

Ball, D. G., and R. L. Nichols
1960 Saline lakes and drill-hole brines, McMurdo Sound, Antarctic. Bull. Geol. Soc. Am., *71:* 1703–1708.

Barnes, D. G.
1959 Preliminary report of Lake Peters, Alaska, ice studies. GRD Res. Notes, *29:* 102–110.

British Museum
1921–1964 Natur. Hist. Rep. Brit. Antarctic (Terra Nova) Exped. 1910. London.

Cherry-Gerrard, A.
1930 The worst journey in the world. Antarctic 1910–1913. xiv + 585 pp. Dial, New York.

Clarke, F. W., and H. S. Washington
1924 The composition of the earth's crust. U.S. Geol. Surv. Prof. Pap. 127: 117 pp.

Daly, R.
1914 Igneous rocks and their origin. 563 pp. McGraw-Hill, New York.

Debenham, F.
1921 Recent and local deposits of McMurdo Sound region. Nat. Hist. Rep. Br. Antarct. Terra Nova Exped., *1:* 63–100.

Dougherty, E. C., and L. G. Harris
1963 Antarctica micrometazoa: Freshwater species in the McMurdo Sound area. Science, *140:* 497–498.

Fritsch, F. E.
1912 Freshwater algae. Nat. Antarctic Exped. 1901–1904 Natur. Hist. Rep. 6: 42–101. Brit. Mus., London.

Fukushima, H.
1962 Notes on diatom vegetation of the Kasumi Rock ice-free area, Prince Olav Coast, Antarctica. Antarct. Rec., *15:* 1267–1280.
1967 A brief note on diatom flora of antarctic inland waters. JARE Scient. Rep., spec. issue 1: 253–264.

Goldman, C. R.
1963 The measurement of primary productivity and limiting factors in freshwater with carbon-14. *In* M. S. Doty (Ed.), Proceedings of the conference on primary productivity measurement, marine and freshwater. Div. Tech. Inform. Rep. TID 7633: 103–113. U.S. Atomic Energy Commission, Washington, D. C.
1964 Primary productivity studies in antarctic lakes. *In* R. Carrick, M. W. Holdgate, and J. Prevost (Eds.), Biologie antarctique, pp. 291–299. Hermann, Paris.
1968 The use of absolute activity for eliminating serious errors in the measurement of primary productivity with C^{14}. J. Cons. Perm. Int. Explor. Mer, *32:* 172–179.
1970 Antarctic freshwater ecosystems. *In* M. W. Holdgate (Ed.), Antarctic ecology. *2:* 609–627. Academic, New York.

Goldman, C. R., and D. T. Mason
1962 Inorganic precipitation of carbon in productivity experiments utilizing carbon-14. Science, *136:* 1049–1050.

Goldman, C. R., D. T. Mason, and B. J. B. Wood
1963 Light injury and inhibition in antarctic freshwater phytoplankton. Limnol. Oceanogr., *8:* 313–322.

Goldman, C. R., D. T. Mason, and J. E. Hobbie
1967 Two antarctic desert lakes. Limnol. Oceanogr., *12:* 295–310.
1969 Variations in photosynthesis in two shallow antarctic lakes. Verh. Inst. Verein. Theor. Angew. Limnol., *17:* 414–418.

Goodwin, T. W.
1957 The nature and distribution of carotenoids in some blue-green algae. J. Gen. Microbiol., *17:* 467–473.

Greenfield, L. J., and F. A. Kalber
1954 Inorganic phosphate measurement in sea water. Bull. Mar. Sci. Gulf Caribb., *4:* 323–335.

Harvey, H. W.
1951 Micro-determination of nitrogen in organic matter without distillation. Analyst, Lond., *76:* 657–660.

Haxo, F. T., and D. C. Fork
1959 Photosynthetically active accessory pigments of cryptomonads. Nature, *184:* 1051–1052.

Hustedt, F.
1930a Bacillariophyt (Diatomeae). *In* A. Pascher (Ed.), Die Süsswasser-Flora Mitteleuropas. *10:* 466 pp. Schweizerbart'sche, Stuttgart.
1930b Die Kieselagen. *In* L. Rabenhorst (Ed.), Kryptogamen-Flora von Deutschland, Osterreich und der Schweiz. *7:* 1–3. Schweizerbart'sche, Stuttgart.

Jensen, H. I.
1916 Report on antarctic soils. 6. Contributions to paleontology and petrology of south Victoria Land, British antarctic expedition 1907–1909. Rep. of Sci. Invest. Geol. 2: 89–92.

Kear, J.
1963 The agricultural importance of wild goose droppings. Wildfowl Trust Annu. Rep. 14: 72–77.

Kelly, W. C., and J. H. Zumberge
1961 Weathering of a quartz diorite at Marble Point, McMurdo Sound, Antarctica. J. Geol., *69:* 433–446.

Ko-Bayashi, T.
1963 Variations in some pennate diatoms from Antarctica, 1. JARE Scient. Rep., ser. E, *18:* 1–20.

Lindstrom, F., and H. Diehl
1960 Indicator for the titration of Ca plus Mg with (ethylenedinitrilo) tetra acetate. Analyt. Chem., *32:* 1123–1127.

Llano, G. A.
1959 Antarctic plant life. IGY Bull., *24:* 10–13.

Margalef, R.
1968 Perspectives in ecological theory. 111 pp. University of Chicago Press, Chicago.

Mason, D. T.
1963 An exercise in lake morphometry. Turtox News, *41:* 226–229.

Mawson, D.
1916 A contribution to the study of ice structures, British antarctic expedition 1907–1909. Rep. Sci. Invest. Geol. 2: 1–24.

Meyer, G. H., M. B. Morrow, O. Wyss, T. E. Berg, and J. L. Littlepage
1962 Antarctica: The microbiology of an unfrozen saline pond. Science, *138:* 1103–1104.

Mullin, J. B., and J. P. Riley
1955 The spectrophotometric determination of nitrate in natural waters with particular reference to sea water. Analytica Chim. Acta, *12:* 464–480.

Murray, J.
1909 Biology. *In* E. H. Shackleton (Ed.), The heart of the Antarctic. *2:* xvi + 449 pp. Lippincott, Philadelphia.
1910 Tardigrada. Sci. Rep. Br. Antarct. Surv., ser. 1, *1:* 81–185.

Parsons, T. R.
1961 On the pigment composition of eleven species of marine phytoplankters. J. Fish. Res. Bd Can., *18:* 1017–1025.

Parsons, T. R., and J. D. H. Strickland
1963 Discussion of spectrophotometric determination of

marine plant pigments, with revised equations for ascertaining chlorophylls and carotenoids. J. Mar. Res., *21:* 155–163.

Patrick, R., and C. W. Reimer
1966 The diatoms of the United States, 1. Monogr. Acad. Nat. Sci. Philad., *13:* 688 pp.

Peragallo, M.
1921 Diatomées d'eau. *In* Expédition antarctique français, 1908–1910.

Péwé, T. L.
1960 Multiple glaciation in the McMurdo Sound region Antarctica, a progress report. J. Geol., *68:* 498–514.

Ponting, H. G.
1923 The great white south. With Scott in the Antarctic. xxvi + 306 pp. McBride, New York.

Rudolph, E. D.
1967 Environmental factors. *In* Terrestrial life of Antarctica, Antarctic Map Folio Ser., folio 5, pp. 2–3. Amer. Geogr. Soc., New York.

Scott, R. F.
1913 Scott's last expedition. *1:* xxiv + 443 pp. Dodd, Mead, New York.

Simpson, G. C.
1923 Meteorology. *3*, tables. Harrison and Sons, London.

Smith, W. C.
1954 The volcanic rocks of the Ross Archipelago. Nat. Hist. Rep. Br. Antarct. Terra Nova Exped., *2:* 1–107.

Strickland, J. D. H.
1960 Measuring the production of marine phytoplankton. Bull. Fish. Res. Bd Can., *122:* 1–172.

Strickland, J. D. H., and T. R. Parsons
1960 A manual of seawater analysis. Bull. Fish. Res. Bd Can., *125:* vi + 185 pp.

Sugawara, K.
1961 Effect of sea breeze on the chemical composition of coastal freshwater lakes. Verh. Inst. Verein. Theor. Angew. Limnol., *14:* 889–892.

Taylor, G.
1916 With Scott: The silver lining. xv + 464 pp. Dodd, Mead, New York.

Tedrow, J. C. F., F. C. Ugolini, and H. Janetschek
1963 An antarctic saline lake. N. Z. Jl Sci., *6:* 150–156.

Thompson, D. C., and W. J. P. MacDonald
1959 Radiation balance at Scott Base. Nature, 184: 541–542.

Treves, S. B.
1962 The geology of Cape Evans and Cape Royds, Ross Island, Antarctica. *In* H. Wexler, M. J. Rubin, and J. E. Caskey, Jr. (Eds.), Antarctic research, Geophys. Monogr. Ser., *7:* 40–46. AGU, Washington, D. C.

Umbreit, W. W., R. H. Burris, and J. F. Stauffer
1957 Manometric techniques. 338 pp. Burgess, Minneapolis.

U.S. Naval Weather Research Facility
1961 Climatology of McMurdo Sound. vii + 66 pp. Washington, D. C.

van Heurck, H.
1880 Synopsis des diatomées de Belgique. 235 pp. Antwerp.

West, W., and G. S. West
1911 Freshwater algae, British antarctic expedition, 1907–1909. Rep. of Sci. Invest. Biol. 1: 263–298.

PRIMARY PRODUCTIVITY AND ASSOCIATED PHYSICAL, CHEMICAL, AND BIOLOGICAL CHARACTERISTICS OF LAKE BONNEY: A PERENNIALLY ICE-COVERED LAKE IN ANTARCTICA

DERRY D. KOOB[1] AND GEOFFREY L. LEISTER[2]

Department of Botany and Institute of Polar Studies, Ohio State University, Columbus, Ohio 43210

Abstract. Limnologic factors in Lake Bonney were measured during the austral summer of 1965–1966. These factors included chlorophyll *a*, bacterial and green cell concentrations, carbon fixation, conductivity, temperature, light, dissolved oxygen, *p*H, and alkalinity. The ice cover at the study area was 4 meters thick, and the lake including the ice was 30 meters deep. *Chlamydomonas subcaudata* was found between 7 and 16 meters, the maximum concentration being 1.5×10^4 cells/l in the chemocline, 12 meters below the ice surface. An unidentified unicellular coccoid alga was found between 4 and 13 meters, the maximum concentration being 7.9×10^4 cells/l at 6 meters. Numerous unidentified pigmented rod-shaped organisms were found at 4 meters, just under the ice. Large populations of bacteria were found associated with the algal populations. It is inferred from alkalinity, *p*H, and chlorophyll *a* data that zones of photosynthetic activity were present at 4, 6, and 12 meters. Maximum concentrations of unicellular coccoid algal cells and *C. subcaudata* cells occurred at 6 and 12 meters, respectively. The additional photosynthetic zone at 4 meters was populated by uncounted numbers of small rod-shaped organisms possibly belonging to the Cyanophyta. Primary productivity was measured by the in situ ^{14}C method. These data were expressed in both counts per minute and milligrams of carbon per cubic meter hour. Carbon fixation maximums occurred at 9 and 13 meters, secondary peaks occurring at 4 and 6 meters. There is little doubt that the carbon fixation at 4 meters is due to pigmented rod-shaped organisms, the fixation at 6 meters is due to unicellular coccoid algae, and the fixation at 13 meters is due to *C. subcaudata*. The carbon uptake at 9 meters was completely unexpected and coincides neither with any phytoplanktonic populations detected in this investigation nor with any photosynthesis-related chemical parameter. A limited study was made of nonphotosynthetic organic uptake by using sodium acetate-2-^{14}C. Heterotrophic use of ^{14}C-labeled acetate was high in the mixolimnionlike region and low in the monimolimnion, but definite correlations between acetate uptake rates and planktonic populations were impossible because of the paucity of data. Suggestions for future work are presented.

Studies of Lake Bonney, Antarctica, were carried out during the austral summer of 1965–1966. The objectives of these studies were to identify the phytoplanktonic populations, to estimate the densities of the phytoplanktonic populations, to measure physicochemical parameters such as conductivity, temperature, oxygen, hydrogen ion concentration, and alkalinity, to ascertain the photosynthetic rates of the phytoplanktonic populations by the in situ ^{14}C method, and to correlate the physicochemical and biological data.

Lake Bonney is a two-lobed perennially ice-covered lake located in the upper Taylor valley in Victoria Land, Antarctica. Its length is about 4.8 km, and its width is about 0.77 km at its widest point. The western lobe of the lake lies at the foot of Taylor Glacier and is connected to the larger eastern lobe through a narrow channel (Figure 1). Studies were carried out in the center of the eastern lobe approximately 1.6 km from the eastern end of the lake; the ice was 4 meters thick and the lake including the ice was 30 meters deep. This depth is the approximate maximum depth of Lake Bonney.

Detailed physical and chemical characteristics of Lake Bonney were reported in papers by Armitage and House [1962], Angino and Armitage [1963], Angino et al. [1964], Hoare et al. [1964], Ragotzkie and Likens [1964], Shirtcliffe [1964], Shirtcliffe and Benseman [1964], Goldman et al. [1967], Yamagata et al. [1967], and Torii et al. [1967]. The geo-

[1] Now at Department of Wildlife Resources, Utah State University, Logan, Utah 84321.

[2] Now at Department of Botany, Duke University, Durham, North Carolina 27706.

Fig. 1. Aerial photograph of the eastern lobe of Lake Bonney.

logic features of Taylor valley were discussed by McKelvey and Webb [1961], Gunn and Warren [1962], and Armstrong et al. [1968].

Until recently only those organisms confined to the littoral areas of Lake Bonney and to a few freshwater melt ponds in Taylor valley have received attention from biologists. Holm-Hansen [1964] has isolated and cultured the following algae from collections in Taylor valley: *Chlorella* sp., *Nostoc commune*, and *Schizothrix* sp. He also found *Oscillatoria* sp., *Phormidium* sp., *Binuclearia* sp., *Chlorella* sp., *Prasiola* sp., and diatoms in preserved collections from the same area. Cameron [1966] identified *Nostoc commune*, *Microcoleus lyngbyaceus*, and *Schizothrix calcicola* in collections made by T. E. Berg from the Lake Chad vicinity of Taylor valley. Armitage and House [1962] reported finding a '*Chlorella*-like' alga in Lake Bonney as well as the rotifer *Philodina* spp. from the lake's edge. Other micrometazoa were reported from Taylor valley by Dougherty and Harris [1963]. Goldman et al. [1967] found *Ochromonas* sp. and *Chlamydomonas* spp. as well as two unidentified 'green coccoid forms' below the ice of Lake Bonney. They also found bacterial populations associated with the phytoplankton.

MATERIALS AND METHODS

An ice auger 20 cm in diameter was employed to penetrate the 4 meters of ice on the surface of Lake Bonney. A 3-liter Kemmerer bottle for chemical and physical measurements and a 9-liter opaque Van Dorn sampler for primary productivity samples were used to obtain water from below the ice. Oxygen measurements were obtained in the field originally by the unmodified Winkler method as described in American Public Health Association [1965] and later by a Precision galvanic cell oxygen analyzer. Oxygen probe correction factors due to ionic interference were calculated for each depth by using data from Winkler titrations of water from each level. The temperature was measured with a thermistor sensor attached to the oxygen analyzer. A Beckman model N *p*H meter was used in the field to measure the hydrogen ion concentration. Total alkalinity was ascertained in the field by the American Public Health Association [1965] method and was expressed in milligrams per liter of $CaCO_3$. The conductivities of water samples collected on January 21, 1966, at 1-meter intervals from 4 to 30 meters were measured with a model RC-1B conductivity bridge (Industrial Instruments, Inc.) at McMurdo Station.

On December 13, 1965, just prior to our return from the field to the laboratory at McMurdo Station, water samples were collected with a 3-liter Kemmerer bottle at 1-meter intervals from 4 to 30 meters and were placed in acid-cleaned polyethylene bottles that had been rinsed with demineralized water. The samples were analyzed at McMurdo Station for chlorophyll *a* concentration as well as for phytoplanktonic concentration. Chlorophyll *a* concentrations were measured with a Beckman model DU spectrophotometer by the method of Strickland and Parsons [1961] as follows: The milligrams of pigment per cubic meter equals C/V, where $C = 11.6E_{6650} - 1.31E_{6450} - 0.14E_{6300}$ and V is the volume in liters of the filtered sample (E is the extinction value for a 1-cm light path at the wave length indicated by the subscript).

The densities of the phytoplanktonic populations were calculated from both colony counts and direct cell counts. Two sets of sterile plastic Millipore (disposable) petri dishes were inoculated with 1 ml of water from each sample. One set contained an autoclaved agar medium similar to that of Provasoli and Pintner [1953], and the other a high inorganic salt medium (appendix). The 1 ml of water was spread over the agar surface, and the plates were left undisturbed in a continuously lighted New Brunswick controlled environment chamber at 4°C. Five weeks later the plates were examined, and the number of colonies of algae and bacteria was counted. Algal cell counts were also made from membrane filters by the methods of deNoyelles [1968]. Then 50-ml subsamples of lake water from the 4- to 9-meter levels containing living organisms were filtered in duplicate through 13-mm 0.45-μ HA Millipore filters at a vacuum not exceeding 38 cm of Hg. Owing to clogging of the filters, only 25 ml was used at the 10-, 11-, and 12-meter levels, and 10 ml at the 13-, 14-, and 15-meter levels. The organisms were preserved on the surface of the filters by rinsing them with Craf's solution. The filters were cleared with immersion oil and permanently affixed to glass slides with a suitable mounting medium and a cover slip. Cell counts were made at 1000× magnification with a binocular phase microscope. One hundred microscope fields were counted per filter, and the counts were expressed in cells per liter. Direct algal cell counts were also made from samples collected on January 21, 1966, by the same procedures described above.

Primary productivity studies were carried out on January 18, 1966, by the in situ ^{14}C method described by Steemann-Nielsen [1952] with modifications taken from Strickland and Parsons [1961] and Goldman [1963]. Water samples were collected with an opaque plastic Van Dorn water sampler. Collections were taken between 1000 and 1100 hours local time at 1-meter intervals from 4 to 15 meters and at 5-meter intervals from 15 to 30 meters. Water samples were dispensed into transparent 300-ml productivity bottles. Samples were continuously shaded to prevent damage by the high light intensity at the surface. Opaque 300-ml productivity bottles were filled with water sampled from 5, 10, 15, 20, 25, and 30 meters. All opaque productivity bottles were painted white so that the 'blackbody effect' would not raise the temperature of their contents during the incubation period.

Glass ampoules, each ampoule containing 1.00 ml of $Na_2{}^{14}CO_3$, were prepared in the laboratory to contain 10 μc (microcuries) of $Na_2{}^{14}CO_3$. The $Na_2{}^{14}CO_3$ solution was prepared in one large lot from $Na_2{}^{14}CO_3$ of high specific activity so that the resulting solution was essentially 'carrier free.'

In the field the $Na_2{}^{14}CO_3$ solution was removed from the ampoule by using a 5-cm^3 syringe equipped with a 14-gage 10-cm laboratory cannula and was carefully deposited at the bottom of the productivity bottle. The ampoule was rinsed once with 2 ml of water drawn into the syringe from the top of the productivity bottle. The rinse solution was redeposited into the bottle from which it was removed. The bottle was capped, shaken, and placed in a light-tight box. At 1230 hours local time the productivity bottles were suspended in situ at the same depths from which their contents had initially been collected. While the bottles were in place, oxygen, temperature, *p*H, alkalinity, and light measurements were made at 1-meter intervals from 4 to 30 meters through a second ice hole previously drilled approximately 7 meters away from the hole through which the productivity bottles were suspended. Four hours later the productivity bottles were brought to the surface, 10 ml of sample was removed, and 10 ml of 40% neutral formalin was added to each bottle to stop all carbon assimilation.

In another set of experiments sodium acetate-2-^{14}C was employed to ascertain whether the plankton could assimilate an organic carbon source. Sodium acetate-2-^{14}C having a specific activity of 403 μc/mg was prepared in one large lot to contain 10 μc/ml by dilution with demineralized and deionized water. Glass ampoules, each ampoule containing 1.00 ml of this solution, were sealed and autoclaved to prevent bacterial use of the acetate-2-^{14}C.

Water samples were collected at depths of 5, 10, 15, and 30 meters on January 21, 1966, between 1100 and 1200 hours local time. Water from each depth sampled was dispensed into three opaque white 300-ml productivity bottles. Before the addition of sodium acetate-2-^{14}C one of the three bottles from each depth was treated with 10 ml of 40% neutral formalin. These bottles served as controls to ascertain the amounts of acetate-2-^{14}C adsorbed to the cell surfaces of nonliving plankton. The contents of one ampoule containing sodium acetate-2-^{14}C were added to each productivity bottle in a manner similar to that used for ^{14}C carbonate. At 1220 hours local time the bottles were suspended at the depth from which their contents had initially been obtained. Four hours later the bottles were brought to the surface, and the samples were preserved with neutral formalin.

On our return to the laboratory at McMurdo Station the contents of each productivity bottle were filtered through a 25-mm 0.45-μ HA Millipore filter at a vacuum not exceeding 38 cm of Hg. Since there was some evidence of precipitation in a few of the productivity bottles to which $Na_2{}^{14}CO_3$ had been added, all filters containing organisms exposed to $Na_2{}^{14}CO_3$ were rinsed with about 5 ml of 0.003 N HCl to remove precipitated $Na_2{}^{14}CO_3$. All filters were air dried and stored in a desiccator until they were assayed for ^{14}C at Ohio State University.

Primary productivity was calculated in milligrams of carbon per cubic meter hour from the following formula modified from Saunders et al. [1962]: $P = (r \times C \times f)/(R \times hr)$, where

- P, carbon fixation in mg C/m³ hr;
- r, assimilated ^{14}C in counts/min;
- R, total available ^{14}C in counts/min;
- C, total available stable inorganic carbon in mg/m³;
- f, isotope correction factor;
- hr, total incubation time in hours.

The counts per minute of $Na_2{}^{14}CO_3$ assimilated by the plankton r were recorded with a Beckman CPM-100™ liquid scintillation system. Each Millipore filter was dissolved in a screw top scintillation vial containing 1.5 ml of methyl cellosolve (2-methoxy-ethanol) and 0.5 ml of dimethyl sulfoxide. The dimethyl sulfoxide was added to aid in the dissolution of the Millipore filters as well as possibly to aid in the penetration of the scintillation fluor into the cells. At the end of 12 hours 10 ml of 1,4-dioxane (p-dioxane) containing 7 grams of 2,5-diphenyloxazole per liter and 100 grams of naphthalene per liter was added to each vial. A suitable amount of the thixotropic gel powder 'Cab-O-Sil' was added, and the vials were thoroughly mixed on a Vortex-Genie mixer. The gel kept the particulate matter (cells) in suspension. All samples were counted to a 2-σ statistical error.

To ascertain the amount R of available $Na_2{}^{14}CO_3$ injected into each productivity bottle, the contents of an ampoule prepared from the original lot were added to a 250-ml volumetric flask containing a buffer solution at pH 8. The procedure and the equipment used in transferring the contents of each ampoule to productivity bottles in the field were also used in transferring the contents of the ampoule to be assayed. The resulting volume was brought to 250 ml. Then ½ ml of this solution was pipetted into each of three scintillation vials containing the scintillation cocktail. The counts per minute were averaged, and the total counts per minute per ampoule were computed.

The total available stable inorganic carbon C in milligrams per cubic meter was calculated from the data on temperature, pH, and total alkalinity, and a table of conversion factors was constructed by Saunders et al. [1962]. The isotope correction factor was given as 1.06 by Saunders et al. [1962], and the incubation time hr was 4 hours.

Millipore filters containing organisms exposed to sodium acetate-2-^{14}C were assayed for ^{14}C in the same scintillation cocktail used in assaying for $Na_2{}^{14}CO_3$, and the results were expressed in counts per minute. The amount of sodium acetate added was not computed because no measure of in situ sodium acetate was made and no absolute measurement of sodium acetate assimilation was possible.

RESULTS

Armitage and House [1962] discovered that the water in Lake Bonney is density stratified. In our study conductivity was measured in water samples collected on January 21, 1966. From these data it was concluded that the mixolimnionlike region at that time extended from 4 to 10 meters, the chemocline from 10 to 14 meters, and the monimolimnion from 14 to 30 meters.

Lake Bonney is also thermally stratified throughout the antarctic summer. This inverse thermal stratification was first reported by Armitage and House [1962] and has been discussed in detail by Ragotzkie and Likens [1964], Hoare et al. [1964], Shirtcliffe

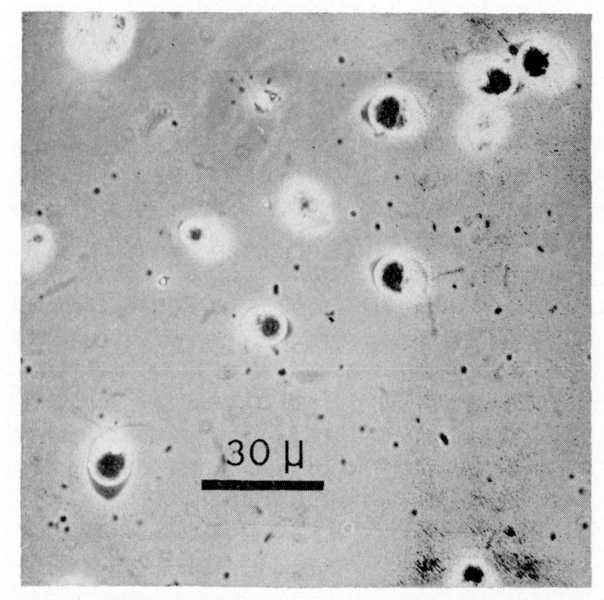

Fig. 2. *Chlamydomonas subcaudata* cells from the 14-meter level cultured on an organically enriched agar medium. Leitz phase contrast microscopy.

[1964], and Shirtcliffe and Benseman [1964]. There is now little doubt that solar radiation is absorbed in the region of the chemocline and causes the reported temperature profiles. In our study, temperature measurements were made at depth intervals of 0.5 meter on November 13 and 15, 1965, December 11 and 13, 1965, and January 18 and 20, 1966. The temperature maximums recorded for these dates were 7.2°C at 13.5 meters, 6.6°C at 13.0 meters, and 7.4°C at 14 meters in November, December, and January, respectively. The temperature data for all dates are essentially the same except for the unexplained temperature decrease between November and December in the region of the chemocline. Since the elevated temperatures in this lake are generally accepted to be due to the absorption of solar radiation, continuous measurements of solar radiation and water temperature would be necessary to investigate such anomalies further.

The alkalinity minimum (100 mg of $CaCO_3$ per liter) and maximum (960 mg $CaCO_3$ per liter) occurred on January 18, 1966, at 6 meters and at the lower boundary of the chemocline, respectively. A maximum *p*H of 9 at 5 meters and a minimum *p*H of 6.2 at 16 meters were found on January 18, 1966. The vertical concentration of dissolved oxygen (DO) was measured galvanometrically on November 13 and December 13, 1965, and on January 18, 1966. A maximum of approximately 39 mg of DO per liter occurred in the mixolimnionlike region in December, and a minimum of <1.8 mg of DO per liter occurred in the monimolimnion on all three dates. In general there was an increase in the concentration of DO in the mixolimnionlike region in December 1965 over that in November 1965, there being a subsequent decrease in January 1966.

Chlorophyll *a* concentrations were measured on January 21, 1966. Two discrete chlorophyll *a* concentration maximums were obtained: 0.73 mg/m³ at 12 meters and 1.29 mg/m³ at 4 meters.

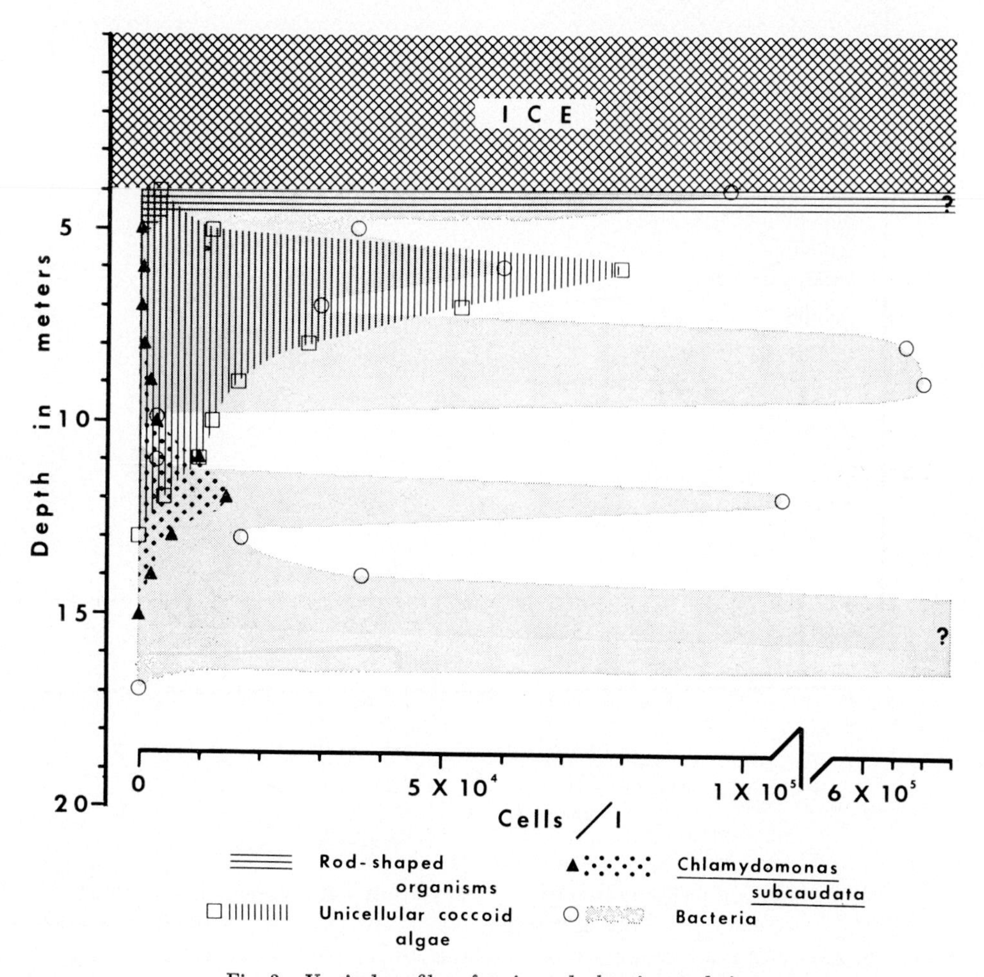

Fig. 3. Vertical profiles of various planktonic populations.

Armitage and House [1962] found a *Chlorella*-like alga during the first limnologic investigation of Lake Bonney. More recently Goldman et al. [1967] found populations of *Ochromonas* sp., *Chlamydomonas* spp., and green coccoid forms as well as concentrations of bacteria below the zones of algae.

A unicellular flagellated chlamydomonad is one biologically important phytoplankter that we found in Lake Bonney. The organism was isolated and maintained in a unialgal culture. Although the taxonomy of the genus *Chlamydomonas* is confused, we have relegated the organisms found in Lake Bonney to the species *C. subcaudata* Wille [Wille, 1903] (Figure 2). It differs from *C. caudata*, the closest related species, in that (1) the papilla is usually small and is not wart-like, (2) the hyaline caudal region is mostly broadly tapered and is at times merely rounded, but is never long and thin, (3) the protoplast is broadly rounded and never has a basal conical protrusion, and (4) the flagella of the parent cell often remain active after production of the four motile daughter cells within the parent cell wall. These characters were consistent for specimens cultured on both normal and high salt media for a period of 2 years. *Chlamydomonas subcaudata* has previously been reported in antarctic collections from Cape Adare by Fritsch [1912], from Hallett Station by E. Schofield (personal communication, 1964), and from numerous ponds and lakes on Ross Island by D. D. Koob (unpublished data, 1964). A quantitative vertical distribution of *C. subcaudata* cells was graphed from membrane filter counts containing organisms sampled on December

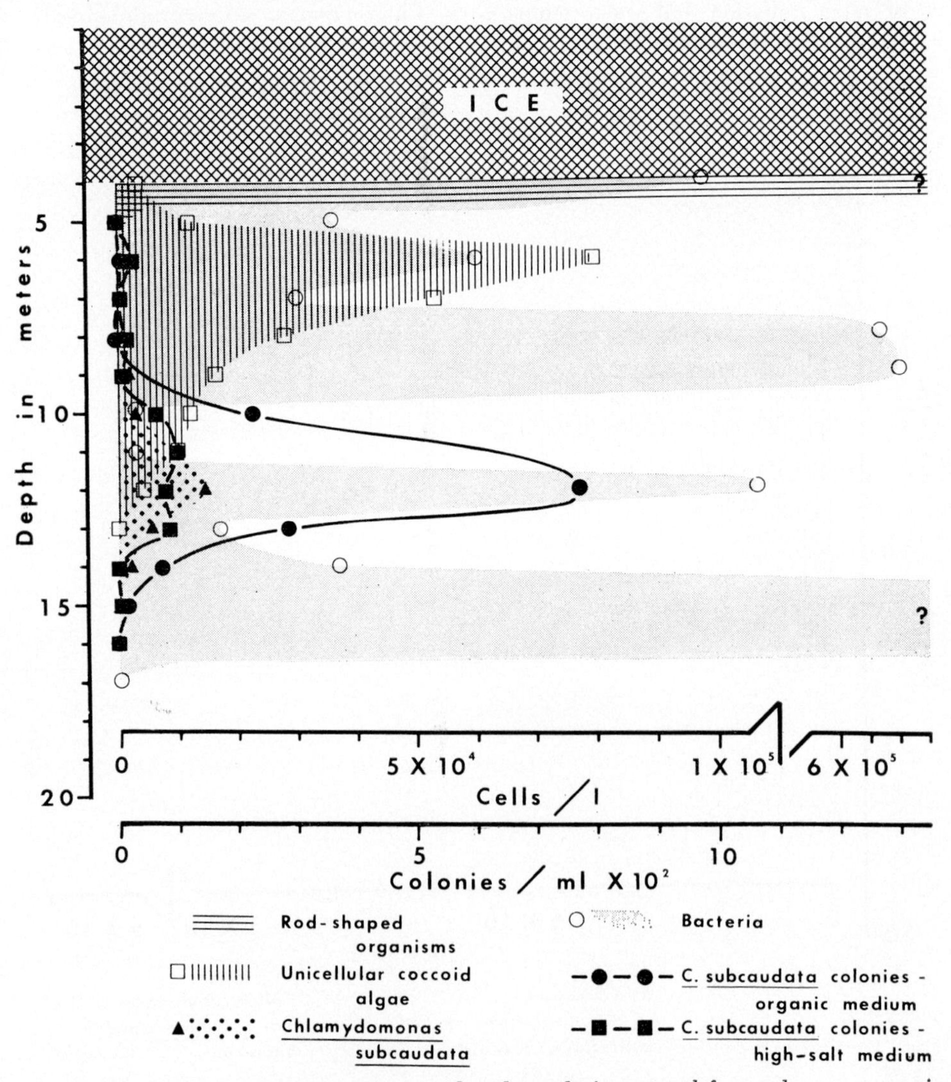

Fig. 4. Vertical profiles of *Chlamydomonas subcaudata* colonies counted from cultures on organic and high-salt media and of other planktonic populations.

13, 1965 (Figure 3). Maximums of approximately 1.6×10^4 cells/l and approximately 1.5×10^4 cells/l occurred at 12 meters in December and January, respectively. The vertical distribution patterns are nearly identical for both months. The vertical distribution of *C. subcaudata* was also ascertained from counts of colonies cultured on modified Provasoli and Pintner [1953] organic media and on high inorganic salt media. A maximum of 797 colonies of *C. subcaudata* grew when 1 ml of inoculum sampled from 12 meters on December 13 was placed on a medium containing organics, and a maximum of 86 colonies of the same organism grew on a high inorganic salt medium. When the colony count values for both media were graphed, the resulting curves were similar. The peaks in both curves occurred at the same depth (Figure 4). Since *C. subcaudata* grew on both media and was initially collected from in situ environments composed of various mineral salt concentrations, the highly salt tolerant population was inseparable from the other in a discrete depth relationship. The vertical profiles constructed from cultured colony counts and membrane filter cell counts were similar in that all three curves had the same general peak at 12 meters. The discrepancy between the Millipore counts and the culture colony counts is probably due to asexual cell division on the culture plates. The 1 ml of lake water was spread on the surface of the petri dish, and thus the organisms could swim actively until the water evaporated. During this time asexual reproductive cycles could have taken place. These new cells would have been released from the parent cell walls to begin independent motile existences. With the evaporation of the lake water these cells would have been deposited on the agar surface, where each would have subsequently grown into a macroscopic green colony.

Another important planktonic population found in Lake Bonney is a yet unidentified unicellular coccoid alga (Figure 5). This form did not grow on the culture media used. Counts were obtained only from membrane filters. Similar organisms were found by Goldman et al. [1967] from their collections at 5-meter intervals. They described 10-μ and 5-μ green coccoid forms whose maximum populations at 15 meters below the ice surface were 1.6×10^6 and 2.5×10^6 cells/l, respectively. Maximum cell counts for the coccoid forms in our study regardless of cell diameter totaled 7.9×10^4 cells/l at 6 meters in the mixolimnionlike region. The vertical profile for the distribution of this organism is included in Figure 3.

The bacteria are the most numerous plankters found in Lake Bonney. Counts were made from the same culture plates containing the *C. subcaudata* colonies. These organisms were identified only as to their characteristic pigmentation (pink or white). Five discrete bacterial populations were found at 4, 6, 8–9, 12, and 15–16 meters (Figure 3). The pink organisms were found only at the ice–water interface (4 meters). Goldman et al. [1967] reported a maximum population of 1.2×10^8 bacterial cells/l at 20 meters below the ice surface in acid-Lugol-preserved water samples examined by inverted microscopic techniques at 400× magnification. This 20-meter zone of bacteria may consist of anaerobes that would not have grown in the aerobic conditions used in culturing *C. subcaudata*. Thus the presence of this deep bacterial population may have been inadvertently overlooked in the present study.

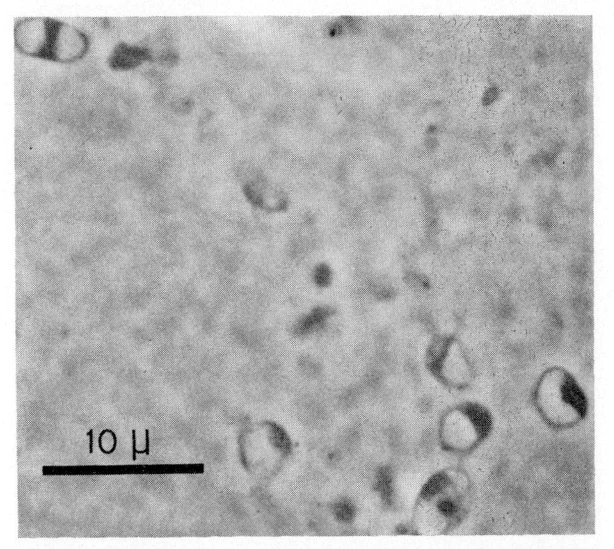

Fig. 5. Unicellular coccoid algae from the 7-meter level on a membrane filter. Leitz phase contrast microscopy magnified 2930×.

Numerous rod-shaped structures were observed on the surface of membrane filters from water sampled in January at the 4-meter ice–water interface. Their size (approximately 1.6 μ × 0.6 μ) and apparent pigmentation would imply that they were blue-green algae. They were so numerous that no cell counts were possible. Since they were found in large numbers immediately beneath the ice, they either may grow on the ice surface and be washed off at the time of sampling or may be truly planktonic forms.

Goldman et al. [1967] reported *Ochromonas* spp. in their samples from Lake Bonney. In the present investigation *Ochromonas* was found neither on the surface of membrane filters nor on the two culture media used. Some *Ochromonas* spp. are characterized by thin delicate cell walls, and thus the cells may have been destroyed by membrane filtration. Hutner et al. [1953] and Hutner and Provasoli [1964] reveal that some *Ochromonas* spp. have complex nutritional requirements including needs for specific vitamins and amino acids. The media used in culturing *C. subcaudata* did not support *Ochromonas* spp. growth.

Primary productivity was estimated by the Steemann-Nielsen ^{14}C uptake method. Determination of the absolute values of carbon fixation depends on a knowledge of the availability of inorganic carbon. The tables of Saunders et al. [1962] were employed to calculate the concentration of inorganic carbon in Lake Bonney. These tables are unfortunately of unknown applicability to water of high salinity. Therefore data in counts per minute are also presented.

On January 18, 1966, there were carbon fixation maximums of 265 counts/min and 0.36 mg C/m^3 hr at 9 meters and 134 counts/min and 0.73 mg C/m^3 hr at 13 meters. Two additional apparent carbon fixation peaks of 112 counts/min at 4 meters and 113 counts/min at 6 meters appeared in the data expressed in counts per minute.

Heterotrophic use of ^{14}C-labeled acetate was ascertained from one experiment in January 1966 at depths of 5, 10, 15, and 30 meters. No absolute values of acetate use rates were possible since the in situ acetate concentration was not measured. These data were therefore expressed as total acetate uptake minus nonbiological acetate uptake in counts per minute. The maximum uptake (14,435 counts/min) of ^{14}C-labeled acetate occurred in samples at the 5-meter level. Uptake rates decreased with increasing depth.

DISCUSSION

Lake Bonney has been studied by numerous investigators since the initial limnologic survey by Armitage and House in 1962, but few biological data have been collected. The present study disclosed a discrete population of *Chlamydomonas subcaudata* near the upper boundary of the chemocline. A large population of unicellular coccoid algae was found in the mixolimnionlike region. Five spatially distinct bacterial populations were found, two in the mixolimnionlike region of the lake, one each at the upper and lower boundaries of the chemocline, and one at the center of the chemocline. An interpretation of the interrelationships of these organisms with the nonbiotic environment follows.

Conductivity. One of the most significant deductions from our data was the extreme stratification of the planktonic populations in the water column of Lake Bonney. The vertical profiles of planktonic concentration exhibit five discrete bacterial communities and at least two phytoplanktonic populations. When one compares the vertical depth profile of conductivity with that of planktonic distribution, some of these populations appear to be related to pycnoclines, as Goldman et al. [1967] have suggested. This apparent relationship is especially true of the bacteria. The large bacterial bands above 10 meters and below 14 meters coincide with the upper and lower boundaries of the chemocline, respectively (Figure 6). The unicellular coccoid algae do not appear to be associated with a discontinuity layer. These organisms are found throughout the mixolimnionlike region and into the chemocline. This distribution is not the case with *C. subcaudata.* Photosynthetic flagellated organisms may be expected to migrate to different levels in the water column as there are changes in the quality and the quantity of light. However, the *C. subcaudata* population density maximums occurred at 12 meters in both December and January. In Lake Bonney this organism appears to be restricted to the region of the chemocline even though it grows in the laboratory on osmotically diverse agar media.

The extreme chemical density stratification in Lake Bonney provides a variety of discrete stable environments in which specific organisms exist. The salinity stratification and the associated segregation of the planktonic population may remain throughout the year, since there is no wind mixing in this lake.

Dissolved oxygen. High concentrations of DO were present throughout the region above the chemocline from the middle of November through the middle of January (Figure 7). Such an accumulation of DO may be considered to be due to the diffusion of oxygen through the ice into the water and throughout this mixolimnionlike region only if one assumes a very low rate of biological use of oxygen. However, this hypothesis seems unlikely because of the presence of the three large bacterial bands at 4, 6, and 8–9 meters. Therefore the respiratory rate of oxygen uptake by the populations in this region above the chemocline must be exceeded by the rate of photosynthetic oxygen production. The distinct population of unicellular coccoid algae that extends from the ice–water interface down to the 13-meter level (the maximum concentration occurring at 6 meters) may be

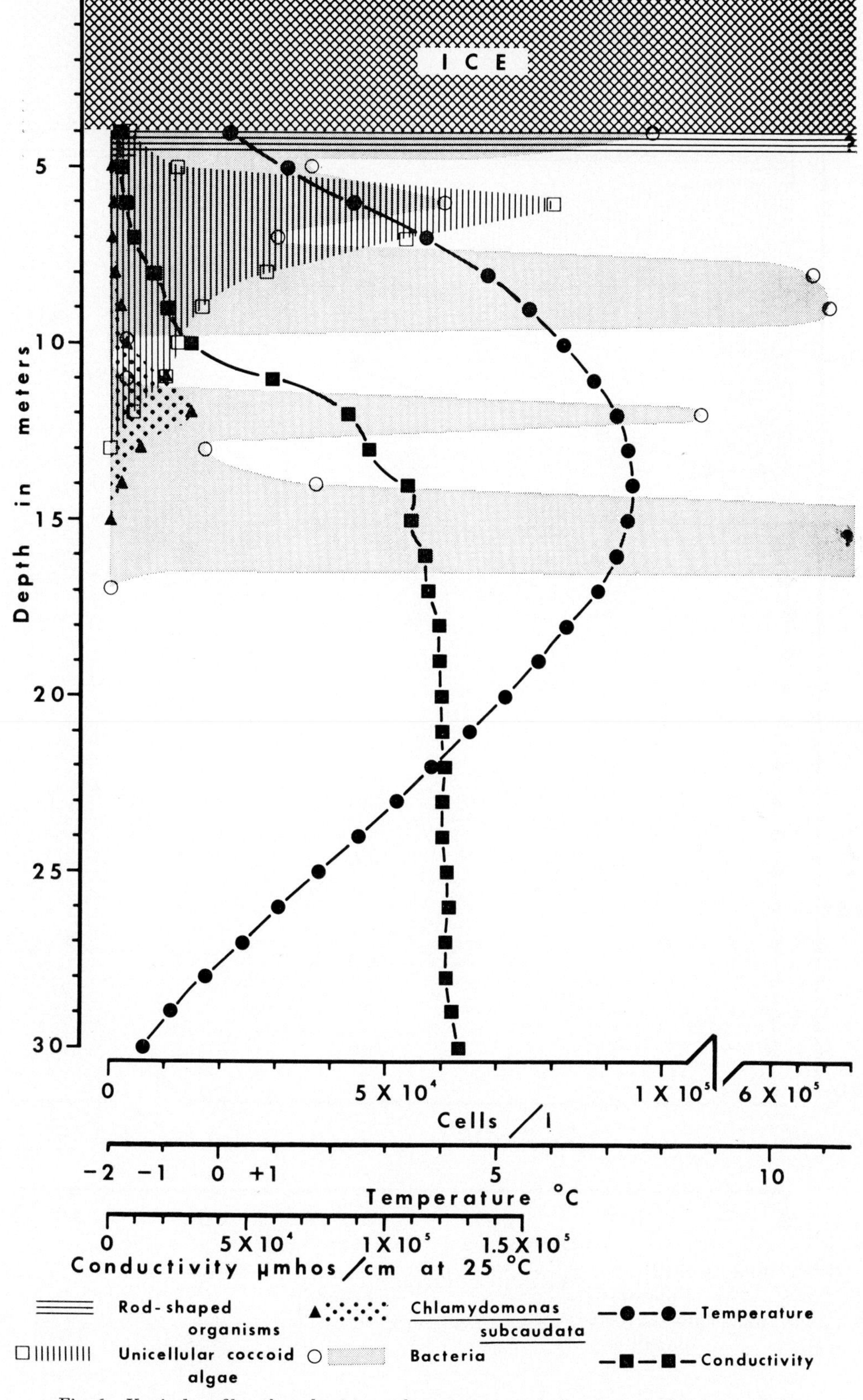

Fig. 6. Vertical profiles of conductivity and temperature and of various planktonic populations.

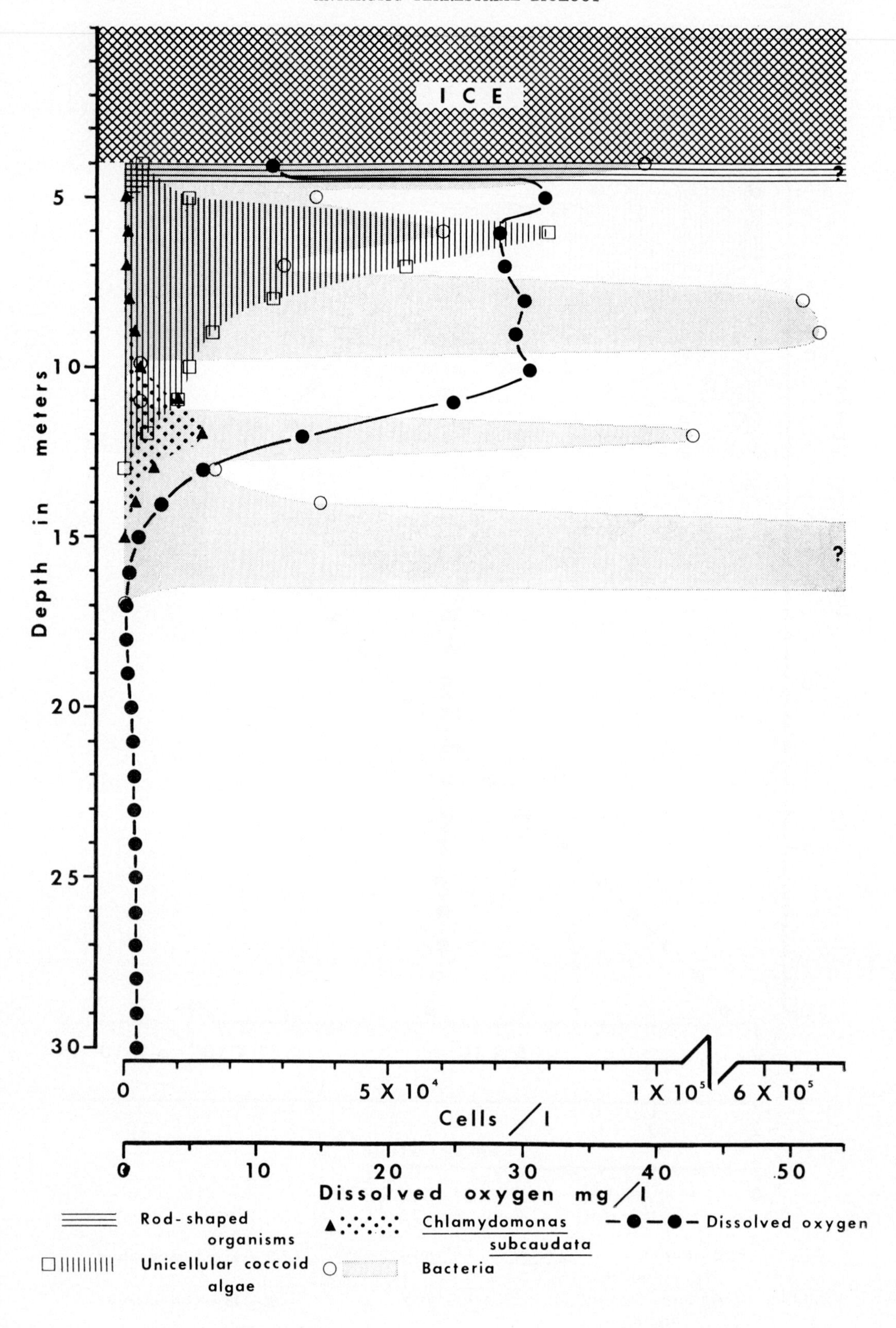

Fig. 7. Vertical profiles of dissolved oxygen and of various planktonic populations.

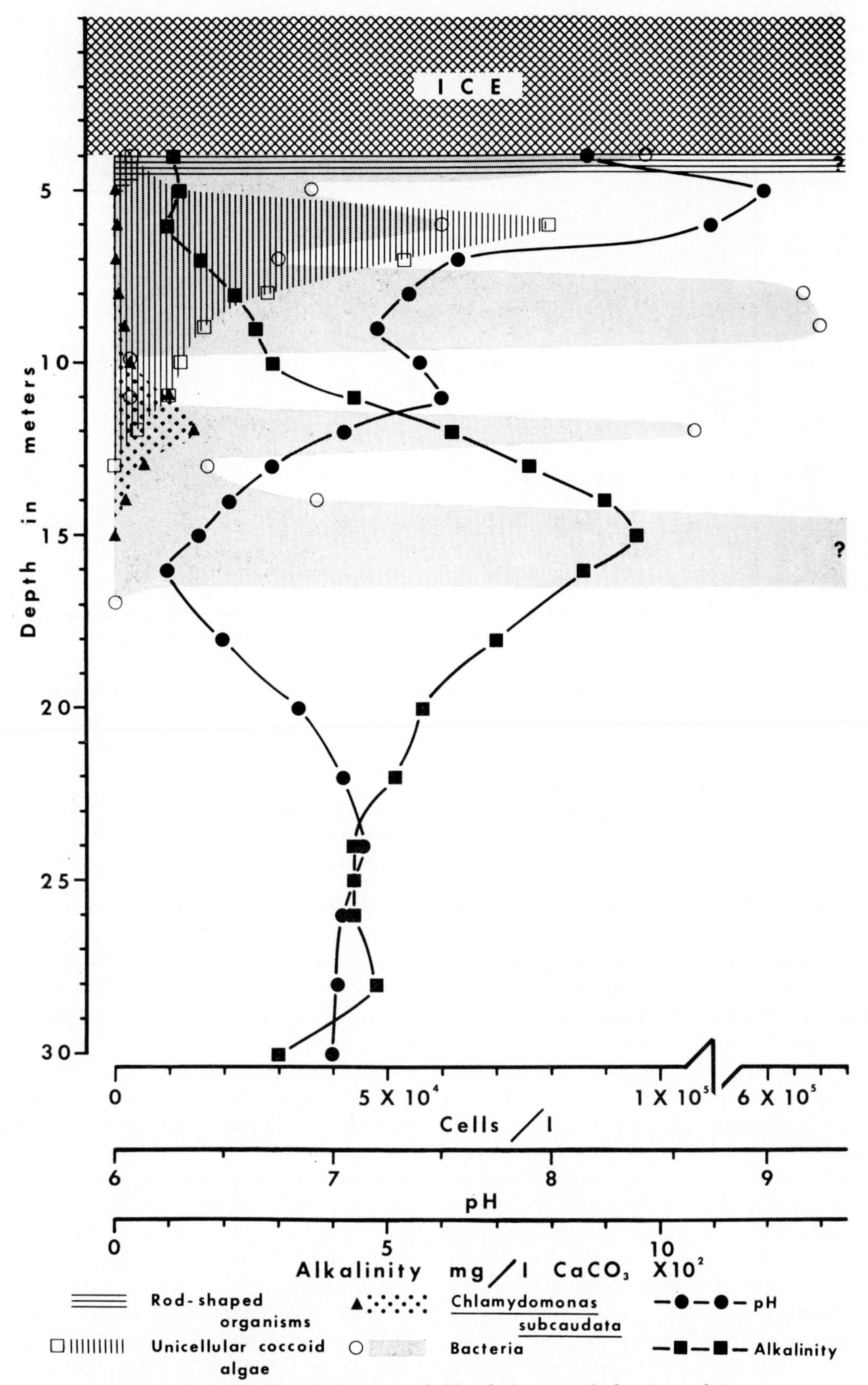

Fig. 8. Vertical profiles of alkalinity and *p*H and of various planktonic populations.

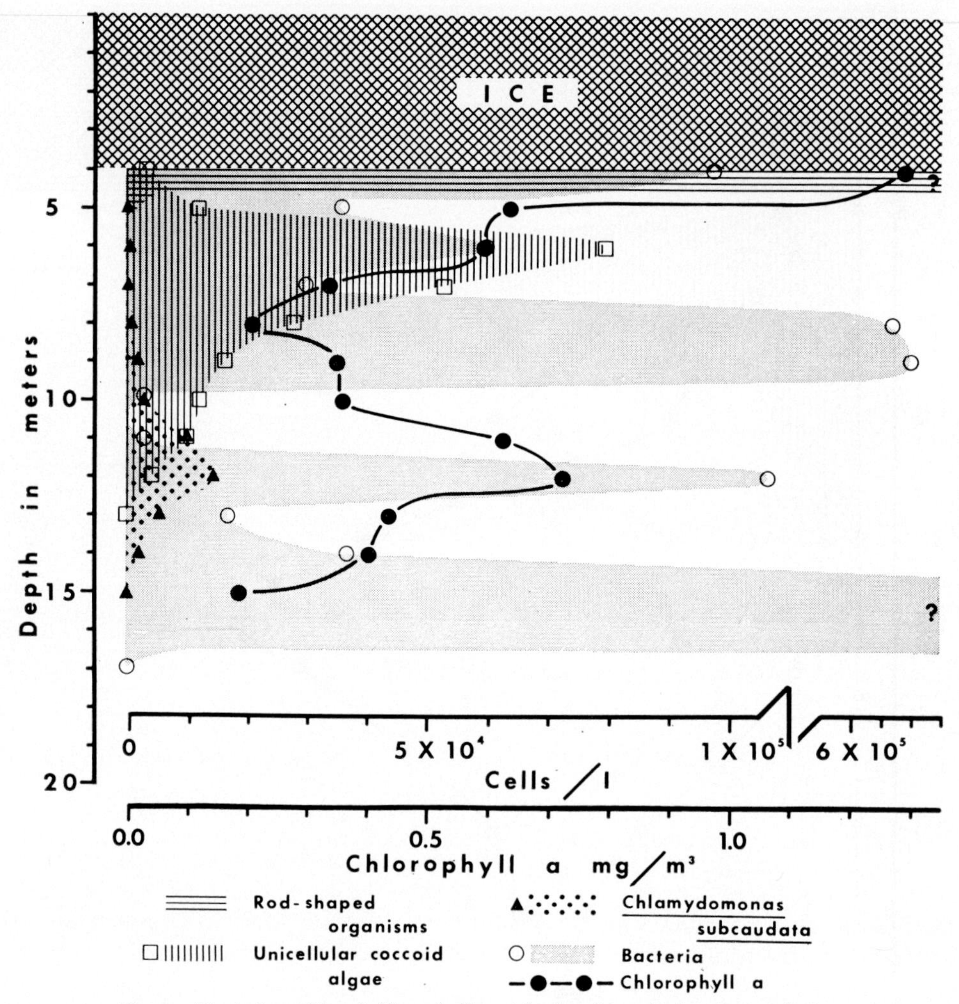

Fig. 9. Vertical profiles of chlorophyll *a* and of various planktonic populations.

responsible for the high concentration of DO in the mixolimnionlike region.

Unique evidence supporting the conclusion of a photosynthetically active mixolimnion comes from observations made when Lake Bonney ice was penetrated with the ice drill on December 11, 1965. A gusher of gas-saturated water spurted from the hole, and the water within the ice hole effervesced for several hours after drilling. No gushing or effervescing of the water was noted in November and January. The pressure and the resulting effervescence observed in December suggest that the water was at that time supersaturated with oxygen. From the preceding observations and from the DO data it is concluded that the highest rates of photosynthesis occurred during early December.

A population of *C. subcaudata* was found in the region of the chemocline in which there was a transition from the oxygen rich mixolimnionlike region to the oxygen deficient monimolimnion. The respiratory rate of oxygen uptake appears to exceed the photosynthetic rate of oxygen evolution by the populations in the oxygen deficient region. The effect of *C. subcaudata* on the concentration of DO appears to be negligible.

Although no biomass data were taken for the bacterial populations, their extreme numerical abundance supports the inference that these organisms are an important part of the Lake Bonney ecosystem. The bacterial peaks at 4, 6, and 9 meters are directly correlated with decreases in DO concentrations in these zones. The vertical distribution of these populations seems to be stable. The minimum oxygen concentrations at 4, 6, and 9 meters were stable; i.e., they were present on November 13 and 15 and on December 11 at both 0330 and 1730 hours local time and on December 13. The phenomenal decrease in DO in the chemocline may be due in part to the

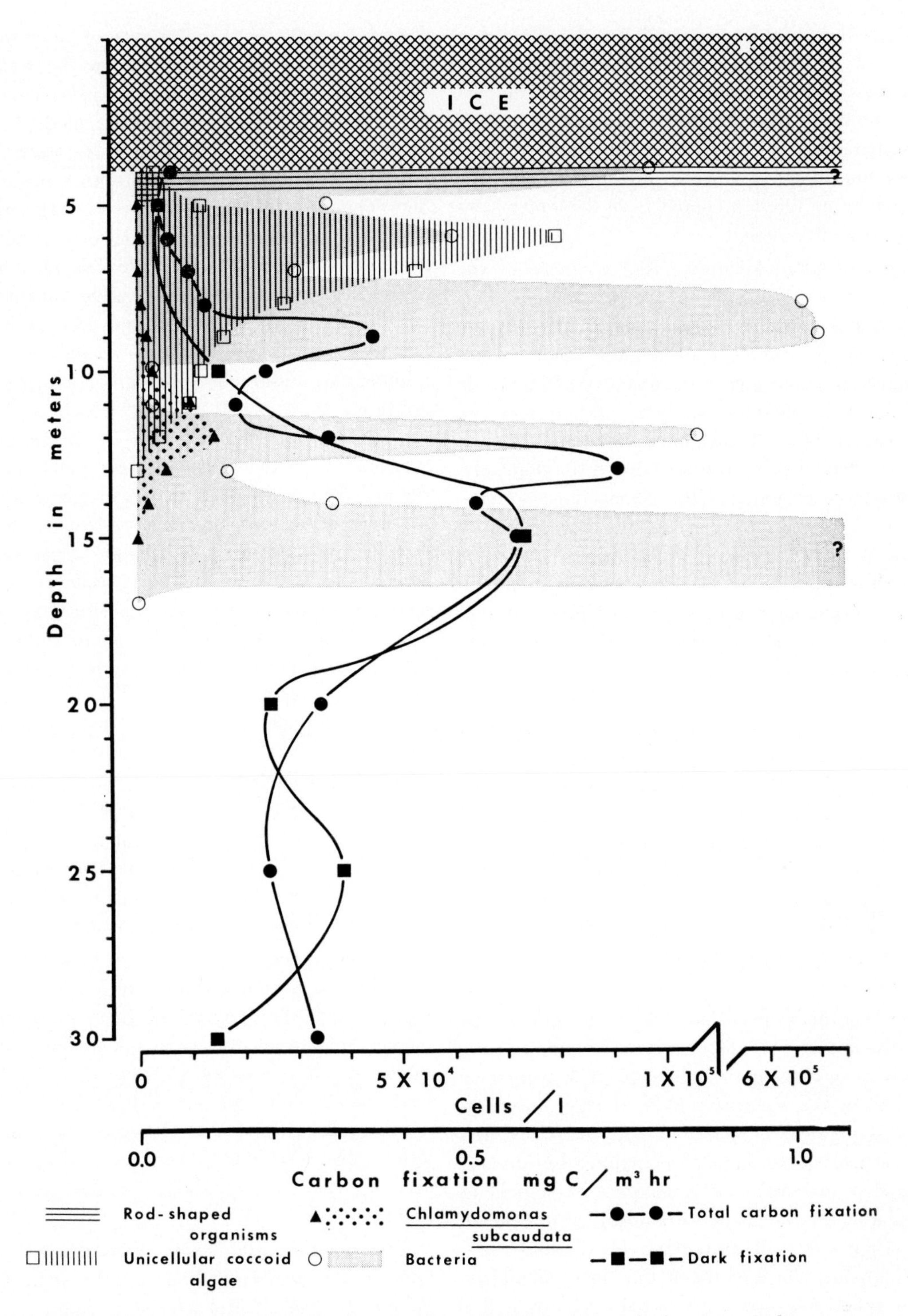

Fig. 10. Vertical profiles of total carbon fixation and dark fixation and of various planktonic populations.

respiratory activity of the large bacterial population between 15 and 16 meters.

The 16- to 30-meter (monimolimnion) portion of the DO vertical profile warrants additional remarks. Measurements of DO in this region may be erroneous, owing to the unknown influence of high ion concentrations on the Winkler test and on the Precision galvanic cell oxygen analyzer. The entire monimolimnion may indeed be anaerobic. Further support for this hypothesis comes from the absence of growth of phytoplanktonic colonies on aerobic culture plates and from the absence of green phytoplanktonic cells on the membrane filters prepared from the water samples taken from this region.

Alkalinity and pH. Alkalinity is a measure of the capacity of a water sample to accept protons. This property is due to the concentrations of bicarbonate, carbonate, and hydroxide ions in most waters. Alkalinity values indirectly reflect the quantity of inorganic carbon available for photosynthetic carbon fixation. The vertical profile of alkalinity for Lake Bonney closely parallels that of conductivity throughout the mixolimnionlike region and the chemocline (Figures 6 and 8). Two discrepancies are the low alkalinity values at 4 and 6 meters. These values could be caused by the rapid use of inorganic carbon by phytoplankton. Maximum numbers of unicellular coccoid forms and rod-shaped unicells did occur at the 6-meter and 4-meter levels, respectively.

Changes in *p*H values may also reflect biological activity. High photosynthetic activity is inferred from increases in *p*H, and high respiratory rates from decreases in *p*H. The vertical *p*H profile in Lake Bonney exhibits maximums at 5 and 11 meters and minimums at 9 and 16 meters (Figure 8). These *p*H maximums and minimums are associated with the maximum populations of unicellular coccoid algae at 6 meters and of *C. subcaudata* at 12 meters and with the bacterial populations at 9 and 15–16 meters, respectively.

Pigments. Another criterion for ascertaining the spatial distribution of photosynthetic populations is the relative concentration of chlorophyll *a* at various depths. When the vertical profile of the concentration of this pigment is superimposed over the graph for the vertical distribution of plankton, the obvious chlorophyll *a* maximum at 4 meters corresponds to the population of rod-shaped organisms just beneath the ice (Figure 9). Measurements of alkalinity and DO also support the hypothesis that this population is photosynthetically active. A possible chlorophyll *a* peak at 6 meters is partially obscured by the pigment peak at 4 meters and most likely corresponds to the maximum population of unicellular coccoid algae at 6 meters. The third pigment peak at 12 meters corresponds exactly with the location of the population of *C. subcaudata.*

Primary productivity. The rate of inorganic carbon assimilation by phytoplankton can be estimated by associated oxygen evolution, by *p*H changes, or by the Steemann-Nielsen ^{14}C uptake method. The last method was employed in the present study because it is the most sensitive. Two plots were made from the data obtained by the ^{14}C method. The results are expressed in counts per minute in one plot and in milligrams of carbon per cubic meter hour in the other. There were carbon fixation maximums at 9 and 13 meters (Figures 10 and 11) in both cases. On the graph expressing the results in counts per minute, two additional apparent photosynthetic peaks appeared at 4 and 6 meters. There is little doubt that the carbon fixation peak at 13 meters reflects the activity of the *C. subcaudata* population in that region. The small peak in counts per minute at the ice–water interface may be correlated with the presence of the pigmented rod-shaped organisms in that zone. The small productivity peak recorded at 6 meters is correlated with the unicellular coccoid population. The strong peak in photosynthetic carbon fixation at 9 meters coincides neither with any phytoplanktonic populations detected in this investigation nor with any photosynthesis-related chemical parameter. It may reflect the presence of a phytoplanktonic population completely missed in this investigation. Goldman [1964] found a peak in photosynthetic carbon fixation at 9 meters, but his sampling depths were 4, 5, 6, 7, 9, and 14 meters. Thus it is not possible to state the actual depth of the mixolimnetic photosynthetic maximum from his data.

It is unfortunate that, although in our investigation light bottles were suspended at 1-meter depth intervals through the chemocline, dark inorganic carbon uptake measurements were taken only at 5-meter intervals. It is now evident that dark bottles should be included with each set of light bottles. The dark uptake rates may be as discretely stratified as the light rates. Without these data, conclusions concerning true 'photosynthetic' uptake rates are much more tenuous.

The apparent high rates of inorganic carbon uptake in the monimolimnion may be artifacts or may be due to anaerobic bacterial populations, although we have no evidence to support either hypothesis. All bacterial counts were made on aerobically incubated petri plates and would not be expected to reveal populations in the anaerobic zone below the chemocline. Future studies to elucidate the biological activity in these bottom waters should prove fruitful.

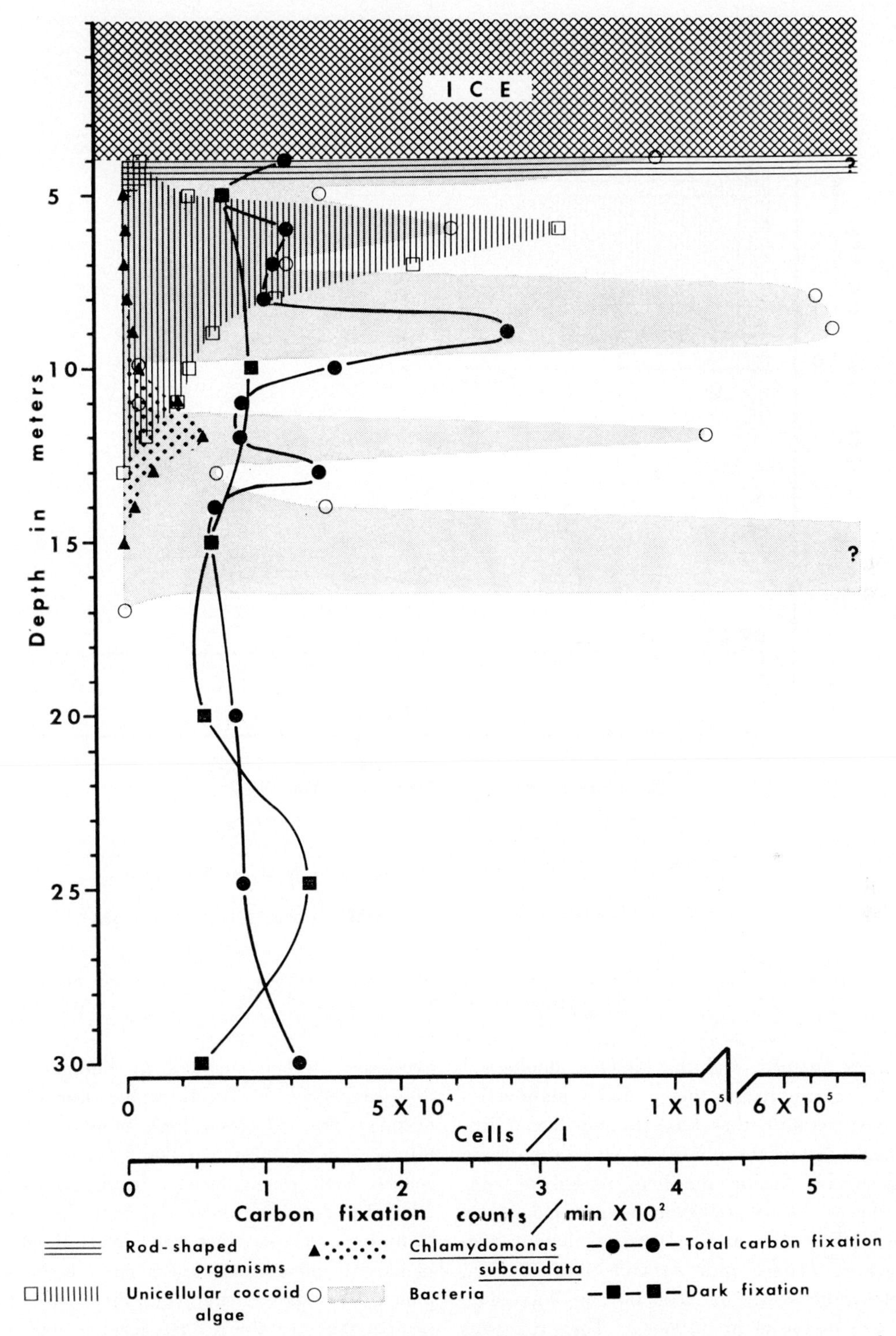

Fig. 11. Vertical profiles of total carbon fixation and dark fixation in direct counts per minute and of various planktonic populations.

Heterotrophy. Most biological research in Antarctica is conducted during the period of continuous illumination. The antarctic winter with its period of continuous darkness places severe but not insolvable limitations on man's ability to conduct field research. Biological studies of Lake Bonney during the winter

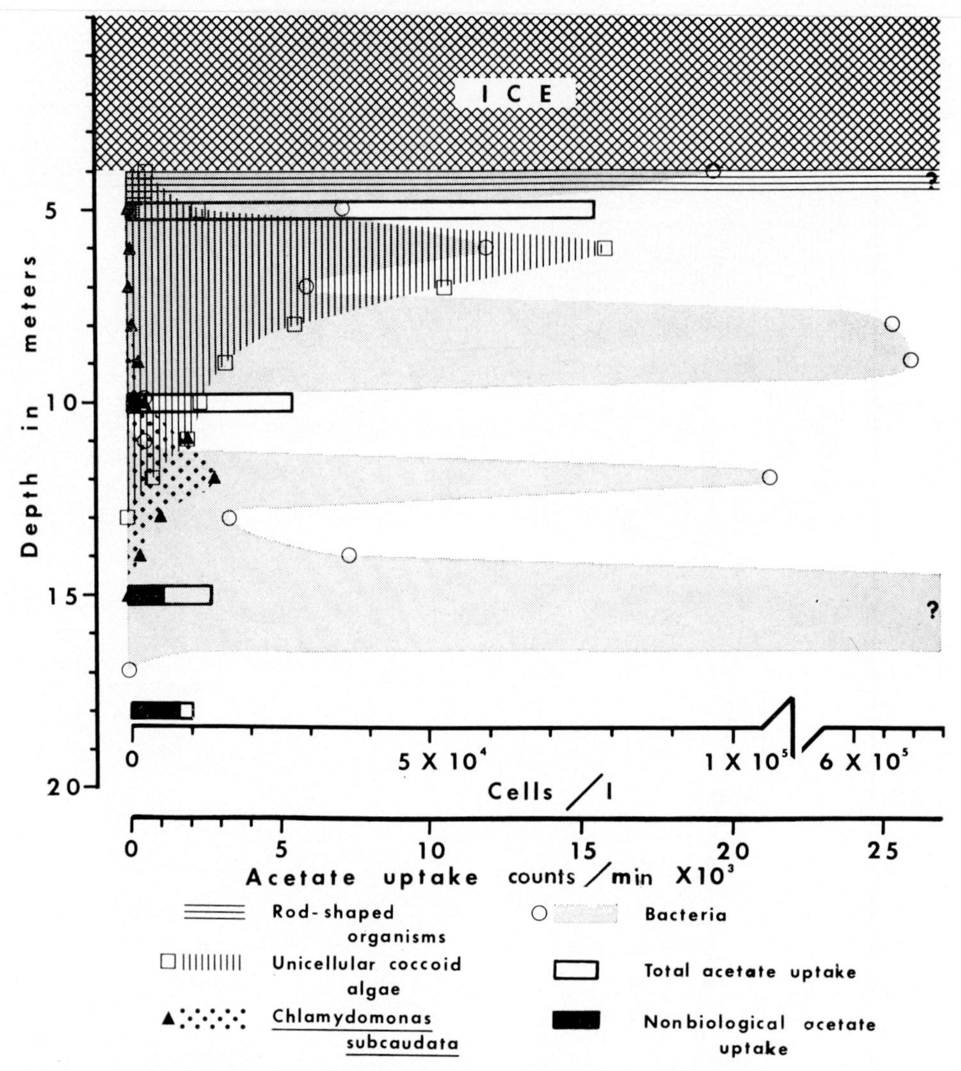

Fig. 12. Vertical profiles of acetate uptake and nonbiological uptake in counts per minute and of various planktonic populations.

have not been pursued, and the kinds of biological activities that occur during this period remain enigmatic. If one assumes that Lake Bonney algae live autotrophically during the summer, then how do these organisms survive during the long period of continuous darkness? A suggestion made most eloquently by Rodhe [1955] for arctic lakes is that heterotrophic carbon fixation may explain their survival when photosynthetic carbon reduction is impossible (however, see Rodhe et al. [1966]). The organisms may also become encysted and simply remain dormant. The survival of bacterial populations can easily be explained by heterotrophic use of organic compounds. Although this question can be answered only by experimentation in the winter, one brief experiment was conducted in January to ascertain the occurrence of heterotrophic use of ^{14}C-labeled acetate. The extremely high number of counts per minute recorded for samples from the 5-meter level reflects high rates of heterotrophic activity (Figure 12). Uptake rates decreased with increasing depth. Definite correlations between acetate uptake rates and planktonic populations are impossible because of the paucity of data. However, there are populations present that are able to use acetate. There is a need for more intensive experimentation in an effort to separate and to measure bacterial and possibly algal heterotrophic activity. The significance of heterotrophic activity during the winter by the plankton in Lake Bonney remains unknown.

FUTURE STUDY

Lake Bonney is one of the simplest large ecosystems in the world today. The constant ice cover prevents wind mixing and the development of circulation currents. The absence of fish and zooplankters results in an exceedingly simplified food web. The isolation of Lake Bonney has exempted it from common pollutants so characteristic of almost all other natural ecosystems on earth. These conditions provide the aquatic biologist with unique opportunities for study. The inverse thermal stratification as well as the meromictic conditions in this lake present unprecedented opportunities to study (1) the migrations of populations to different levels in the water column according to changes in the quality and the quantity of light, (2) the relation of plankton to pycnoclines, and (3) the problems of the cycling of nitrogen, iron, sulfur, and autochthonous organic substances. Because of the extreme stratification of the planktonic populations in Lake Bonney, sampling at depth increments of <1 meter would likely prove advantageous.

Another area of research for which the Lake Bonney ecosystem is ideal involves the interrelationships among the bacterial and phytoplanktonic populations. There appear to be distinct correlations between the vertical distributions of bacterial and algal populations. Are the bacteria directly dependent on the algae for an organic energy source? Are the algae dependent on the bacteria for organic growth factors? What effect would nutrient enrichment have on the densities of the phytoplanktonic and bacterial populations? What is the relative importance of photosynthesis, chemosynthesis, and heterotrophic uptake of organic materials to the various populations? What changes occur in these populations during the long winter months when 24 hours of darkness prevail?

APPENDIX: MEDIA FOR THE ISOLATION OF ALGAE FROM LAKE BONNEY

Organic medium. The following concentrations of constituents were used (all values except that of agar are in milligrams per liter):

K_2HPO_4	20.00
$MgSO_4 \cdot 7\ H_2O$	20.00
Sodium citrate 2 H_2O	20.00
Ferric citrate	1.00
Citric acid	1.00
Neopeptone	600.00
Trypticase	160.00
Yeast extract	50.00
Glycolic acid	0.50
Trace elements	
H_3BO_3	1.00
$ZnSO_4 \cdot 7\ H_2O$	1.00
$MnSO_4 \cdot 4\ H_2O$	0.40
$CoCl_2 \cdot 6\ H_2O$	0.20
$NaMoO_4 \cdot 2\ H_2O$	0.20
$CuSO_4$	0.04
Agar, grams	15.00

High-salt medium. The following concentrations of high-salt constituents were used (all values except that of agar are in milligrams per liter):

NaCl	20,000
$MgSO_4 \cdot 7\ H_2O$	3,200
$Ca(NO_3)_2$	500
$CaCO_3$	400
K_2HPO_4	200
Ferric citrate	100
Citric acid	100
Agar, grams	15

Acknowledgments. The financial assistance provided by National Science Foundation grant GA-123 (awarded to D. D. Koob) is gratefully acknowledged. We also wish to thank the members of U.S. Navy Task Force 43, who provided excellent logistics for this work. Special appreciation for field assistance during the first year is given to Mr. Paul W. Richard. Contribution 213 of the Institute of Polar Studies, Ohio State University, Columbus, Ohio.

REFERENCES

American Public Health Association
1965 Standard methods for the examination of water and wastewater. 12th ed., 769 pp. New York.

Angino, E. E., and K. B. Armitage
1963 A geochemical study of Lakes Bonney and Vanda, Victoria Land, Antarctica. J. Geol., *71:* 89–95.

Angino, E. E., K. B. Armitage, and J. C. Tash
1964 Physiochemical limnology of Lake Bonney, Antarctica. Limnol. Oceanogr., *9:* 207–217.

Armitage, K. B., and H. B. House
1962 A limnological reconnaissance in the area of McMurdo Sound, Antarctica. Limnol. Oceanogr., *7:* 36–41.

Armstrong, R. L., W. Hamilton, and G. H. Denton
1968 Glaciation in Taylor valley, Antarctica, older than 2.7 million years. Science, *159:* 187–189.

Cameron, R. E.
1966 Soil studies—Desert microflora. 13. Identification of some algae from Antarctica. Space Programs Sum. 37-40, *4:* 123–133. Jet Propul. Lab., Calif. Inst. of Technol., Pasadena.

deNoyelles, F., Jr.
1968 A stained organism filter technique for concentrating phytoplankton. Limnol. Oceanogr., *13:* 562–565.

Dougherty, E. C., and L. G. Harris
1963 Antarctic micrometazoa: Fresh-water species in the McMurdo Sound area. Science, *140:* 497–498.

Fritsch, F. E.
1912 Freshwater algae. National antarctic expedition 1901–1904. *6:* 66 pp.

Goldman, C. R.
1963 The measurement of primary productivity and limiting factors in freshwater with carbon-14. *In* M. S. Doty (Ed.), Proceedings of the conference on primary productivity measurement, marine and freshwater, Hawaii, 1961, pp. 103–113. U.S. Atomic Energy Commission, Washington, D. C.
1964 Primary productivity studies in antarctic lakes. *In* R. Carrick, M. W. Holdgate, and J. Prevost (Eds.), Biologie antarctique, pp. 291–299. Hermann, Paris.

Goldman, C. R., D. T. Mason, and J. E. Hobbie
1967 Two antarctic desert lakes. Limnol. Oceanogr., *12:* 295–310.

Gunn, B. M., and G. Warren
1962 Geology of Victoria Land between the Mawson and Mulock glaciers, Antarctica. Bull. Geol. Surv. N. Z., *71:* 157 pp.

Hoare, R. A., et al.
1964 Lake Bonney, Taylor valley, Antarctica: A natural solar energy trap. Nature, *202:* 886–888.

Holm-Hansen, O.
1964 Isolation and culture of terrestrial and freshwater algae of Antarctica. Phycologia, *4:* 43–51.

Hutner, S. H., and L. Provasoli
1964 Nutrition of algae. A. Rev. Pl. Physiol., *15:* 37–56.

Hutner, S. H., L. Provasoli, and J. Filfus
1953 Nutrition of some phagotrophic fresh-water chrysomonads. Ann. N. Y. Acad. Sci., *56:* 852–862.

McKelvey, B. C., and P. N. Webb
1961 Geological reconnaissance in Victoria Land, Antarctica. Nature, *189:* 545–547.

Provasoli, L., and I. J. Pintner
1953 Ecological implications of *in vitro* nutritional requirements of algae flagellates. Ann. N. Y. Acad. Sci., *56:* 839–851.

Ragotzkie, R. A., and G. E. Likens
1964 The heat balance of two antarctic lakes. Limnol. Oceanogr., *9:* 412–425.

Rodhe, W.
1955 Can plankton production proceed during winter darkness in subarctic lakes? Verh. Inst. Verein. Theor. Angew. Limnol., *12:* 117–122.

Rodhe, W., J. E. Hobbie, and R. T. Wright
1966 Phototrophy and heterotrophy in high mountain lakes. Verh. Inst. Verein. Theor. Angew. Limnol., *16:* 302–313.

Saunders, G. W., F. B. Trama, and R. W. Bachmann
1962 Evaluation of a modified C^{14} technique for shipboard estimation of photosynthesis in large lakes. Publ. 8: 61. Great Lakes Res. Div., Univ. of Mich., Ann Arbor.

Shirtcliffe, T. G. L.
1964 Lake Bonney, Antarctica: Cause of the elevated temperatures. J. Geophys. Res., *69:* 5257–5268.

Shirtcliffe, T. G. L., and R. F. Benseman
1964 A sun-heated antarctic lake. J. Geophys. Res., *69:* 3355–3359.

Steemann-Nielsen, E.
1952 The use of radioactive carbon (C^{14}) for measuring organic production in the sea. J. Cons. Perm. Int. Explor. Mer, *18:* 117–140.

Strickland, J. D. H., and T. R. Parsons
1961 A manual of seawater analysis. J. Fish. Res. Bd Can., *125:* 1–185.

Torii, T., N. Yamagata, and T. Cho
1967 Report of the Japanese summer parties in dry valleys, Victoria Land, 1963–1965. 2. General description and water temperature data for the lakes. Antarct. Rec., *28:* 1–14.

Wille, N.
1903 Algologische Notizen 11. Nytt Mag. Naturvid., *41:* 118–120.

Yamagata, N., T. Torii, and S. Marata
1967 Report of the Japanese summer parties in dry valleys, Victoria Land, 1963–1965. 5. Chemical composition of lake waters. Antarct. Rec., *29:* 53–75.

FRESH-WATER ALGAE OF THE ANTARCTIC PENINSULA
1. SYSTEMATICS AND ECOLOGY IN THE U.S. PALMER STATION AREA

BRUCE C. PARKER AND GENE L. SAMSEL

Department of Biology, Virginia Polytechnic Institute and State University, Blacksburg, Virginia 24061

G. W. PRESCOTT

University of Montana Biological Station, Flathead Lake, Big Fork, Montana 59911

Abstract. A great variety of fresh-water planktonic and Aufwuchs algae from small melt pools and year-round ponds near the U.S. Palmer Station, Anvers Island, Palmer Archipelago, are described. When it is possible, this preliminary survey attempts to correlate algal distributions with geologic, physical, and chemical environmental features measured during the collecting. The total picture is still incomplete, but from the large number of species reported here for the first time in Antarctica we feel that a much greater diversity of nearly all major groups of fresh-water antarctic algae occurs at more southerly latitudes than has been supposed previously. In the Palmer Station area, fresh-water habitats exhibit the complete range of trophic levels from extreme oligotrophy with a paucity of species and low organic production to hypereutrophy with blooms of a single or a few species. Eutrophication is enrichment induced during the antarctic summer by penguin rookeries, elephant seal colonies, and other resident bird populations.

According to Hirano [1965] few systematics, distributional, and ecologic studies of algae from the Antarctic Peninsula or subantarctic islands have been published since the early works of de Wildeman [1900, 1935], Hariot [1908], West and West [1911], Fritsch [1912], Gain [1912], and Frenguelli [1943]. Such information from this region, the portion of the antarctic continent closest to other global land masses, is vital for understanding dispersal mechanisms, environmental adaptation, and algal evolution, as well as community structure, function, and energy flow in antarctic fresh-water ecosystems. This report constitutes the first in a series of surveys of the fresh-water algae of the Antarctic Peninsula that correlates the physical, geologic, chemical, and biological features of selected major algal habitats.

Approximately 50 fresh-water habitats ranging from simple melt water pools to lakes 1500 m^2 in surface area were visited during the 1970–1971 austral summer (January 10–26). Seventeen of these sites contained well-developed algal communities considered to be of probable taxonomic and/or ecologic interest. The approximate locations of these sites are shown in Figure 1. Seven habitats were selected for detailed study, including investigations of their physical, geologic, and chemical characteristics, on the basis of their size, apparent permanency, and/or biological content. We summarize the characteristics of these seven habitats as determined during the sampling period and list the algae collected from our 17 sample sites; we especially note the algae not collected previously in Antarctica.

MATERIALS AND METHODS

The collections included plankton net (70-μ pores) samples, scrapings from rocks, and mud samples. All the plankton were examined with a Nikon field microscope; sketches, notes, and some photomicrographs were made prior to preservation of the samples in 5–10% formalin. Fresh collections from some sites were inoculated into various culture media (Table 1). Preliminary comparisons of these media with those used by previous workers for the isolation and the culture of antarctic algae [Thornton, 1922; Holm-Hansen, 1964; Cameron, 1966; Koob, this volume] carried out on fresh-water samples taken from near the Virginia Polytechnic Institute and State University campus showed that our media resulted in the recovery of more species. Mixed cultures were incubated under illumination and ambient temperature conditions outside the

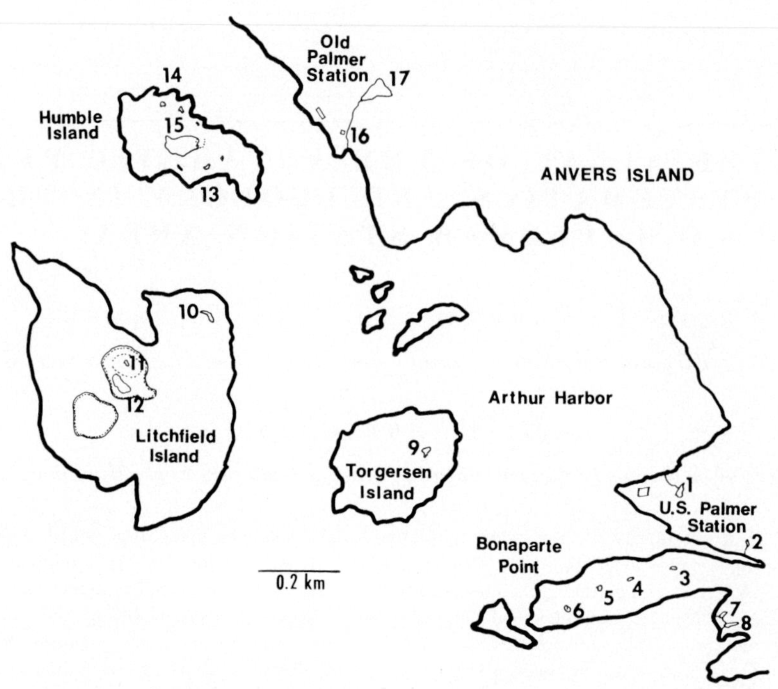

Fig. 1. Map of Palmer Station area, Anvers Island, sketched from aerial photographs that shows the approximate locations of the fresh-water habitats collected during the austral summer of 1970–1971. The habitats were assigned the following unofficial names to facilitate communication (asterisks indicate the seven most intensively studied habitats): 1, CB pond*; 2, Commander Frost's pool*; 3, Bonaparte Point pool 1; 4, Bonaparte Point pool 2; 5, Bonaparte Point pool 3; 6, Bonaparte Point pool 4*; 7, Elephant seal pool 1 (green)*; 8, Elephant seal pool 2 (red); 9, Torgersen Island penguin pool*; 10, Litchfield Island penguin mud; 11, Litchfield Island mountaintop pool; 12, Litchfield Island intermontane valley; 13, Humble Island pools, southeast rim; 14, Humble Island pools, northeast rim; 15, Humble Island Lake*; 16, Old Palmer stream, Norsel Point; and 17, Old Palmer Lake, Norsel Point.*

Palmer Station laboratory and were subsequently transported (under refrigeration whenever possible) to the university for cultivation under approximately 5000 lux of continuous cool white fluorescent illumination at 5°C.

The physical, geologic, chemical, and biological examination of the seven most thoroughly studied habitats included (1) morphometric measurements, (2) identification of the parent rock type composing the lake basin, (3) analyses of select inorganic ions, dissolved O_2, and *p*H with the Hach Chemical Company's direct reading portable engineer's laboratory, the duplicate determinations always showing $<10\%$ variation, (4) radiation measurements, both incident and subsurface (20 cm), with a Yellow Springs Instrument Company model 68 pyrheliometer to 5% precision and ±0.1 ly/min accuracy, (5) ambient air and water temperature measurements with a Yellow Springs Instrument Company model 41 thermistor and probe to ±0.1°C accuracy, (6) chlorophyll measurements via extraction in acetone of 0.22-μ Millipore GS membranes through which water had been filtered and subsequent colorimetric determination according to the Scor (Scientific Committee on Oceanographic Research)/Unesco formula and the methods of Strickland and Parsons [1968], and (7) primary productivity measurements by using the ^{14}C method of Steeman-Nielsen [1952] modified by Strickland and Parsons [1968] and employing 250-ml screw cap pharmaceutical bottles, which we have found superior in optical properties to many other types of glass containers [Parker and Samsel, 1970]. The values reported here for total chlorophyll and for carbon fixation represent the means of the results for duplicate samples having $<5\%$ variation.

The identifications of taxa were accomplished with the aid of classical systematics works, such as Frémy [1930], Fritsch [1911, 1917], Fukushima [1959],

TABLE 1. Four Culture Media Used for the Isolation and the Growth of Antarctic Algae

Solute	Modified Bristol's Medium [Bold, 1942]	Bozniak's Community Medium [Bozniak, 1969]	Synthetic 'Old Palmer Lake' Medium	Synthetic 'Humble Lake' Medium
$NaNO_3$	250			
$CaCl_2$	25			27.7
$MgSO_4 \cdot 7H_2O$	75	10.2		
K_2HPO_4	75	6.0		
KH_2PO_4	175	3.6		
Na_2SiO_2		59.0	2.3	3.05
$Ca(NO_3)_2 \cdot 4H_2O$		144		
$NaHCO_3$		60.0		
$MgCl_2$			5.2	11.1
Na_2CO_3			13.45	18.0
NaCl	25			
EDTA		10.0		
NH_4NO_3				10
$Ca(OH)_2$			8.0	
Na_2HPO_4			0.14	6.4
$KHCO_3$				2.6
$FeSO_4 \cdot 7H_2O$	1			
H_2SO_4 (1:1000 dilution), ml/l			2.5	2.5
H_3BO_3		2.8		
$NaNO_2$				2.01
$ZnSO_4 \cdot 7H_2O$				
$MnCl_2 \cdot 4H_2O$		0.890	0.410	0.410
MoO_3		0.0075		
Ferric citrate		1.0		
Citric acid		1.0		
$FeCl_2$			0.417	0.417
HCl (1:1000 dilution), ml/l			1.5	1.5
KCl			2.2	
NH_4Cl			0.25	
$CuSO_4 \cdot 5H_2O$		0.000125		
K_2CrO_4		0.0037		
$CoCl_2 \cdot 6H_2O$		0.02		
$ZnCl_2$		0.0104		
$VOSO_4 \cdot 2H_2O$		0.0039		
Thiamin hydrochloride*		1.0×10^{-6}		
Cyanocobalimine*		1.0×10^{-6}		
Total (approximate)	626	298.5+	37.66	86.97

All values are in milligrams per liter unless otherwise indicated.

*Vitamin.

Geitler [1932], Kol [1968], Kol and Flint [1968], Komarek and Ruzicka [1965], and de Wildeman [1935]. In general the arrangement of chlorophytan genera follows the scheme advanced by Bourrelly [1966]. We fully recognize the weaknesses of this method compared to detailed culture study methods [e.g., Herndon, 1958; Bischoff and Bold, 1963; Trainor and McLean, 1964; Shihira and Krauss, 1965; Cox and Deason, 1968; Groover and Bold, 1968; Uhlik and Bold, 1970; Thomas and Brown, 1970; Archibald and Bold, 1970] for establishing the range of morphologic variations within a single genotypic species. Nonetheless we present our list of genera, species, and perhaps some ecophenes for comparison with other classical algal taxonomic investigations in Antarctica, particularly because most of the algae collected did not grow in the various culture media used, including modified Bristol's solution, which was employed most extensively by the authors cited above. Ultimately we hope to extend and to reexamine some of these organisms by culture techniques; until this work is accomplished, however, our approach must remain relatively classical and therefore comparable to other algal systematics studies in Antarctica [Hirano, 1965; Koob, 1967].

RESULTS AND DISCUSSION

The appendix lists the algae collected and identified from the 17 fresh-water habitats indicated in Figure 1. Among the dominant algae we found species of *Chlamydomonas* and *Chlorella* most frequently representing the phytoplankton. *Oscillatoria, Chroococcus,* and *Trochiscia* were dominant, especially as Aufwuchs algae, and *Prasiola* was frequently a dominant macroscopic alga in habitats where bird excrement had enriched the water. Only at site 6 (Figure 1), which had a dissolved silica content of >1.0 ppm, did we find significant numbers of fresh-water diatoms.

Blue-green algal genera having proven nitrogen-fixing capabilities appeared only rarely in our collections; this result suggests that aerobic photosynthetic nitrogen fixation is probably not important in these fresh-water habitats. Our laboratory microecosystem studies (G. L. Samsel and B. C. Parker, unpublished data, 1971) further suggest that nitrate reduction by aquatic microorganisms is also insignificant or absent. Thus the main sources of nitrogen for fresh-water algae appear to be ammonia and ammonium in the air, rain, and runoff.

That relatively few algae grew in our culture media justifies the necessity for field identifications and for the use of preserved material in a complete floral analysis. Especially noteworthy in this respect is the unidentified dinoflagellate, apparently the first reported for Antarctica. This organism formed a dense bloom with *Chlamydomonas* sp. but was extremely delicate and disintegrated shortly after collection and photomicrography. It was destroyed in preservation and failed to grow in any of our culture media.

Hirano [1965] has summarized the literature on

fresh-water algae from Antarctica and the subantarctic regions. He lists 229 species and varieties from Antarctica belonging to 64 genera. We have recognized approximately 103 species and varieties in the Palmer Station area, not all of which have been fully named. Of these we report about 60 species and varieties and 25 genera apparently for the first time in Antarctica exclusive of the subantarctic islands. The rather chaotic nature of the algal taxonomic literature no doubt partially explains our recognition of so many species and genera of algae not previously reported from Antarctica. Nevertheless this rather sizable number of algae new to Antarctica also points to the incomplete status of our present knowledge and further substantiates the need for additional, more comprehensive collections throughout the entire continent.

Hirano [1965] reviews the numbers of algal components for major groups at 48°–50°S latitude (Îles Kerguelen), 50°–60°S latitude (South Georgia, Falkland Islands, Tierra del Fuego), 60°–65°S latitude (South Orkney Islands, South Shetland Islands, Graham Land), 66°–70°S latitude (Wilhelm II Coast), and 70°–80°S latitude (Victoria Land). His data reveal (1) no obvious latitudinal trends in Cyanophyceae or diatoms, (2) an apparent reduction in numbers of species of all groups at the higher latitudes, (3) a reduction in the numbers of Heterokontae, filamentous Conjugatae including desmids, and Rhodophyceae with increasing latitudes and the total elimination of members of these groups, except one desmid, at latitudes of >66°S. Although our data are still insufficient to modify these trends, our discovery of many new species and genera, some of which were heterokonts and desmids, suggests that greater numbers and varieties of fresh-water algae penetrate to more southerly latitudes along the Antarctic Peninsula than has been assumed previously.

Unfortunately too little of the physiology of desmids and other Conjugatae is known to explain their poor representation on the antarctic continent. We do know, however, that these algae as a group typically dominate in acid waters, few of which occur in Antarctica, and that they are not tolerant of saline conditions; a few members apparently survive transmission through avian guts [Proctor, 1959]. Similarly Conjugatae apparently have not been reported among the nearly 200 viable algae isolated from the atmosphere [Schlichting, 1969]. These points suggest that these algae may be poorly represented in the Antarctic because they have had little opportunity to reach that continent via water, air, and/or biological dispersal mechanisms; however, there is also a trend in the Arctic toward a reduction in the numbers of species of some of these groups as one progresses northward without interruption by the sea.

In our opinion the question of algal endemism remains complicated and unsolved. At present we can only imagine the mechanisms of distribution (wind, water, migratory animals, and so forth) of algae to and from Antarctica. The species of fresh-water algae in Antarctica seem quite common throughout the world. We wish to note, however, that much of our difficulty in naming some species derives from slight differences in their morphology from that of taxa already described. One can view these differences as evolutionary trends, if not environmental responses, of algae recently having reached Antarctica or of algae yet to be distributed from Antarctica. Thus, in the light of our current knowledge about global algal distributions, the endemism of microscopic algae may well be an illusion caused by inadequate data.

Table 2 summarizes our data on chemical analyses, extractable chlorophylls, and primary productivity for seven of the fresh-water habitats visited during our field studies. Among the notable findings were that (1) dissolved oxygen was saturated or supersaturated (≥13 mg/l) at all sites during this period of greatest solar radiation; (2) there were striking differences between the two most eutrophic highly productive habitats (i.e., sites 7 and 9) and the other habitats in environmental parameters including *p*H, total alkalinity, nitrite, phosphate, and fluoride; (3) the total dissolved solids, based on the limited spectrum of analyses, were significantly higher in habitats 7, 9, and 15 than in the other habitats; (4) primarily productivity measured by the ^{14}C method revealed a wide range of trophic levels for the fresh-water habitats of the U.S. Palmer Station area. This range appears consistent with that recorded for Alaskan tundra lakes by Howard and Prescott [1971].

The numbers of algal species also ranged widely among the 17 habitats (appendix). Each of the distinctly eutrophic habitats (i.e., habitats 7, 8, and 9 in Figure 1) contained about five species, whereas habitat 1, which was one of the more extremely oligotrophic fresh-water bodies, contained only one algal species. The greatest number of species (40) occurred in habitat 15, which was intermediate in its trophic level and collected somewhat more extensively than the other habitats.

Except for habitat 9 all the sites collected resembled each other geologically. They were underlain by

TABLE 2. Results of Analyses of Select Chemicals, Chlorophyll, Primary Productivity, and Other Features of Seven Fresh-Water Habitats near the U.S. Palmer Station

Determination	Habitat 1	Habitat 2	Habitat 6	Habitat 7	Habitat 9	Habitat 15	Habitat 17
Maximum depth, meters	1.5	0.6	0.3	0.6*	0.2	0.6	6.2
Secci disc visibility, meters	>1.5	>0.6	>0.3	0.1	0.2*	>0.6	>6.2
pH	6.4	6.4	6.4	8.9	8.3	6.3	6.4
Dissolved O_2, mg/l	14.0	14.0	14.0	14.0	14.0	13.0	14.0
Total alkalinity, mg/l	15.0	10.0	10.0	70.0	80.0	12.0	10.0
Total hardness, mg/l	5.0	5.0	10.0	80.0	25.0	15.0	10.0
Calcium hardness, mg/l	5.0	5.0	5.0	50.0		10.0	5.0
Ammonium (as N), mg/l	0.01	0.02	0.35	0.65	1.05	2.5	0.1
Nitrate (as N), mg/l	0.095	1.99	1.8	3.2		1.75	1.0
Nitrite (as N), mg/l	0.005	0.01	0	3.8	3.0	0.05	0
Orthophosphate (as P), mg/l	0.01	0.02	0.47	0.069	0.11	1.0	0.02
Total phosphate (as P), mg/l	0.02	0.06	0.47	2.0	7.0	1.1	0.03
Dissolved silica, mg/l	0.10	0.05	1.3	2.5	19.0	1.0	0.75
Iron (as Fe^{++}), mg/l	0.01	0.01	0	1.35	0.7	0.02	0.02
Sulfate, mg/l	2.0	2.0	2.0	0	0.01	6.0	4.0
Chloride, mg/l	2.5	10.0	25.0	17.0	12.0	35.0	7.5
Fluoride, mg/l	0.28	15.0	0.38	117.5		0.95	0.65
Total phytoplanktonic chlorophyll, mspu/m^3	26.0	814.0†	683.0	2194.0		112.0	37.0
Surface radiation, langleys	0.5	0.5	0.45	0.38	0.40	0.5	0.38
Primary productivity, mg C/m^3/hr	1.2	340.0‡	3.0	830		72.0	0.78

Habitats correspond to numbered locations in Figure 1; mspu, million standard pigment units.
* Approximate.
† Benthic, attached; glacial melt water flowed through quite rapidly and thereby precluded phytoplankton.
‡ Not comparable to other values based on phytoplankton; this sample was one of concentrated attached algae.

quartz-diorite (tonalite) and had <5% acid feldspars (e.g., biotite, pyroxene) and traces of sodium plagioclase. Habitats 6–8 had somewhat higher amounts of silicates in the rocks than the other habitats; this finding correlated with the higher dissolved silica. Habitat 9 contained a volcanic, acidic type of granite, probably porphyritic tuff (i.e., riolite), and some sodium plagioclase and traces of other minerals. Habitats 1, 2, and 17 had ≤5 cm of inorganic rock-derived sediment, there being little or no accumulation of organic matter. Habitat 6 contained some organic sediment derived from a nearby skua nest. Habitats 15, 7, and 9 had sediments rich in organic matter derived from aquatic mosses, elephant seal excrement, and penguin guano, respectively.

An interesting observation was the apparent trend toward larger-celled algae in habitats having a higher content of total dissolved solids. This trend was especially obvious for the widespread genus *Chlamydomonas*. Prescott [1962] has observed a similar phenomenon in other fresh-water habitats, especially for desmids. One may speculate that this difference of cell size relates to the conservation of energy for osmoregulation in some fresh-water algae. Thus, for a eukaryotic green alga like *Chlamydomonas*, small-celled species would have a relatively high ratio of surface area to volume and thus more effective absorption of solutes from the dilute aqueous environment surrounding them. In contrast larger-celled species having lower ratios of surface area to volume might thrive in aqueous habitats having higher solute concentrations, which would be more readily absorbed. If this hypothesis is correct, one may ask further whether the cytological differences between the *Chlamydomonas* species are genetically based and hence constitute legitimate taxonomic criteria or whether we may merely be observing phenotypic variations in one or more species.

APPENDIX: FRESH-WATER ALGAE OF THE U.S. PALMER STATION AREA AND NOTES ON THEIR PREVIOUS COLLECTION FROM ANTARCTICA

(Asterisks and daggers indicate species and genera, respectively, to the authors' knowledge not previously reported from the antarctic continent.)

Cyanophyta: Chroococcales, Chroococcaceae

Aphanocapsa grevillei (Haas.) Rab.*

Collection site: Humble Island pools, southeast rim.

Specimens: preserved.

Previous collection: *Aphanocapsa montana* Cramer described for Antarctica [Hirano, 1965].

Aphanothece microspora (Menegh.) Rab.*
Collection site: Commander Frost's pool.
Specimens: alive in field and preserved.
Previous collection: *Aphanothece prasina* A. Br. described [Hirano, 1965].

A. microscopica Naeg.*
Collection site: Torgersen Island penguin pool.
Specimens: preserved.
Previous collection: reported from Îles Kerguelen [Hirano, 1965].

Chroococcus limneticus var. *carneus* (Chodat) Lemm.*
Collection site: Old Palmer stream, Norsel Point.
Specimens: preserved.
Previous collection: none.

C. minor (Kuetz.) Naeg.
Collection site: Bonaparte Point pool 2.
Specimens: preserved.
Previous collection: none.

C. minutus (Kuetz.) Naeg.
Collection sites: Bonaparte Point pool 2; Humble Island pools, southeast rim; Humble Island Lake; Old Palmer stream, Norsel Point; Old Palmer Lake, Norsel Point.
Specimens: Bonaparte Point specimens alive in field and preserved; Old Palmer stream specimens in culture and preserved; all other specimens preserved.
Previous collections: none.

C. pallidus Naeg.
Collection site: Humble Island pools, southeast rim.
Specimens: preserved.
Previous collection: none.

C. varius A. Braun*
Collection site: Elephant seal pool 2 (red).
Specimens: in culture.
Previous collection: none.

Dactylococcopsis irregularis G. M. Smith*
Collection site: Elephant seal pool 1 (green).
Specimens: preserved.
Previous collection: reported from Tierra del Fuego and south Patagonia [Hirano, 1965].

D. raphidioides Hangrig
Collection site: Humble Island Lake.
Specimens: preserved.
Previous collection: none.

Gloeocapsa sp.
Collection site: ? Humble Island Lake.
Specimens: preserved.
Previous collection: three *Gloeocapsa* species described from Antarctica [Hirano, 1965].

Merismopedia elegans A. Braun*
Collection site: Humble Island Lake.
Specimens: preserved.
Previous collection: none.

M. sp.
Collection site: Humble Island Lake.
Specimens: preserved.
Previous collection: *M. punctata* Meyen and *M. tenuissima* described from Antarctica [Hirano, 1965].

Entophysalidaceae

Heterohormogonium sp.†
Collection site: Litchfield Island mountaintop pool.
Specimens: preserved.
Previous collection: none.

Chamaesiphonales, Pleurocapsaceae

Myxosarcina concinna Printz†
Collection site: Bonaparte Point pool 4.
Specimens: preserved.
Previous collection: none.

Oscillatoriales, Oscillatoriaceae

Lyngbya spp.
Collection site: ? Humble Island Lake.
Specimens: preserved.
Previous collection: 17 *Lyngbya* species reported from Antarctica [Hirano, 1965].

Oscillatoria angusta Koppe*
Collection sites: Commander Frost's pool, Bonaparte Point pool 2, and Humble Island Lake.
Specimens: Commander Frost's pool specimens alive in field and preserved; Bonaparte Point and Humble Island Lake specimens preserved.
Previous collection: none.

O. angustissima West and West*
Collection sites: Humble Island pools, southeast rim, and Old Palmer stream, Norsel Point.
Specimens: preserved.
Previous collection: none.

O. brevis (Kuetz.) Gomont
Collection sites: Litchfield Island mountaintop pool and Humble Island Lake.
Specimens: preserved.
Previous collection: none.

O. granulata Gard.*
Collection sites: Bonaparte Point pool 4; Humble Island Lake; and Old Palmer stream, Norsel Point.
Specimens: preserved.
Previous collection: none.

O. granulata Gard. var. nov.*
Collection sites: Humble Island Lake and Old Palmer stream, Norsel Point.
Specimens: preserved.
Previous collection: none.

O. hormogenea Fremy*
Collection site: Torgersen Island penguin pool.
Specimens: preserved.
Previous collection: none.

O. meslinii Fremy*
Collection site: Old Palmer stream, Norsel Point.
Specimens: preserved.
Previous collection: none.

O. profunda Kirch.*
Collection site: Bonaparte Point pool 4.
Specimens: alive in field and preserved.
Previous collection: none.

O. splendida Greville*
Collection site: Humble Island Lake.
Specimens: preserved.
Previous collection: also reported from the South Orkney Islands [Hirano, 1965].

O. subbrevis Schmidle*
Collection site: Humble Island pools, northeast rim, and Old Palmer Lake, Norsel Point.
Specimens: preserved.
Previous collection: none.

O. tenuis var. *levis* Gard.
Collection sites: Commander Frost's pool, Elephant seal pool 1 (green), and Humble Island Lake.
Specimens: Commander Frost's pool specimens alive in field and preserved and other specimens preserved.
Previous collections: *O. tenuis* also reported from the South Orkney Islands, Tierra del Fuego, and south Patagonia [Hirano, 1965].

O. terebriformis Agardh*
Collection site: Elephant seal pool 1 (green).
Specimens: preserved.
Previous collection: none.

O. spp.
Collection site: Humble Island Lake.
Specimens: alive in field.
Previous collection: none.

Phormidium mucicola Huber-Pestalozzi et Naumann*
Collection site: Humble Island Lake.
Specimens: preserved.
Previous collection: 14 species reported from Antarctica [Hirano, 1965].

P. tenuissimum Woronichin*
Collection site: Humble Island Lake.
Specimens: preserved.
Previous collection: none.

P. spp.
Collection site: Humble Island Lake.
Specimens: alive in field.
Previous collection: from several locations in Antarctica [Koob, 1967].

Schizothrix calcicola (Ag.) Gomont
Collection site: ? Humble Island Lake.
Specimens: preserved.
Previous collection: reported by Cameron and Benoit [1970] from Deception Island.

Nostocales, Nostocaceae

Anabaena sp.
Collection site: Bonaparte Point pool 2.
Specimens: alive in field.
Previous collection: two *Anabaena* species reported from Antarctica [Hirano, 1965].

Rivulariaceae

Calothrix sp.
Collection site: Humble Island pools, southeast rim.
Specimens: alive in field.
Previous collection: four *Calothrix* species reported from Antarctica [Hirano, 1965].

? Mastigocladaceae

Mastigocladus-like†
Collection site: Humble Island Lake.
Specimens: alive in field.
Previous collection: none.

Chlorophyta: Volvocales, Chlamydomonadaceae

Chlamydomonas basimaculata Pascher et Jahoda*
Collection site: Humble Island Lake.
Specimens: preserved. (For application of the culture method to *Chlamydomonas* spp., see Uhlik and Bold [1970].)
Previous collection: none.

C. caudata Wille
Collection site: Torgersen Island penguin pool.
Specimens: alive in field and preserved.
Previous collection: *C. caudata* reported from the South Orkney Islands [Hirano, 1965] and Lake Bonney [Koob, 1967].

C. elliptica Korschikoff*
Collection site: Humble Island pools, southeast rim.
Specimens: preserved.
Previous collection: none.

C. gracilis Snow*
Collection sites: Elephant seal pools 1 (green) and 2 (red).
Specimens: preserved.
Previous collection: none.

C. snowiae Printz.
Collection site: Humble Island pools, southeast rim.
Specimens: preserved.
Previous collection: none.

C. spp. (vegetative cells and zygotes)
Collection sites: Commander Frost's pool; Bonaparte Point pools 1, 3, and 4; Elephant seal pools 1 (green) and 2 (red); Torgersen Island penguin pool; Humble Island pools, southeast and northeast rims; and Humble Island Lake.

Specimens: Bonaparte Point pools 1 and 4 specimens in culture, alive in field, and preserved; Commander Frost's pool specimens alive in field; Bonaparte Point pool 3 specimen in culture; all other specimens preserved.

Previous collection: *Chlamydomonas* is by far the dominant and most widely distributed genus of algae in the fresh-water plankton of the western Antarctic Peninsula [Hirano, 1965; Koob, 1967].

Polytoma sp.†
Collection site: Old Palmer Lake, Norsel Point.
Specimens: in culture.
Previous collection: none.

Volvocaceae

Gonium sp.†
Collection site: ? Litchfield Island intermontane valley.
Specimens: alive in field.
Previous collection: none.

Phacotaceae

Pteromonas sp.†
Collection site: ? Humble Island Lake.
Specimens: alive in field.
Previous collection: *Pteromonas nivalis* (Shuttlew) Chodat reported from the South Orkney Islands [Hirano, 1965].

Tetrasporales, Gloeocystaceae

Palmellopsis gelatinosa Korschikoff†
Collection site: Humble Island Lake.
Specimens: preserved.
Previous collection: none.

Palmellaceae

Sphaerocystis schroeteri Chodat†
Collection sites: Bonaparte Point pool 2 and Humble Island pools, southeast rim.
Specimens: preserved.
Previous collection: also reported from the South Orkney Islands [Hirano, 1965].

Chlorococcales, Chlorococcaceae

Chlorococcum infusionum (Schrank) Meneghini*
Collection site: Humble Island Lake.
Specimens: preserved. (*Chlorococcum* is a form species and genus as used here without the culture method [Thomas and Brown, 1970; Archibald and Bold, 1970].)
Previous collection: none.

C. sp. (resembling *C. botryoides* Rab.)*
Collection sites: Bonaparte Point pools 1, 3, and 4; Humble Island Lake; and Old Palmer Lake, Norsel Point.
Specimens: Bonaparte Point pools 1 and 3, Humble Island Lake, and Old Palmer Lake specimens in culture; Bonaparte Point pool 2 specimens preserved.
Previous collection: *C. humicola,* another form species, reported from Deception Island by Cameron and Benoit [1970].

Neochloris sp.
Collection site: Bonaparte Point pool 3.
Specimens: preserved.
Previous collection: none.

Spongiochloris sp.†
Collection site: Bonaparte Point pool 4.
Specimens: in culture. (Also, perhaps *Spongococcum* [Trainor and McLean, 1964; Cox and Deason, 1968].)
Previous collection: none.

Trebouxia sp.†
Collection site: Commander Frost's pool.
Specimens: alive in field.
Previous collection: none.

Oocystaceae

Also perhaps *Chlorococcum* [Archibald and Bold, 1970].

Ankistrodesmus setigerus (Schroed.) G. S. West†
Collection site: Old Palmer stream, Norsel Point.
Specimens: preserved.
Previous collection: none.

Chlorella vulgaris
Collection sites: CB pond; Bonaparte Point pool 1; Elephant seal pool 2 (red); and Old Palmer stream, Norsel Point.
Specimens: Old Palmer stream specimens in culture and preserved; all other specimens in culture.
Previous collection: Second in importance to the genus *Chlamydomonas, Chlorella* is a dominant and widespread member of the fresh-water phytoplankton in the western Antarctic Peninsula area. *Chlorella vulgaris* also reported from Deception Island by Cameron and Benoit [1970].

C. sp.
Collection sites: Elephant seal pool 2 (red); Humble Island Lake; and Old Palmer Lake, Norsel Point.
Specimens: Humble Island Lake specimens in culture and alive in field; other specimens in culture. (*Chlorella* is used here as a form species and genus only [Shihira and Krauss, 1965].)
Previous collection: three *Chlorella* species reported in Antarctica [Hirano, 1965].

Oocystis borgei Snow†
Collection site: Bonaparte Point pool 4.
Specimens: preserved. (*Oocystis* is used here as a form genus and species [Groover and Bold, 1968].)
Previous collection: several *Oocystis* species reported from the subantarctic regions [Hirano, 1965].

O. lacustris Chodat†
Collection site: Bonaparte Point pool 1.
Specimens: in culture.
Previous collection: none.

O. spp. (pitted wall)†
Collection site: Humble Island Lake.
Specimens: preserved.
Previous collection: none.

Scotiella antarctica Fritsch
Collection sites: Humble Island pools, southeast rim, and Humble Island Lake.
Specimens: preserved.
Previous collection: several locations on the Antarctic Peninsula [Koob, 1967].

S. sp.*
Collection sites: Bonaparte Point pools 3 and 4.
Specimens: in culture.
Previous collection: none.

Trochiscia granulata (Reinsch) Hansgrig*
Collection site: Humble Island Lake.
Specimens: preserved.
Previous collection: also reported from Tierra del Fuego and south Patagonia [Hirano, 1965].

T. zachariasi Lemmermann*
Collection site: Humble Island Lake.
Specimens: preserved.
Previous collection: none.

T. spp.
Collection site: Humble Island Lake.
Specimens: preserved.
Previous collection: four species of *Trochiscia* reported from Antarctica.

Hormotilaceae

Palmodictyon varium (Naeg.) Lemmermann†
Collection site: Bonaparte Point pool 2.
Specimens: preserved.
Previous collection: none.

Coccomyxaceae

Diogenes bacillaris (West) Bourrelly (formerly *Nannochloris bacillaris* Naumann)†
Collection sites: Bonaparte Point pools 3 and 4 and Old Palmer Lake, Norsel Point.
Specimens: in culture.
Previous collection: none.

Dictyosphaeriaceae

Westella botryoides (West) de Wildeman†
Collection site: Old Palmer Stream, Norsel Point.
Specimens: preserved.
Previous collection: none.

Ulotrichales, Ulotrichaceae

Binuclearia tatrana Wittrock
Collection site: Bonaparte Point pool 2.

Specimens: preserved.
Previous collection: none.

Stichococcus flaccidus (Kuetz.) Gay*
Collection sites: Commander Frost's pool and Old Palmer stream, Norsel Point.
Specimens: Commander Frost's pool specimens alive in field and Old Palmer stream specimens in culture. (For application of culture method to *Stichococcus*, see Arce [1971].)
Previous collection: three *Stichococcus* species reported from Antarctica [Hirano, 1965].

Ulothrix sp.
Collection site: Old Palmer Lake, Norsel Point.
Specimens: in culture.
Previous collection: Seven species of *Ulothrix* are known to Antarctica.

Chlorosarcinaceae

Chlorosarcina elegans Gerneck†
Collection sites: Bonaparte Point pool 2 and Humble Island Lake.
Specimens: Bonaparte Point specimens preserved and Humble Island Lake specimens in culture and preserved. (*Chlorosarcina* used here as only a form species and genus since the culture method is not used [Herndon, 1958; Cox and Deason, 1968]. *Chlorosphaera* reported near Showa Base [Koob, 1967].)
Previous collection: none.

C. lacustris (Snow) Lemmermann†
Collection sites: Humble Island Lake and Old Palmer stream, Norsel Point.
Specimens: Humble Island Lake specimens preserved and Old Palmer stream specimens in culture.
Previous collection: none.

Chaetophoraceae

? *Gongrosira* sp. nov.†
Collection site: Bonaparte Point pool 4.
Specimens: in culture (may be a new genus if not a new *Gongrosira* species).
Previous collection: none.

Ulvales, Prasiolaceae

Prasiola crispa (Lightf.) Meneghini
Collection sites: Bonaparte Point pools 1 and 3; Litchfield Island intermontane valley; Humble Island pools, southeast rim; Humble Island Lake; Old Palmer stream, Norsel Point; and Old Palmer Lake, Norsel Point.
Specimens: preserved.
Previous collection: *P. crispa* is widespread and dominant among the attached algae and is located throughout Antarctica and the subantarctic regions.

P. crispa (Lightf.) Menegh. with parasite (formerly *Schizogonium crispa* Gay)
Collection sites: Humble Island Lake.
Specimens: preserved.
Previous collection: none.

P. fluviatilis (Sommer.) Areschoug
Collection site: Old Palmer stream, Norsel Point.
Specimens: preserved.
Previous collection: none.

P. tesselata Kuetzing*
Collection site: Litchfield Island intermontane valley.
Specimens: preserved.
Previous collection: none.

P. spp. (filamentous stages)
Collection sites: Bonaparte Point pool 4; Humble Island Lake; and Old Palmer stream, Norsel Point.
Specimens: Bonaparte Point and Humble Island Lake specimens preserved, and Old Palmer stream specimens alive in field.
Previous collection: several locations for *P.* spp. and *P. calophylla* (Carm.) Menegh. [Koob, 1967].

Ulvaceae

Monostroma sp.†
Collection site: Humble Island Lake.
Specimens: alive in field and preserved.
Previous collection: dominated in biomass during December 1970 here.

Zygnematales, Mesotaeniaceae

Cylindrocystis brebissonii Meneghini*
Collection site: Old Palmer stream, Norsel Point.
Specimens: preserved.
Previous collection: reported from Îles Kerguelen, the South Orkney Islands, Tierra del Fuego, and south Patagonia [Hirano, 1965], and *C.* spp. from two locations on the Antarctic Peninsula [Koob, 1967].

Desmidiaceae

Only five desmids have been described previously for the Antarctic Peninsula area [Gain, 1912], and Hirano [1965] lists only eight species for the entire antarctic continent.

Cosmarium spp.
Collection site: Old Palmer stream, Norsel Point.
Specimens: alive in field.
Previous collection: Antarctic Peninsula [Koob, 1971].

Staurastrum disputatum West and West (formerly *S. dilatatum* var. *insignis* Raciborski) *
Collection site: Old Palmer stream, Norsel Point.
Specimens: preserved.
Previous collection: none.

Chrysophyta: Xanthophyceae, Pleurochloridaceae

Ellipsoidion solitare (Geitler) Pascher†
Collection site: Humble Island Lake.
Specimens: preserved.
Previous collection: none.

E. sp. nov.†
Collection site: Humble Island Lake.
Specimens: preserved.
Previous collection: none.

Trachychloron ellipsoideum Pascher†
Collection site: Humble Island Lake.
Specimens: preserved.
Previous collection: none.

Chrysophyceae, Chrysococcales, Chrysococcaceae

Chrysococcus tesselatus Fritsch†
Collection sites: Bonaparte Point pool 4 and Humble Island Lake.
Specimens: preserved.
Previous collection: none.

Chrysococcus sp. cysts†
Collection sites: Bonaparte Point pools 2 and 4 and Humble Island Lake.
Specimens: Bonaparte Point pool 2 and Humble Island Lake specimens preserved; Bonaparte Point pool 4 specimens in culture and preserved.
Previous collection: none.

Chrysomonadales, Chromulinaceae

Chromulina minima Doflein†
Collection site: Bonaparte Point pool 4.
Specimens: in culture.
Previous collection: none.

C. sp.†
Collection site: Bonaparte Point pool 1.
Specimens: in culture.
Previous collection: none.

Chrysapsidaceae

Chrysapsis sp.†
Collection site: Humble Island Lake.
Specimens: in culture.
Previous collection: none.

Bacillariophyceae, Pennales, Achnanthaceae

Achnanthes lanceolata (Bréb.) Grun.
Collection site: Old Palmer stream, Norsel Point.
Specimens: preserved.
Previous collection: reported from Îles Kerguelen [Hirano, 1965].

Cocconeis sp.
Collection site: Bonaparte Point pool 4.
Specimens: alive in field.
Previous collection: two species reported from Antarctica [Hirano, 1965].

Pinnularia sp.
Collection site: Bonaparte Point pool 4.
Specimens: alive in field.
Previous collection: four species reported from Antarctica.

Fragilariaceae

Fragilaria virescens Ralfs
Collection site: Old Palmer stream, Norsel Point.
Specimens: preserved.
Previous collection: five species of *Fragilaria* reported from Antarctica [Hirano, 1965].

Tabellaria spp.
Collection site: Bonaparte Point pool 4. (Note that this site was the only noneutrophic habitat with a dissolved silica content of more than 1.0 ppm. The rocks were highest in SiO_2 at this location.)
Specimens: alive in field.
Previous collection: *Tabellaria flocculosa* (Roth) Kutz is the only species described from Antarctica [Hirano, 1965].

Naviculaceae

Diploneis sp.
Collection site: Bonaparte Point pool 4.
Specimens: alive in field.
Previous collection: none.

Navicula sp.
Collection sites: Commander Frost's pool; Bonaparte Point pools 2 and 4; Litchfield Island intermontane

valley; Humble Island pools, southeast rim; Humble Island Lake; and Old Palmer stream, Norsel Point.

Specimens: Old Palmer stream specimens preserved; Bonaparte Point pool 4 specimens alive in field and preserved; Humble Island Lake specimens in culture and preserved; all other specimens alive in field.

Previous collection: *N.* spp. and *N. multicopsis* Van Heurck from several locations in Antarctica [Koob, 1967].

Surirellaceae

Surirella sp.

Collection site: Bonaparte Point pool 4.

Specimens: alive in field.

Previous collection: *Surirella angusta* Kutz is the only species described from Antarctica [Hirano, 1965].

Centrales, Coscinodiscaceae

Coscinodiscus was also found in pools near Old Palmer stream, Norsel Point, which received sea spray, was not strictly fresh water, and was high in chloride.

Melosira borreri Greville*

Collection site: Old Palmer stream, Norsel Point.

Specimens: preserved.

Previous collection: three *Melosira* species described from Antarctica [Hirano, 1965].

Pyrrhophyta: Dinophyceae, Dinococcales

Unidentified athecate dinoflagellate†

Collection site: Elephant seal pool 2 (red). (This pool had only 120 ppm of chloride and hence was regarded as fresh water.)

Specimens: in culture.

Previous collection: *Peridinium umbonatum* Stein var. *inaequale* Lemm. reported from Îles Kerguelen [Hirano, 1965].

Acknowledgment. We are grateful to the National Science Foundation for grant GA-16768 in support of this research, to Drs. Eugene Bozniak and Richard Smith for reviews of this manuscript, and to Dr. Leroy Sharon for confirming our geologic observations.

REFERENCES

Arce, G.
1971 *Stichococcus sequoieti* sp. nov. *In* B. C. Parker and R. M. Brown (Eds.), Contributions in phycology, pp. 25–30. Allen, Lawrence, Kans.

Archibald, P. A., and H. C. Bold
1970 Phycological studies. 9. The genus *Chlorococcum meneghini.* Publ. 7015: 115 pp. Univ. of Tex., Austin.

Bischoff, H. W., and H. C. Bold
1963 Phycological studies. 4. Some soil algae from Enchanted Rock, and related algal species. Publ. 6318: 95 pp. Univ. of Tex., Austin.

Bold, H. C.
1942 The cultivation of algae. Bot. Rev., *8:* 69–138.

Bourrelly, P.
1966 Les algues d'eau douce. Algues vertes. 511 pp., 117 pls. Boubée et Cie, Paris.

Bozniak, E. G.
1969 Laboratory and field studies of phytoplankton communities. Ph.D. thesis, 106 pp. Dep. of Bot., Wash. Univ., St. Louis, Mo.

Cameron, R. E.
1966 Soil studies—Desert microflora. 13. Identification of some algae from Antarctica. Publ. 37-40, *10:* 123–133. Jet Propul. Lab., Calif. Inst. of Technol., Pasadena.

Cameron, R. E., and R. E. Benoit
1970 Microbial and ecological investigations of recent cinder cones, Deception Island, Antarctica—A preliminary report. Ecology, *51* (5) : 802–809.

Cox, E. R., and T. R. Deason
1968 *Axilosphaera* and *Heterotetracystis,* new chlorosphaeracean genera from Tennessee soil. J. Phycol., *4:* 240–249.

de Wildeman, E.
1900 Expédition antarctique belge. Note préliminaire sur les algues rapportées par M. E. Racovitza, naturaliste de l'expédition antarctique. Bull. Acad. R. Belg. Cl. Sci., *1900:* 558–569.
1935 Observations sur les algues. Expédition antarctique belge. Résult. Voyage S. Y. Belgica, *45:* 12 figs.

Frémy, P.
1930 Les Myxophycées de l'Afrique équatoriale française. Thèse, 507 pp., 362 figs. Fac. Sci., Paris, Caen.

Frenguelli, J.
1943 Diatomeas de las Orcadas del Sur. Revta Mus. La Plata, *5:* 221–265.

Fritsch, F. E.
1911 Freshwater algae collected in the South Orkneys. J. Linn. Soc., *40:* 293–366, pls. 10, 11.
1912 Algae of the South Orkneys. Rep. Scient. Results Scott. Natn. Antarct. Exped., *3:* 95–134.
1917 Nat. Hist. Rep. Br. Antarct. Terra Nova Exped., part 1: 1–16, pl. 1.

Fukushima, H.
1959 General report on fauna and flora of the Ongul Island, Antarctica, especially on freshwater algae. J. Yokohama Munic. Univ., ser. C-31, *112:* 1–10, pls. 1–10.

Gain, L.
1912 La flore algologique d'eau douce de l'Antarctique. *In* Deuxième expédition antarctique française, 1908–1910. La flore algologique des régions antarctiques et subantarctiques, pp. 156–218. Masson, Paris.

Geitler, L.
1932 Cyanophyceae. Rabenhorst's Kryptogamen-Flora. *14:* 1–1196, 780 figs.

Groover, R. D., and H. C. Bold
1968 Phycological notes. 1. *Oocystis polymorpha* sp. nov. SWest. Nat., *13:* 129–135.

Hariot, J.
1908 Algues. *In* Expédition antarctique française, 1903–1905, pp. 1–9. Masson, Paris.

Herndon, W. R.
1958 Studies on chlorosphaeracean algae from soil. Am. J. Bot., *45:* 298–308.

Hirano, M.
1965 Freshwater algae in the antarctic regions. *In* J. Van Mieghem and P. Van Oye (Eds.), Biogeography and ecology in Antarctica, pp. 127–193. Junk, The Hague.

Holm-Hansen, O.
1964 Isolation and culture of terrestrial and freshwater algae of Antarctica. Phycologia, *4:* 43–51.

Howard, H. H., and G. W. Prescott
1971 Primary production in Alaskan tundra lakes. Am. Midl. Nat., *85:* 108–123.

Kol, E.
1968 Algae from Antarctica. Annls Hist.-Nat. Mus. Natn. Hung., *60:* 71–77, 11 figs.

Kol, E., and E. A. Flint
1968 Algae in green ice from Balleny Islands, Antarctica. N. Z. J. Bot., *6*(3) : 249–261, 4 figs.

Komarek, J., and J. Ruzicka
1966 Freshwater algae from a lake in proximity of the Novolazarevskaya Station, Antarctic. Preslia, *38:* 237–244, pls. 11, 12.

Koob, D. D.
1967 Algal distribution. *In* S. W. Greene et al., Terrestrial life of Antarctica, Antarctic Map Folio Ser., folio 5, pp. 13–15, pls. 6, 7. Amer. Geogr. Soc., New York.

Koob, D. D., and G. L. Leister
1972 Primary productivity and associated physical, chemical, and biological characteristics of Lake Bonney: A perennially ice-covered lake in Antarctica. *In* G. A. Llano (Ed.), Antarctic terrestrial biology, Antarct. Res. Ser., *20:* 51–68. AGU, Washington, D. C.

Parker, B. C., and G. L. Samsel
1970 The 'container-effect' in ^{14}C primary productivity. J. Phycol., *6*(suppl.) : 6.

Prescott, G. W.
1962 Algae of the western Great Lakes area. 2nd ed., 977 pp., 7 pls. Brown, DuBuque, Iowa.

Proctor, V. W.
1959 Dispersal of fresh-water algae by migratory water birds. Science, *130:* 623–724.

Schlichting, H. E., Jr.
1969 The importance of airborne algae and protozoa. J. Air Pollut. Control Ass., *19:* 946–951.

Shihira, I., and R. W. Krauss
1965 Chlorella. Physiology and taxonomy of forty-one isolates. 95 pp. University of Maryland, College Park.

Steeman-Nielsen, E.
1952 The use of radioactive carbon (C^{14}) for measuring organic production in the sea. J. Cons. Perm. Int. Explor. Mer, *18:* 117–140.

Strickland, J. P. U., and R. T. Parsons
1968 A practical handbook of seawater analysis. Bull. Fish. Res. Bd Can., *167:* 311 pp.

Thomas, D. L., and R. M. Brown, Jr.
1970 New taxonomic criteria in the classification of *Chlorococcum* species. 3. Isozyme analysis. J. Phycol., *6:* 293–299.

Thornton, H. G.
1922 On the development of standardized agar medium for counting soil bacteria, with a special regard to the repression of spreading colonies. Ann. Appl. Biol., *9:* 241–274.

Trainor, F. R., and R. J. McLean
1964 A study of a new species of *Spongiochloris* introduced into sterile soil. Am. J. Bot., *51:* 57–61.

Uhlik, D. J., and H. C. Bold
1970 Two new species of *Chlamydomonas*. J. Phycol., *6:* 106–110.

West, W., and G. S. West
1911 Freshwater algae. *In* British antarctic expedition, 1907–09. Biology. *1:* 263–298. Heinemann, London.

PHOTOSYNTHESIS OF LICHENS FROM ANTARCTICA

Otto L. Lange and Ludger Kappen
Botanisches Institut II, Universitaet Wuerzburg, Germany

Abstract. Laboratory studies of lichens from Antarctica (Hallett Station, Victoria Land) revealed that their adaptation to the cold and dry conditions of this continent is a result of special features of their frost resistance and photosynthetic capabilities. Specifically our findings have shown the following: (1) The desiccation and the cold resistance of antarctic lichens were extraordinarily high. Some species recovered their photosynthesis and dark respiration fully even after the hydrated thalli were cooled rapidly to —196°C. Consequently frost impairment in nature appears unlikely. Efficiency would not have been influenced because only a few aftereffects of the extremely low temperatures were found. (2) The optimal thallus temperatures for net photosynthesis of all the tested species were <15°C. In some cases optimal CO_2 uptake occurred near 0°C. Thus high photosynthetic production of antarctic lichens in nature during brief periods of hydration by melting snow can occur despite the low temperatures. No temperature adaptation was found in the gross photosynthesis. Like other species from cold regions, the antarctic lichens had a higher rate of dark respiration (related to their net photosynthesis) at raised temperatures than desert or tropical species. (3) A small amount of CO_2 assimilation occurred even at thallus temperatures of —12.5°C to —18.0°C. This finding indicates that antarctic lichens can achieve a photosynthetic gain during the long periods in which they are frozen. (4) The light compensation point was low at low temperatures. Thus lichen productivity is possible when the lichens are snow covered. (5) Desiccated thalli were able to reactivate their photosynthetic activity after the uptake of water vapor from the air. Therefore CO_2 assimilation by lichens need not be restricted to periods of liquid water hydration. This ability to absorb water vapor would especially benefit the lichens of the coastal regions of Antarctica. In addition to these findings the different behaviors of the investigated species are discussed in relation to ecologic considerations. Ecophysiological findings along with the results of field studies will help to explain the existence of lichens in the world's coldest habitats and to understand their role as primary colonizers.

Lichens are a dominant part of the terrestrial plant life of Antarctica. In addition to the 350 lichen species that grow in Antarctica, there are 75 moss species, two liverwort species, 75 fungus species, and two higher plant species [Ahmadjian, 1970]. The degree of covering exhibited by lichens is a better expression of their dominance among the types of antarctic vegetation than the number of their species. The organization of a lichen thallus appears to be well suited to the hostile conditions of a cold environment. The question of how far this organization expresses itself in the photosynthetic capability of antarctic lichens has been considered in the present laboratory study.

The habitats of antarctic lichens are characterized by intense and prolonged periods of cold and by a high degree of dryness [Llano, 1965; Rudolph, 1966a; Ahmadjian, 1970]. The periods of productive photosynthetic activity for poikilohydric plants such as lichens are very much restricted because of the preceding conditions. The lichen thalli along the coastal regions of Antarctica remain covered with snow throughout most of the winter. During the dark period of midwinter photosynthetic activity is impossible. In late winter, when the sun reappears, a small amount of light must reach the snow-covered lichens, which are still exposed to low temperatures. In early spring, i.e., September in the coastal region of Victoria Land, the snow begins to thaw locally. The lichens are in contact with liquid water only during this melting, sometimes while they are under snow. Their major metabolic activity must take place during this period even though it is interrupted daily by freezing and thawing. As a rule these periods of hydration occur in conjunction with relatively low temperatures. In the antarctic summer the lichen thalli can be heated considerably by direct exposure to solar radiation. According to Rudolph [1966b] values of >32°C were recorded for rock surfaces. The lichens on these rocks dried and went into a latent state of activity. During the summer, liquid water may be imbibed occasionally from the melting of fallen snow.

The existence of lichens in antarctic habitats may be explained by their resistance to unfavorable climatic periods and their ability to undergo sufficient rates of photosynthesis even under conditions of periodic cold and dry inactivation [Billings and Mooney, 1968]. Experimental investigations of these problems were stimulated by Ahmadjian. His coworker Gannutz carried out field measurements on the CO_2 gas exchange of lichens in relation to their microclimatic conditions in the vicinity of Hallett Station in Victoria Land, Antarctica [Gannutz, 1969]. Otto L. Lange was invited to participate in these investigations in an advisory capacity. During the investigations lichen material was collected from Antarctica, and CO_2 gas exchange and cold resistance were examined in the laboratories of the Institute for Forest Botany, University of Goettingen.

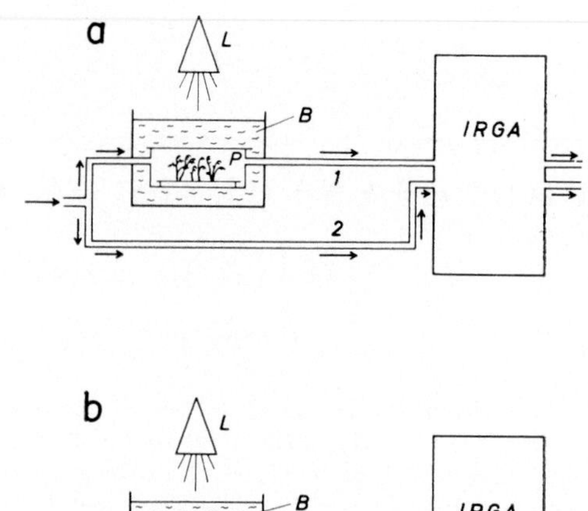

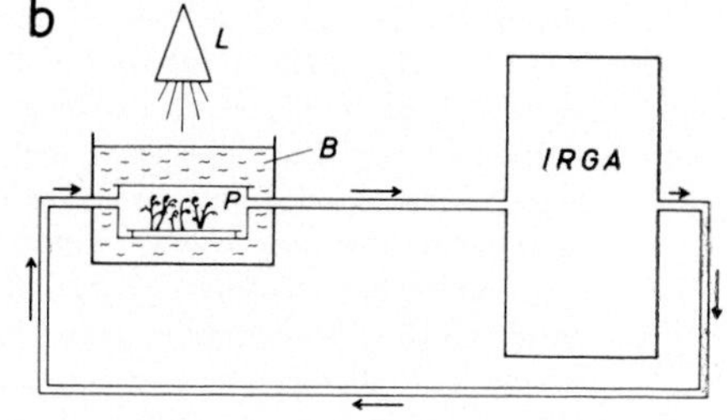

Fig. 1. Gas exchange measurements in (*a*) an open system and (*b*) a closed system (IRGA, infrared gas analyzer; L, light source; B, temperature-controlled bath; P, enclosed plants).

LICHEN MATERIAL

The lichen samples [Murray, 1963; Rudolph, 1963] were collected near Hallett Station (Cape Hallett, Seabee Hook), which is located at 72°18'S, 170°18'E. *Buellia frigida* Darb., *Lecanora melanophthalma* (Ram.) Ram., *Xanthoria elegans* (Link) Th. Fr., and *Xanthoria mawsoni* Dodge grew on stones and rocks of the talus slope above an extensive penguin colony in the immediate vicinity of the station. *Neuropogon acromelanus* (Stirt.) Lamb f. *picatus* Lamb, and *Umbilicaria decussata* (Vill.) Zahlbr. were collected from a slope of boulders about 11 km west-southwest of Hallett Station across the Edisto Inlet at the southern base of point 1070. The lichens were transported to Germany in the desiccated state and stored at −20°C in a deep freezer. Previous investigations [Lange, 1969a; Lange et al., 1970] have shown that the metabolic capability of frost and drought resistant lichens is not changed by such treatment [Feige, 1967]. Before testing, the thalli were placed in a temperature-controlled plant chamber where for 10 days they were exposed to 12-hour periods of alternating light and darkness at +4°C and were moistened twice a day.

METHODS

The photosynthesis and the respiration of the lichens were determined by measuring their CO_2 gas exchange with an infrared gas analyzer (Hartmann und Braun, Frankfurt). Specimens were enclosed in small plexiglass chambers submerged in water or in ethylene glycol in which the temperature was held constant to an accuracy of ±0.1°C. The temperature of the lichens was recorded by means of thermistores attached to the tips of medical injection needles [Lange, 1965a]. Most gas exchange measurements were taken in an open air system according to the gas difference procedure (Figure 1). An air stream with a CO_2 concentration of about 330 ppm was divided into two parts: (1) The measured air was passed over the plant at a constant flow rate. (2) The reference air was not influenced by the plant. The difference in the CO_2 concentrations of both streams was recorded by the gas analyzer. The product of the flow rate and this difference gave the rate of CO_2 exchange of the enclosed material. When the reactions of the lichens became weaker, as they did at temperatures of <-8°C, the sensitivity of the analyses was increased by the effect of accumulation through measurement in a closed system (Figure 1). In the present case the air was recirculated in a closed volume including the cuvette and the sampling tube of the analyzer. The rate of increase or decrease of the CO_2 concentration within the system is a measure of the intensity of the CO_2 exchange of the enclosed plant. A quantitative evaluation was dispensed with for these last experiments; the conclusions were based on the presence or the absence of CO_2 uptake. (Therefore a quantitative comparison between the described findings and the results of Scholander et al. [1953] for the respiration of frozen lichens obtained by another method is not possible.)

As the detailed descriptions in previous publications [Lange, 1965b; pp. 2–4, 13–14] show, the validity of the gas exchange measurements, especially those at low temperatures, has been examined carefully. Blank runs (without lichens) on calibrated air mixtures proved the tightness of the systems with respect to CO_2 and showed no material enclosed with CO_2 uptake or elimination. Control measurements of the dead lichen thalli verified that CO_2 gas exchange was restricted to living material. The temperature of the algal layer of the lichens was controlled during the experiments by means of thermocouples 15–20 μm thick. Repetition of the experiments at low temperatures by another method, i.e., the incorporation of radioactive $^{14}CO_2$, confirmed the gas exchange measurements just described [Lange and Metzner, 1965].

The lichens enclosed in the plant chambers were illuminated by mercury high-pressure lamps (400- or 1000-watt, Osram, Berlin); measurements of light intensity were made with selenium photoelements (produced by B. Lange, Berlin). The light intensity was varied by shading with gray filters, which at the same time avoided any change in the relative spectral energy dispersion. The CO_2 gas exchange in relation to temperature and light intensity and the location of the light compensation points were determined for completely hydrated thalli.

To determine their cold resistance (fully imbibed), the lichen thalli, after 10 days of preliminary culture in climatized chambers at 4°C and 10 klux, were cooled for a given period. Experiments were conducted in duplicate; each experiment was done on several specimens of one species. From the results of preliminary experiments with other species the cold resistance was expected to be very high. Thus the lichen thalli were cooled to −30°C by means of an electronic control with a temperature gradient of 15°C/hr. Next the material was kept at −50°C for a half hour and then maintained for 6 hours at the temperature of liquid nitrogen (−196°C). The lichen thalli were rewarmed stepwise in the same manner. This treatment is called gradual cooling in contrast to rapid cooling, in which the thalli are plunged directly into liquid nitrogen and then rewarmed in air at a temperature of +4°C.

The CO_2 gas exchange of the samples was measured before and after the cold treatment under standard conditions (4°C, 10 klux) and several times during a recovery period of about 3 weeks' 'cultivation' in the climatized chambers. The behavior of the lichens after the cold treatment was expressed as a percentage of their activity before the treatment.

By means of the distinction between gross photosynthesis and respiration, the timing and the extent of the phycobiont recovery could be indicated [Ried, 1960]. The calculation of the relative gross photosynthesis was accurate, because the values had the same base and errors before and after the cold treatment.

To examine their photosynthesis and respiration during water vapor uptake, the lichens were first dried for several weeks in desiccators (silica gel). The dehydrated thalli were then installed in a cuvette at 4°C, illuminated with 10 klux, and exposed to nearly water-saturated air (98% relative humidity at 4°C). The photosynthesis and dark respiration were observed as the thalli absorbed water vapor. Great care was taken to avoid condensation.

The amounts of gas exchange of *Buellia frigida* and the *Xanthoria* species were related to the thallus-covered rock surface. That of the remaining species was related to the dry weight of the thalli (24 hours at 105°C). More detailed information concerning our methods can be obtained from Lange [1965a, 1969a] and Kappen and Lange [1972].

RESULTS

Temperature relations of the CO_2 gas exchange. The temperature dependence of the apparent CO_2 gas exchange of four lichen species at different light intensities is represented in Figures 2a–d. In all cases, optimal net photosynthesis occurred only at temperatures of <15°C. *Neuropogon acromelanus* and *L. melanophthalma* exhibited especially low temperature optimums. Except for *L. melanophthalma* (for which there were only a few readings at high light intensities), as the illumination decreased, the temperature for optimum CO_2 uptake also decreased. At low light levels, optimal CO_2 absorption occurred at temperatures near or below 0°C. In this respect the results are similar to those obtained previously with another antarctic lichen (*Parmelia coreyi*) and other lichens from cold and temperate regions. Tropical lichens show a different behavior [Lange, 1965b, also unpublished data, 1969].

The low temperature compensation points of the gas exchange were noteworthy; for example, *L.*

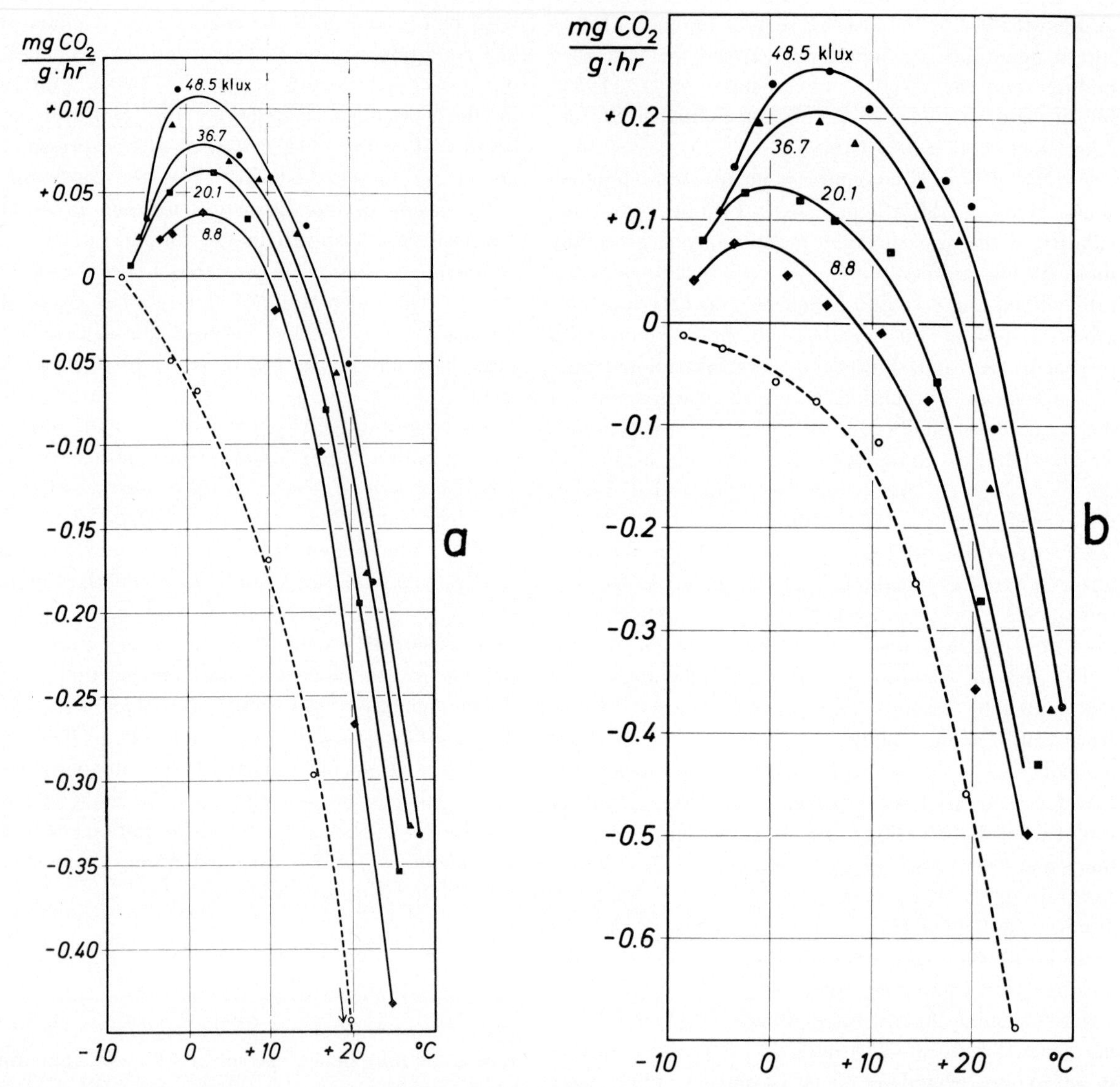

Fig. 2. Apparent CO_2 gas exchange (CO_2 uptake and output) in relation to thallus temperature (the solid lines represent the reaction curves for different levels of light intensity in klux, and the dotted line the dark respiration). (*a*) *Lecanora melanophthalma.* (*b*) *Neuropogon acromelanus.* (*c*) *Buellia frigida.* (*d*) *Xanthoria mawsoni.*

melanophthalma even at a light intensity of 48.5 klux did not have positive net photosynthesis at temperatures of >17°C.

Although the adaptation of antarctic lichens to low temperatures is evident from their net photosynthesis, it is not evident from their gross photosynthesis. The gross value was the sum of the amounts of apparent CO_2 absorption and dark respiration. The temperature dependence of the gross photosynthesis for the same four lichens is illustrated in Figures 3a–d. The optimal temperature for gross photosynthesis at high light intensities was about 20°C for *X. mawsoni* and higher for the other species. In this respect the behavior of the antarctic lichens is similar to that of species from hot desert regions [Lange, 1969b]. The photosynthetic activity of the phycobiont inside the thallus showed no relation to the general temperature regime of the antarctic habitat. This finding is in agreement with the results of Ahmadjian [1958, also unpublished data, 1970], who found that some *Trebouxia* species isolated from antarctic lichens grew well under low light intensities at temperatures averaging 15°–18°C, the optimal growth temperature range of most lichen algae. The particular adaptation of antarctic phycobionts is their ability to photosynthesize over a wide range of temperatures

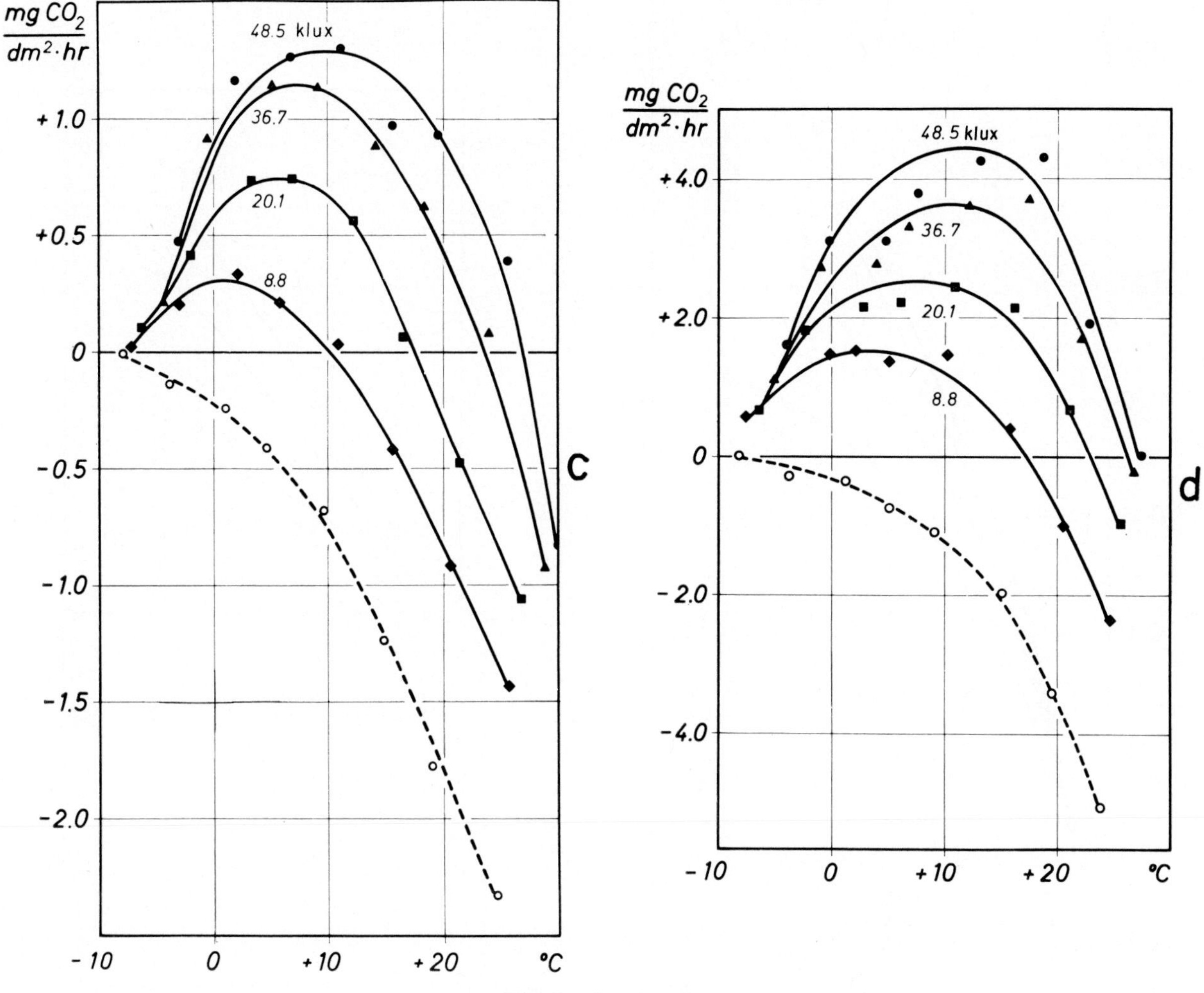

Fig. 2. (continued)

and with particular efficiency even at low temperatures. This ability of the phycobiont coupled with the relatively weak respiratory activity of the mycobiont accounts for the high productivity of the lichen under cold climatic conditions.

The remarkable differences between the temperature relations of net and gross photosynthesis are due to the extremely high respiratory rates of antarctic lichens at increased temperatures (see Figures 2a–d for dark respiration). According to Billings and Mooney [1968] a high dark respiration is also typical of phanerogams from cold regions. However, from a study of arctic and tropical lichens Scholander et al. [1952] concluded that 'oxygen consumption data overlap at all temperatures, and hence show no clear metabolic adaptation to different climates.' The adaptation becomes even clearer if we consider the ecologic aspects of respiration from the point of view of productivity under different climatic conditions and if we express the respiratory loss at increased temperature in terms of the assimilatory capability of a lichen. To do so, we use the quotient of the maximum apparent CO_2 uptake of a lichen at a given light intensity (10 klux) and the dark respiration of the same thallus at a relatively high temperature (20°C). This value is very low for the four antarctic species investigated (the average being 0.32). Thus these lichens have very high respiratory rates in comparison to species from other regions. The highest values of this quotient belong to *Ramalina maciformis* from the Negev Desert (1.80) and to most of the tropical species that we tested (the average for nine species being 0.91; O. L. Lange and L. Kappen, unpublished data, 1969). Alpine lichens also generally showed low values (the average for three species being 0.51). From these results the particular

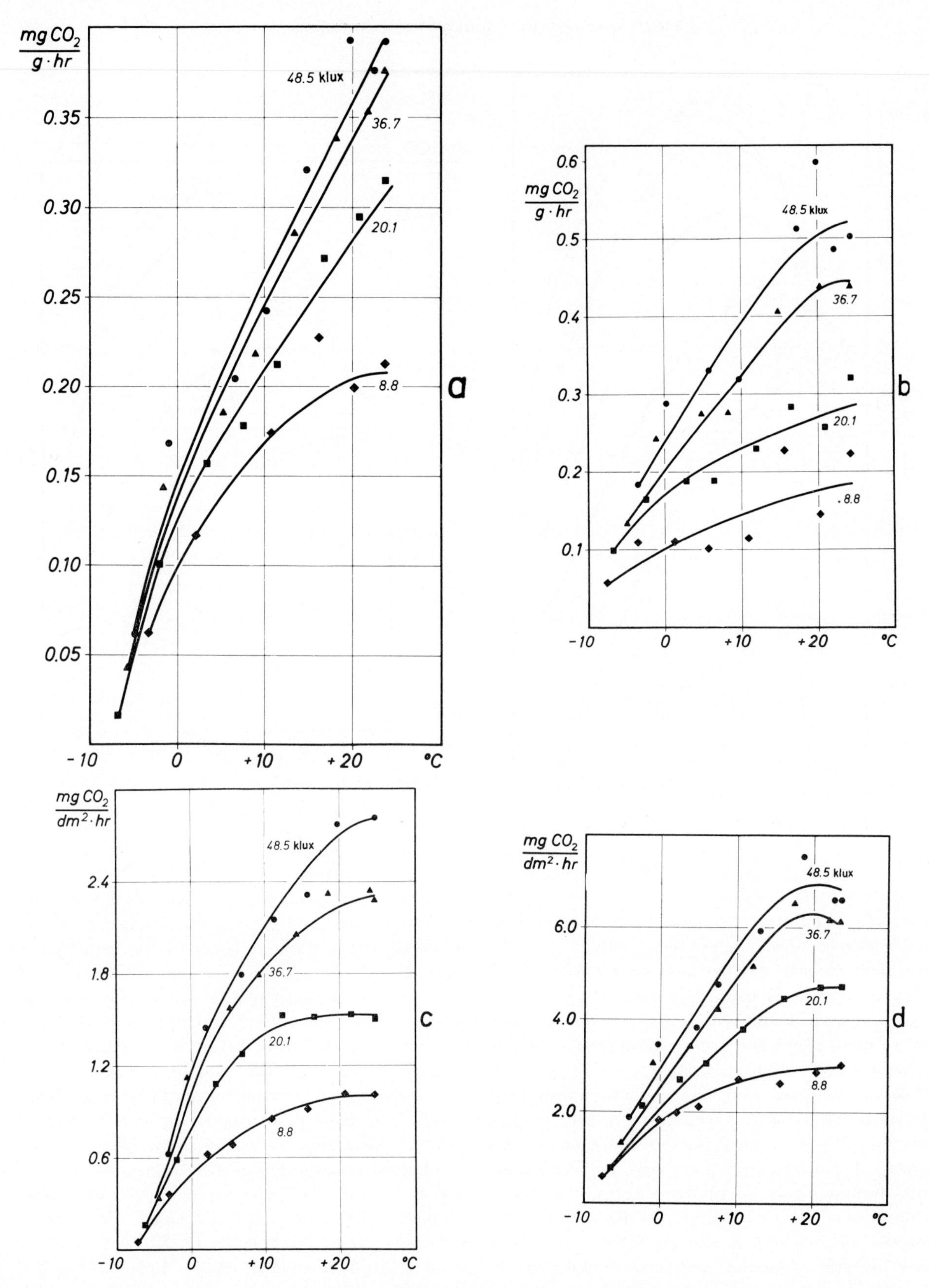

Fig. 3. Gross photosynthesis at indicated light intensities in klux in relation to thallus temperature. (*a*) *Lecanora melanophthalma.* (*b*) *Neuropogon acromelanus.* (*c*) *Buellia frigida.* (*d*) *Xanthoria mawsoni.*

TABLE 1. CO_2 Gas Exchange at Low Thallus Temperatures

Thallus Temperature, °C	*Buellia frigida*	*Lecanora melanophthalma*	*Xanthoria mawsoni*	*Neuropogon acromelanus*
+1.5	+	+	+	+
−0.5	+	+	+	+
−2.5	+	+	+	+
−4.5	+	+	+	+
−6.5	+	+	+	+
−8.5	+	+	+	+
−10.5	+	+	+	+
−12.5	+	+	+	+
−14.5	−	+	+	+
−16.5	−	+	+	+
−18.5	−	−	−	+
−20.5	−	−	−	−

Signs under species headings indicate positive or negative measurable CO_2 uptake.

temperature adaptation of the CO_2 gas exchange of lichens from cold regions is obvious.

Temperature minimum of CO_2 uptake. The antarctic lichens still absorbed CO_2 at below freezing temperatures down to approximately −8°C (Figures 2 and 3). The absolute lower limit was still not reached at that level. The results of our experiments on minimum CO_2 gas exchange recorded in the closed circulatory system are compiled in Table 1. Like lichens from other habitats, the lichens from Antarctica still assimilated CO_2 at temperatures of −12.5°C and −18.5°C. In earlier experiments [Lange, 1965b] *Parmelia coreyi* from Cape Hallett showed similar reactions.

A study of the incorporation of $^{14}CO_2$ by other species revealed that the absorption of CO_2 by lichens at low temperatures does in fact represent photosynthetic carbon fixation. This finding also applies to the antarctic species of the present study [Lange and Metzner, 1965; H. Metzner, W. Koch, and O. L. Lange, unpublished data, 1968].

Photosynthetic CO_2 uptake continues at temperatures below the freezing temperature of the capillary water in the medulla. The ice formed in the cortex must be so loose even at −24°C that CO_2 diffusion to the phycobiont is still possible [Lange, 1965b]. At this temperature the cells of the phycobiont are considerably dehydrated by freezing [Levitt, 1956], and their water potential is correspondingly decreased. Bertsch [1966] and Lange [1969a] have shown that lichen algae can conduct photosynthesis at normal temperatures at a water potential as low as −280 atm. The present study shows that photosynthetic enzymatic reactions also remain intact to a certain extent at temperatures far below freezing under conditions of comparably decreased water potential.

The absolute amounts of CO_2 assimilated at low temperatures are small. The net assimilation of *X. mawsoni* at −10°C corresponded to one-twentieth to one-fiftieth of that at +10°C (48.5 klux). However, the yield resulting from the minute CO_2 absorption at low temperatures becomes significant when it is considered in the light of the long periods in which antarctic lichens are exposed to temperatures below freezing.

Photosynthesis after periods of heavy frost. The cold resistance of antarctic lichens proved to be extraordinarily high. Like those of tested species from other regions, the desiccated thalli were practically unaffected by low temperatures. Immediately after a treatment at −196°C, rewarming, and rewetting, they again showed their original CO_2 gas exchange. Thalli in the fully hydrated state were only slightly impaired by gradual cooling to the temperature of liquid nitrogen. The calculated relative gross photosynthesis after this treatment was as high as or higher than before (Figure 4a). In most cases the response to the cold treatment was a remarkable increase in dark respiration that reached >300% in *B. frigida.* This response might be explained as mostly a reaction of the mycobiont due to a sort of restitution process. It shows that, although the fungus was affected by cooling, there was no indication of irreversible injury.

The initial rise of the respiration of all the tested species was higher after rapid cooling than after gradual cooling to −196°C. The gross photosynthesis of *U. decussata* after rapid cooling remained <10% of the original value (Figure 4b). Despite their high respiratory capacity the thalli were practically killed. *Buellia frigida* and *X. mawsoni* attained 30% of their original gross photosynthesis. No recovery could be achieved in the postcultural period of these species. On the other hand, *X. elegans* and *L. melanophthalma* recovered fully. Their frost resistance seems to be without limit even under fully hydrated conditions. They are the most frost resistant lichens known [Kappen and Lange, 1972].

Photosynthesis following the absorption of water vapor. It has been shown in the laboratory that the photosynthetic mechanism of the desiccated thalli of Middle European and desert lichens can be reactivated solely through the absorption of water vapor [Lange

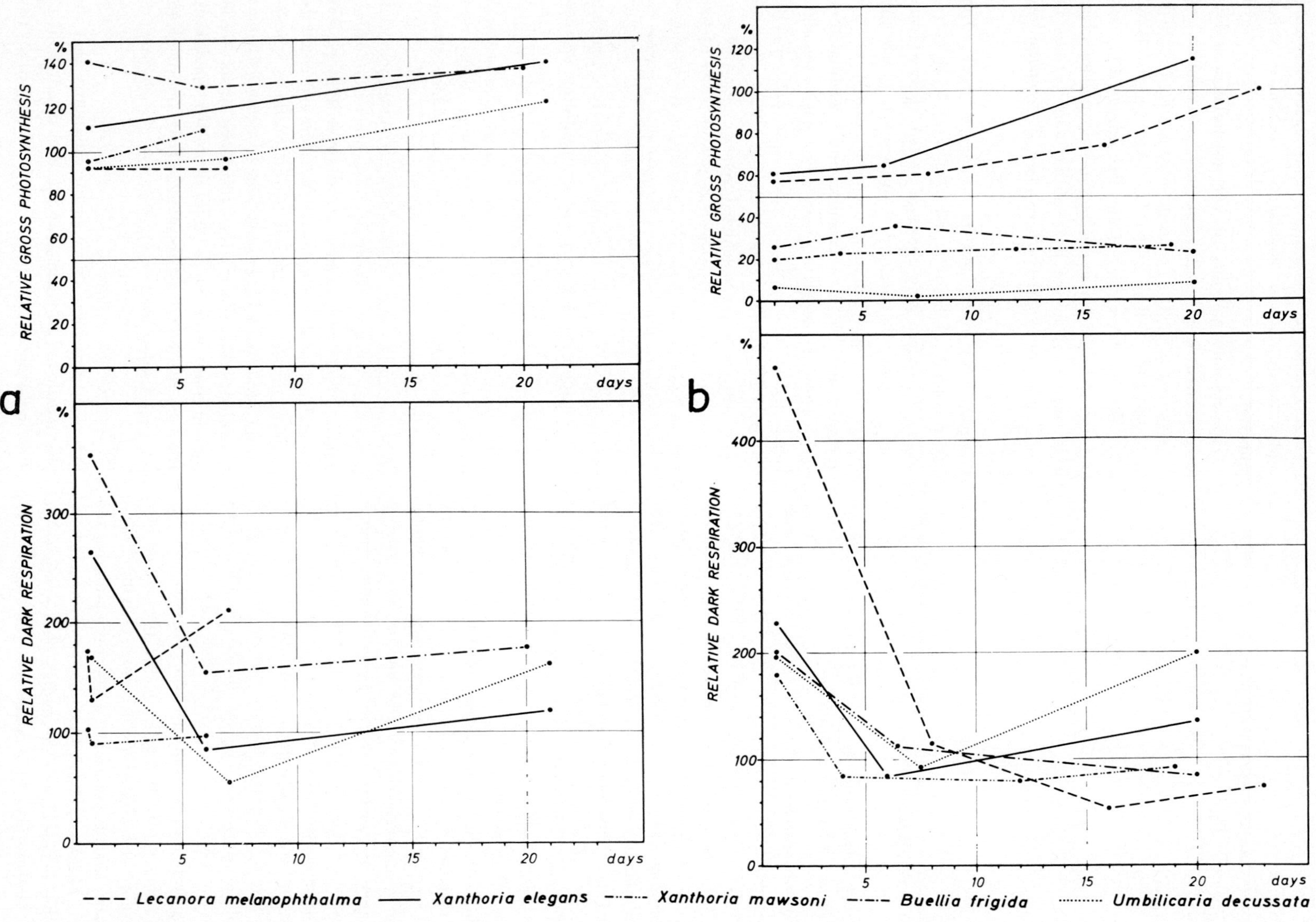

Fig. 4. Gross photosynthesis and dark respiration (in per cent of value before treatment) of antarctic lichens after cooling to the temperature of liquid nitrogen in relation to the time (in days) of cultivation after the frost treatment. (*a*) Gradual cooling. (*b*) Rapid cooling.

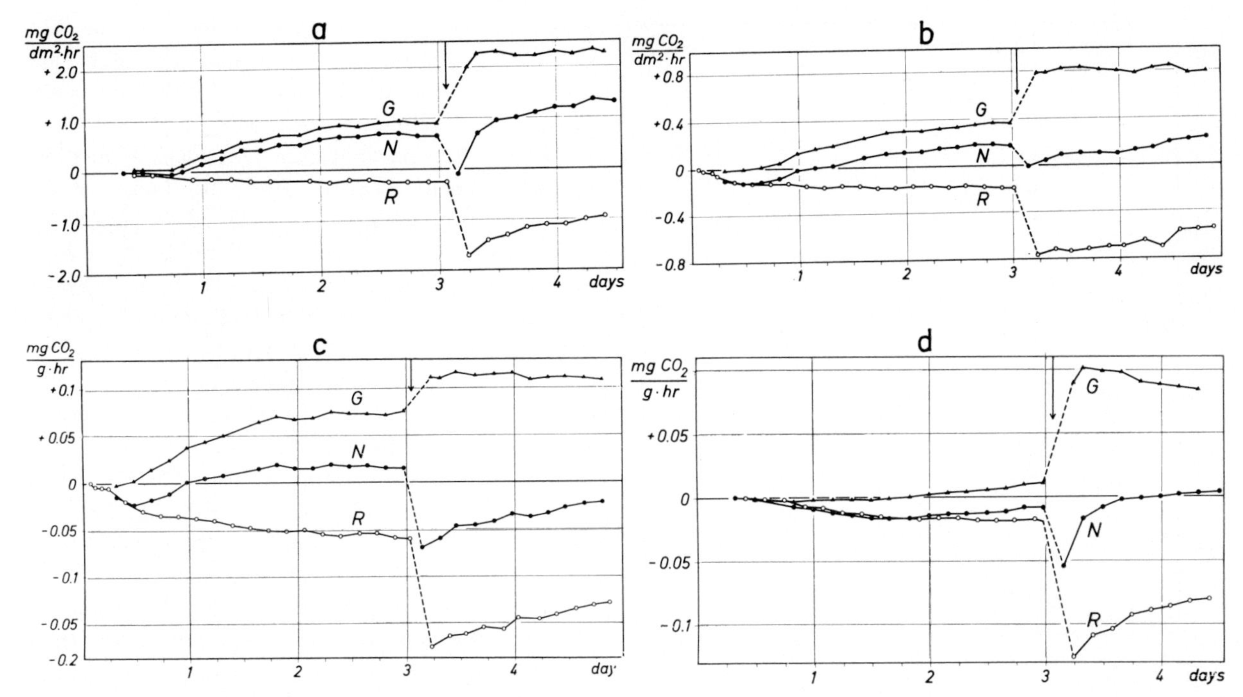

Fig. 5. Gross photosynthesis G, net photosynthesis N, and dark respiration R of desiccated thalli during uptake of water vapor from air (the ordinate is CO_2 uptake and output, and the abscissa is length of treatment with air of high water potential; the arrow indicates the point at which the thalli were sprayed with water). (*a*) *Xanthoria mawsoni.* (*b*) *Buellia frigida.* (*c*) *Neuropogon acromelanus.* (*d*) *Lecanora melanophthalma.*

and Bertsch, 1965; Bertsch, 1966; Lange, 1969a]. Considerable rates of photosynthesis are possible without hydration by liquid water. This capability is used by the lichens to attain a positive metabolic balance in their natural desert habitats after cold nights without dew condensation [Lange et al., 1970].

Experiments have illustrated that antarctic lichens have the same capability (Figures 5a–d). After a dry period of several weeks the desiccated thalli of *X. mawsoni* that had been installed in assimilatory chambers showed the first signs of apparent CO_2 absorption after 24 hours of exposure to nearly saturated air. Gross CO_2 uptake was first recognizable after 10 hours of exposure. In the course of further water vapor absorption the net photosynthesis rose considerably. It reached 50% of the value recorded for the lichens after hydration with liquid water (Figure 5a). The capability of the lichens in relation to the gross photosynthetic rate amounted to 41% of the value recorded at full hydration. After the uptake of liquid water the respiration of a lichen thallus was more strongly stimulated than its photosynthesis. The possible causes for this phenomenon, which has also been observed in species from other regions, are beyond the scope of the present report.

The ability to reactivate net photosynthesis through the absorption of water vapor was not as pronounced in *N. acromelanus*, *B. frigida*, and *L. melanophthalma.* Nevertheless they reached 65%, 44%, and 12%, respectively, of their maximum gross photosynthetic potential after 3 days of water vapor absorption under experimental conditions. This result was less evident from the apparent gas exchange because of the concurrently higher respiration. Note that a considerable reactivation of the metabolic processes takes place in all species following water vapor hydration.

Light compensation point. The light compensation point of the CO_2 gas exchange, i.e., the light intensity at which the photosynthetic CO_2 absorption equals the respiratory loss, is relatively high in lichens in comparison to that in leaves of higher plants. It increases markedly at high temperatures mostly because of increased respiration. Figure 6 illustrates this behavior in *B. frigida* and *N. acromelanus.* An intensity of 17–26 klux (about one-fifth of the maximum environmental light intensity) is needed for compensation at a thallus temperature of 16°C. This behavior can hardly be considered a disadvantage in the antarctic environment, where higher temperatures usually occur in connection with high light intensities.

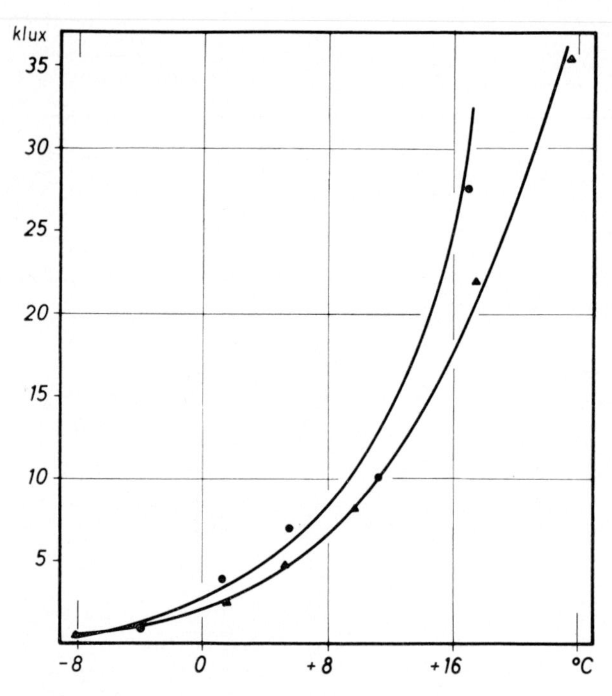

Fig. 6. Light compensation point in relation to thallus temperature for *Neuropogon acromelanus* (circles) and *Buellia frigida* (triangles).

However, note that from an ecologic point of view the compensation point at temperatures at and below the freezing point decreases to a few hundred lux.

DISCUSSION AND ECOLOGIC CONCLUSIONS

The experiments that have been described illustrate some characteristics of the photosynthetic activity of antarctic lichens and provide information concerning their physiological adaptations. This information allows hypothetical conclusions about the ecologic properties of lichens that may help to interpret the measurements made in the field. We do not agree with Gannutz [1969, pp. 169–170], who feels that 'a movement of the (lichen) material, such as into the laboratory for experimentation, cannot be tolerated' and who sees basic problems in comparing laboratory studies with field studies. We recently conducted experiments with *Ramalina maciformis* and other lichens under controlled conditions in the laboratory as well as under natural conditions in the Negev Desert. For the field investigations we used completely climate-conditioned plant chambers to insure measurements under ambient conditions reproduced by follow-up control [Koch et al., 1971] and to avoid the well-known 'cuvette effect.' The results showed an exact correspondence between the CO_2 gas exchange of the lichens in the laboratory and those in the field under comparable climatic conditions. The combination of both findings allows relevant conclusions about the functional adaptations of desert lichens [Lange, 1969a, b; Lange et al., 1970]. We feel that, in the field of ecophysiology, studies made under exact and reproducible conditions provide information about the basic behavior of a plant in relation to individual internal and external factors. Such work generally requires an experimental accuracy that is difficult or even impossible to achieve under the frequently varying microclimatic conditions in the field.

The lowest air temperature measured in Antarctica was −88°C [Rubin and Weyant, 1965]. Lichens can withstand far lower temperatures. Thus cold injury is impossible in their natural environment. Although all the lichens tested were unimpaired by gradual cooling to the temperature of liquid nitrogen, there were differences from species to species in resistance to rapid cooling to −196°C. This temperature is not ecologically relevant, but the behavior of the lichens after the rapid treatment may indicate physiological differences among the species. For example, the fully imbibed active thalli of *X. elegans* and *L. melanophthalma* were the most tolerant to freezing stresses, whereas those of *U. decussata* were the most sensitive.

Especially because of their resistance to heavy freezing, lichens can potentially colonize any area of the Antarctic that is at least temporarily free of snow and ice. However, in addition they need some minimal growth conditions, as the results of measurements of the CO_2 metabolism of the lichens show. Such conditions appear to be absent in some snow- and ice-free areas, e.g., the 'dry valleys' such as Taylor valley near McMurdo Sound in Antarctica, where over large areas there are almost no lichens. The permanent lack of water excludes nearly all lichens from these regions.

In the coastal area of the antarctic continent, the periods of metabolic activity after hydration of the lichens under natural conditions are interrupted by periods of desiccation and/or low temperatures. Like most terrestrial lichens, the antarctic species proved to be well adapted to sudden changes from inactivation to activation caused by frost or desiccation. After cold periods, within the ecologic temperature range of our experiments, photosynthesis showed practically no phase of regeneration, and each hour of suitable hydration and thallus temperature can be used fully, with no lag phase for photosynthetic productivity. Lichens can also survive extended frost periods, as experiments with *Cladonia alcicornis* from a German habitat [Lange, 1966] and with other lichen material from an antarctic habitat have shown.

The ecologic significance of the ability of antarctic lichens to assimilate CO_2 at high rates during freezing is obvious. All periods of hydration from melting snow and ice can be successfully used to attain a positive metabolic yield. Their low light compensation point enables them to reach a considerable metabolic balance even under snow cover if water is available. In general the temperatures encountered by lichens hydrated with melt water were between 0° and +5°C (temperature measurements taken on lichen thalli in the vicinity of Hallett Station from October to November 1966; for methods see Lange [1965a]). Only infrequently was the temperature higher. The maximum temperature recorded in a thallus of *X. elegans* fully imbibed with melt water was about +15°C. The rock on which the thallus grew was located in a shallow snow hole and was covered by a thin transparent sheet of ice formed by sintered snow (the 'glasshouse effect'). Our laboratory experiments showed that the net photosynthesis of the antarctic lichens is in general optimally adapted to this temperature range.

There were significant differences in the temperature curves of the CO_2 gas exchange of the four species treated (Figures 2a–d). The metabolic activity of *L. melanophthalma* and *N. acromelanus* seemed to be best adapted to low temperatures. The quotients of their maximum net photosynthesis and their dark respiration at 20°C were 0.08 and 0.17, respectively. Even at high illumination these species showed optimal rates of CO_2 uptake near 0°C. Both of these lichens are restricted to regions of cold climate. *Neuropogon acromelanus* is a circumpolar subantarctic species [Lamb, 1964]. The experimental material for this lichen was collected at a scree slope. The thalli grew 10–15 cm under the surface of the slope within the small gaps between and under stones. Apparently in this type of location they profited from the snow blown in by the strong wind that is typical of this open habitat. Sun radiation heats the dark colored stones, and the snow gradually melts and moistens the lichens, which are also protected from being quickly desiccated. Since the thalli are never heated by direct solar radiation, metabolic activity in the field must always be restricted to low temperatures and to low light intensities. Thus the temperature relations of the net photosynthesis might help explain the ability of *N. acromelanus* to colonize this particular microhabitat. *Lecanora melanophthalma*, on the other hand, grew on fully exposed rocks; thus it clearly preferred places with sufficient snow cover to supply water after melting. The light, whitish-green color of its thalli suggests that during irradiation it remains at lower temperatures than species with dark thalli. This behavior may agree with the differences in the temperature curves of the CO_2 exchange between *L. melanophthalma* and *B. frigida*, which grow together on the same substrate. *Buellia frigida* is a typical dark, black, or gray crustaceous antarctic lichen. It is a common endemic species of circumpolar distribution [Lamb, 1968]. In keeping with the higher thallus temperatures expected in this lichen during periods of sunshine, we found that the optimum temperature at the highest tested illumination shifted to about +10°C. The calculated quotient of CO_2 gas exchange was 0.19. The highest quotient (0.46) was found for *X. mawsoni*, which in this respect seems to be the lichen least adapted to colder regions in spite of its distribution only on the antarctic continent. This lichen is closely related to *X. candelaria*, which is very common in temperate areas.

The periods of active photosynthesis of antarctic lichens are not restricted to occasions when liquid water is available. They are also able to assimilate CO_2 at temperatures far below the freezing point. It is conceivable that lichens exposed to temperatures of $\leqslant -15$°C can undergo a small amount of photosynthetic metabolism as long as the light intensity and the hydration are sufficient. These conditions will occur in late winter and spring more frequently and for longer periods than the exposure to liquid water. Although the absolute assimilation rates below freezing are rather small, the balance may be significant to the antarctic lichens. In what form the photosynthetic products are stored is still unknown, since transport in the frozen thallus is hardly imaginable.

The assimilates accumulated during cold periods may be an energy source for the lichens during periods of reactivation. The assimilates (sugars and polyols [Feige, 1970]) may increase the frost resistance of the lichens [Sakai, 1961; but Lewis and Smith, 1967]. Another possibility has been proposed by Levitt [1967]. In the light of his $SH \leftrightharpoons SS$ hypothesis of freezing injury, he supposes that the survival value of the photosynthesis of lichens below freezing is due not to the accumulation of carbohydrates but perhaps to the maintenance of a high reduction capacity.

All the antarctic lichens tested could reactivate photosynthesis solely through the absorption of water vapor. Like that of hot desert lichens, this ability may be significant to the antarctic species in their natural habitats. During summer the snow-free lichen thalli dry out. The water vapor absorption during periods of high air humidity suggests that metabolic activity and photosynthetic production can be achieved occasionally even during times without snow. The findings of Gannutz [1969] confirm this possibility in the field, at least for *X. mawsoni*. In our laboratory experiments this species showed the most complete recovery of its net photosynthesis after moistening with water vapor. This recovery is apparently promoted by the small lobate shape and thin cortex of the thallus. The enlarged surface of *X. mawsoni* enables it to take up water vapor very quickly. This capability, which is demonstrated well by *N. acromelanus*, must be particularly important ecologically along the coastal areas of Antarctica, where air humidity is periodically high. Perhaps this humidity is one of the reasons for the abundance of this particular lichen along the seashore. *Xanthoria mawsoni* is one of the dominant species near Hallett Station [Rudolph, 1963]. It is typical for a nitrophilous lichen community, which occurs near penguin rookeries (O. L. Lange and O. Klement, unpublished data, 1971).

The periods of water vapor reactivation, which are significant for the productivity of antarctic lichens, may not be restricted to periods during which the lichens are free of snow. The degree of water vapor saturation is probably high under snow cover, especially in the small spaces around the lichen thalli. Photosynthesis may be possible under snow without liquid water. This property may be of advantage to *N. acromelanus* because of the nature of its habitat in openings inside the scree slope. Note that even *B. frigida* can reactivate by means of water vapor absorption. This lichen probably also profits from this ability at its coastal habitats. There is only a very small CO_2 uptake by *L. melanophthalma* after water vapor moistening. The compactness of its cushionlike thalli appears to prevent sufficient water vapor uptake. This species apparently depends exclusively on liquid water.

Ecophysiological laboratory studies result in findings about the functional characteristics of plants and allow assumptions about the ecologic consequences. The significance of such characteristics for the existence of a plant in its natural environment must be determined through measurements in the field. Thus a combination of field and laboratory work helps us to understand and to explain the ecologic behavior of plants. Thus the results of Ahmadjian and Gannutz concerning the CO_2 gas exchange of lichens in their antarctic habitats are awaited with great interest.

Acknowledgments. The research in the field was financed by the National Science Foundation under the U.S. Antarctic Research Program. I would like to express my gratitude to Dr. V. Ahmadjian, Clark University, Worcester, Massachusetts, for the invitation to take part in his investigations (NSF grant LA-227) and to Dr. G. Llano, National Science Foundation, Washington, D. C., for his support. I would also like to thank Dr. Gannutz, Dr. Frishmann, and Dr. Gless for their help during the joint research. Professor Ahmadjian kindly assisted in drafting this paper into English. We are greatly obliged to Prof. Dr. A. Henssen, Marburg, Germany, Dr. O. Klement, Kreuzthal, Germany, Dir. Dr. I. M. Lamb, Cambridge, Massachusetts, and Prof. Dr. J. Poelt, Graz, Austria, for their help in identifying the lichen material.

REFERENCES

Ahmadjian, V.
1958 Antarctic lichen algae. Carolina Tips, *21:* 17–18.
1970 Adaptation of antarctic terrestrial plants. *In* M. W. Holdgate (Ed.), Antarctic ecology. *2:* 801–811. Academic, New York.

Bertsch, A.
1966 Über den CO_2-Gaswechsel einiger Flechten nach Wasserdampfaufnahme. Planta, *68:* 157–166.

Billings, W. D., and H. A. Mooney
1968 The ecology of arctic and alpine plants. Biol. Rev., *43:* 481–529.

Feige, B.
1967 Untersuchungen zum Kohlenstoff- und Phosphatstoffwechsel der Flechten unter Verwendung radioaktiver Isotope. Thesis, 179 pp. Univ. Wuerzburg, Wuerzburg.
1970 Untersuchungen zur Stoffwechselphysiologie der Flechten unter Verwendung radioaktiver Isotope. *In* Deutsche Botanische Gesellschaft (Ed.), Flechten symposium 1969, Vorträge aus dem Gesamtgebiet der Botanik. *4:* 35–44. Fischer, Stuttgart.

Gannutz, T. P.
1969 Effects of environmental extremes on lichens. Mém. Soc. Bot. Fr., 169–179.

Kappen, L., and O. L. Lange
1972 Die Kälteresistenz einiger Makrolichenen. Flora, *161:* 1–29.

Koch, W., O. L. Lange, and E.-D. Schulze
1971 Ecophysiological investigations on wild and cultivated plants in the Negev Desert. 1. Methods: A mobile laboratory for measuring carbon dioxide and

water vapour exchange. Oecologia (Berl.), *8:* 296–309.

Lamb, I. M.
1964 Antarctic lichens. 1. The genera *Usnea, Ramalina, Himantormia, Alectoria, Cornicularia.* Sci. Rep. Br. Antarct. Surv., *38:* 34 pp.
1968 Antarctic lichens. 2. The genera *Buellia* and *Rinodina.* Sci. Rep. Br. Antarct. Surv., *61:* 129 pp.

Lange, O. L.
1965a Leaf temperatures and methods of measurement. *In* F. E. Eckardt (Ed.), Proceedings of the Montpellier Symposium, pp. 203–209. Unesco, New York.
1965b Der CO_2-Gaswechsel von Flechten bei tiefen Temperaturen. Planta (Berl.), *64:* 1–19.
1966 CO_2-Gaswechsel der Flechte *Cladonia alcicornis* nach langfristigem Aufenthalt bei tiefen Temperaturen. Flora, Jena, ser. B, *156:* 500–502.
1969a Experimentell-ökologische Untersuchungen an Flechten der Negev-Wüste. 1. CO_2-Gaswechsel von *Ramalina maciformis* (Del.) Bory unter kontrollierten Bedingungen im Laboratorium. Flora, Jena, ser. B, *158:* 324–359.
1969b Die funktionelle Anpassung der Flechten an die ökologischen Bedingungen arider Gebiete. Ber. Dt. Bot. Ges., *82:* 3–22.

Lange, O. L., and A. Bertsch
1965 Photosynthese der Wüstenflechte *Ramalina maciformis* nach Wasserdampfaufnahme aus dem Luftraum. Naturwissenschaften, *52:* 215–216.

Lange, O. L., and H. Metzner
1965 Lichtabhängiger Kohlenstoff-Einbau in Flechten bei tiefen Temperaturen. Naturwissenschaften, *52:* 8.

Lange, O. L., E.-D. Schulze, and W. Koch
1970 Experimentell-ökologische Untersuchungen an Flechten der Negev-Wüste. 2. CO_2-Gaswechsel und Wasserhaushalt von *Ramalina maciformis* (Del.) Bory am natürlichen Standort während der sommerlichen Trockenperiode. Flora, Jena, *159:* 38–62.

Levitt, J.
1956 Hardiness of plants. 278 pp. Academic, New York.
1967 The mechanism of hardening on the basis of the SH⇋SS hypothesis of freezing injury. *In* E. Asahina (Ed.), Cellular injury and resistance in freezing organisms, Proceedings of the international conference on low temperature science. *2:* 51–61. Hokkaido University, Hokkaido, Japan.

Lewis, D. H., and D. C. Smith
1967 Sugar alcohols (polyols) in fungi and green plants. 1. Distribution, physiology and metabolism. New Phytol., *66:* 143–184.

Llano, G. A.
1965 The flora of Antarctica. *In* T. Hatherton (Ed.), Antarctica, pp. 331–350. Methuen, London.

Murray, J.
1963 Lichens from Cape Hallett area, Antarctica. Bot. Trans. R. Soc. N. Z., *2:* 59–72.

Ried, A.
1960 Stoffwechsel und Verbreitungsgrenzen von Flechten. 2. Wasser- und Assimilationshaushalt, Entquellungs- und Submersionsresistenz von Krustenflechten benachbarter Standorte. Flora, Jena, *149:* 345–385.

Rubin, M. J., and W. W. Weyant
1965 Antarctic meteorology. *In* T. Hatherton (Ed.), Antarctica, pp. 375–401. Methuen, London.

Rudolph, E. D.
1963 Vegetation of Hallett Station area, Victoria Land, Antarctica. Ecology, *44:* 585–586.
1966a Terrestrial vegetation of Antarctica: Past and present studies. *In* J. C. F. Tedrow (Ed.), Soils and soil forming processes, Antarctic Res. Ser., *8:* 109–124. AGU, Washington, D. C.
1966b Lichens ecology and microclimate studies at Cape Hallett, Antarctica. Biometeorology, *2:* 900–910.

Sakai, A.
1961 Effect of polyhydric alcohols on frost hardiness in plants. Nature, *189:* 416–417.

Scholander, P. F., W. Flagg, V. Walters, and L. Irving
1952 Respiration in some arctic and tropical lichens in relation to temperature. Am. J. Bot., *39:* 707–713.

Scholander, P. F., W. Flagg, R. J. Hock, and L. Irving
1953 Studies on the physiology of frozen plants and animals in the Arctic. J. Cell. Comp. Physiol., *42:* 1–56.

FIELD OBSERVATIONS AND LABORATORY STUDIES OF SOME ANTARCTIC COLD DESERT CRYPTOGAMS

EDMUND SCHOFIELD

Department of Botany and Institute of Polar Studies, Ohio State University, Columbus, Ohio 43210

VERNON AHMADJIAN

Department of Biology, Clark University, Worcester, Massachusetts 01610

Abstract. Field observations and laboratory studies suggest that the type of nitrogen compounds and the concentration of water-soluble salts determine the distribution of macroscopic terrestrial cryptogams (algae, mosses, and lichens) in continental Antarctica. In laboratory experiments the isolated mycobiont of *Lecanora tephroeceta* Hue, a 'nitrophilous' lichen collected in a snow petrel rookery in West Antarctica, grew best in pure culture on reduced nitrogen compounds, particularly on ammonia and ammonia-yielding compounds. Uric acid, xanthine, urea, three amino acids with nonpolar side chains (proline, alanine, and leucine), and two amino acid amides (asparagine and glutamine) in addition to organic and inorganic ammonium salts sustained significant growth, whereas ribosides, pyrimidines, peptides, allantoin, and nitrates did not. The mycobiont hydrolyzed urea in a test medium and probably deamidated the two amides before absorbing their nitrogen as ammonia. The observed pattern of nitrogen source use suggests that neutral molecules are absorbed much more readily than charged molecules; this pattern is an important attribute for an organism growing in saline habitats. The distribution of *L. tephroeceta*, as determined from all available information on previous collections, indicates that the important determining factor is ornithogenic nitrogen in the environment. In related experiments *Prasiola crispa* (a green alga long known from its association with bird rookeries to be a highly ornithocoprophilic species) grew twice as well in unialgal culture on uric acid as on any other nitrogen source tested. A blue-green alga often associated with *P. crispa*, '*Phormidium autumnale*,' appears from herbarium studies to have a distribution pattern that is also correlated with the presence of ornithogenic nitrogen. A moss (*Bryum algens*), on the other hand, grew well in pure culture on all nitrogen compounds supplied; thus its distribution is probably not limited by the availability of any particular class of nitrogenous compounds. Two lichen algae (*Trebouxia* spp.) grew best on a peptide (peptone) but hardly used ammonium. They also did not use urea and nitrates. All the lichen phycobionts were typical obligate psychrophiles, their temperature optimums being near 15°C and their maximums near 20°C. The two mycobionts tested displayed two separate temperature optimums under certain conditions, one below 10°C and one above 10°C. This result may indicate that the species evolved in a warmer climate than that of present-day Antarctica. The growth responses of the lichen symbionts to *p*H, light, and osmotic conditions were also examined in the laboratory, and the concentration of reduced nitrogen in soil, guano, and plant samples was determined. The soluble salt concentration in soils was found to be a critical determinant of plant distribution in addition to the type of nitrogen compounds. At Cape Royds, Ross Island, the potentially promotive influence on lichen and moss growth of reduced nitrogen was apparently negated by toxic salt levels. Ammonia was considered to be an ideal source of nitrogen for plants in the saline cold deserts of continental Antarctica.

Antarctica is divided conveniently into two botanical zones: (1) maritime or oceanic Antarctica (the areas between the southern limit of extensive closed phanerogamic vegetation and the southern limit of extenrogamic vegetation and the southern limit of extensive relatively rich cryptogamic vegetation) and (2) continental Antarctica (the areas south of the southern limit of extensive closed cryptogamic vegetation) [Holdgate, 1970, part 12]. Continental Antarctica can be subdivided into three zones: (1) coastal Antarctica, (2) the antarctic slopes, and (3) the antarctic ice plateau [Holdgate, 1970; Weyant, 1966]. Holdgate [1970] has briefly characterized the vegetation of these subdivisions of continental Antarctica.

Ugolini [1970] has recently summarized the knowledge of the soils of both maritime and continental

TABLE 1. Electrical Conductivities of Some Soils from Southern Victoria Land and Ross Island in Relation to Plant Growth

Soil Category	Conductivity at 25°C, μmhos · cm^{-1} Minimum	Maximum	Number of Samples
All soils	14	83,400	119
Soils not associated with plant growth	66	83,400	68
Soils associated with plant growth			
With all plant growth	14	5,270	48
With Cyanophyta (alga) growth	86	5,270	11
With Chlorophyta (*Prasiola crispa*) (alga) growth	186	3,290	6
With moss growth	14	844	24
With lichen growth	14	432	18

Samples consisted of the top 5 cm of soil from an area of <1 m^2 wherever possible; measurements were performed at room temperature on material sifted through 20-mesh screens; 1:2 aqueous extracts (200 grams of soil in 400 ml of distilled water stirred for 30 min and allowed to settle) were used.

Antarctica with reference to the occurrence of plants and animals; Cameron et al. [1970] have correlated the distribution of microorganisms with soil factors. The discussions in both of these papers are pertinent to the results reported herein, as is that in Syroechkovskii [1959].

The climatic factors that impinge directly on plants in the cold deserts of Antarctica (intense cold, extreme aridity, windiness, and so forth) and severely limit their distributions further restrict their occurrence by inducing additional, secondary unfavorable conditions (e.g., an elevated salt content in soils) (Table 1). In the cold deserts of continental Antarctica favorable plant habitats often appear to be merely lucky combinations of otherwise unfavorable factors, as on Cape Royds, Ross Island, where permanent and semipermanent snowdrifts ameliorate the soil and microenvironmental conditions inhibiting plant growth nearby (Figures 1–3). Thus, because edaphic factors have very significant effects on plant distribution in continental Antarctica, pedological data are as necessary as climatic and micrometeorologic data for interpreting plant distributions.

NITROGEN IN THE ENVIRONMENT

Where ice, snow, toxic levels of salts, extreme aridity, and wind abrasion do not prevent plant growth, the lack of a suitable nitrogen source may. There is observational evidence that nitrogenous compounds of animal origin (e.g., penguin guano) permit the occurrence of terrestrial lichens and algae in Antarctica, Filson's [1966] disclaimer notwithstanding [Dodge, 1964, 1965; Follmann, 1965; Lamb, 1948, 1968; Rudolph, 1963; Siple, 1938; Syroechkovskii, 1959]. Although none of these authors proved by experiment that nitrogen (and not, for example, phosphorus) was responsible for the correlation that they noted between the occurrence of bird life and that of certain plants, Holdgate et al. [1967] have shown that nitrogen is the only major nutrient likely to limit plant growth on Signy Island, South Orkney Islands (maritime Antarctica). Table 2 shows that the nitrogen contents of cold desert soils may also be very low.

Lichens and algae. Certain lichens (e.g., *Xanthoria mawsoni* Dodge, *Caloplaca elegans* var. *pulvinata* (D&B) Murray, and *Mastodia tesselata* (Hook. f. & Harv.) Hook. f. & Harv.) are well-known nitrophilous species. In West Antarctica *Buellia latemarginata* Darb. [Lamb, 1968] and *Lecanora tephroeceta* Hue (Appendix 2) are others. Follmann [1965] and Redón [1969] have described ornithocoprophilic lichen associations in West Antarctica. In East Antarctica, where the field work was carried out, *Lecanora lavae* Darb., *L. griseomarginata* D&B, *Omphalodiscus ex-*

TABLE 2. Nitrogen Contents of Soils and Plants from the Granite Harbor Area of Victoria Land (76°55'S, 162°28'E)

Material	Average Concentration of Reduced Nitrogen,[a] %	Accumulation Factor[b]
Soils		
Not associated with plant growth	0.021	1.00
Associated with plant growth	0.055	2.6
Sifted from plant samples[c]	0.017	5.1
From surfaces and crevices of rocks in association with plants	0.106	5.1
Plants		
Blue-green algae	2.119	105.7[d]
Mosses	0.480	22.9
Lichens[e]		
Neuropogon sp.	0.310	14.8
Omphalodiscus spp.	1.203	57.3

[a] Determined by microkjeldahl analysis. A small percentage of any nitrate in the samples is apparently reduced by the method employed, as determined with nitrate controls.

[b] Based on the value for soils not associated with plants.

[c] Small pieces of plants were probably present.

[d] This high value undoubtedly reflects atmospheric nitrogen fixed by the heterocystous types.

[e] The wide difference in the two genera tested was consistent for all samples.

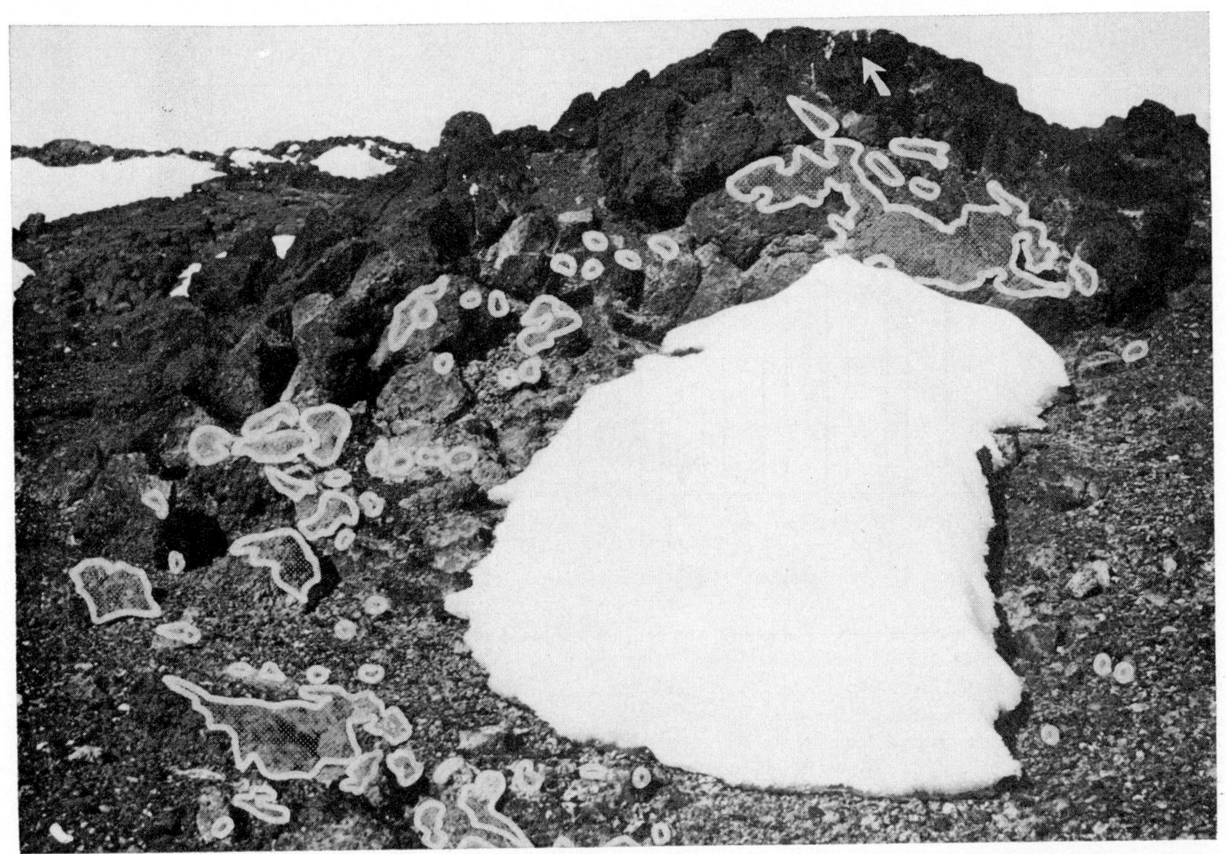

Fig. 1. Distribution of lichens (primarily *Caloplaca elegans* var. *pulvinata* (D&B) Murray, a nitrophilous species) in relation to a permanent snowdrift near Cape Royds similar to that for which nitrogen content and conductance values were determined (Figure 2). All areas of significant lichen growth are outlined in white. The arrow points to a skua perch and droppings.

sulans (Th. Fr.) Dodge, and *Parmelia coreyi* D&B appear to be more or less nitrophilous or ornithocoprophilous (cf. Appendix 1).

The green alga *Prasiola crispa* (Lightf.) Menegh., one of the most nitrophilous organisms known, is strongly ornithocoprophilic wherever it is found. In polar regions it becomes lichenized where nitrogen levels are lowered and constitutes the phycobiont (i.e., the algal partner) of *Mastodia tesselata* [Lamb, 1948]. '*Phormidium autumnale*' (according to Drouet [1962] an aquatic ecophene of the blue-green alga *Microcoleus vaginatus* (Vaucher) Gomont) has a distribution similar to that of *P. crispa* (Appendix 2).

The degree of correlation between most of the lichen species and the organic nitrogen deposits (guano) or bird nesting sites is not so strong for most lichen species as it is for *P. crispa* and '*P. autumnale*,' no doubt because lichens grow far more slowly and can thus subsist on smaller amounts of nitrogen.

Mosses. Data on herbarium specimens and in the literature suggest that the unusual endemic antarctic moss *Sarconeurum glaciale* (C. Müll.) Card. et Bryhn, which is primarily a coastal species, may be nitrophilous [Greene et al., 1970]. The other two principal mosses of extreme southern Victoria Land (*Bryum argenteum* Hedw. and *B. antarcticum* Hook. f. & Wils.) display no consistent preference for habitats rich in organic nitrogen so far as one can determine from field notes on herbarium specimens, although Horikawa and Ando [1967] have characterized *B. argenteum* as a weedy nitrophilous species. Their distributions appear to be controlled more by water and salt relations. For example, McCraw's [1967] map of soil types in Taylor valley, Victoria Land, reveals that mosses (most likely *Bryum* spp.) are found almost exclusively near the feet of glaciers, along intermittent streams, and at the edges of lakes in the coastal two-thirds of the valley, which is more subject than the western one-third to marine influences [Jones and Faure, 1969].

Kohn et al. [1971] found about 10 nesting pairs

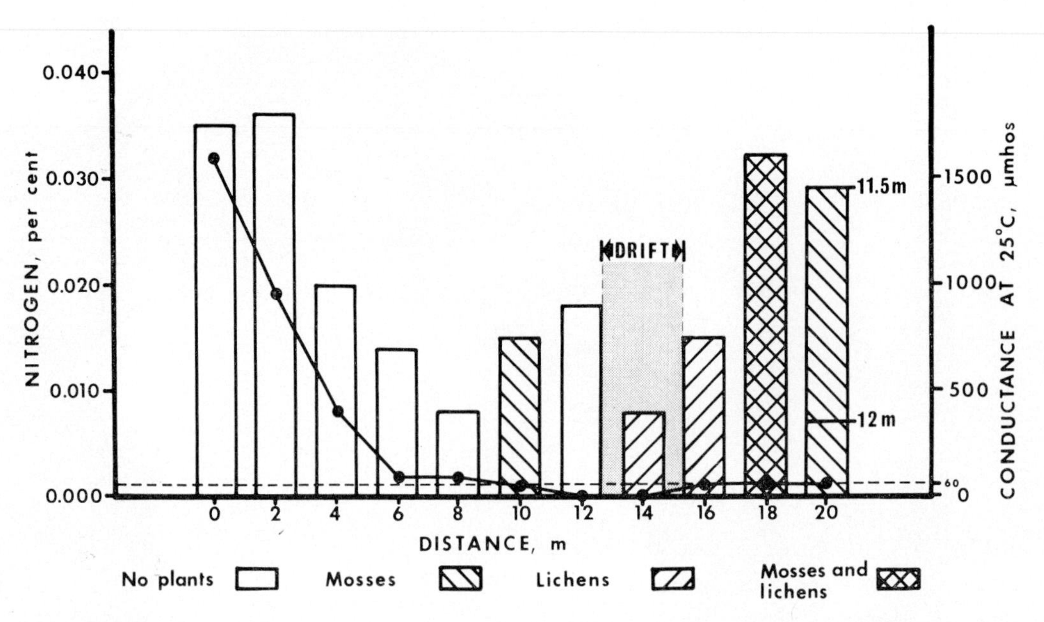

Fig. 2. The effects of snow cover, salt content, and reduced nitrogen on the distribution of mosses and lichens near Cape Royds, Ross Island. The bars represent the amount of nitrogen in soil samples collected at 2-meter intervals through the middle of a quadrat, and the points on the solid line represent the conductance due to water-soluble salts in the samples. The permanent snowdrift appears to reduce the content of salts in the substrate; this reduction creates conditions favorable to plant growth. Nitrogen appears to encourage growth only where the conductance is below the critical level of 60 μmhos (dashed horizontal line); plants cannot grow where the salt content is higher despite adequate amounts of nitrogen. (The two values for nitrogen at 20 meters are for samples collected 0.5 meter apart at the points at 11.5 × 20 and 12 × 20 meters. There were mosses at the former point, but there were none at the latter point. No conductance measurement was made for the point at 12 × 20 meters. See Figure 1.)

of skuas (*Catharacta maccormicki*) near the mouth of Taylor valley and 40–50 mats of moss along melt water streams and 'under stones at altitudes greater than about 270 m,' mostly in extremely wet environments in lower (i.e., coastal) Taylor valley. They concluded that the distribution of mosses is directly controlled by the available water. They also found the remains of penguins in a number of ice-free valleys, and others have found the preserved carcasses of seals. There is therefore some transport of organic nitrogen inland by living animals. The amount of this 'diffused' organic nitrogen must decrease very rapidly with distance from the coast, however.

Greene [Greene et al., 1967] has characterized the distribution patterns of *B. argenteum* and *B. antarcticum* as follows: '*Bryum argenteum* . . . is particularly abundant in southern Victoria Land and Ross Island . . . [where] . . . it is often abundant in open situations in the *wettest areas* of drainage channels or seepage slopes, on scree, on morainic detritus, and on sand and gravel of outwash fans . . . , while *B. antarcticum* . . . is . . . locally abundant in southern Victoria Land and Ross Island, where it forms a conspicuous and important element in the sparse bryophyte vegetation along the *drier sides* of drainage channels and in the *drier parts* of seepage areas on a variety of substrates' [emphasis ours].

Greene's field observations appear to be substantiated by the interesting pattern of salt crusts on herbarium specimens of the two species. Only one of the 21 specimens of *B. argenteum* examined had an obvious salt crust, whereas at least six and possibly nine of the 15 specimens of *B. antarcticum* examined had apparent salt crusts.

The field and herbarium observations considered together suggest an inverse relationship between soil moisture content and soil salt concentration, at least in some habitats. It is suggested that the distinctive distributional patterns of these two related species of moss reflect soil moisture and salinity relationships.

These facts imply that nonnitrophilous or merely facultatively nitrophilous plants are restricted by environmental factors other than organic nitrogen. The

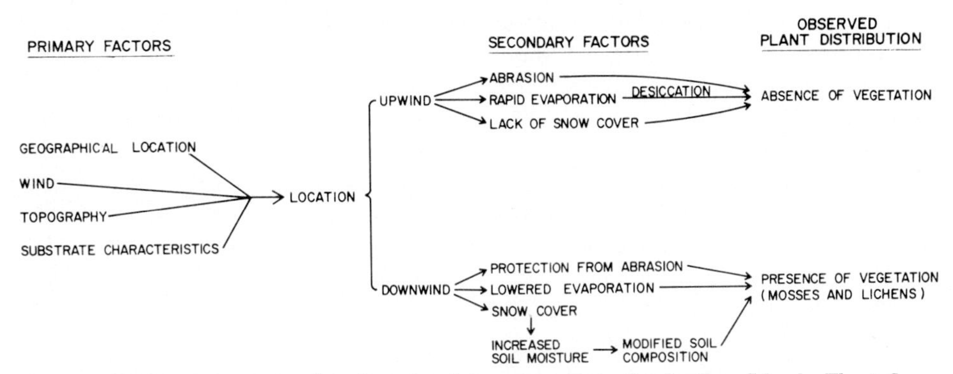

Fig. 3. Suggested sequence of environmental factors at Cape Royds, Ross Island. The influence of the primary factors is modulated by the secondary factors in a cause and effect chain of events that culminates in either the presence or the absence of vegetation. At Cape Royds the persistence of snow leads to increased soil moisture, which in turn leads to decreased soil salts and thus to conditions suitable for the growth of mosses and lichens. The pattern of persistent snow cover is determined by wind direction and speed and by topographic features. The concentration of soil salts is determined by geographical relationships, wind, topography, and substrate characteristics. (See Figure 2.)

studies of Rudolph [1963, 1967] are especially pertinent in this regard.

'Nitrophily'

Nitrophily, according to Räsänen [Barkman, 1958], is really a question of 'ammoniophily.' That is, ammonia is responsible for the presence of the so-called nitrophilous plants. As Barkman [1958] points out, however, nitrophily is a very complex phenomenon that is far from fully explained. The only sure way to settle the question is to test the nitrogen requirements and preferences of the plant species of interest laboriously in pure culture. Some of the results reported below may contribute to the ultimate resolution of the question.

Ornithocoprophily and guanotrophy. The terms 'ornithocoprophilic' or 'guanotrophic' can be applied not only to species occurring exclusively in areas with thick obvious coverings of guano ('guanotrophic habitats' [Leentvaar, 1967]), but also to species occurring in skua rookeries, snow petrel rookeries, rendezvous sites, bathing areas, and perches, with a small but continual supply of certain organic nitrogenous compounds in amounts sufficient for slow growing plants like lichens. Siple and Lindsey [1937], for example, described two small ponds about 50 km from the sea to which bathing skuas contributed sufficient nitrogen to support a remarkably lush biota. A similar fertilizing influence occurs on Kar Plateau in southern Victoria Land. There plant growth is much more abundant near a small skua bathing pool than anywhere else on the plateau, yet the only evidence of bird life, aside from the appearance of the birds themselves after the pond has melted in midsummer, are nitrophilous components of the relatively lush lichen flora.

Uric acid. In Antarctica uric acid, the principal nitrogenous compound in bird urine [Sturkie, 1964; Bose, 1944; Levine et al., 1947] or an early appearing component [Folk, 1969a, b; but see Lonsdale and Sutor, 1971] may be long lived [Jones and Walker, 1964] or short lived. (A report by Winsnes [1969] indicates that the organic deposit from the Tottan Hills described by Jones and Walker [1964] is actually 'mumiyo,' which is the spittle of the snow petrel, *Pagodroma nivea.* The antarctic material consisted of 57% lipoids, 12% uric acid, 7% rock and inorganic material, and 24% unidentified material. Because the uric acid was in the oily spittle and not in the guano, the long-term persistence of the uric acid appears to be a special case. In any event the uric acid of 'mumiyo' is probably almost entirely unavailable to plants and is thus of little ecologic importance.) In some locations it and its derivatives affect the compositions of soils for only relatively short periods after nesting sites have been abandoned by birds [Campbell and Claridge, 1966; F. C. Ugolini, unpublished data, 1964].

Under the proper conditions uric acid gives rise to urea, ammonia, and nitrate and therefore to a flora with a wide diversity of nitrogen source preferences. In areas where birds are common, plants requiring nitrogen in one of these forms can exist if other factors do not inhibit them. At progressively greater distances from the influence of birds, certain species will drop out as the forms of nitrogen on which they depend disappear from the environment. Breaks are probably not sharp because (1) a species may be able to use other, less suitable nitrogen sources when its preferred nitrogen source disappears and (2) wind may distribute nitrogenous compounds long distances, as several workers have suggested [e.g., Swan, 1963, 1968; Llano, 1962].

Uric acid, which is only slightly soluble in water at neutral and acid *p*H values, would probably be dispersed as particles [cf. Folk, 1969a, b] and thus would not be found as far from nesting sites as urea, ammonia, or ornithogenic nitrate, all of which are quite soluble in water. The *p*H of guano tends to be acid, whereas that of almost all other soils tends to be very basic.

The liquid fraction of fresh skua droppings collected near Cape Crozier, Ross Island, contained only 0.23% reduced nitrogen, whereas the solid fraction contained 24.6%. Two older samples of penguin guano from a nearby Adélie penguin rookery contained 0.53 and 0.87 mg of urea nitrogen per gram, whereas two samples of water from the periphery of the rookery contained 0.20 and 0.40 mg of urea nitrogen per milliliter.

At moist localities near the coast uric acid is probably a more reliable source of nitrogen; at arid inland sites ammonium and nitrate could accumulate [e.g., Johannesson and Gibson, 1962; Gibson, 1962; Claridge and Campbell, 1968; Mueller, 1968]. Coastal species dependent on the nearly water-insoluble organic compounds could not exist at the inland sites. The ultimate case would be species capable of using oxidized (mainly nitrate) or molecular nitrogen. In such cases nitrate-assimilating lichens and mosses and nitrogen-fixing blue-green algae would occur.

Since both the absorption and the assimilation of nitrate nitrogen and the reduction of molecular nitrogen require metabolic energy (whereas the absorption and assimilation of ammonium nitrogen may not because the latter can exist as neutral NH_3 and NH_4OH molecules [Jennings, 1963, chapter 10] and is already in the reduced oxidation state), atmospheric nitrogen and nitrate nitrogen are apparently less suitable for plants of the harsh cold deserts of continental Antarctica.

The distribution patterns of plants in the McMurdo Sound region appear to reflect the operation of the preceding environmental factors (i.e., dry, saline, and often nitrogen poor conditions inland, the more lush and varied flora being restricted to the coast, to sheltered locations, and to places where ornithogenic nitrogen, even in small quantities, supports the growth of certain species). This pattern is strongly modified by salt levels and wind patterns. The field and laboratory studies described below were attempts to clarify and to substantiate speculations about and preliminary explanations of plant distribution in continental Antarctica.

Other Potential Nitrogen Sources

Three additional nitrogen sources for plants in Antarctica should be mentioned. In their probable order of effectiveness and availability, they are (1) nitrogen-fixing organisms, (2) the atmosphere (in precipitation and from oxidized nitrogenous gases), and (3) mineral substrates.

Excreted nitrogen. Holm-Hansen [1963] and Fogg and Stewart [1968] have demonstrated that *Nostoc commune* Vaucher fixes atmospheric nitrogen in situ under field conditions in Antarctica, and Fogg and Stewart [1968] have demonstrated that the cyanophycean phycobionts of lichens in maritime Antarctica also fix nitrogen. However, since virtually all species of lichens in continental Antarctica contain chlorophycean phycobionts, nitrogen fixation would probably have to be an indirect, external source of the element for most antarctic lichens.

There are numerous reports that nitrogen-fixing blue-green algae excrete nitrogenous compounds into the environment [e.g., Fogg, 1952; Mayland and McIntosh, 1966; Shields, 1957; Shields et al., 1957; Stewart, 1963, 1967]. In some places in Antarctica nitrogen originating from the activities of heterocystous blue-green algae must therefore have a significant influence on patterns of plant distribution. In southern Victoria Land such an influence appears to be operating in the Penny Lake–Walcott Glacier area (Appendix 1), on The Flatiron in Granite Harbor, and at Marble Point (Figure 4), where more or less lush stands of nitrophilous lichens occur. This influence is especially probable in the Penny Lake–Walcott Glacier area, where there is no apparent source of organic or reduced nitrogenous compounds other than blue-green algae.

In some places where *Nostoc* and other possible nitrogen fixers occur, high salt concentrations may inhibit vegetation despite the presence of fixed nitrogen (cf. Tables 1 and 2 and Figure 2).

Mosses are sometimes intimately associated with blue-green algae, which cling tightly to their protone-

Fig. 4. *Nostoc* near Marble Point, Victoria Land. This nitrogen-fixing blue-green alga is found in the greatest abundance in depressions, where both the moisture content and the salt concentration are high.

mata. The alga (*Nostoc* sp.) may obtain the necessary moisture from this association, and the mosses nitrogen. Such a phenomenon occurs in the tundra of arctic Alaska (V. Alexander and D. Schell, personal communication, 1972). However, Horikawa and Ando [1967] and Matsuda [1968] report that blue-green algae are detrimental to antarctic mosses.

Nitrogen-fixing bacteria may play a role in plant distribution as well. Dodge [1964, 1965], for instance, reports that *Azotobacter* occurs in the thalli of some antarctic lichens and thus gives rise to more luxuriant specimens. Boyd and Boyd [1962] detected *Azotobacter* in soil from the Windmill Islands off the Budd Coast in Wilkes Land.

Precipitation. Another potential source of nitrogen for antarctic plants is precipitation, which in the present case is almost exclusively snow [Wilson, 1959; Wilson and House, 1965; Claridge and Campbell, 1968; Jones and Faure, 1969; Lorius et al., 1969]. This source may be adequate for lichens, which grow extremely slowly, but it would be far from adequate for mosses and algae, unless they also grew very slowly or unless the nitrogen tended to accumulate. Where nitrates accumulate, conditions are usually dry [Claridge and Campbell, 1968]; they are so dry in fact that few macroscopic plants, including lichens, can survive.

The substrate. The third possible source of nitrogen is the substrate itself, including clays and rocks. Eugster and Munoz [1966] have discussed this possibility, and Stevenson [1959] has pointed out that there is fixed ammonium in rocks. Mueller [1968] has suggested that nitrate deposits in Chile and Antarctica may be due in part to compounds leached from rocks and subsequently oxidized. It is therefore significant that Ugolini [1970] has recently reported that *Xanthoria mawsoni* and *Polycauliona pulvinata* (< *Caloplaca elegans* var. *pulvinata*) deplete silica and iron from basaltic and volcanic rocks [cf. Silverman and Munoz, 1970], since bound nitrogenous compounds may be mobilized in the process (see also Jackson and Keller [1970]).

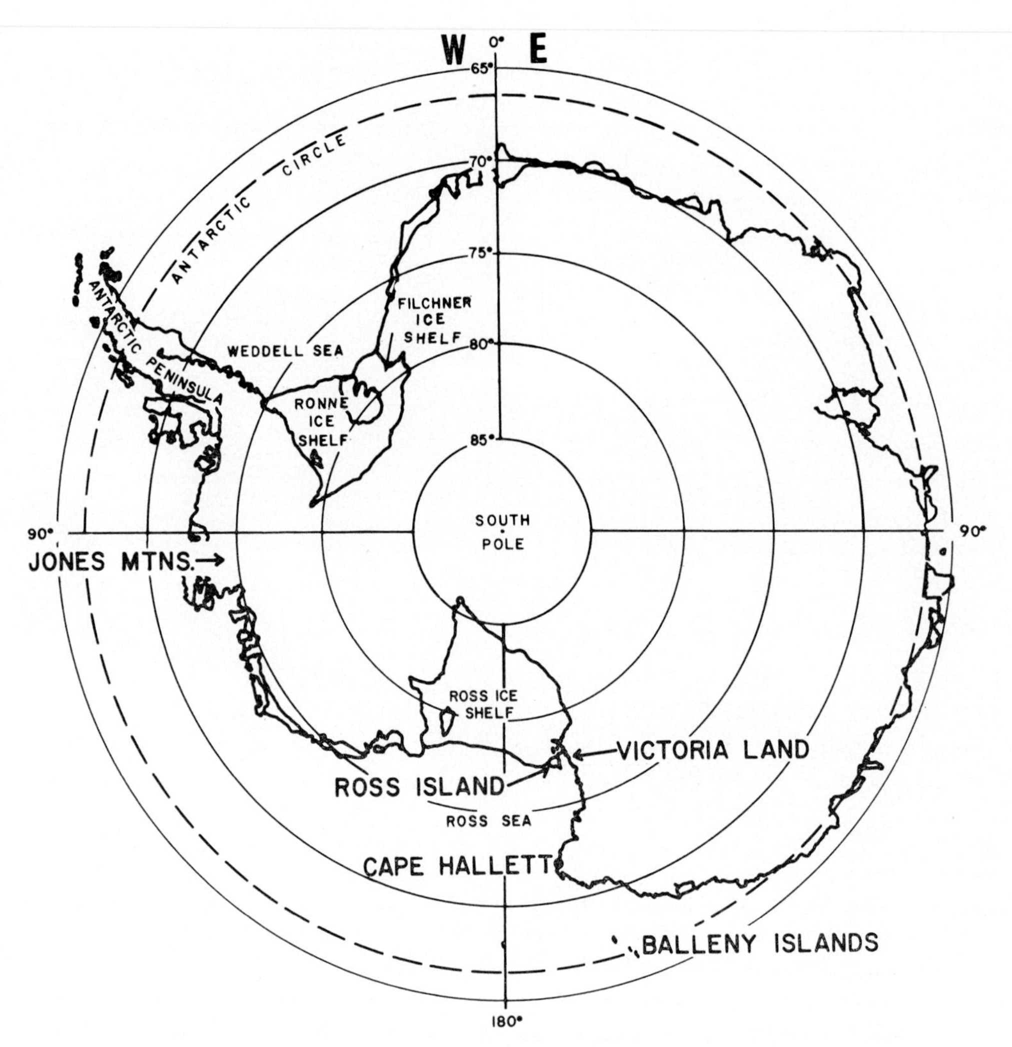

Fig. 5. Map of Antarctica showing localities where organisms used in laboratory studies were collected and where other collections mentioned in the text were obtained.

In the experiments described below, only substrate (primarily soil and guano) nitrogen was considered. The other potential sources of nitrogen may be available in amounts adequate for a few species of lichens but certainly not for most other macroscopic terrestrial plants in continental Antarctica. For example, the very small quantities of nitrogen in precipitation at the South Pole, 5 ppb (parts per billion) deposited at the rate of 5×10^{-8} g/cm²/yr [Wilson and House, 1965], and the relatively unobtainable nitrogen bound in rocks could sustain only the very slowest growing lichens.

FIELD WORK

The field work consisted primarily of collecting samples of plants, soils, and water for use in laboratory investigations. Numerous collecting trips were made for this purpose by Edmund Schofield in Victoria Land and on Ross Island during the austral summers of 1963–1964, 1967–1968, and 1968–1969. The lichens collected during the first of these seasons are enumerated in Appendix 1. Dr. Carroll W. Dodge, University of Vermont, Burlington, very kindly provided identifications for them. He has described five new species from among the lichens in the collection [Dodge, 1968].

In addition to the collecting trips in Victoria Land and on Ross Island, a visit was made to the Balleny Islands on March 9, 1964. A number of collections were made, including a sample of green ice from which *Prasiola crispa* was isolated (see discussion below).

A number of lichen and moss samples collected by

Mr. Kelvin P. Rennell of New Zealand in the Jones Mountains, Eights Coast, West Antarctica, were kindly provided by him for study. Among his samples was a new lichen [Dodge, 1968]. From another of his samples a moss (*Bryum algens*) was isolated into pure culture and used in a study of its nutritional characteristics (see discussion below).

The areas in which plants were collected are shown in Figure 5.

EXPERIMENTAL DATA

Lichen Fungi

Introduction. Because lichen thalli are difficult or impossible to maintain in the laboratory for any length of time and because determining the responses to compounds of intact lichen thalli is difficult, it is often convenient to isolate the symbionts of lichens into pure culture. In this isolation the mycobiont (fungal symbiont) is separated and grown independently of the phycobiont (algal symbiont). There is no definite information for judging whether the results obtained in vitro from separated symbionts can be applied to lichens growing in their natural habitats; nevertheless it is probably safe to assume that some of the findings obtained in the laboratory can be extrapolated to field conditions.

During the antarctic summer of 1963–1964 approximately 200 single-ascospore isolates of lichens from Victoria Land (in the Cape Hallett, McMurdo Sound, and ice-free valley areas), Ross Island (primarily Hut Point Peninsula, Cape Royds, and Cape Crozier), and other scattered regions of the continent were obtained on various media by the isolation methods described by Ahmadjian [1961]. These 200 successful isolations were <10% of the 2500 isolations attempted. Standard microbiological media modified to provide different *p*H values, vitamins (biotin, thiamine, and so forth), and other constituents (glucose, asparagine, and yeast extract) were employed for the initial growth of the isolates. There was no discernible correlation between successful spore germination or subsequent colony growth and cultural conditions, date of collection, or species (Appendix 1). Ascospores from one collection of a species may have displayed a very high or a very low percentage germination, whereas those from another collection of the same species from a different area may have behaved to the contrary.

Materials and methods. Two mycobionts were selected for studies of temperature and *p*H optimums and nitrogen source preference. One isolate (from the Jones Mountains, Eights Coast, West Antarctica) was selected from among the isolates of one of the mycobionts (lichen A), whereas the other (from Ross Island, East Antarctica) had been in culture for approximately 8 years [Ahmadjian, 1958, 1961] (lichen B). Lichen A mycobiont was isolated from *Lecanora tephroeceta* Hue, and lichen B mycobiont from *Lecidea* sp. These two isolates were chosen because they grew much more rapidly than most of the other isolates and had been in culture for different lengths of time (<1 year versus 8 years, respectively).

The diameter of lichen A mycobiont was 7 mm after 9 months in a screw cap culture tube on a modified malt extract–yeast extract agar medium (20 g/l of malt extract, 2 g/l of yeast extract, 200 μg/l of pyridoxine, 10 μg/l of inositol, 200 μg/l of thiamine, 5 μg/l of biotin, 20 g/l of agar, and a final *p*H adjusted with 0.1 *N* NaOH of 6.8). The growth of the other isolates of this species from the same collection site was similar on all four of the isolation media used, Sabouraud's dextrose agar, nutrient agar, corn meal agar, and the modified malt extract–yeast extract agar. The viability of the ascospores of *L. tephroeceta* based on a total of 20 spores obtained from the same specimen was approximately 90%. (Hue [1915] states that this species is very fertile, its thallus soon disappearing under the quantity of apothecia. All known collections of *L. tephroeceta* are described in Appendix 2.) The isolates, all of which were from single ascospores, were originally pink; with age the cultures turned yellow.

Lichen B (isolate 11 in Ahmadjian [1961]) was collected from the ground on Ross Island in 1957. The mycobiont used in this study (a monospore isolate) developed very slowly at first: after 9–12 months of growth at 8°C it was only 1–2 mm in diameter. After several years in culture its growth rate increased, and at 18°C its growth was relatively good. Initially the colonies were white; later they became partly yellow and then light brown.

Lichen A was used in the studies of nitrogen source preference. Thirty-eight nitrogen sources were supplied separately at concentrations of 0.035 g/l (ammonium and nitrate salts) or 0.018 g/l (amino, amide, purine, pyrimidine, riboside, ureide, and miscellaneous nitrogenous compounds) to 0.2–0.3 mg of washed, homogenized mycelium taken from cultures grown in Lilly and Barnett synthetic medium plus biotin (5 μg/l) and thiamine (100 μg/l) [Ahmadjian, 1967, p. 21] for 3 weeks at 15°C and then stored in polyethylene bottles for 1 month at 5°C in a few milliliters of nitrogen-free medium. (Molybdenum should have been added to the media because reduction of nitrate requires trace quantities of the element. It was not supplied. However, it has been shown many times that water, reagents, and other materials contain sufficient quantities of molybdenum to support the growth of even rapidly growing fungi such as *Aspergillus* and *Penicillium*. Molybdenum can be eliminated from media only by painstaking procedures. Because lichen fungi, especially antarctic lichen fungi, grow hundreds or thousands of times more slowly than *Aspergillus* and *Penicillium* and because the constituents of the media were not treated to remove contaminating molybdenum, it can be assumed with confidence that there was sufficient molybdenum for enzymatic reduction of nitrate by *L. tephroeceta*.)

The experimental media were sterilized with Millipore (0.45-μm pores), Morton ultrafine sintered glass or Seitz filters and dispensed in 45-ml aliquots to 125-ml cotton-stoppered Erlenmeyer flasks previously cleaned with sodium dichromate–sulfuric acid solution. The medium was buffered with 10 ml of M/15 sodium, potassium phosphate buffer (*p*H 5.15) per flask. The water used was once pyrex distilled. (A smaller pored Millipore filter should have been employed. Nevertheless there was no pattern of contamination that could be traced to the use of the larger sized pores. Fortunately only the solutions with aspartic and glutamic acids were sterilized with Millipore filters; all other solutions were sterilized with either Morton or Seitz filters.)

Inoculum (0.2 mg of macerated mycelial fragments in 5 ml of distilled water according to 50 dry weight determinations) was added to each flask with a sterile Cornwall syringe pipette. These additions brought the total volume of medium per flask to 60 ml. The *p*H of the medium plus buffer was 5.3 in all except six cases, whose *p*H values fell between 5.1 and 5.6, the *p*H range for all nitrogen sources being 0.5.

The inoculated flasks were incubated at 20°C. They were not shaken. The mycelial dry weights of at least four flasks were determined for each nitrogen source after 21, 78, and 99 days. All the mycelium from each flask was rinsed on filter paper with distilled water, dried overnight on the filter paper at 60°–70°C, and weighed to the nearest 0.1 mg.

Two experiments were performed to determine the optimal temperatures for growth of the mycobionts of lichens A and B. In the first of these experiments, which involved both of the mycobionts, the incubation temperatures were 3°, 7°, 11°, 15°, and 19°C. The dry weights of mycelium were determined after 9 and 18 weeks of growth in a medium consisting of 20 g/l of malt extract (Difco), 2 g/l of yeast extract (Difco), 5 μg/l of biotin, and 100 μg/l of thiamine made to volume with distilled water (the initial *p*H was approximately 6.3). The mycelia were incubated in 125-ml Erlenmeyer flasks containing 25 ml of medium and 2 ml of inoculum. The initial dry weight of the inoculum of lichen A was 1.42 mg per flask, and that of lichen B 0.08 mg.

In the second temperature experiment, which involved only lichen A, incubation temperatures were −1°, +4°, +9°, +14°, +19°, and +24°C. The medium consisted of 5 g/l of glucose (Difco), 2.5 g/l of peptone (Difco), 1.0 g/l of malt extract broth (Difco), and 2.5 g/l of casein hydrolysate (Nutritional Biochemicals) (the initial *p*H was 6.4). Each of the 125-ml flasks used in this experiment contained 50 ml of filter-sterilized medium plus 4 ml of inoculum. The inoculum dry weight was 0.4 mg.

The responses of both mycobionts to *p*H were determined by similar cultural techniques. The study of *L. tephroeceta* was not as thorough as that of *Lecidea* sp., however.

Lecidea sp. mycobiont (lichen B) was grown in liquid Lilly and Barnett synthetic medium adjusted to and buffered at a number of *p*H values between 4.0 and 8.0. The inoculum, which had been grown at 15°C in a liquid medium of *p*H 6.5 for 14 weeks, was prepared by suspending the combined yield from 10 flasks (about 340 mg) in 100 ml of sterile distilled water plus buffer (1:2), blending it in a Waring blender for 1 min, and pipetting the resulting suspension in 10-ml aliquots to nine Erlenmeyer flasks, each flask containing 250 ml of Teorell's universal buffer [Cavanaugh, 1956] prepared to yield a different *p*H. Each buffered mycelial suspension contained 0.15–0.17 mg of mycelium per milliliter, as three dry weight determinations performed on each sample showed.

The experimental media were prepared in two distinct steps: (1) a 10× strength glucose and asparagine solution was sterilized by filtration, and a minerals solution, plus each of nine buffer solutions, was autoclaved; (2) the resulting sterile solutions plus the nine buffered fungal inocula were combined in sterile 125-ml Erlenmeyer flasks, the final concentrations for each constituent being equivalent to those for Lilly and Barnett medium (biotin and thiamine

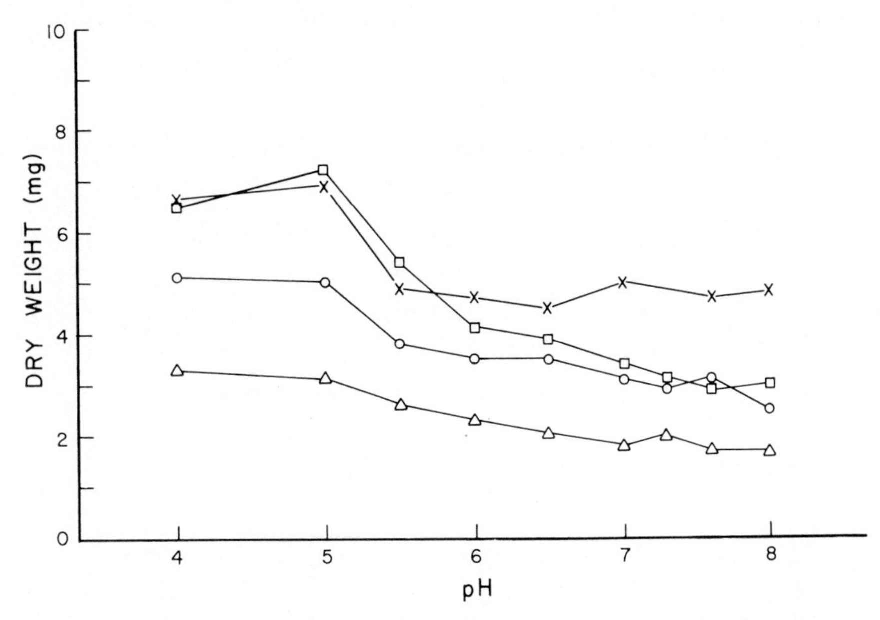

Fig. 6. Growth of mycobiont *Lecidea* sp. (Ahmadjian 32) in liquid cultures at various *p*H values after 24 (triangles), 49 (circles), 74 (squares), and 99 (crosses) days.

were not added) and the inoculum being 0.95 mg per flask. The results were nine series of 25 ml per flask of vitamin-free Lilly and Barnett medium plus macerated fungal mycelium.

After inoculation the *p*H values of some of the buffer solutions changed; the intended and actual (in parentheses) values were as follows: 4.0 (4.0), 5.0 (5.0), 5.5 (5.4–5.5), 6.0 (6.0), 6.5 (6.5), 7.0 (7.0), 7.5 (7.3), 8.0 (7.6), and 9.0 (7.9–8.1). The incubation temperature was 20°C. The results, which are plotted as the means of mycelial dry weight for three to five flasks, are presented in Figure 6.

The mycobiont of *L. tephroeceta* (lichen A) was grown at 12 *p*H values between 4.0 and 8.4 in 50 ml of the following liquid medium: 2.5 g/l of glucose, 2.5 g/l of malt extract broth, 1 g/l of yeast extract, and 2.5 g/l of casein hydrolysate. The *p*H of the medium, which at first was approximately 6.4, was adjusted with either 1 *N* NaOH or 1 *N* HCl. The incubation temperature was 20°C. The inoculum dry weight was 0.2 (±0.1) mg per flask.

Mycelium grown at 14°–15°C for 2 months in the preceding liquid medium at a *p*H of about 6.4 was harvested, washed twice with 700 ml of sterile distilled water, macerated in a Waring blender for 1 min, and dispensed in 5-ml aliquots to 125-ml Erlenmeyer flasks with a Cornwall syringe pipette. The dry weight determinations made after 21 days are shown in Table 3.

Results. The results were as follows.

Nitrogen source preference: The results of the growth studies on nitrogen source preference of lichen A are presented in Table 4. The table shows that the control flasks contained 4.2 mg dry weight of mycelium after 72 days but <1 mg after 98 days; thus autolysis probably took place when the carry-over nitrogen was depleted. The growth curves are plotted in Figure 7.

Ammonium nitrogen sources supported excellent growth. Dry weight readings of about 16–18 mg per flask were obtained after 78 days and of 17–23 mg

TABLE 3. Growth Attained by the Mycobiont *Lecanora tephroeceta* Hue at Various *p*H Values

Initial *p*H	Final *p*H (58 Days)	Dry Weight (21 Days), mg
4.00	4.09	2.8*
4.20	4.14	2.7
5.24	5.20	1.3
5.70	5.89	1.0
6.12	6.29	0.9
6.43	6.56	1.1
7.00	6.95	1.3
7.30	7.20	0.7
7.45	7.44	0.8
7.90	7.79	0.8
8.20	8.12	0.6
8.40	8.32	1.0

Values are the means for five replicate flasks.
* At both 20° and 10°C.

TABLE 4. Growth Attained by the Mycobiont *Lecanora tephroeceta* Hue on Various Sources of Nitrogen in a Buffered Medium

	Dry Weight Yield, mg			*p*H*		
Nitrogen Source	At 21 Days	At 78 Days	At 98 Days	Initial	At 78 Days	Change after 78 Days
Glutamine	3.2	23.9	24.9		5.2	—0.1
Urea	2.1	18.4	22.3		5.8	+0.5
Ammonium tartrate	2.8	17.6	19.5		5.2	—0.1
NH_4Cl	2.5	17.5	18.0		5.1	—0.1
$(NH_4)_2SO_4$	2.3	16.6	23.0		5.2	—0.1
Proline	2.6	16.4	14.2		5.4	
Alanine	3.1	16.2	...		5.4	+0.1
NH_4NO_3	2.9	15.9	16.8		5.1	—0.2
Xanthine	...	15.3	12.9	no measurement	5.6	?
Leucine	3.6	15.1	...			
Asparagine	3.3	14.3	14.7		5.2	—0.1
Uric acid	3.3	13.2	...	5.6	5.6	
Casein hydrolysate	2.3	8.2	5.4			
Arginine	2.8	7.3	3.6			
KNO_3	2.7	7.2	5.7			
Glutamic acid	3.3	7.1	...	5.1	5.1	
$Ca(NO_3)_2$	3.0	6.6	6.9			
Serine	3.1	6.2	4.8			
Methionine	2.4	5.9	...			
Cysteine	3.0	5.8	3.1	5.1	5.1	
Cytosine	3.5	5.6	3.4			
Tryptophane	3.1	5.5	2.4			
Allantoin	2.3	5.3	...	5.5	5.3	—0.2
Adenosine	3.5	5.2	...			
Phenylalanine (DL)	2.5	5.1	2.2			
Thymidine	2.8	5.0	2.1			
Cytidine	2.3	4.8	...			
Histidine	3.7	4.7	2.1			
Aspartic acid	2.1	4.6	3.0	5.1	5.1	
Glutathione (DL)	2.4	4.6	...		5.2	—0.1
Thymine	3.0	4.6	2.6			
Peptone	2.5	4.4	...		5.4	
Adenine	2.5	4.3	2.6	5.1	5.1	
Control	2.2	4.3	0.8			
Tyrosine	3.1	4.0	2.7			
Uracil	2.1	3.8	2.8			
Uridine	2.2	3.7	2.7			
$NaNO_3$	1.8	3.0	0.8			
Threonine	2.7	2.3	2.1			

Nitrogen sources are listed in descending order of dry weight yield of fungal mycelium at 78 days. Unless it is indicated otherwise, the L stereoisomer of the amino acid was used.

* Where no value is given, the initial *p*H was 5.3–5.4, the *p*H at 78 days was 5.3, and the change in *p*H was <0.1.

after 98 days; thus at 98 days the mycelium was still in the growth phase.

Nitrate nitrogen sources supported much less growth than ammonium nitrogen sources. Of the three nitrate sources used (except for NH_4NO_3) maximum growth was obtained with KNO_3 at 78 days and with $Ca(NO_3)_2$ at 98 days. Between the 78- and 98-day readings the mycelium in the KNO_3 had decreased by nearly 2 mg; this finding suggests that autolysis had set in. In comparison to the control medium sodium nitrate inhibited growth. This result may be related to the well-known and widespread process of active transport of sodium ions out of the cells of many organisms; the potassium ions in the flasks with KNO_3, on the other hand, may have been actively accumulated. The fate of the accompanying cation may have influenced the rate of uptake of nitrate in these two cases.

The *p*H values of the filtered medium obtained from the combined flasks of each nitrogen source are presented in Table 4. Despite the high concentration of buffer and the small amount of fungal growth achieved, there was a measurable and consistent drop in *p*H in the media containing ammonium. For all

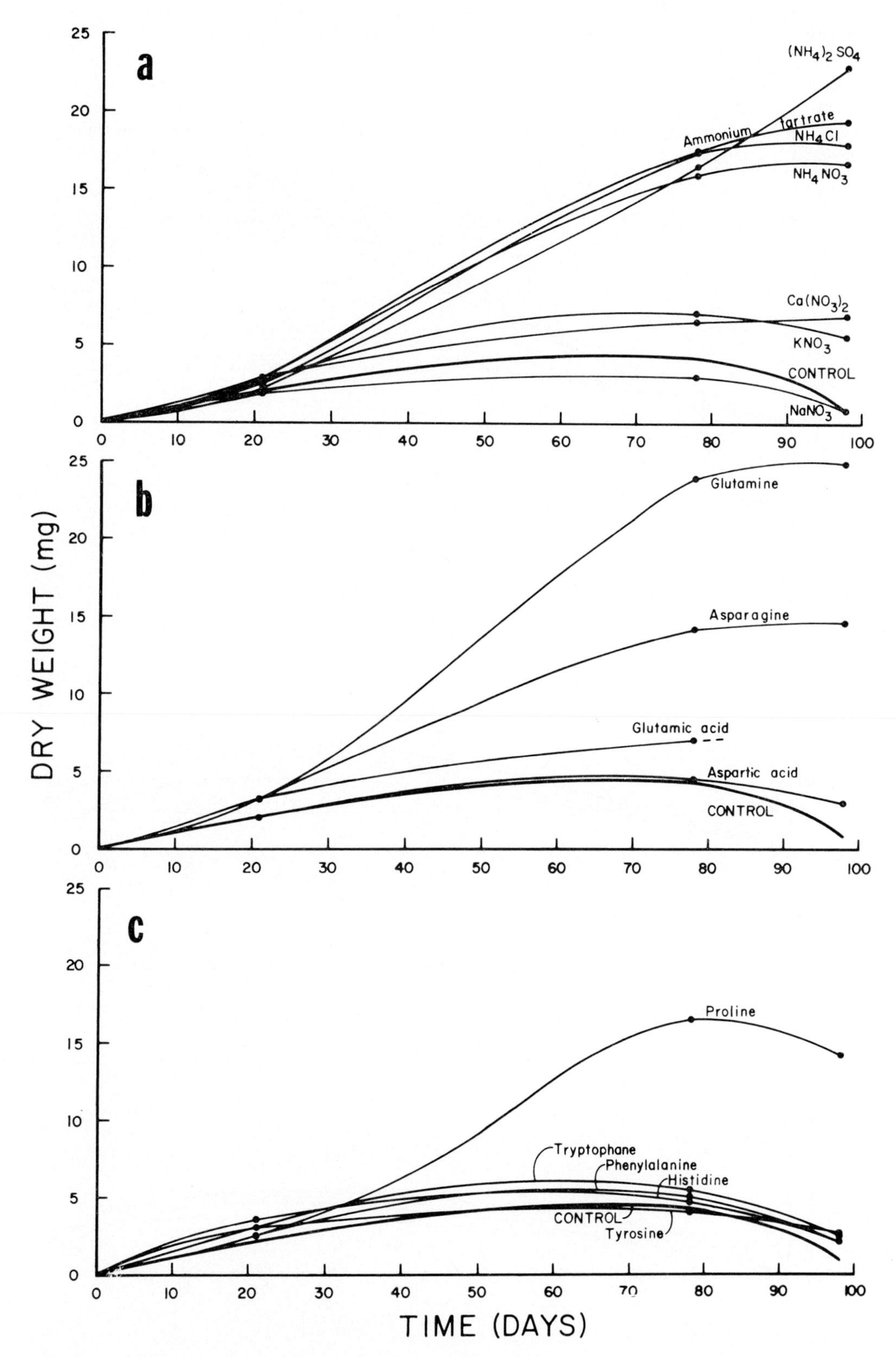

Figs. 7*a–c*. Growth of the mycobiont *Lecanora tephroeceta* Hue on (*a*) nitrate and ammonium salts, (*b*) dicarboxylic amino acids and their amides, and (*c*) aromatic and heterocyclic amino acids.

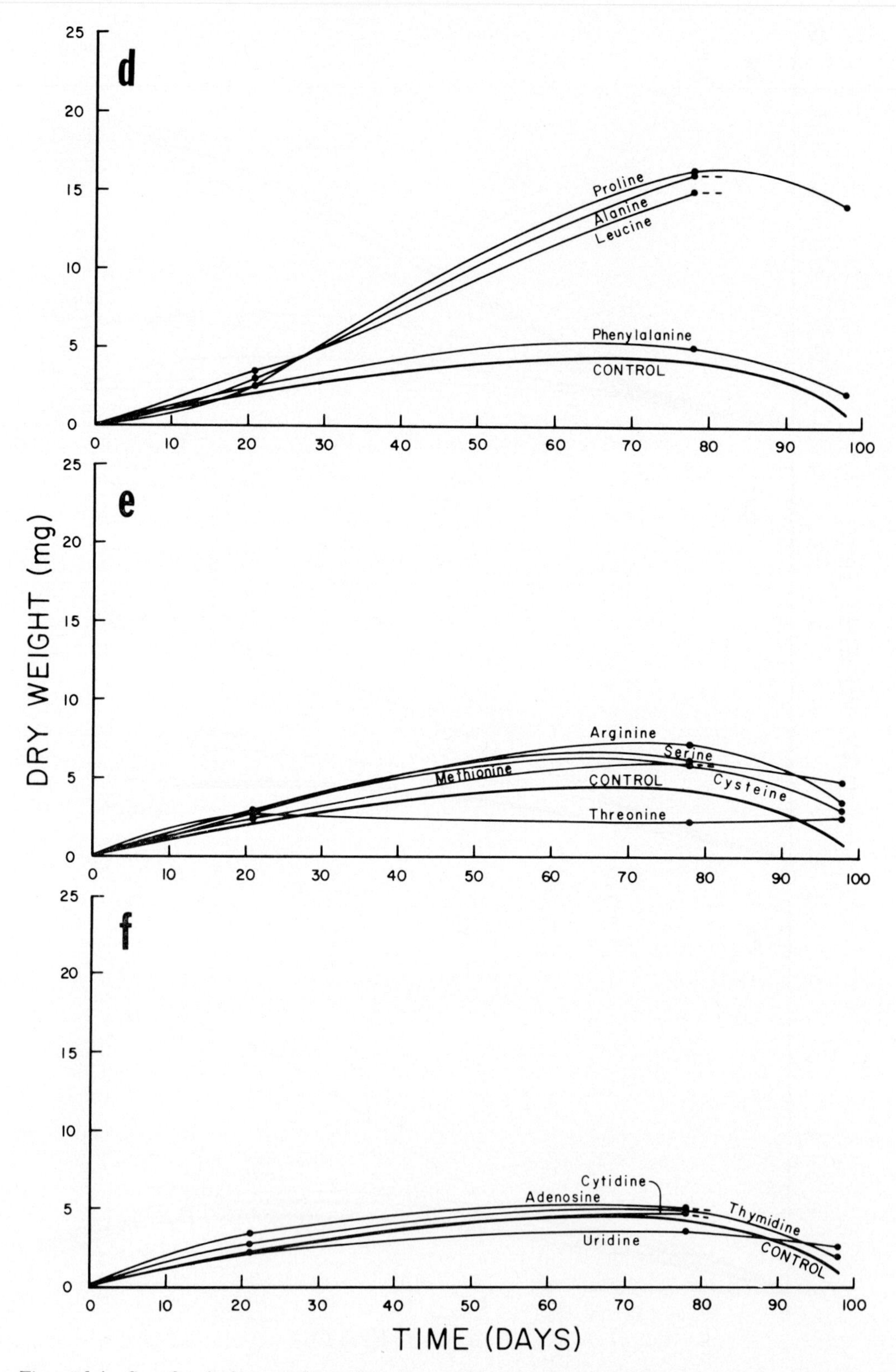

Figs. 7*d–f*. Growth of the mycobiont *Lecanora tephroeceta* Hue on (*d*) amino acids with non-polar side chains, (*e*) other amino acids, and (*f*) ribosides.

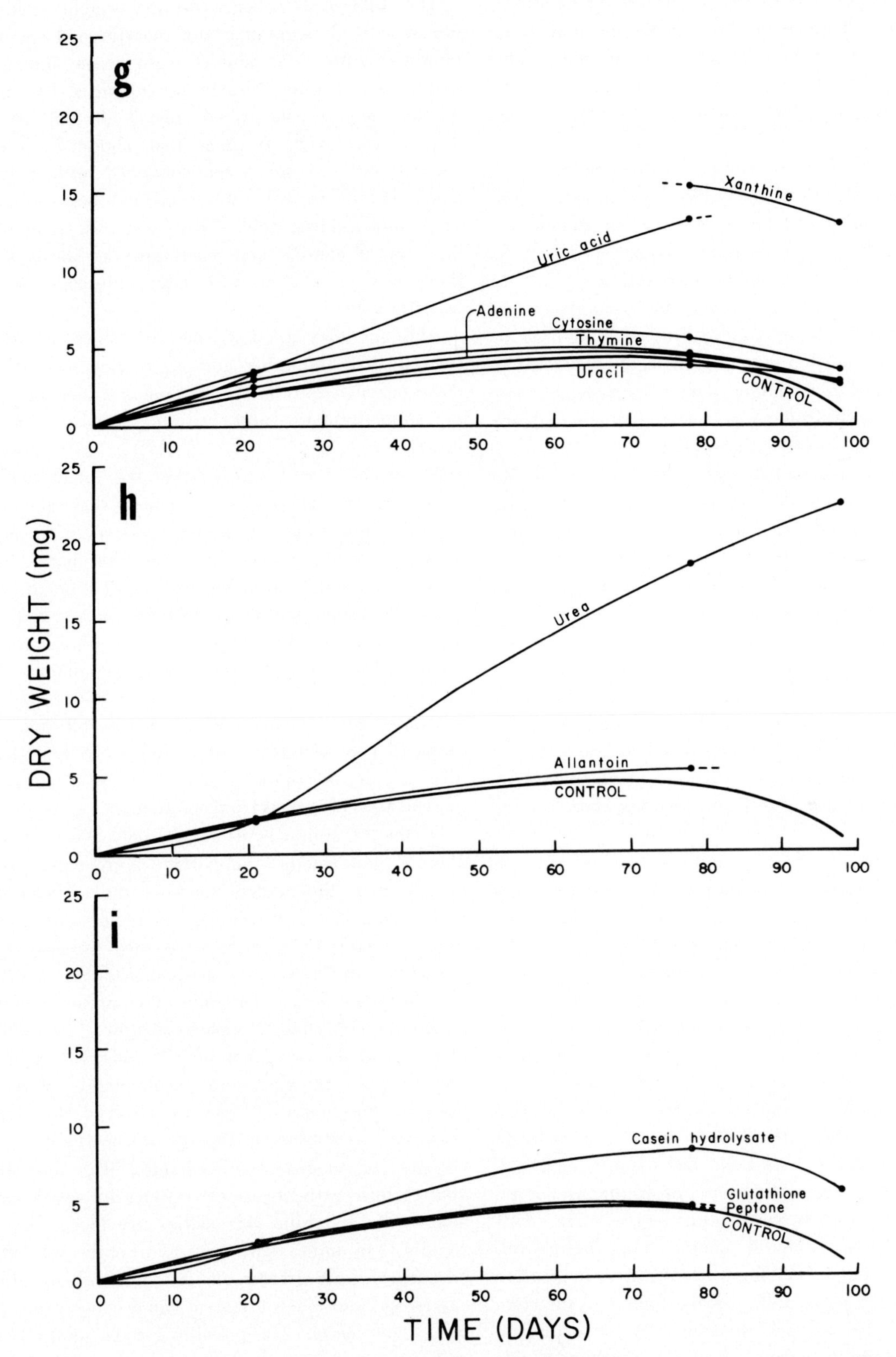

Figs. 7*g–i*. Growth of the mycobiont *Lecanora tephroeceta* Hue on (*g*) purines and pyrimidines, (*h*) ureides, and (*i*) miscellaneous nitrogen sources.

four ammonium sources the *p*H decreased by 0.1 or 0.2 *p*H units because of the uptake of ammonium ions from the medium. There was no measurable change in *p*H in the control and nitrate flasks. Significant uptake of nitrate would have caused a rise in *p*H.

The growth of the mycobiont in the medium with urea was excellent (higher at 78 days than that produced on all other nitrogen sources except glutamine). At 98 days the growth on urea was exceeded only by that on glutamine and ammonium sulfate.

The change in *p*H of +0.5 in the urea flasks was probably due to the enzymatic hydrolysis of urea by extracellular fungal urease, which would have led to the production of ammonia and to a consequent rise in *p*H. This *p*H change was greater than that of any of the other 37 sources of nitrogen.

A standard microbiological test was carried out with the mycobiont of lichen A to detect the production of urease. The test, which employs Difco urea agar, was carried out at 19°C. Within 18 hours the ammonia released by hydrolysis of the urea had produced a discernible color change in the medium due to the higher *p*H. Similar results had previously been obtained with nearly all lichen fungi tested for production of urease, including the mycobiont of *Rinodina frigida* (Darb.) Dodge, another antarctic species. It is therefore concluded that the urea in the experimental medium was hydrolyzed to ammonia by extracellular urease before it was absorbed by the fungus.

Of the riboside, purine, and pyrimidine nitrogen sources employed, the fungus used only xanthine and uric acid to any extent. Allantoin, which is the oxidation product of uric acid and which is degraded by some organisms to urea and glyoxalate, was not used. These results are interpreted to show that uric acid and xanthine were absorbed as intact molecules rather than as ammonia produced (via allantoin and urea) from them. The fungus *Candida utilis* absorbs uric acid and xanthine by two specific carrier systems [Quetsch and Danforth, 1964]; such a mechanism may account for the selective use of uric acid and xanthine by *L. tephroeceta* and for its nonuse of related substances such as allantoin and adenine. This interpretation assumes of course that the growth rates of the fungus on the various nitrogen sources reflect both the permeability of the fungus to the compounds and the enzyme complement of the organism. It is impossible in the present case to distinguish one influence from another.

The differences between the dry weight yields obtained with glutamic acid and aspartic acid and those obtained with their amides (glutamine and asparagine) are striking. Glutamine supported by far the greatest amount of growth of all the 38 nitrogen sources employed: 6 times that obtained with the control and >3 times that obtained with glutamic acid. The same 3:1 ratio also held between asparagine and aspartic acid. There was a drop in *p*H of 0.1 with glutamine and asparagine at 78 days, but there was no change with either glutamic acid or aspartic acid.

Although they are far from conclusive, the results suggest that glutamine and asparagine were deamidated enzymatically in the medium and that the resulting ammonium was absorbed and thus led to a drop in *p*H.

Moiseeva [1961] showed that the intact thalli of nearly all the 41 species of lichens that she tested, including two antarctic species, produced extracellular asparaginase and urease. On the other hand, Smith [1960] concluded from his own studies on discs cut from the intact thalli of *Peltigera polydactyla* that asparagine was either taken up as the whole molecule or adsorbed and only very slowly deamidated. Thus the mechanism by which glutamine and asparagine were used by *L. tephroeceta* is open to question. Our tentative conclusions are that they were first deamidated by the mycobiont and that the resulting ammonia was then absorbed from solution.

Of the 14 amino acids tested, only three (proline, alanine, and leucine) supported significantly greater growth than the control medium did. These three amino acids have in common nonpolar side chains; this characteristic may have resulted in more rapid absorption of these amino acids than of the less lipid-soluble amino acids. The only other amino acid with a nonpolar side chain was phenylalanine. Since phenylalanine and the two other amino acids with aromatic constituents (tyrosine and tryptophane) were not used by the fungus, it can be assumed that amino acids with aromatic constituents are not used.

Some organisms (e.g., carrots [Birt and Hird, 1956]) have active-transport systems by which amino acids with lipophilic side chains are taken up more rapidly than amino acids with hydrophilic side chains because of their greater affinity for a cell membrane carrier system. Such a system could conceivably exist in *L. tephroeceta*. The possibly greater solubilities of such amino acids in the cell membrane or their lower net electrostatic charge (at least on their side chains)

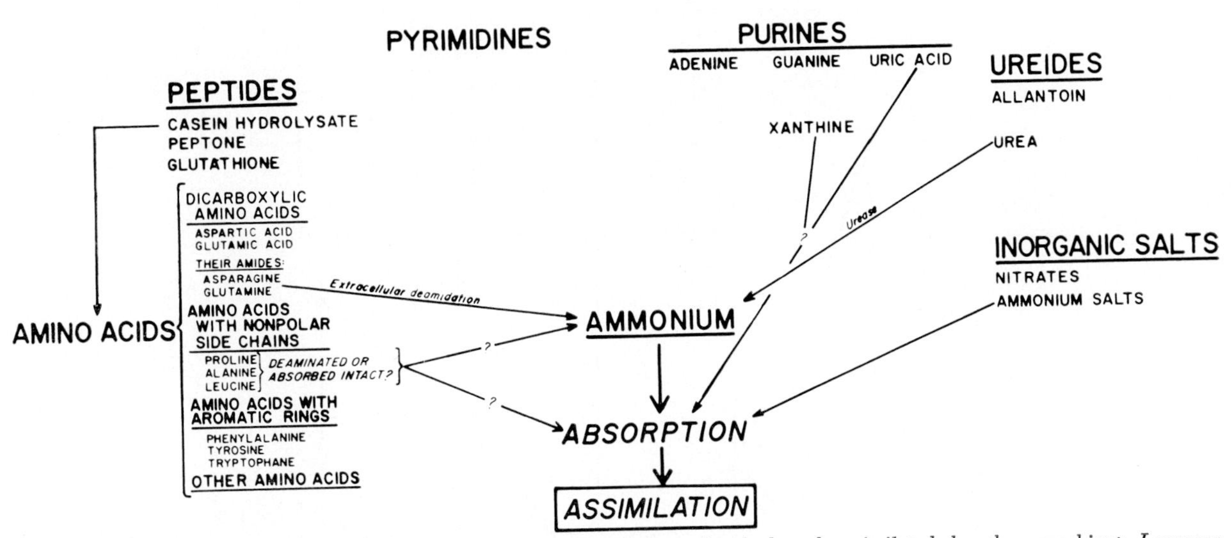

Fig. 8. Suggested pathways by which the nitrogen sources were absorbed and assimilated by the mycobiont *Lecanora tephroeceta* Hue in laboratory experiments. Arrows indicate compounds that were used and the probable routes by which they were absorbed (compounds or classes of compounds that supported little or no growth have no arrows). The results indicate that urea was hydrolyzed and that glutamine and asparagine were deamidated by extracellular enzymes. Only those amino acids with nonpolar side chains were used by the fungus to any significant degree. A special absorption mechanism may exist for the two related compounds, uric acid and xanthine.

may also have enhanced their absorption by the mycobiont.

Glutathione and peptone supported about the same amount of growth as the control medium, and casein hydrolysate supported only about twice as much growth as the control. The results obtained with these three nitrogen sources indicate (1) that peptides are not used by *L. tephroeceta* (i.e., that enzymes that will hydrolyze peptides to their amino acid constituents are not produced) and (2) that very few amino acids are used, either because they cannot be absorbed or because there are no enzymes within the cells for converting them to other, essential amino acids.

Figure 8 summarizes in diagrammatic form the results and conclusions of the nitrogen source studies with *L. tephroeceta.* The compounds used are arranged in groups roughly according to their biochemical relationships.

From the results and conclusions it is possible to infer that nitrogen compounds are used by *L. tephroeceta* in either of two ways: (1) as ammonia or (2) by intact absorption. In the first case the compounds are converted extracellularly to yield ammonia either by hydrolysis (urea) or by deamidation (asparagine and glutamine). In the second case the compounds are lipid soluble (proline, alanine, and leucine) or are actively absorbed (uric acid and xanthine).

If these inferences are correct, the mycobiont of *L. tephroeceta* and possibly those of other antarctic lichens found in similar habitats display selectivity in using nitrogen sources; cell permeability is limited to ammonium ions (or ammonia molecules), lipophilic amino acids, and specific compounds such as uric acid and xanthine. If the mycobiont is in fact permeable to nitrate, the nitrate is not reduced in the cells to ammonia because the appropriate enzymes are lacking.

Temperature experiments: The results of the two temperature experiments are presented in Figure 9, in which it is clear that in the first experiment the temperature curves for both mycobionts were bimodal. Growth was less at 11°C than at higher or lower temperatures. The curves are remarkably similar, especially when the great difference in inoculum dry weights is considered. The two optimums for lichen B (*Lecidea* sp.) were close to 7° and 15°C, and those for lichen A (*L. tephroeceta*) were close to 3° and 19°C. The results of the second experiment with lichen A (inset in Figure 9) are puzzling in comparison with those of the first; the peak in the temperature curve at 3°C is absent at 33, 53, and 84 days; at 84 days, however, growth had greatly accelerated at 9° and 14°C but had virtually ceased at 4°C. This disparity between the results of the

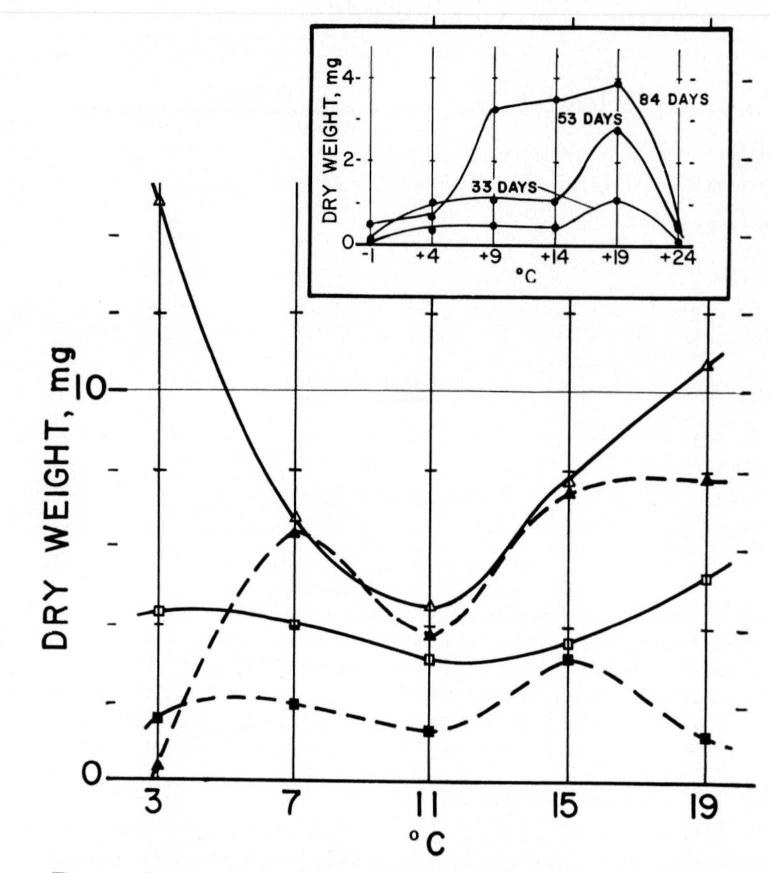

Fig. 9. Growth of mycobionts *Lecanora tephroeceta* Hue (lichen A) after 9 (open squares) and 18 (open triangles) weeks and *Lecidea* sp. (Ahmadjian 32, lichen B) after 9 (solid squares) and 18 (solid triangles) weeks. Inset shows the growth of lichen A at five temperatures in a different medium.

first and second experiments with lichen A may be due to the presence of vitamins (biotin and thiamine) and other growth factors in the medium used for the first experiment. However, there are too few data to determine the factors involved, and further experiments are necessary.

The points for 18 weeks in Figure 9 represent the means for the dry weights of mycelium from at least five flasks and in all but one case (lichen A at 3°C) those from 10–15 flasks. An examination of the frequency distributions of the dry weights showed that the means accurately reflect the true weights attained at each temperature. For example, at 3°C lichen A yielded a mean of 14.9 mg of mycelium, values in the five flasks being 13.3–16.7 mg. The means and the standard deviations (in parentheses) for experiment 2 on lichen A at 84 days (inset in Figure 9) in milligrams were 0.97 (0.79) at −1°C, 1.12 (0.40) at +4°C, 3.83 (1.42) at +9°C, 4.0 (1.88) at +14°C, 4.31 (1.51) at +19°C, and 0.96 (0.89) at +24°C. (The weight of the inoculum, 0.4 mg, was subtracted before the points were plotted for Figure 9.)

Thus the results of both experiments on temperature appear to be valid, and the reasons for the differences must lie with the two media employed. This conclusion underscores the need for care in interpreting the results; nutritional factors apparently play a significant role in the response of *L. tephroeceta* to temperature.

If the mycobionts of antarctic lichens do have two temperature optimums under some conditions, there may be two enzyme complements, one of which evolved under warmer environmental conditions than the other. There are two possible interpretations of this situation: (1) the higher temperature optimums may have evolved outside Antarctica, and the lower optimums in Antarctica after the lichens had migrated from warmer regions, or (2) the higher temperature optimums may have evolved in indigenous lichens during warmer periods in Antarctica itself. This type of speculation is unsafe and must be verified by careful experimentation. If it turns out that many or most lichen mycobionts display the same type of double temperature optimums, such speculation would carry more weight. On the other hand, double temperature optimums may be a necessary characteristic of antarctic lichens; i.e., the double optimums may have evolved in direct response to the present environment of Antarctica. Comparative physiological studies of nonantarctic and antarctic lichen mycobionts would be an effective way of resolving the question.

pH optima: Both mycobionts displayed a preference for acid media, at least during their initial periods of growth. In this property they resemble other lichen mycobionts.

Figure 6 shows that higher *p*H values did not prevent growth but merely slowed it down; growth was more rapid at acid *p*H but also reached the stationary phase sooner. This effect is reflected in the lower dry weights at 99 days for *p*H 5.0 and 5.5 and in the leveling off of growth at *p*H 4.0; there was considerable growth at 74–99 days at *p*H 6.0 and above.

The data in Table 3 show only that in the early stages of growth acid media are best for *L. tephroeceta;* since there are no data for the later stages, nothing further can be concluded.

Lichen Algae

Nitrogen nutrition. Nitrogen nutrition was studied as follows.

Materials and methods: Two lichen algae (both were strains of *Trebouxia,* the most common genus of lichen algae) were used in a similar study of nitrogen nutrition. One of them (the phycobiont of *Buellia pernigra* Darbishire) was isolated by Mrs. Jen-rong Wang Yang in February 1964 with a micropipette, using the procedure described by Ahmadjian [1967] from a specimen collected at Hallett Station by Dr. E. D. Rudolph and identified by Dr. C. W. Dodge. The resulting cultures were maintained at 4°–5°C.

Buellia pernigra occurs along the coasts of the Ross Sea, Marie Byrd Land, and the Princess Martha Coast [Greene et al., 1967]. Dodge [1948] cites a collection from George V Coast; Lamb [1968] does not mention the species in his monograph of the genera *Buellia* and *Rinodina* from peninsular Antarctica, however. The type specimen was collected on 'Mount Erebus,' Ross Island [Darbishire, 1923], a designation that could easily include nearby Cape Royds and Cape Evans, where penguins and/or skuas breed, since collection data on older specimens tend to be vague.

The culture used for the experiment was grown in Trebouxia liquid medium [Starr, 1964], NH_4NO_3 being the nitrogen source. The inoculum was washed in sterile distilled water, concentrated by centrifugation, and resuspended in sterile distilled water. It was then homogenized in a Waring blender.

The basal medium used in the experiment was nitrogen-free Trebouxia liquid medium without soil water. Eight nitrogen sources were tested; the basal medium without an added nitrogen source served as the control. The experiment was carried out with cotton-stoppered 125-ml Erlenmeyer flasks, each flask containing 25 ml of the medium. The average dry weight of the inoculum (determined for five aliquots) was 0.8 mg per flask. The flasks were inoculated with the algal suspension on December 8, 1964, and were then placed in an incubator at 10°C with constant illumination of 260–390 lux.

The dry weights of four flasks were determined for each nitrogen source after 44 days. The results are presented in Table 5.

The second alga (the Trebouxioid phycobiont of *Polycauliona citrina* Dodge), which was also from a specimen collected at Cape Hallett by Dr. Rudolph and identified by Dr. Dodge, was isolated with a micropipette by Mrs. Yang. *Polycauliona citrina* is a coastal species in the Ross Sea sector, Marie Byrd Land [Greene et al., 1967], and Queen Mary Coast (type collection) [Dodge, 1948].

Two concentrations of each nitrogen source were employed with this phycobiont: (1) the concentration given for each nitrogen source in Table 5 and (2) one-half of each concentration given in Table 5. Dry weights were determined after 31, 66, and 84 days. In other respects the experimental procedure was similar to that of the previous study.

The flasks were inoculated on February 18, 1965, with 0.5 mg of algal cells. The results are presented in Figure 10.

Temperature optimums. The temperature optimums were studied as follows.

Materials and methods: The phycobionts of six antarctic lichens were grown in darkness and in light at 10°, 15°, and 20°C to determine their responses to temperature. All phycobionts were species of *Trebouxia* obtained as single-cell isolates. The lichen species from which they were obtained and the collection sites are listed below (species marked with

TABLE 5. Use of Nitrogen Sources by the *Trebouxia* Phycobiont of *Buellia pernigra* Darbishire

Nitrogen Source	Concentration of Nitrogen Source, g/l	Dry Weight of Alga,* mg	*p*H	
			Initial	Final
Peptone	10.00	30.5	6.25	5.72
NH_4Cl	5.72	4.7	5.19	4.55
NH_4NO_3	4.28	4.2	5.31	3.80
KNO_3	10.82	4.0	4.97	5.48
$(NH_4)_2SO_4$	7.08	3.7	4.37	4.03
Asparagine	7.08	3.4	4.92	4.58
Control	...	3.3	4.89	5.69
Arginine	4.66	3.1	6.75	3.65
Urea†	3.22	0.9	6.43	6.25

The alga was grown for 44 days in 125-ml flasks containing 25 ml of Trebouxia liquid medium (without soil water) at 10°C and constant illumination of 260–390 lux. The inoculum dry weight was 0.8 mg.

* Values are averages of values for four flasks.

† Filter sterilized.

asterisks were collected by Dr. Rudolph and isolated by Mrs. Yang in February 1964 and identified by Dr. Dodge):

Species	Collection Site
Caloplaca sp.	Ross Island
Polycauliona citrina Dodge*	Cape Hallett
Buellia pernigra Darbishire*	Cape Hallett
Parmelia coreyi Dodge & Baker	Ross Island
Rinodina frigida (Darbishire) Dodge*	Cape Hallett
Xanthoria mawsoni Dodge*	Cape Hallett

The algae from the Ross Island specimens were isolated in 1958 and maintained in culture at 20°C [Ahmadjian, 1958]. Those from Cape Hallett were maintained at 4°–5°C. Several to numerous clones, each clone derived from a single cell, were obtained from all of the Cape Hallett specimens except *Polycauliona citrina,* from which only one clone was obtained.

This experiment employed 50-ml Erlenmeyer flasks, each flask containing 20 ml of Trebouxia liquid medium (no soil water), in which peptone is the source of nitrogen. The initial *p*H of the medium was 6.28.

The inoculum was taken from colonies grown at 10°C on slants of Trebouxia agar. A portion of a colony was homogenized in sterile distilled water in a Waring blender, from which equal aliquots containing an average of 0.6–0.8 mg of algal cells were pipetted into the experimental flasks. The inoculated flasks were placed in the light (about 430 lux at the illuminated surface of the flask) or in the dark. Dry weight determinations were made after 28 days of growth. The values obtained (Figure 11) are averages of the weights of the algae from four flasks.

Discussion. The preceding results are discussed below.

Nitrogen studies: Although fewer nitrogen compounds were employed in the studies with the phycobionts than in the study with the mycobiont *L. tephroeceta,* some interesting comparisons can nevertheless be made. The generalizations about nitrogen source use by antarctic lichen fungi appear to be inapplicable to the algae of lichens from similar (i.e., nitrogen rich) habitats. For example, peptone was by far the preferred nitrogen source for the phycobionts, which were isolated from lichens collected near the Cape Hallett penguin and skua breeding areas, and neither the ammonium compounds nor the urea was a better nitrogen source for the algae than nitrates (in sharp contrast to the results for the mycobiont *L. tephroeceta*).

Asparagine was a very good source of nitrogen for the phycobiont of *P. citrina* but was a poor one for that of *B. pernigra.* The results obtained with arginine suggest a strong *p*H effect that can be seen directly in *B. pernigra,* for which a final *p*H reading was made (Table 4), and indirectly in *P. citrina,* which attained 3 times as much growth in the medium with one-half the concentration of arginine and therefore with presumably less adverse *p*H conditions (Figure 10). The two algae would probably have achieved much greater growth in a buffered medium of lower *p*H.

The results of these nutritional studies suggest that the use of many types of nitrogenous compounds by the symbionts of antarctic lichens may be complementary. Nitrates and in one case asparagine are exceptions to this generalization; neither the algae nor the fungus used nitrate to any significant extent. The generalization must be restricted of course to the compounds on which the algae were grown. Only further research with other symbionts and other compounds will disclose the degree to which 'complementarity' may be the rule. (In intact lichen thalli, nitrate may be reduced symbiotically, as molecular nitrogen is in the root nodules of legumes; there is absolutely no information on this point, however.)

The results may be interpreted in another way. The algae tested appear to be more fastidious in their nitrogen requirements than the fungus: a complex nitrogen compound not likely to be abundant in the environment (peptone) was by far the best nitrogen source for both of the algae. On the basis of the experimental results obtained, it may be suggested

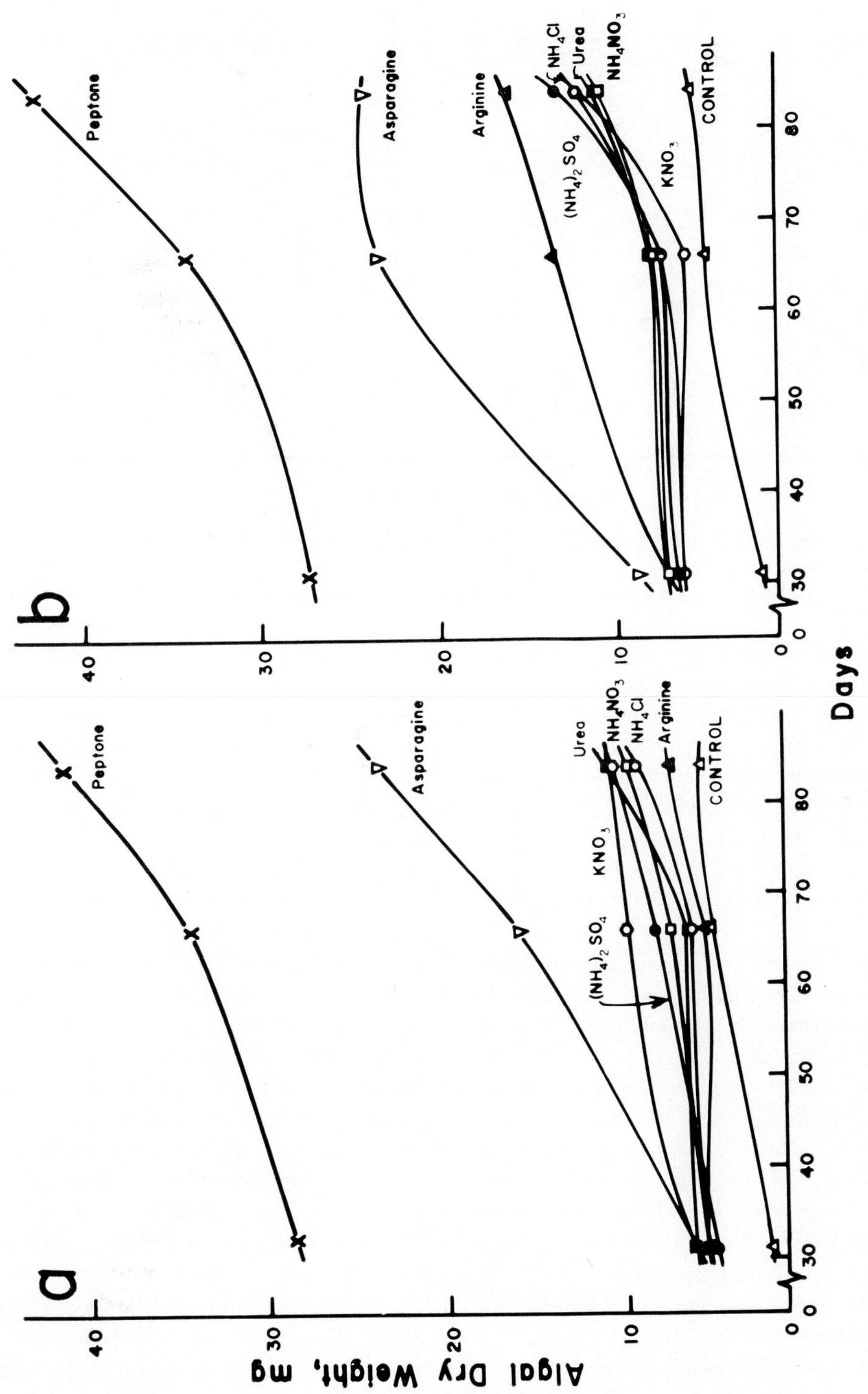

Fig. 10. Nitrogen source use by the phycobiont (*Trebouxia* sp.) of the lichen *Polycauliona citrina* Dodge. (*a*) Full concentration. (*b*) Half concentration.

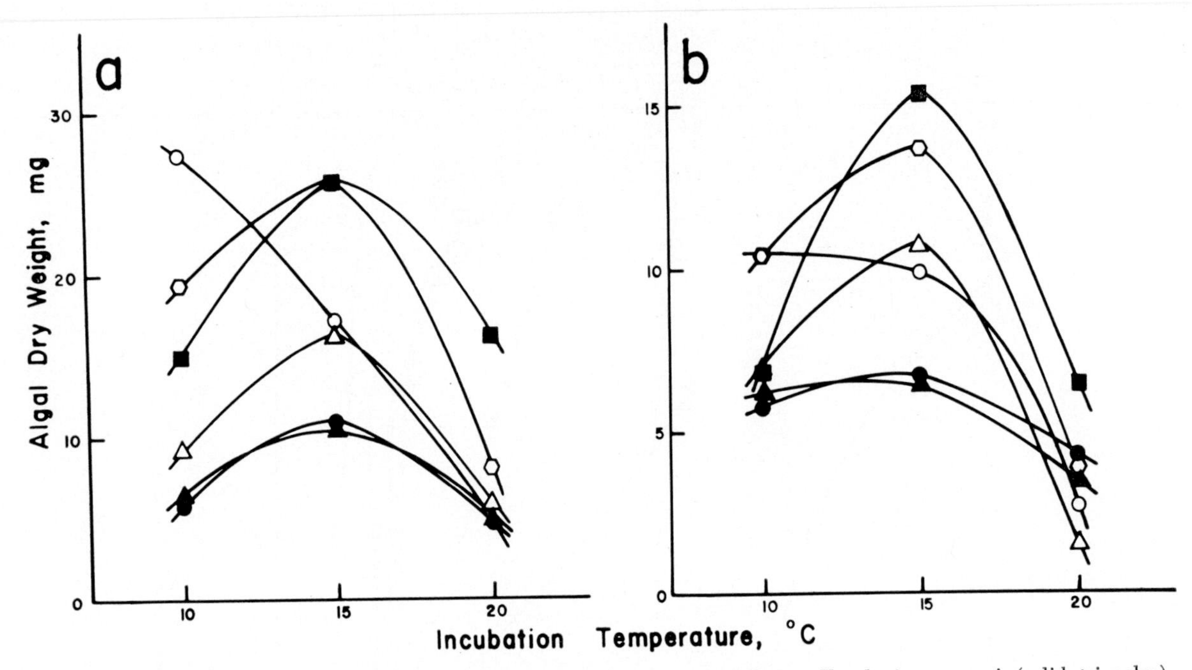

Fig. 11. Growth of the phycobionts (*Trebouxia* spp.) of six antarctic lichens: *Xanthoria mawsoni* (solid triangles), *Caloplaca* sp. (open triangles), *Rinodina frigida* (solid circles), *Polycauliona citrina* (open circles), *Parmelia coreyi* (squares), and *Buellia pernigra* (hexagons) at 10°, 15°, and 20°C. (*a*) In the light. (*b*) In the dark.

that (1) antarctic lichen algae are largely dependent on their cosymbionts for a source of nitrogen (except of course for the few antarctic lichens with blue-green phycobionts) and (2) they have stricter requirements for nitrogen sources than their cosymbionts. Since the sampling of species and compounds was small and since the symbionts being compared came from different species of lichens, these suggestions must be taken with caution. More research is clearly called for.

A possible route for nitrogen in lichen nutrition is as follows. Environmental nitrogen (e.g., ammonium, nitrate, urea, purines, amino acids) is absorbed and assimilated by the mycobiont, which then synthesizes and excretes more complex nitrogenous compounds (e.g., polypeptides) into the thallus or in some other way passes them to the phycobiont. This scheme is consistent with the results. Such an arrangement would protect the phycobiont from detrimental factors in the environment yet would constitute a means of transporting nitrogen from the environment to the phycobiont via the (presumably) less sensitive mycobiont.

Temperature studies: The results of the temperature studies strongly suggest that antarctic lichen algae grow best at about 15°C; a very pronounced decrease in dry weight yield occurs above and below this temperature. Of the six algae tested only one had a lower (undetermined) optimum.

There was a remarkable consistency between the light and dark experiments in the grouping of the algae according to dry weight yield. This consistency was most obvious at 15°C, where the algae fell into three discrete groups (in order of decreasing dry weight): (1) *P. coreyi* and *B. pernigra*, (2) *P. citrina* and *Caloplaca* sp., and (3) *R. frigida* and *X. mawsoni*. The dry weights at 15°C of the algal cells in the light experiment were almost precisely 1.67 times those in the dark experiment; thus the three groups retained their relative positions under both light and dark experimental conditions. The maverick alga (from *P. citrina*) was stimulated by light at 10°C far more than the other algae, and it appears from the data that its optimum temperature is $<10°C$; at $\geqslant 15°C$ it behaved exactly like the other algae. The temperature curves for the most slowly growing pair (*R. frigida*–*X. mawsoni*) have less well defined optimums than those for the other algae.

The growth of the two lichen fungi (Figure 9) and of these six lichen algae was somewhat similar between 10° and 20°C; except for the phycobiont of *P. citrina* both the algae and the fungi grew less near 10° than near 15°C. However, the growth of the fungi was not depressed at 20°C, whereas that of the

Fig. 12. The orange lichen *Caloplaca elegans* var. *pulvinata* growing on dark kenyte rocks and pebbles on Cape Royds, Ross Island. The photograph was taken with high speed infrared-sensitive film through a Kodak Wratten no. 89B filter, which passed only the near-infrared wave lengths. The two arrows are 6 cm apart.

algae was. In addition the results do not indicate whether the algae have a second temperature optimum below 10°C. One fact is most clear, however; antarctic lichen algae grow best below 20°C and are thus obligate psychrophiles according to the definitions of Stokes [1963], Morita and Haight [1964], Baxter and Gibbons [1962], and Jacobs et al. [1957].

Although the sample of species used in these simple growth experiments is too small to serve as a basis for generalizations, the results obtained indicate that the temperature tolerances of antarctic lichen algae are narrower than those of antarctic lichen fungi and that the growth characteristics of lichen algae may therefore decrease the ecologic amplitudes of the composite lichen organisms in which they are indispensable partners.

Interrelations between light intensity and temperature in the field: It is impossible to duplicate exactly in the laboratory the environment that organisms encounter in their natural habitats. This difficulty is true with respect to light characteristics and temperature interrelationships.

The 'light intensity' (i.e., illumination) range for the optimal growth of *Trebouxia* phycobionts is about 1600–2700 lux [Ahmadjian, 1967]; however, illuminations as high as 50,000 lux frequently occur during midday in Antarctica and may exceed 60,000 lux. Ahmadjian [1967] reports that *Trebouxia decolorans*, the phycobiont of *Buellia punctata* and of *Xanthoria parietina* (nonantarctic species), ceases to grow and begins to lose its chlorophyll at illuminations above 1000 lux.

Although it is not safe to assume that the physiological states of phycobionts grown in liquid culture are at all like those of phycobionts in intact lichen thalli growing in their natural habitats and although data obtained for one strain or species of alga cannot be used to interpret observations made for another, it is possible to make some broad comparisons. Unless antarctic *Trebouxiae* are completely different from nonantarctic *Trebouxiae*, the laboratory observations on isolated phycobionts reported by Ahmadjian and the measurements of field conditions under which antarctic lichens grow are at first glance irreconcilable.

Figure 12 suggests one possible explanation for the disparity between the laboratory and field observations. The near-infrared wave lengths of about 700–900 nm (the maximum sensitivity being 770–840 nm) are recorded in the photograph, which reveals that the lichen thallus absorbs much less solar radiation in this range than the darker substrate. This difference in near-infrared albedo should result in a difference in the surface temperature of the thallus and should also greatly reduce the amount of sunlight reaching the *Trebouxia* phycobiont in the thallus. Table 6 indicates that there is indeed a measurable difference in temperatures. The important fact is that the temperatures of the lichens remained <15°C (the temperature optimum for all the algae

TABLE 6. Lichen Thallus and Adjacent Rock Surface Temperatures

Date and Time Species Observed	Location	Rock Color	Weather	Illumination, klux	Solar Radiation Intensity, ly min^{-1}	Temperature,[c] °C	
						Lichen	Rock
		C. elegans var. *pulvinata*					
Dec. 28, 1968, 1430 hours	Cape Royds, Ross Island	dark gray to black	partly cloudy	...	0.6–1.0	15.3[d]	16.3[d]
Jan. 2, 1968, 1200 hours	Cape Royds, Ross Island	dark gray to black	sunny	...	...	12.0(n); 16.0(s)	16.0(n); 15.0(s)
Jan. 23, 1968, 1330 hours	Cape Crozier, Ross Island	dark brick red	sunny	...	...	13.8–15.7	18.0–18.2
		Buellia frigida					
Dec. 14, 1968, 1920–1950 hours	Cape Crozier, Ross Island	dark brick red	sunny	2.2–2.8	0.43–0.47	12.3–13.8	14.1–15.5
Dec. 16, 1968, 1538–1610 hours	Cape Crozier, Ross Island	dark brick red	cloudy	1.5–1.7	...	12.6–13.3	13.2–14.2
Jan. 27, 1969, 1000 hours	Kar Plateau, Victoria Land	dark red	foggy	1.0[b]	0.31–0.49	−1.0 to +1.8[d]	−1.0 to +2.1[d]
Dec. 3, 1968, 1330–1406 hours	Miers valley area, Victoria Land (770 meters elevation)	light gray	cloudy, sun visible[a]	...	0.29–0.82	1.4–4.0	1.6–5.0
Jan. 17, 1969, 2222–2224 hours (west facing surface)	Miers valley area, Victoria Land (770 meters elevation)	light gray	cloudy, sun visible	0.3	0.1	−0.2 to −0.3[d]	−0.8[d]
Jan. 17, 1969, 2300 hours (south facing surface)	Miers valley area, Victoria Land (770 meters elevation)	light gray	cloudy, sun visible	0.3	0.09	−0.9 to −1.9[d]	−2.4 to −1.7[d]

Three center dots indicate that no measurement was taken.

Illumination was determined with a Gossen Tri-Lux foot-candle meter (P. Gossen & Co., Erlangen, West Germany) with a 20× opal multiplier disk over the receiving surface.

Solar radiation intensity was estimated from the trace of a recording pyrheliometer (Belfort Instrument Co., Baltimore, Maryland, catalog no. 5-3850). The instrument was mounted in a horizontal position within 300 meters of the surfaces being studied.

[a] There were also periods of bright sunlight and of snow flurries.

[b] Value is for sunlight falling on a nearby horizontal surface; all other illumination values are for sunlight measured perpendicular to the surface being studied.

[c] The 'n' and 's' distinguish temperatures of north and south facing surfaces, respectively.

[d] Value obtained with a Yellow Springs Instruments telethermometer, model 44-TE, with a probe 408; all other temperatures were measured with a Wallac Thermex thermoanemometer, model GGA2B, with surface temperature probe NI-103.

under most conditions of light and darkness) most of the time, whereas the temperatures of the rocks were often >15°C. The fact that this difference in temperature was smaller or nonexistent under cloudy conditions is consistent with the image produced on the infrared sensitive film (Figure 12), since the infrared portion of the solar spectrum (which accounts for about half the energy that impinges on the surface of the earth) is absorbed by clouds.

The significance of this mechanism, if it can be so called, is revealed by the fact that *Xanthoria mawsoni* at Cape Hallett photosynthesized maximally between −2° and +14°C, whereas it photosynthesized at an extremely low rate outside this temperature range [Gannutz, 1969].

From this preliminary and admittedly fragmentary evidence, antarctic lichens appear to have reflectance properties similar to those of nonantarctic lichens [cf. Gates, 1965] and of many other plants. For numerous reasons high reflectance in the infrared would be advantageous to plants growing in the arid cold deserts of Antarctica. High reflectance of near-infrared light would shield the algal symbiont from excessive light, the photosynthetic wave lengths would not be interfered with, the thallus temperature would be kept within the optimal range for the photosynthesis and growth of the alga, and moisture would presumably be conserved because it would evaporate more slowly from the cooler thallus.

Prasiola crispa

Introduction. *Prasiola crispa* was studied as follows.

Balleny Islands reconnaissance expedition, 1964: On March 9, 1964, a helicopter landing was made on Sabrina Island, the Balleny Islands (66°54′S, 163°20′E) from the USS *Glacier* for the collection of plants and other scientific purposes [Dawson et al., 1965]. Terrestrial algae were found on the western slopes of the isthmus between sea level and about 50 meters. Myxophyceae and Chlorophyceae (*Prasiola crispa* (Lightf.) Kütz. spp. *antarctica* Kütz.) were collected near the top of the slope. *Prasiola* was common on most of the slope; it appeared to be lichenized at the top of the ridge. Lichens were very rare, possibly because of high concentrations of salts or because of the scarcity of suitable substrate; in fact there were only a few very depauperate sterile specimens of a *Xanthoria*-like genus near the top of the ridge. No mosses could be found; previous reports of 'moss' are probably based on the abundant *Prasiola* there. Ice colored green by microscopic algae was collected and returned frozen to the United States, as were soil samples from the top of the ridge. Kol and Flint [1968] have reported on some algal isolates from the green ice, including a new species of *Chlamydomonas*, and Wise [1964] has reported on three species of mites that he isolated from algal and soil samples collected at the time.

Prasiola crispa is one of the most common algae in Antarctica [e.g., Skottsberg, 1905; Hirano, 1965]. In places it may cover large expanses. For example, Rudolph [1963] found that it covered approximately 12.8% (the range was 0.1–89.0%) of a quadrat of about 4200 m^2 on Seabee Hook near Cape Hallett, an area greater than that covered by any other species of plant. Rudolph attributed its abundance to moisture and to nitrogen from penguin and skua guano. Syroechkovskii [1959] arrived at a similar conclusion for Haswell Island.

Prasiola crispa (alga AA-115) collected frozen from the large Adélie penguin rookery near Cape Crozier, Ross Island, contained 5.50% reduced nitrogen, as we determined by microkjeldahl analysis, a proportion greater than that contained by any other type of plant analyzed (cf. Table 2).

Experimental data. The results of our experiments were as follows.

Isolation: Prasiola crispa was isolated from a subsample of the green ice collected on Sabrina Island in a liquid medium containing ammonium oxalate. It was used in a study of nitrogen source preference. The first isolation attempt was made on April 19 from the melted subsample. The *p*H of the melted ice was about 5.5 some 20 hours after it was placed in constant light at 12°–13°C on April 18. It had a distinctly fishy odor.

On April 19 loopfuls of the resulting algal suspension were placed on agar slants consisting of soil extract, NH_4NO_3, urea, and Bristol's solution. The inoculated slants were then incubated in constant light at 12°C. No growth occurred. In a solution consisting only of uric acid, however, growth did occur.

The remaining sample was refrozen on April 20. On June 1 a second attempt was made to isolate algae on the following media: (1) a saturated filter-sterilized aqueous solution of uric acid, with and without Bristol's solution; (2) a filter-sterilized 0.1 *M* solution of urea, with and without Bristol's solution; (3) an autoclaved 0.5% solution of ammonium oxalate, with and without Bristol's solution; (4) a filter-sterilized 0.5% solution of ammonium oxalate, with and without Bristol's solution; (5) Cyanophycean medium (liquid) [Starr, 1964]; and (6) Beneche's nutrient solution [Cavanaugh, 1956].

The basic media were dispensed in 50-ml aliquots to 125-ml Erlenmeyer flasks. Five ml of Bristol's solution were added to half of the flasks of media 1–4, there being a total of 55 ml of medium per flask.

Growth experiments: On October 27 pieces of *Prasiola crispa* approximately 0.5–1.0 cm in diameter, which had developed in medium 3 (with 5 ml of Bristol's solution), were transferred to 25-ml flasks containing 15 ml of a medium consisting of equal amounts of soil extract diluted 1:3 with distilled water and of Bristol's solution minus $FeCl_3$. By the time of this transfer the formerly bright green algae had begun to turn yellow.

The inoculated flasks were incubated at about 5°C in continuous light for 2 days, after which 2 ml of the following solutions were added: (1) uric acid (filter-sterilized Bristol's solution, 0.0665 g/l of uric acid replacing the $NaNO_3$), (2) xanthine (filter-sterilized Bristol's solution, 0.0602 g/l of xanthine replacing the $NaNO_3$), and (3) urea (filter-sterilized Bristol's solution, 0.475 g/l of urea replacing the $NaNO_3$).

In each case 20 ml/l of phosphate buffer at *p*H 7.0 were used instead of the prescribed phosphate salts, and one drop of a 1% aqueous solution of $FeCl_3 \cdot 6\,H_2O$ was added to each liter of the resulting solution. An equal number of flasks containing the above media plus 2% agar were also inoculated, but growth was poor because of contaminating microorganisms.

The flasks were incubated at about 5°C in constant light of approximately 1500 lux. Growth was excellent in all three liquid media; numerous long, narrow, bright green thalli arose from the original dying inoculum.

The algal thalli resulting from growth in the preceding three media were combined, washed with sterile distilled water, added to about 350 ml of nitrogen-free Bristol's solution, and homogenized in a Waring blender for about 15 sec. On March 18, 1965, the resulting algal suspension was pipetted in 2-ml aliquots to the following media: (1) equimolar (3.5×10^{-4} *M*) filter-sterilized solutions of uric acid, xanthine, and allantoin made up in Bristol's solution minus the $NaNO_3$ and containing equal amounts (0.0196 g/l) of nitrogen (0.0588 g/l of uric acid, 0.0532 g/l of xanthine, or 0.0553 g/l of allantoin; the uric acid and the xanthine were first dissolved in a small amount of 0.5 *N* NaOH); (2) solutions of urea, NH_4Cl, NH_4NO_3, and $NaNO_3$ made up in Bristol's solution minus the $NaNO_3$ and containing approximately 0.1648 gram of nitrogen per liter, the amount of nitrogen in 1 gram of $NaNO_3$ (0.353 g/l of urea, 0.629 g/l of NH_4Cl, 0.4708 g/l of NH_4NO_3, 1.000 g/l of $NaNO_3$); (3) a double strength (0.9416 g/l) solution of NH_4NO_3 made up in nitrogen-free Bristol's solution; (4) a solution consisting of autoclaved nitrogen-free Bristol's solution plus 0.5 g/l NH_4NO_3, the *p*H being adjusted to 6.0, and filter-sterilized nitrogen-free Bristol's solution plus 0.0266 g/l xanthine and 0.1765 g/l urea, the *p*H being adjusted to 6.0; the first part of the solution was added to the second part in the ratio 1:28, the final xanthine concentration being about 0.9 mg/l and the final urea concentration about 5.9 mg/l; (5) medium 1 with xanthine plus 0.0121 g/l of gibberellic acid (75% potassium salt); (6) medium 2 with NH_4NO_3 plus 0.0121 g/l of gibberellic acid (75% potassium salt); (7) medium 2 with $NaNO_3$ plus 0.0121 g/l of gibberellic acid (75% potassium salt); (8) medium 2 with $NaNO_3$ plus 0.0753 g/l of kinetin (dissolved first in 25 ml of 0.1 *N* HCl); and (9) Bristol's solution plus 100 μg/l of thiamine and/or 5 μg/l of biotin.

The algae, except for those on medium 4, were incubated in continuous fluorescent light for 51 days at 15°C in 25-ml Erlenmeyer flasks containing 20 ml of the medium. For medium 4, 50-ml flasks and 30 ml of the medium were used. Each flask was inoculated by pipette with 2 ml of the blended algal suspension. The average dry weight of the inoculum was <0.1 mg per flask (from 21 dry weight determinations). The results are presented in Table 7.

A separate series was set up at the same time for a rough comparison of growth at 5°, 10°, 15°, and 20°C. For medium 4, 50-ml flasks and 30 ml of medium were used. Since it was impossible to provide the identical illumination for all flasks because of technical limitations, only visual estimates of growth were made. No growth occurred at 20°C; the thalli were white and presumably dead. The thalli that grew at 10°C were yellowish green, and there was little apparent growth. The thalli that grew at 5°C were light green; they appeared to be healthy and to be growing actively.

Discussion: The results clearly indicate that *P. crispa* grows best on uric acid, which is the principal form in which nitrogen is excreted by birds. With uric acid as the nitrogen source the alga was large, healthy, and bright green. At the time of collection there were numerous moulting penguins on Sabrina Island, and there was a rookery of some 1500–2000 birds on part of the island [Dawson et al., 1965]. Cape pigeons (*Daption capensis*) and snow petrels (*Pagodroma nivea*) were also nesting on the island. These birds were undoubtedly a source of uric acid.

Prasiola crispa is found almost exclusively in moist places near bird rookeries. An examination of specimens in a number of herbaria confirmed the occurrence of this species in guanotrophic habitats, as a review of the literature on *P. crispa* from antarctic, arctic, and north temperate locations also did. The present experimental results indicate that uric acid is one important factor responsible for the observed correlation between bird life and the distribution of *P. crispa.*

The compounds most closely related to uric acid (xanthine and allantoin) supported significantly less growth than uric acid. Allantoin, in particular, was a poor source of nitrogen. Urea, ammonium salts, and nitrate salts supported approximately the same amount of growth as xanthine. The plant growth substance gibberellic acid had a stimulatory effect in combination with xanthine, an inhibitory effect in combination with $NaNO_3$, and no significant effect in combination with NH_4NO_3. Kinetin had a very definite growth-inhibiting effect in combination with $NaNO_3$. Thalli grown in the presence of kinetin were small and yellowish green to brownish green.

TABLE 7. Growth of *Prasiola crispa* from the Balleny Islands on Various Nitrogen Sources and with Added Growth Factors

Nitrogen Source and Additions	Medium	Concentration of Nitrogen, g/l	Dry Weight of Alga,* mg	Final *p*H†	Remarks
Uric acid	1	0.019	3.5 (3.3–3.6)	7.0	bright green large thalli
Xanthine	1	0.019	1.5 (1.4–1.6)	6.2	
Allantoin	1	0.019	0.9 (0.8–0.9)	6.2	light green
Urea	2	0.165	1.4 (1.1–1.6)	6.2	
NH_4NO_3	2	0.165	1.3 (1.2–1.4)	4.5	
NH_4NO_3	3	0.330	1.2 (1.1–1.5)	4.9	
$NaNO_3$	2	0.165	1.7 (1.3–2.2)	6.7	
NH_4Cl	2	0.165	1.2 (1.0–1.5)	4.5	
Xanthine (0.9 mg/1), urea (6.0 mg/1), and NH_4NO_3 (0.5 g/1)	4	0.172	1.5 (1.4–1.5)	6.0	
Xanthine and gibberellic acid	5	0.019	2.3 (1.7–2.8)	6.1	
NH_4NO_3 and gibberellic acid	6	0.165	1.4 (1.1–1.5)	4.5	
$NaNO_3$ and gibberellic acid	7	0.165	1.2 (1.1–1.4)	6.5	
$NaNO_3$ and kinetin	8	0.165	0.4 (0.4–0.6)	6.1	yellowish to brownish green; least growth
$NaNO_3$ and biotin	9	0.165	2.2 (1.9–2.4)	...	contaminated with a white yeast
$NaNO_3$ and thiamine	9	0.165	3.1 (2.8–3.7)	7.0	contaminated with a white yeast
$NaNO_3$, biotin, and thiamine	9	0.165	3.5 (2.8–4.1)	7.1	contaminated with a white yeast

Prasiola crispa spp. *antarctica* from the Balleny Islands (66°53′S, 163°19′E) was originally isolated in a medium containing ammonium oxalate. The original sample was in the form of green ice [Kol and Flint, 1968]. The alga was grown in a liquid medium for 51 days at 15°C and 970 lux of continuous fluorescent light.

* Values are the averages for three flasks. The ranges of weights are given in parentheses. The dry weight of the inoculum (average of 21 aliquots) was <0.01 mg per flask.

† Values are the *p*H values of the combined filtered medium from three flasks.

The results obtained with the vitamins thiamine and biotin are unreliable because the vitamins stimulated the growth of what appeared to be a single species of yeast that was associated with the alga. Although the dry weight was significantly greater when vitamins were present, the conclusion that vitamins per se stimulated the growth of the alga on nitrate is untenable because it is quite likely that the yeast transformed the nitrate to a more suitable nitrogen compound. That possibility cannot at least be ruled out in this case. Microscopic examination of the contaminated growth medium indicated that the same yeast may have been involved in each case and that the contamination seemed to be essentially a pure culture of one yeast, although a definite identification was not attempted. As seen through a microscope, the yeast was white, and some of the cells had longitudinal divisions. There was some pseudomycelium having yeastlike conidia at the tips. Because the standard mycologic techniques necessary for identifying yeasts were not employed, it cannot be stated that only one species was present. Since the inoculum used in this study came from material first isolated in a medium containing oxalate, other organic acids (particularly glyoxylic acid, which is produced by the enzymatic hydrolysis of allantoin, and glycolic acid, a well-known extracellular compound related chemically to oxalic acid) should be employed in future studies of *P. crispa*. Oxalic acid (COOH-COOH) is produced in a single metabolic step from glyoxylic acid (CHO-COOH). Guano and soils from rookery areas should be tested for the presence of these and other organic acids.

The results of this experiment could be rejected because the material was merely unialgal and nonaxenic. The following facts indicate that except for the results for the media containing vitamins the results can be accepted with relative confidence. (1) The basic medium was essentially a mineral solution (i.e., it contained no carbohydrate or carbon source); thus only autotrophic organisms could grow in it. (2) No visible growth of contaminating organism(s) occurred except in the media containing vitamins. (3) Neither xanthine nor allantoin supported as much growth of *Prasiola* as uric acid; thus uric acid was probably absorbed by the alga intact rather than after transformation by contaminating organisms to some other nitrogen compound. In addition, neither urea nor ammonium supported as much growth as uric acid. It is safe to state that contaminating microorganisms were present only in negligible amounts and had a negligible influence on the results, except in media containing biotin and/or thiamine.

Holm-Hansen [1964] and Kol and Flint [1968] failed to isolate *P. crispa* from samples from the McMurdo Sound area and the Balleny Islands, although *Prasiola* was seen in preserved samples from Cape Crozier (probably *P. crispa*) and from Taylor valley (possibly *P. calophylla*) by Holm-Hansen and was isolated by the authors from a subsample of the green ice used by Kol and Flint. None of the media used by Holm-Hansen and Kol and Flint contained uric acid, urea, or organic ammonium salts. Their failure to obtain *Prasiola* can reasonably be attributed to their use of unsuitable nitrogen sources.

Ecology of P. crispa: There is already a great deal of information on the ecology of *P. crispa* in both antarctic and nonantarctic regions. Virtually all observations support the conclusion that this very interesting species is highly nitrophilous and usually ornithocoprophilic, e.g., Petersen [1928, 1935], Grønlie [1948], Letts [1913], and Knebel [1935] in nonantarctic areas and Lamb [1948], Rudolph [1963], Longton [1967], and Gimingham and Lewis Smith [1970] in antarctic areas. *Prasiola crispa* can accumulate large quantities of nitrogen: $\leqslant$8.94% in Northern Ireland [Letts, 1913] and $\geqslant$5.50% in Antarctica (E. Schofield, unpublished data, 1969). Its ecology can be further characterized as (1) hydrophilous and (2) halotolerant. Justification for the first characterization is provided by the observation that *P. crispa* is found exclusively in standing (Seabee Hook, Cape Hallett) or flowing (Cape Crozier) water, in locations where there is frequent precipitation or blowing snow (Cape Crozier), or on soil that is otherwise very moist (Cape Crozier and Cape Royds). Further confirmation is found in Gannutz' discovery that *Prasiola* requires 5–7 days in the wet condition after the winter season to become photosynthetically active [Ahmadjian, 1970].

Justification for the second characterization is supplied by Gimingham and Lewis Smith [1970], who reported that it is the only macroscopic plant capable of colonizing the mud of penguin colonies and elephant seal wallows, where trampling, excavating, and extremely high concentrations of marine and organic salts make conditions unsuitable for most organisms. Justification is also supplied by data showing that penguin guano has very high concentrations of water-soluble salts (the mean electrical conductivity of guano extracts from two penguin colonies on Ross Island was 16,000 μmhos $\cdot$ cm^{-1}) (Table 1). Skuary soils were less saline by 1 order of magnitude.

Bryum algens

Introduction. A specimen of moss collected in the Jones Mountains, Eights Coast, West Antarctica, by Mr. Kelvin P. Rennell and referred by Dr. Stanley W. Greene on the basis of cultured material to *Bryum algens* Cardot was isolated into axenic culture and used in laboratory studies of its nutritional characteristics.

The type specimen of *B. algens* was collected in the Granite Harbor area of southern Victoria Land in 1902. The species has since been collected from (among other localities) McMurdo Sound (islet in old ice) in East Antarctica, Deception Island and Petermann Island in maritime West Antarctica, and Signy Island, where it forms compact cushions along the courses of melt water streams, in sheltered rock crevices, and in other moist or permanently wet habitats influenced by basic rocks and soils in the Moss Hummock subformation of the maritime Antarctic [Gimingham and Lewis Smith, 1970; Longton, 1967].

Dixon [Dixon and Watts, 1918] and Clifford [1957] have reduced *B. algens* to synonymy with *B. antarcticum*, from which they claim it is indistinguishable. Greene, Horikawa, and Ando [Horikawa and Ando, 1967] do not concur in this judgment, however.

Isolation. Eight clones of *B. algens* were isolated into axenic culture on March 1, 1964, by cutting the newly elongated tips of about 20 gametophytic green shoots from a clump of moss, rinsing them in ethanol diluted with sterile distilled water, and placing them on Trebouxia agar [Ahmadjian, 1967] with biotin (5 μg/l) and thiamine (100 μg/l). The eight contamination-free isolates were then transferred to Knudson's medium in Erlenmeyer flasks covered with metal caps and polyethylene film [Ward, 1960]. Figure 13 is a photograph of one of the clones in axenic culture.

Experimental data. The following results were obtained for *Bryum algens.*

Introduction: Three preliminary studies were made

Fig. 13. Clonal isolate of *Bryum* cf. *algens* from Jones Mountains, Eights Coast, West Antarctica, in axenic culture [cf. Gressitt et al., 1964].

of the response of the isolates to nitrogen sources, growth factors, vitamins, and adsorbing and chelating agents. The objects were to (1) survey the nitrogen source preferences of *B. algens*, (2) determine whether growth factors induce the sporophytic stage of the organism, which apparently is sterile in Antarctica, (3) detect any requirements for vitamins that may play a role in limiting its distribution, (4) determine whether mineral deficiencies or mineral toxicity influences its distribution, and (5) assess the range of morphologic variation that the preceding factors may produce.

In the first study the eight clones were grown on one of a number of media in an attempt to determine the over-all variability of the species. Three basic media were used, and were modified by adding so-called microelements and urea. In the second study five of the eight clones were transferred to a medium containing soil extract with or without a minerals solution, carbon source, and/or adsorbing or chelating compound. In the third study seven nitrogen sources, three plant growth substances, two water-soluble vitamins, a chelating agent, and a vitamin-adsorbing compound were added to the basic medium separately and in combination. In addition the moss was cultured with an imperfect fungus that had occurred in large numbers in moss communities on Seabee Hook, Cape Hallett, to detect any mutual effects on the growth of the two organisms. Since other workers [e.g., Sironval, 1947; von Maltzahn and MacQuarrie, 1958] had found that fungi (e.g., *Penicillium*) stimulated the growth of mosses, it was possible that the fungal isolate from Antarctica was influencing the growth of the associated moss at Cape Hallett.

First experiment: In this preliminary experiment the eight clones were grown on one of three basic media that had been either unmodified or modified by the addition of certain compounds or elements. Clone 1 was grown on Knudson's medium [Ward, 1960] without additions, as were clones 2–4. Clone 5 was grown on Knudson's medium plus microelements [Ward, 1960]. Clone 6 was grown on Bristol's solution agar with 0.001 *M* urea (added as a filter-sterilized solution), and clone 7 was grown on unmodified Bristol's solution agar. Finally clone 8 was grown on soil water agar. All the media except the Bristol's solution and the Bristol's solution plus urea were solidified with 15 grams of agar per liter of final medium; for the two exceptions 20 grams of purified agar were used.

The moss was first incubated at a temperature of 11°C under continuous cool white fluorescent light. During an unavoidable interim period of about 9 days it was grown at various temperatures, under fluctuating light intensities, and in different photoperiodic regimes (for 4 days in the dark at 6°C or under fluctuating illumination at 15°C and then for 5 days with an approximately working day diurnal photoperiod at room temperature). The moss was then incubated at 12°–14°C under continuous cool white illumination.

Second experiment: In the second experiment clones 1, 2, 3, 4, and 6 were grown in four flasks of each of the following five media: (1) soil water agar made according to the directions in Ahmadjian [1967, p. 121] (medium 1a) and soil water agar darker in color than medium 1a (medium 1b); (2) soil extract agar (soil water plus distilled water, 3:4) plus Bristol's solution, 10:1; (3) medium 2 plus 1 g/l of glucose; (4) medium 2 plus 10 g/l of activated carbon (Norit-A); and (5) medium 2 plus about 0.01 g/l of gibberellic acid (75% potassium salt). To each of these media except the last, 15 grams of purified agar (Difco) were added per liter of the medium. Ordinary agar (Difco Bacto-agar) was used with medium 5. The inoculated flasks were placed in continuous cool white fluorescent light at a temperature of approximately 14°C.

Third experiment: Clone 6 was used in a third study of nutritional response. Uric acid, xanthine, allantoin, urea (in three concentrations), NH_4Cl, $NaNO_3$, NH_4NO_3 (in two concentrations), gibberellic acid, kinetin, thiamine, and biotin were supplied in the amounts given for the nine experimental media employed in the study of *Prasiola crispa.* Indole-3-acetic acid (0.06131 g/l, filter sterilized), Norit-A (5 g/l), and ethylenediaminetetraacetic acid (EDTA), the EDTA being dissolved first in KOH in a 50:31 ratio, were added to Bristol's solution alone and in various combinations. Urea was supplied in concentrations of 0.021, 0.042, and 0.353 g/l, and NH_4NO_3 in concentrations of 0.4708 and 0.9416 g/l. A fungus (AF-132) isolated on malt extract agar from soil under moss near Hallett Station was inoculated onto Bristol's solution agar (solidified with 20 grams of purified agar per liter), and the moss was added 2 days later. The fungus was very common in the vicinity of the Adélie penguin rookery on Seabee Hook at Cape Hallett and had been isolated from various substrates on a number of occasions. It resembled most closely members of the form genus *Phoma.*

The inoculated flasks were incubated at 15°C for the first 55 days in continuous unilateral cool white fluorescent light. The illumination was not measured and was not exactly the same in every flask, but it was most likely 250–1000 lux in all cases. After the first 55 days the incubation temperature was 10°C; diffuse cool white fluorescent light was used.

In all three experiments the media were inoculated with the moss by transferring small pieces of the eight clones with a sterile inoculating spear. Since the usual procedures for measuring dry weight were unsuitable for *B. algens,* which had to be grown on a solid medium, yield was estimated qualitatively by noting the color and the conditions of the shoots and protonemata and semiquantitatively by measuring the diameter of growth, counting the number of new shoots, and estimating the sizes of the 'leaves' and shoots.

Results and discussion: The results indicate that all the nitrogen compounds employed were more or less suitable for *B. algens,* although ammonium salts were somewhat superior. Since urea, uric acid, nitrates, and to a lesser extent allantoin and xanthine supported growth as well, it appears that the type of nitrogen compounds in the habitat of *B. algens* is not a critical factor for its distribution, as it is for that of *L. tephroeceta* and *P. crispa.* It is interesting, however, that the type of nitrogen source did affect the gross morphology of the moss [cf. Burkholder, 1959]; this effect suggests but by no means proves that some so-called subspecies and even species may be no more than ecophenes attributable to the type of nitrogen source(s) in the habitat.

For example, *Bryum siplei* Bartram, which is practically indistinguishable from the cosmopolitan *B. argenteum* Hedwig (the difference being primarily in the production by *B. siplei* of axillary gemmae), seems to be found almost without exception where birds occur [cf. Siple, 1938, p. 498; Bartram, 1938, 1957]. *Bryum argenteum,* which is considered by Horikawa and Ando [1967] to be a weedy nitrophilous species, is not found so exclusively in close association with birds. Greene [Greene et al., 1967, p. 12] has reduced *B. siplei* to synonymy with *B. argenteum,* as Horikawa and Ando [1961] had suggested. Only further experiments will show unequivocally whether nitrogenous or other kinds of compounds do in fact modify the morphology of antarctic mosses.

For some reason (perhaps the presence of sucrose in rather high concentration) Knudson's medium, both with and without microelements, was unsuitable for *B. algens.* The inocula (clones 1–5) did not grow out onto the new medium but remained completely on the small remnants of the old medium (Trebouxia agar plus biotin and thiamine). This difference in the suitability of the two media does not seem to be attributable to osmotic factors because the glucose in the Trebouxia agar was present in approximately the same molar concentration as the sucrose in Knudson's medium (0.055 *M* versus 0.058 *M*). It is therefore reasonable to infer that for some reason the sucrose itself inhibited the *B. algens.*

Bristol's solution agar with 0.001 *M* filter-sterilized urea (solidified with 20 grams of purified agar per liter) supported good protonematal growth, but only three green shoots developed. However, on Bristol's solution without urea, about 50 developed. Growth was also more restricted with the urea (0.6–0.7 cm in diameter versus 1.5 cm in diameter).

Of the three basic media employed in experiment

1, soil water agar supported the best growth. There were numerous green shoots about 1 cm tall, and the diameter of the moss, which was healthy, green, and spreading, extended some 3.5 cm.

Soil water agar was employed in experiment 2. The two different batches of soil water agar without additives (media la and lb) yielded somewhat different amounts of growth. Soil water plus Bristol's solution sustained very good growth of all five clones; with 1 gram of glucose per liter of medium there were a few more shoots (an average of 35 with glucose compared to 25 without). Growth was best on soil water plus Bristol's solution and 10 grams of activated charcoal per liter. On this medium, growth was healthy and green, and there were almost 90 shoots per clone, about 2.5 times the number on the next best medium.

Gibberellic acid (GA) did not markedly stimulate growth with NH_4NO_3, although the protonemata produced in its presence were greener and bore more shoots than those produced in its absence. Protonematal growth with xanthine was somewhat stimulated in the presence of GA. The results obtained with $NaNO_3$ and GA are difficult to explain. Perhaps the inoculum was insufficient or died. However, the results of experiment 2 suggest that GA in the presence of $NaNO_3$ may in some cases reduce the amount of growth produced.

Kinetin stimulated protonemata and somewhat inhibited shoots. Aerial protonemata developed in the centers of the colonies grown on kinetin; this characteristic appeared otherwise only with ammonium salts. The effects of indole-3-acetic acid cannot be evaluated because the compound was added along with other substances.

The protonemata on the medium with $NaNO_3$ and xanthine did not grow straight out from the center along radii but tended to grow outward in very definite arcs. This phenomenon was not seen in any other case.

Light had very pronounced effects on both shoots and protonemata. With certain media (e.g., NH_4Cl, uric acid, allantoin, $NaNO_3$, xanthine, and urea at 0.042 and 0.021 g/l but not at 0.353 g/l), these effects were marked. Although there was no discernible pattern in the occurrence of these photic responses, they can be classified into a number of groups: (1) photophobic growth (protonemata), (2) photophilic growth (shoots), (3) growth perpendicular to the light, (4) growth or initiation inhibited on the side toward the light source (shoots on protonemata), (5) pronouncedly acentric growth of the whole plant, and so forth. None of these responses was exhibited on every medium; the photic responses sometimes occurred together on the same media but not always. It is impossible to determine whether the photic responses were due to differences in the media or in incubation conditions such as temperature and illumination. Further studies will be necessary to clarify this point.

The three concentrations of urea used in experiment 3 had remarkably different effects on growth, especially on photic responses. With 0.042 gram of urea per liter there was a marked photic response; over-all growth was acentric, and shoots were definitely photophilic; however, they were produced primarily on the side shielded from the light. With 0.353 gram of urea per liter, over-all growth was perfectly circular. With 0.021 gram of urea per liter, growth was more spreading and extensive than that with 0.042 gram, and shoots were long, chlorotic, and more numerous on protonemata shielded from direct light. With 0.353 gram of urea per liter protonemata were green, with 0.042 g/l they were brownish red, and with 0.021 g/l they were almost completely brownish. Preliminary tests of *B. algens* on urease agar indicate that this species may hydrolyze urea with extracellular urease (more critical tests must be performed to confirm this finding). Since the hydrolysis of urea raises the *p*H and yields NH_3 and CO_2, the observed differences in protonematal growth could have been due to either *p*H or nutritional differences among the media. As a case in point Gimingham and Lewis Smith [1970] report that *B. algens* occurs on basic rocks and soils. A further possibility is that the lower concentrations of urea limited growth because of a deficiency of nitrogen. Chlorosis, a symptom of nitrogen deficiency in plants, was noted in the shoots of *B. algens* grown on 0.021 gram of urea per liter. It is interesting that Holdgate et al. [1967] concluded that nitrogen was the only element likely to limit plant growth in the maritime Antarctic. If this conclusion is indeed true, the results point up the extent to which the concentration of nitrogen is critical to the distribution of mosses in Antarctica.

The vitamins biotin and thiamine appeared to have different effects on the growth of *B. algens.* Biotin induced a compact growth habit and inhibited the production of shoots, whereas thiamine did not seem to induce a compact growth habit. Thiamine supported good protonematal growth; large shoots developed in the area of inoculation, but only numerous very small shoots developed on the new protonemata. When both vitamins were in the medium, shoots were numerous in the inoculated area, but after 2 months there were no shoots on the protonemata. Biotin seems to have inhibited growth, since, even though the protonemata were green, no shoots had developed on them 2 months after inoculation. Since Boyd et al. [1966] detected three vitamins in relatively high quantities near the Adélie penguin rookery on Cape Royds and found detectable amounts of the same vitamins in the ice-free valleys of Victoria Land (they apparently did not test for biotin and thiamine, however), it is conceivable that mosses growing in the vicinity of nesting areas could be morphologically different from specimens of the same species growing in areas where there are no free vitamins in the environment.

The most consistent and convincing morphologic effects were produced by Norit-A and EDTA. Norit-A unquestionably stimulated the production of green shoots and apparently suppressed the formation of protonemata. (Since the medium containing Norit-A was black and opaque, it was impossible to discern whether significant protonematal growth had in fact occurred; this problem was compounded by the photophobic nature of the protonemata.) Since Norit-A is merely activated charcoal, an adsorbing compound routinely used to render media vitamin free [cf. Lilly and Barnett, 1951, p. 432], the Norit-A in the medium may have adsorbed substances such as biotin that suppress the initiation of shoots.

The adsorption of biotin is only one plausible explanation for the results obtained with Norit-A, however, since activated charcoal readily adsorbs all hydrophobic (nonpolar) substances. Furthermore the stimulation by Norit-A may have been due to light effects, since the photophobic reaction of the protonemata could have been reduced by the opaque black medium that resulted when the charcoal was added.

The EDTA, which was added in a higher than customary concentration to the medium (1.0 g/l versus the more common 0.2 g/l), apparently chelated some essential microelements such as molybdenum, which is required for the enzymatic reduction of nitrate. Since they were not specifically added to the medium, any microelements present would have to have been background contamination in the reagents used for the medium. The effectiveness of chelation is revealed by the virtual absence of growth in the flasks containing EDTA. Such complete sequestration or lack of microelements probably would not occur in the field in Antarctica, but it might in places. An examination of field samples would reveal the extent to which microelements are available to terrestrial plants there. Conversely and probably more significantly the extent to which elements either singly or in combination inhibit growth should be determined, especially in the cold deserts of continental Antarctica, where saline soils are the rule.

There are, however, other plausible explanations for the results. For example, EDTA readily binds many ions, especially calcium ions, a deficiency of which would have inhibited the growth of the moss. The EDTA may even have been absorbed by moss cells. Intracellular EDTA could cause problems.

Although these studies of *B. algens* were preliminary in nature and yielded inconclusive results, they nevertheless revealed the potential usefulness of laboratory experiments for interpreting plant distribution and speciation in Antarctica.

GENERAL DISCUSSION

The experimental results suggest that some antarctic lichens may subsist on the uric acid in guano and on compounds derived from it but that they cannot use inorganic or oxidized nitrogen. Studies in progress may reveal a relationship between the presence of uric acid and its derivatives urea and ammonia and the distribution of some lichen species. Galinou [1954] attempted such experiments in France and found that the production of allantoinase, allantoicase, and urease by lichens was correlated with the occurrence in the environment of allantoin, allantoic acid, and urea, respectively. In England, Smith [1960] found that the absorptive capacities of *Peltigera polydactyla* coincided with the most prevalent class of nitrogenous compounds in its habitat, i.e., the amino acids.

The problem of attributing the distribution of lichens to a single factor, such as a suitable nitrogen source, is illustrated by the nonantarctic species *Acarospora fuscata* (Nyl.) Arn. In a study of lichens occurring on rocks frequented by birds Hakulinen [1962] concluded that *A. fuscata* is an ammoniophilous rather than, for example, an ornithocoprophilous species. Laboratory studies of the nitrogen preferences of this species have been made by Hale [1958], Ahmadjian [1961, 1964], and Gross and Ahmadjian [1966]. In most cases the results of these studies are consistent, undoubtedly because identical or very similar media and incubation conditions were employed in all the studies.

Overall, *A. fuscata* appears to resemble *L. tephroeceta* in most aspects of nitrogen source use. Ammonium tartrate supported excellent growth [Hale, 1958], and alanine, proline, and (in one of the studies) asparagine all supported relatively large amounts of growth [Ahmadjian, 1961, 1964; Gross and Ahmadjian, 1966] at 18°–20°C in the presence of biotin and thiamine, whereas peptone, casein, and arginine did not. In contrast to *L. tephroeceta*, however, *A. fuscata* grew very well on KNO_3 and very poorly on inorganic ammonium salts and filter-sterilized urea [Ahmadjian, 1964].

The apparent contrast in the growth of the two mycobionts on nitrate, inorganic ammonium salts, and urea is best ascribed to unfavorable *p*H conditions; the final *p*H values of the media containing the inorganic ammonium salts were 2.52–2.65, of those containing urea, 7.03, and of those containing KNO_3, 6.30. Since *p*H optimums for lichen fungi lie between 4.5 and 7.4 [Ahmadjian, 1967, p. 54], the apparently poor growth on the ammonium salts and urea was probably due to the use of an unbuffered medium [cf. Cochrane, 1958, pp. 247–248]. The large rise in *p*H with urea indicates that the mycobiont excreted urease into the medium, which yielded ammonium and increased the *p*H to unfavorable levels. That ammonium was absorbed by the mycobiont of *A. fuscata* is revealed by the following observations. (1) The *p*H of the media fell; this drop was a sign that ammonium had been absorbed [cf. Lilly and Barnett, 1951, pp. 102–103]. (2) The ammonium tartrate supported by far the best growth of all the nitrogen sources used in the four studies [Hale, 1958], probably because the tartrate buffered the medium. (3) The NH_4NO_3 in the medium led to a definite drop in *p*H; this drop was a strong indication that ammonium had been absorbed but that nitrate, the absorption of which would have increased the *p*H, had not been absorbed (the *p*H would have remained more or less stable if both compounds had been absorbed concurrently).

Hakulinen's [1962] observations are not therefore contradicted by the experimental data. In Cochrane's [1958] words, 'No claim of failure to utilize ammonium nitrogen can stand unless the *p*H effect is excluded experimentally.' On the other hand, Ahmadjian's [1964] data showing that *A. fuscata* used KNO_3 must be accepted, especially since Furnari and Luciani [1962] found a comparable effect with the same organism. Hakulinen's designation 'ammoniophile' must therefore be modified to 'facultative ammoniophile.'

The presence of purines and purine derivatives in the antarctic environment can be ascribed primarily to excretion by birds, whereas the presence of amino acids must be ascribed principally to the decomposition of vegetation. In continental Antarctica, where virtually all soils are ahumic [Ugolini, 1970], penguin guano constitutes the largest accumulations of organic matter; algal peats in and around small ponds constitute the only other large accumulations. Moss peats, which are frequent in maritime Antarctica, do not occur to any significant degree in continental Antarctica.

Literally nothing is known about the absorptive capacities of antarctic lichens in relation to the compounds in their habitats, and the only previously published data on the enzyme complements of antarctic lichens appear to be those of Moiseeva and her associates [Moiseeva, 1961; Kuprevich et al., 1957], who reported that *Neuropogon antarcticus* (DR) Savicz and *N. sulphureus* (Koenig) Elenk. produce trace amounts of both extracellular and intracellular urease and asparaginase, among other enzymes. It is interesting in this regard that Galinou [1954] obtained negative results for allantoinase, allantoicase, and urease with the closely related species *Usnea ceratina* Ach., *U. comosa* Röhl., *U. rubicunda* Str., and *U. subpectinata* Str.

In France, Massé [1966a, b] found a correlation between the total nitrogen content of lichen thalli and nitrogen in the substrate, depending on whether the phycobiont was a green or blue-green alga. The thalli of lichens with blue-green algae were rich in nitrogen compared to their substrates (3.7–8.4% versus 0.1–2.1%), the thalli of ornithocoprophobic species with green algae contained approximately the same amounts of nitrogen as their substrates (0.9–2.5% versus 1.0–2.0%), and the thalli of ornithocoprophilic species with green algae contained higher amounts of nitrogen than the ornithocoprophobic species but in quantities in the lower third of the total range for their substrates, in this case, bird excrement (3.5–5.5% versus 3.0–12.0%).

Massé's data for French lichens are in general similar to those obtained for antarctic macroscopic terrestrial plants (Table 2). In one group of analyses the nitrogen contents were: soils, 0.02% reduced nitrogen; guano, 0.23 (liquid fraction) to 24.6% (solid fraction); presumably nonornithocoprophilic lichens with green algae, 0.3–1.2%; blue-green algae,

2.1%; and *Prasiola crispa*, an obligate ornithocoprophile, 5.5%. Clearly, the same factors operative on lichens in France are operative on them in Antarctica. In continental Antarctica, where lichens have only chlorophycean phycobionts, nitrogen in the environment must be of paramount importance in both the qualitative and the quantitative senses.

The pattern of nitrogen source use displayed by *L. tephroeceta* suggests that nitrophily exists in lichens because the mycobiont (1) uses very few amino acids, (2) lacks peptidases or for some other reason does not use protein derivatives, (3) does not use nitrates, and (4) uses compounds derived from the excrement of uricotelic animals (i.e., birds). Nitrophily in lichens may thus be considered the manifestation of a limited capacity to use nitrogen compounds; this limitation tends to restrict such lichens to specific habitats: for ornithocoprophilic species, to the neighborhoods of the breeding, bathing, rendezvous, and resting sites of birds.

Lecanora tephroeceta appears to substantiate Räsänen's contention [Barkman, 1958] that nitrophily is actually a question of ammoniophily. *Prasiola crispa*, on the other hand, reveals that another type of nitrophily (uricophily) exists as well. Indeed, since all organisms require nitrogen, all organisms must be considered nitrophilous, even nitrogen-fixing organisms. Thus it is probably more meaningful to categorize organisms on the basis of their nitrogen nutrition, as Robbins [1937] and others have done for the fungi. According to their schemes the fungi were designated as species able to use (1) molecular nitrogen and nitrate, ammonium, and organic nitrogen compounds; (2) nitrate, ammonium, and organic nitrogen compounds but not molecular nitrogen; (3) ammonium and organic nitrogen compounds only; and (4) only organic nitrogen compounds. *Lecanora tephroeceta* apparently belongs in the third category, whereas both *P. crispa* and *B. algens* apparently belong in the second.

A more realistic system of classification could be established on the basis of whether the nitrogen sources are zoogenic, phytogenic, or inorganic. For antarctic terrestrial plants such a system would be far more meaningful. The following scheme is proposed:

A. Nitrogen source organic.
   1. Nitrogen source phytogenic; nitrogen source proteinaceous or derived from the hydrolysis of proteins.
      a. Proteins.
      b. Polypeptides.
      c. Peptides.
      d. Amino acids and related amines.
   2. Nitrogen source zoogenic.
      a. Animals uricotelic (e.g., birds): (1) purines (uric acid) or (2) ureides (allantoin or urea).
      b. Animals ureotelic (e.g., mammals): urea.

B. Nitrogen source inorganic (biogenic or nonbiogenic).
   1. Nitrogen source solid and more or less soluble in water.
      a. Nitrogen source reduced: ammonia and ammonium.
      b. Nitrogen source oxidized: (1) nitrites or (2) nitrates.
   2. Nitrogen source gaseous: molecular nitrogen.

This scheme, which like any other is subject to modification and refinement, is meaningful in an ecologic context for antarctic terrestrial plants. Species are placed in one or more categories that reflect their relationships to and dependence on plants (primarily nitrogen-fixing blue-green algae, which probably excrete nitrogenous compounds, but also mosses and nonnitrogen-fixing blue-green algae when they contribute the organic component of soils) (category A1), on animals (category A2), or on neither (their independence of other plants and animals) (category B).

According to this scheme *L. tephroeceta* belongs in category A, as both *B. algens* and *P. crispa* do; the latter two species also belong in category B, since they use inorganic nitrogen compounds. All the species fit in one or more subcategories, usually in one more strongly than in the others. These subcategories convey a large amount of information about the distribution of the plants placed in them. Thus *L. tephroeceta* fits most readily into categories A2 and Bla; *B. algens* fits strongly into category Bla and somewhat less strongly into categories Blb and A2 (there is no information on whether *B. algens* uses amino acids); and *P. crispa* fits very strongly and almost exclusively into category A2a(1). Of the three species, *P. crispa* is most dependent on biogenic nitrogen, *L. tephroeceta* is less so, and *B. algens* is least so. Nitrogen-fixing blue-green algae, which would fit into the lowest category (B2), would according to the scheme be most independent of biogenic nitrogen (although they undoubtedly use it when it is present). This independence of biogenic nitrogen should be reflected in their patterns of distribution; these patterns would contrast sharply with those of *P. crispa,* which belongs in one of the higher categories of the scheme.

Classification on the basis of nutritional pattern may prove helpful in interpreting and predicting species distribution in Antarctica, where vast areas contain little or no detectable nitrogen in the substrate and where nitrogen enters the terrestrial ecosystem at a few discrete points (e.g., at rookeries). The situation would be far more complicated in more moderate regions of the earth. To be sure, nitrogen is only one of many environmental factors influencing the distribution of plants in Antarctica. It is, however, a very important one. Appropriate laboratory studies properly applied should give meaning to the often puzzling and seemingly haphazard fashion in which plants occur in the Antarctic, especially studies coupled with field observation and experiment.

APPENDIX 1: GEOGRAPHICAL LIST OF IDENTIFIED COLLECTIONS OF ANTARCTIC LICHENS

Lichen specimens were identified by Dr. C. W. Dodge, University of Vermont, Burlington. Specimens, including unidentified collections, have been deposited in the Herbarium of Cryptogamic Botany, U.S. National Museum, Smithsonian Institution, Washington, D. C. Duplicates of identified specimens are in the herbarium of Dr. Dodge in Burlington. Specimens of algae referred to are in the herbarium of the Academy of Natural Sciences, Philadelphia. Moss specimens have been deposited in the herbarium of the New York Botanical Garden, Bronx Park, New York. Collection information, environmental data, and photographs of some of the collection sites appear in Gressitt et al. [1964] (Jones Mountains) and in Wise et al. [1964] (Ross Island and Victoria Land). The asterisks indicate species used for the isolation attempts.

West Antarctica: Ellsworth Mountains

Sentinel Range. The following species was found in Camp Hills (78°57′S, 85°43′W) on nunataks 1.6 km west of Camp Gould in the cracks of rocks. ? *Lecanora fuscobrunnea* D&B: AA-83* (fragmentary).

Collector: K. P. Rennell, December 7, 1963.

Heritage Range. The following species were found at Welcome Nunatak ('The Pimple') (79°06′S, 85°54′W). (1) A 'blasteniaceous thallus' too young to identify: AA-139 (in cracks in rock) and AA-167. (2) *Buellia actinobola* (Hue) Darb., <*B. latemarginata* Darb., *fide* Lamb (1964): AA-108a*. (3) *Lecanora tephroeceta* Hue: ? AA-87* (apothecium moribund, no ascospores; in cracks in rock) and AA-107* (all of these taxa are nitrophilous). (4) ? *Lecidea cancriformis* D&B: AA-84* (fragmentary). (5) AA-85 and AA-108b unidentified.

Collector: K. P. Rennell, November 26, 1963, and January 16, 1964.

West Antarctica: Eights Coast

Jones Mountains (73°32′S, 94°00′W; 730 meters above sea level [cf. Gressitt et al., 1964]). ? *Caloplaca athallina* Darb. (AA-130* on moss and AA-131*) and *Polycauliona prostrata* (Hue) Dodge (AA-124*) were found at Jones Mountains.

Intrusive Spur (73°30′S, 94°25′W): The species found at Intrusive Spur (where snow petrels were present) were as follows. (1) *Buellia racovitsae* Dodge: AA-126*. (2) *Lecanora daltoniana* Hook. *f.*: AA-127* (microthallus and apothecia smaller than type; dark ashy color). (3) *L. tephroeceta* Hue: ? AA-122* (fragment with small apothecia; thallus blackened; on moss) and AA-129* (used in nitrogen nutrition experiments). (4) *Parmelia rennelli* Dodge: AA-155 (type; on decayed moss tufts). (5) *Polycauliona prostrata* (Hue) Dodge: AA-123* and AA-160. (6) *Rhizocarpon argyreum* (Hue) Darb.: AA-125* (ascospores small but apparently mature).

Polycauliona prostrata (Hue) Dodge was found on rocks near a snow petrel nest (AA-118*, AA-119*, and AA-120*).

Collector: K. P. Rennell, February 6, 1964.

East Antarctica: Victoria Land

Borchgrevink Coast. The following species were found in the Cape Hallett area (where Adélie penguins, snow petrels, and South Polar skuas were present).

Top of cape, north end (72°19′S, 170°18′E): (1) *Buellia llanoi* Dodge: AA-39a* and AA-156. (2) ? *Caloplaca mawsoni* (Dodge) Dodge or ? *Blastenia sparsa* Murray: AA-44* (immature; young thecium immersed in thallus). (3) *Lecanora priestleyi* Dodge: AA-39b?*. (4) *Lecanora schofieldi* Dodge: AA-44 (type). (5) *Rhizocarpon flavum* D&B: AA-29*.

West slope of cape near Willett Cove: *Omphalodina exsulans* (Th. Fr.) Dodge (AA-17* on rock in sheltered position in crevice) was found near the snow line (72°19′S, 170°16′E). The following species were found above the skuary (72°19′S, 170°15′E). (1) *Buellia subtegens* Murray: AA-166. (2) *Gasparrinia harrissoni* Dodge: AA-149 (fertile).

(3) *Omphalodina exsulans* (Th. Fr.) Dodge: AA-18* (abundant in wetter, sheltered locations), AA-149, and AA-158 (abundant above the large patch of moss). (4) *Parmelia coreyi* D&B: AA-142 (with traces of *Gasparrinia* sp. too immature to identify) and AA-149. (5) *P. johnstoni* Dodge: AA-143. (6) *Polycauliona pulvinata* D&B: AA-31*, AA-155, AA-159 (on loose pebbles near large patch of moss), and AA-165. (7) *Rinodina frigida* (Darb.) Dodge: AA-27* and AA-30*. (8) *R. stipitata* D&B: AA-28*. (9) *Xanthoria mawsoni* Dodge: AA-28 and AA-166.

Base of cape near Seabee Hook (72°19′S, 170°15′E): (1) A 'blasteniaceous thallus': AA-154 (epiphytic on another lichen). (2) *Lecanora griseomarginata* D&B (on mosses): AA-15* and AA-163. (3) *Polycauliona pulvinata* D&B: AA-3a* (on small stone), AA-19*, and AA-31*. (4) *Rinodina frigida* (Darb.) Dodge: AA-3b*, AA-21*, and AA-150 (very large, nearly black thallus). (5) *R. rudolphi* Dodge: AA-45a–h* (blacker than usual). (6) *R. stipitata* D&B: AA-19. (7) *Xanthoria mawsoni* Dodge: AA-146 (on stone near penguin rookery).

Seabee Hook (72°19′S, 170°13′E) (Adélie penguins and South Polar skuas were present): Parmelia coreyi D&B (AA-142, mostly on moss) was found between the base of Cape Hallett and Willett Cove [cf. Rudolph, 1963]. *Parmelia griseola* D&B (AA-164, very well developed but sterile; on flat stone in moist ground near runoff from snow field) was found in the skuary [cf. Rudolph, 1963]. Specimens AA-218–AA-220 were not identified. Specimens were collected in November and December 1963.

Moubray Bay, west side of Edisto Inlet. The following species were found in the headland northeast of Luther Peak near a glacier (72°21′S, 169°55′E), where snow petrels and South Polar skuas were present (exposure to west). (1) *Blastenia sparsa* Murray: AA-24. (2) *Buellia grisea* D&B: AA-16* (on moss and sand). (3) *B. llanoi* Dodge: AA-34*, AA-35*, and AA-41*. (4) *B. pallida* D&B: AA-153 (top of headland). (5) *B. quercina* Darb.: AA-34a. (6) *Caloplaca darbishirei* (D&B) Dodge: AA-153 (top of headland; fertile but very immature). (7) *Lecanora griseomarginata* D&B: AA-151 (on soil) and AA-162 (with Moss AA-1). (8) *L. lavae* Darb.: AA-34a, AA-38a* (on rock), and AA-38b* (eroded; on rock). (9) *L. siplei* D&B: AA-24*. (10) *Lecidea physciella* Darb. (mostly on upper half of headland): AA-25*, AA-33a*, AA-33b*, AA-40*, AA-42, and AA-43*. (11) *Omphalodina exsulans* (Th. Fr.) Dodge (on boulders and rocks): AA-32*, AA-37*, AA-145a, and AA-145b. (12) *Parmelia variolosa* D&B: AA-147 (thallus grayer than usual; on mosses). (13) *Polycauliona pulvinata* D&B: AA-20* and AA-26*. (14) *Rinodina frigida* (Darb.) Dodge (most collections made near top of headland): AA-20, AA-22*, AA-23*, AA-33b, AA-36*, AA-42*, AA-144, and AA-148. (15) *Usnea picata* (Lamb) Dodge (all collections made at top of headland): AA-33a, AA-36, and AA-144 (more papillate than usual). (16) *Xanthoria mawsoni* Dodge: AA-151 (on soil) and AA-162 (very young sterile lobes). Specimens were collected on November 12, 1963.

Terra Nova Bay area. The following species were found on Vegetation Island (74°47′S, 163°37′E). (1) *Acarospora gwynni* Dodge & Rudolph: AA-50 (very young). (2) *Alectoria congesta* (Zahlbr.) Dodge: AA-50 (very young). (3) *Buellia pallida* D&B: AA-50*. (4) *Omphalodiscus bakeri* Dodge: AA-50. (5) AA-49, AA-256, and AA-257 were unidentified.

Collector: K. A. J. Wise, December 7, 1963.

Scott Coast. The following species were found in the Mawson Glacier area.

Near Nordenskjöld ice tongue: (1) *Lecidea stancliffi* D&B (AA-102, on soil) and an unidentified specimen AA-190 were found at Bruce Point (76°08′S, 162°15′E) on January 30, 1964. (2) *Rhizocarpon flavum* D&B (AA-104), *R. schofieldi* Dodge (AA-104*, type), and an unidentified specimen AA-209 were found on exposed rock ('Fruticose Nunatak') about 4 km west of Bruce Point and 10 km northeast of Mount Murray (approximately 76°07′S, 162°10′E) on January 30, 1964. (3) *Omphalodiscus subcerebriformis* (Dodge) Dodge (AA-75), *Rinodina sordida* D&B (AA-75*), and unidentified specimens AA-174 and AA-226 were found on a nunatak ('Cliff Nunatak') about 5 km northeast of Mount Murray (approximately 76°07′S, 162°00′E) (collectors were K. A. J. Wise and A. Spain, December 31, 1963). (4) *Lecanora lavae* Darb. (AA-105*), *Rhizocarpon flavum* D&B (AA-103*), and unidentified specimens AA-225 and AA-140 were found on Crash Nunatak (75°46′S, 160°38′E) about 1570 meters above sea level on January 30, 1964.

Granite Harbor area (76°53′S, 162°44′E). (South Polar skuas were present.) The following specimens were found. (1) *Acarospora emergens* Dodge: AA-51* (immature). (2) *Biatorellopsis cerebriformis* (Dodge) Dodge: AA-53. (3) *Lecidea stancliffi* D&B: AA-62*. (4) *Omphalodina exsulans* (Th. Fr.) Dodge:

AA-56*. (5) *Parmelia griseola* D&B: AA-51 (sterile). (6) *Rinodina frigida* (Darb.) Dodge: AA-53* (immature), AA-55*, and AA-56. (7) Specimens AA-57–AA-61 and AA-217 were not identified.

Collector: Y. Kobayasi, December 15–17, 1963. (Description in Kobayasi [1967, p. 218] suggests that Kobayasi visited The Flatiron.)

The Flatiron (77°01′S, 162°23′E): (1) *Caloplaca darbishirei* (D&B) Dodge: AA-96* (on moss and pebble). (2) *C. mawsoni* (Dodge) Dodge: AA-138 (on moss; thallus better developed and ascospores somewhat larger than those in type). (3) *Lecanora griseomarginata* D&B: AA-97*, AA-98*, AA-141a, and AA-141b. (4) *L. lavae* Darb.: AA-138* (on lava). (5) *Lecidea (Biatora) acerviformis* Murray: AA-99. (6) *L. stancliffi* D&B: AA-136*. (7) *Omphalodina exsulans* (Th. Fr.) Dodge: AA-116* and AA-138 (on granite). (8) *Omphalodiscus spongiosus* var. *subvirginis* (Lamb & Frey) Dodge: AA-169. (9) *Rinodina frigida* (Darb.) Dodge: AA-136. (10) *Xanthoria mawsoni* Dodge: AA-141c. (11) Specimens AA-210–AA-216 were not identified. Specimens were collected on January 14–15, 1964.

Marble Point area. (South Polar skuas were present.)

Near Wilson Piedmont Glacier (77°26′S, 163°38′E): *Aspicilia glacialis* Dodge (AA-100*) was found on January 12, 1964.

Near abandoned U.S. Navy camp (77°25′S, 163°40′E): *Caloplaca darbishirei* (D&B) Dodge (AA-72), ? *C. schofieldi* Dodge (AA-161; sterile; identification uncertain), and *Lecanora griseomarginata* D&B (AA-72* and AA-161) were found on moss on December 28, 1963.

Wheeler valley (77°12′S, 161°42′E). The following specimens were found about 45 km west of McMurdo Sound. (1) *Lecidea (Biatora) acerviformis* Murray (on small pebbles in crevices of large boulder): AA-64* and AA-65*. (2) *L. coreyi* D&B: AA-67* (chipped from large boulder on floor of valley). (3) *L. ecorticata* D&B: AA-68* (on small rock on floor of valley). (4) *Omphalodina exsulans* (Th. Fr.) Dodge: AA-66* (chipped from stone; depauperate thallus). (5) *Rinodina sordida* D&B: AA-69* (on small rock on floor of valley). (6) Specimen AA-260 (a large, 25- by 5-cm thallus; sterile) was not identified. Specimens were collected on December 26, 1963.

Victoria valley. *Aspicilia glacialis* Dodge (AA-63; type) and unidentified specimens AA-258 and AA-259 were found near the foot of Packard Glacier (77°21′S, 162°10′E), 34 km west of McMurdo Sound. A *Prasiola* species *(P. ? calophylla)* was present (Alga AA-25; in U.S. National Museum). Specimen was collected on December 28, 1963.

Wright valley. *Acarospora emergens* Dodge (AA-132*, on rock), *Lecidea blackburni* D&B (AA-133*), and *Omphalodina exsulans* (Th. Fr.) Dodge (AA-134*) were found on The Dais (77°33′S, 161°15′E) about 60 km west of McMurdo Sound about 850 meters above sea level.

Collector: K. P. Rennell, February 14, 1964.

Taylor valley. *Caloplaca mawsoni* (Dodge) Dodge (AA-70) was found near Lake Fryxell (77°37′S, 163°11′E) about 10 km west of McMurdo Sound on December 26, 1963.

Vicinity of Blackwelder Glacier. *Rinodina frigida* (Darb.) Dodge (AA-110*, thallus nearly or completely eroded) and unidentified specimens AA-242 and AA-243 were found near Lake Péwé (77°56′S, 164°12′E) about 4 km west of McMurdo Sound on January 10, 1964.

Vicinity of Walcott Glacier. (The specimens were collected about 30 km southwest of McMurdo Sound on January 9, 1964.)

Near Penny Lake (78°17′S, 163°21′E): (1) *Lecanora lavae* Darb.: AA-135. (2) *Thelidium minutum* Dodge: AA-106* (perithecia very immature) and AA-135* (type; parasymbiont of *L. lavae*). (3) Unidentified specimens AA-175–AA-177, AA-182–AA-185, and AA-235–AA-241.

Near top of hill south of Penny Lake: (1) *Caloplaca schofieldi* Dodge: AA-93* (type) and ? AA-94 (blackened by *Enterococcus*). (2) *Lecanora siplei* D&B: AA-92*. (3) *Parmelia johnstoni* Dodge: AA-90* (fertile but young).

Base of hillside south of Penny Lake: *Lecidea harrissoni* Dodge (AA-88*).

The following algae were present. (1) *Nostoc commune* Vaucher, (2) *Schizothrix calcicola* (Agardh) Gomont, and (3) *Anacystis montana* (Lightfoot) Drouet & Daily (Alga AA-2; filed as Zaneveld & Simmonds 64-01-0208; in the Academy of Natural Sciences, Philadelphia (Drouet's herbarium); abundant over sandy gravel about 2 meters from east shore of Penny Lake). Determined by F. Drouet and R. E. Cameron.

East Antarctica: Ross Island

Hut Point Peninsula. (South Polar skuas were present.)

Observation Hill (77°51′S, 166°41′E), west slope: (1) *Blastenia grisea* Dodge: AA-46* (abundant in moist places). (2) *Lecanora siplei* D&B: AA-47* (in drier locations). Specimens were collected on December 23, 1963.

Crater Hill (77°51′S, 166°43′E), southeast slope: (1) *Caloplaca darbishirei* (D&B) Dodge: AA-152 (on moss). (2) *Lecanora griseomarginata* D&B: AA-170 (on soil). (3) *L. siplei* D&B: AA-73* (on moss). Specimens were collected on December 30, 1963, and January 2, 1964.

Near Pram Point (approximately 77°51′S, 166° 44′E): (1) *Caloplaca darbishirei* (D&B) Dodge: AA-76* (on soil). (2) *Lecanora griseomarginata* D&B: AA-77*, AA-81*, and AA-82. (3) *Lecidea (Biatora) aceriviformis* Murray: AA-79* and AA-82*. (4) *Rinodina olivaceobrunnea* D&B: AA-78* (on soil and rock). Specimens were collected on January 2, 1964.

Pram Point (77°51′S, 166°46′E): Lecanora griseomarginata D&B (AA-74*) and an unidentified specimen AA-235 were collected near Scott Base on December 30, 1963.

Cape Royds area (approximately 77°33′S, 166° 12′E). (Adélie penguins and South Polar skuas were present.) The following specimens were found on January 20, 1964. (1) *Blastenia succinea* D&B: AA-114* (on rock outcrop). (2) *Lecanora griseomarginata* D&B: AA-115 (on moss). (3) *Lecidea stancliffi* D&B: AA-117* (on rock outcrop). (4) *L. wadei* D&B: AA-114 (on rock outcrop) and AA-115* (on moss).

Vicinity of Cape Crozier (77°31′S, 169°23′E). (Adélie and Emperor penguins and South Polar skuas were present.) *Gasparrinia adarensis* Dodge (AA-54) and *Polycauliona pulvinata* D&B (AA-54*) were found.

Collector: Y. Kobayasi, December 19, 1963 (information in Kobayasi [1967, p. 213] reveals that Kobayasi visited the Adélie penguin rookery on this date; thus the two preceding specimens were probably collected nearby).

Vicinity of Adélie penguin rookeries (77°27′S, 169°14′E): Xanthoria siplei (D&B) Dodge (AA-109*, thallus badly weathered) was found near the U.S. Antarctic Research Program hut Jamesway (collector was R. C. Wood, February 28, 1964).

Peak ('Bryant') (77°27′S, 169°13′E) in skuary: (1) *Acarospora gwynni* Dodge & Rudolph: AA-101*. (2) *Lecanora griseomarginata* D&B: AA-111* and AA-112. (3) *L. siplei* D&B: AA-113* (on rock outcrop). Specimens were collected on January 20–24, 1964.

Unidentified specimens AA-187, AA-188, AA-191–AA-193, and AA-195–AA-206 were collected in the Cape Crozier area.

East Antarctica: Hillary Coast

Darwin Glacier area. The following collection was made.

Brown Hills: Acarospora emergens Dodge (AA-52*, on quartz) was found on Diamond Hill (79° 52′S, 158°52′E).

Collector: K. A. J. Wise, December 24, 1963.

APPENDIX 2: DISTRIBUTION OF *LECANORA TEPHROECETA* HUE AND *'PHORMIDIUM AUTUMNALE'* FROM ALL AVAILABLE COLLECTIONS

Searches in cryptogamic herbaria at the Field Museum of Natural History, Chicago; U.S. National Museum (Smithsonian Institution), Washington, D. C.; Academy of Natural Sciences, Philadelphia; Farlow Herbarium, Harvard University; and the personal herbaria of Drs. Francis Drouet (Philadelphia) and Carroll W. Dodge (Burlington) yielded the information presented in this appendix. The searches were made to determine from collection data on herbarium packets and through examination of specimens themselves whether the distributions of the two species are correlated with the presence of bird life. When collection data were insufficient, descriptions in the antarctic literature of the localities in which the specimens were collected were used. Short of a personal tour of all of Antarctica, this source was the most suitable way of judging whether the habitats were guanotrophic. The evidence indicates that both taxa can with justification be designated 'nitrophilous'; collection data for *L. tephroeceta* coupled with the experimental data are best explained on the basis of a requirement for ornithogenic reduced nitrogen or for ammonium from any source. Since there are no experimental data for *'P. autumnale,'* the collection data are only circumstantial evidence for a similar pattern in this ecophene of *Microcoleus vaginatus;* in this case, however, we are dealing with an environmentally induced variant form of a morphologically plastic species rather than with a true species.

Habitat Data for All Available Collections of Lecanora tephroeceta Hue from West Antarctica

Eights Coast, Jones Mountains, Intrusive Spur (73° 30′S, 94°25′W), 730 meters above sea level. Schofield AA-122 and AA-129. (1) Herbarium or literature citation: Dodge (Burlington). (2) Substrate: moss. (3) Collector: K. P. Rennell. (4) Collection data: none. (5) Collection date: February 6, 1964. (6) Other information: thousands of snow petrels nesting (J. F. Splettstoesser, personal communication, 1970). (7) Remarks: located 100 km from coast; mycobiont from AA-129 used in nitrogen experiments. (8) Habitat type: guanotrophic.

Ellsworth Mountains

Heritage Range, Welcome Nunatak ('The Pimple') (79°06′S, 85°54′W), 14.4 km south of Camp Gould. Schofield AA-87 and AA-107. (1) Herbarium or literature citation: Dodge (Burlington). (2) Substrate: in cracks of rocks. (3) Collector: K. P. Rennell. (4) Collection data: none. (5) Collection date: November 26, 1963. (6) Other information: isolated nunatak 700 km from coast, apparently no bird life; only very infrequent passing skuas (J. F. Splettstoesser, personal communication, 1970; D. A. Coates, personal communication, 1970). (7) Remarks: very small specimen; associated species are known to prefer guanotrophic locations, i.e., 'blasteniaceous' thallus (*Blastenia* sp.), two collections; *Buellia latemarginata* Darb., >*B. actinobola* (Hue) Darb., *fide* Lamb (1968) (Schofield AA-108a, Heritage Range, January 16, 1964, collector, K. P. Rennell); Lamb [1968, pp. 52–53, and Table XXV] states: 'I have found *B. latemarginata* to be a highly nitrophilous species. . . . It occurs preferentially high up or even on the zenith of bird rocks ... [and is] ... often associated with other characteristically nitrophilous species such as *Xanthoria elegans, X. candelaria, Rinodina petermanni,* and *Mastodia tesselata* [the lichenized form of *Prasiola crispa*].' The occurrence together of *L. tephroeceta* (two collections), *B. latemarginata,* and *Blastenia* sp. on Welcome Nunatak strongly suggests that guanotrophic conditions exist there despite the great distance from the coast; however, a source of ammonium of any kind would also explain these collections and is the most plausible explanation. (8) Habitat type: ? nonguanotrophic.

South Shetland Islands

See Lindsay [1971].

Deception Island (62°55′S, 60°38′W). Follmann 13852. (1) Herbarium or literature citation: Dodge (Burlington). (2) Substrate: moss. (3) Collector: G. Follmann. (4) Collection data: none. (5) Collection date: 1963. (6) Other information: at least four penguin rookeries on island [U.S. Hydrographic Office, 1965, chart HO 6796]; numerous birds flying and walking over inner and outer shores (R. E. Cameron, personal communication, 1969); the nitrophilic plant association, *Ramalinetum terebratae,* and *Prasiola crispa* occur [Follmann, 1965] (specimens of *P. crispa* in U.S. National Museum Herbarium of Cryptogamic Botany and Field Museum Herbarium of Cryptogamic Plants). (7) Remarks: none. (8) Habitat data: guanotrophic.

Redón 45. (1) Herbarium or literature citation: Dodge (Burlington). (2) Substrate: wood. (3) Collector: J. Redón. (4) Collection data: none. (5) Collection date: January 1966. (6) See (6) under Follmann 13852. (7) Remarks: overgrown by a blasteniaceous lichen, an indicator of guanotrophic conditions. (8) Habitat type: guanotrophic.

Pendulum Cove (62°55′S, 60°38′W). Gain 76 (type). (1) Herbarium or literature citation: Hue [1915, pp. 89–90]. (2) Substrate: 'cendres agglomerées.' (3) Collector: L. Gain. (3) Collection data: at 20 meters above sea level. (4) Collection date: December 25, 1908. (5) Other information: abundant bird life near Pendulum Cove (R. E. Cameron, personal communication, 1969); associated with *Lecanora (Caloplaca) cinericola* Hue [Hue, 1915, pp. 89–90 and 194]. (6) Remarks: type locality. (7) Habitat type: guanotrophic.

Robert Island (62°24′S, 59°34′W). Redón 8 and 9. (1) Herbarium or literature citation: Dodge (Burlington). (2) Substrate: 'sobre rocas.' (3) Collector: J. Redón. (4) Collection data: none. (5) Collection date: January 1966. (6) Other information: penguin rookeries at Coppermine Peninsula (K. R. Everett, personal communication, 1970); ringed penguin rookery (exact location on island not specified) [Follmann, 1965]. (7) Remarks: spermogonia only, no apothecia (Redón 8); precise location of collections not stated; Lindsay [1971] visited both Mitchell Cove and Edwards Point on Robert Island in January 1966. (8) Habitat type: very probably guanotrophic [cf. Redón, 1969].

Melchior Islands

Omega Island (64°20′S, 62°56′W), northeast landing. Siple 384. (1) Herbarium or literature citation: Dodge ('Lichen Flora of Antarctica,' unpublished manuscript, undated). (2) Substrate: unknown. (3) Collector: P. A. Siple. (4) Collection data: none. (5) Collection date: ? March 1940. (6) Other information: ringed penguin rookery found at Melchior Harbor; many blue-eyed shag, which probably nest there, were observed [Eklund, 1945; Friedmann, 1945]; Siple made at least six collections of *P. crispa* nearby in March 1940 (specimens in U.S. National Museum Herbarium of Cryptogamic Botany and Field Museum Herbarium of Cryptogamic Plants). (7) Remarks: numerous indications of guanotrophic conditions in a small (about 30-km^2) area. (8) Habitat type: guanotrophic.

Omega Island, 'Chain Point.' Siple, Frazier, and Bailey 364c–m. (1) Herbarium or literature citation: Dodge ('Lichen Flora of Antarctica,' unpublished manuscript, undated). (2) Substrate: not known. (3) Collectors: P. A. Siple, R. G. Frazier, and D. K. Bailey. (4) Collection data: none. (5) Collection date: ? March 1940. (6) Other information: see Siple 384. (7) Remarks: see Siple 384. (8) Habitat type: guanotrophic.

Omega Island, water landing. Siple and Richardson 333. (1) Herbarium or literature citation: Dodge ('Lichen Flora of Antarctica,' unpublished manuscript, undated). (2) Substrate: not known. (3) Collectors: P. A. Siple and H. H. Richardson. (4) Collection data: none. (5) Collection date: ? March 1940. (6) Other information: see Siple 384. (7) Remarks: see Siple 384. (8) Habitat type: guanotrophic.

Lambda Island (64°18'S, 63°00'W), northeast landing: Siple 380-g-20. (1) Herbarium or literature citation: Dodge ('Lichen Flora of Antarctica,' unpublished manuscript, undated). (2) Substrate: pale granite. (3) Collector: P. A. Siple. (4) Collection data: none. (5) Collection date: ? March 1940. (6) Other information: see Omega Island, Siple 384. (7) Remarks: see Omega Island, Siple 384. (8) Habitat type: guanotrophic.

Argentine Islands

Uruguay Island (65°14'S, 64°14'W). Tyler L-2-4(a). (1) Herbarium or literature citation: Dodge (Burlington). (2) Substrate: rock protruding from snow and ice. (3) Collector: J. Tyler. (4) Collection data: exposed to sun most of the time. (5) Collection date: April 8, 1959. (6) Other information: 'abundant animal and bird life, and water, are found in this group' (i.e., the Argentine Islands) [U.S. Hydrographic Office, 1960]; South Polar skua and blue-eyed shag occur [Burton, 1970]. (7) Remarks: young specimen. (8) Habitat type: probably guanotrophic.

Adelaide Island (67°46'S, 68°54'W). Follmann 14173. (1) Herbarium or literature citation: Dodge ('Lichen Flora of Antarctica,' unpublished manuscript, undated). (2) Substrate: decaying wood. (3) Collector: G. Follmann. (4) Collection data: at coast. (5) Collection date: ? 1963. (6) Other information: abundant bird life on nearby Avian Island [U.S. Board on Geographic Names, 1956]; *P. crispa* common in region (specimens in U.S. National Museum Herbarium of Cryptogamic Botany and Field Museum Herbarium of Cryptogamic Plants); initial phase of *R. terebratae,* with *Mastodia tesselata* (lichenized *P. crispa*) as dominant species [Follmann, 1965]. (7) Remarks: immature specimen; island is large (about 115 × 40 km), but the wood substrate may have come from buildings on south end of island, only about 400 meters from Avian Island. (8) Habitat type: guanotrophic.

Debenham Islands (68°08'S, 67°05'W). Bryant 26. (1) Herbarium or literature citation: Dodge ('Lichen Flora of Antarctica,' unpublished manuscript, undated). (2) Substrate: rock. (3) Collector: H. M. Bryant. (4) Collection data: at 24 meters above sea level. (5) Collection date: ? December 1940. (6) Other information: South Polar skuas and probably antarctic terns nest on islands [Bryant, 1945, Figures 3 and 4; Sáiz and Hajek, 1968]; numerous collections of *P. crispa* from Marguerite Bay region (specimens in U.S. National Museum Herbarium of Cryptogamic Botany and Field Museum Herbarium of Cryptogamic Plants). (7) Remarks: group of islands in close proximity. (8) Habitat type: guanotrophic.

Occurrence in Antarctica of 'Phormidium autumnale,' an Unsheathed Aquatic Ecophene of Microcoleus vaginatus (Vaucher) Gomont, As Determined from Hebarium Specimens

Siple 332. (1) Location: Melchior Islands, Omega Island (64°20'S, 62°56'W), at 6–10 meters elevation. (2) Collector and expedition: P. A. Siple, U.S. Antarctic Service Expedition (1939–1941). (3) Collection date: March 14, 1940. (4) Associated species: *P. crispa.* (5) Indications of nitrogen rich conditions: *Prasiola crispa* (Lightf.) Ag.; ringed penguins occur on the island [Perkins, 1945, Figures 10 and 11]. (6) Remarks: determined by F. Drouet (?).

Siple 363. (1) Location: Melchior Islands, Anchorage Island (67°36′S, 68°13′W). (2) Collector and expedition: P. A. Siple, U.S. Antarctic Service Expedition (1939–1941). (3) Collection date: March 4, 1940. (4) Associated species: *P. crispa.* (5) Indications of nitrogen rich conditions: *P. crispa,* partly lichenized (*Mastodia* sp.) with same collection number from Anchorage Island. (6) Remarks: determined by F. Drouet.

Bryant 99. (1) Location: Marguerite Bay, Lagotellerie Island (67°53′S, 67°24′W). (2) Collector and expedition: H. M. Bryant, U.S. Antarctic Service Expedition (1939–1941). (3) Collection date: December 22, 1940. (4) Associated species: none. (5) Indications of nitrogen rich conditions: rookery of 1500 Adélie penguins [Eklund, 1945]. (6) Remarks: determined by F. Drouet.

Nutt 54. (1) Location: Knox Coast, Bunger Hills (66°18′S, 100°45′E). (2) Collector and expedition: E. T. Apfel, U.S. Navy Antarctic Expedition (1947–1948), 'Operation Windmill.' (3) Collection date: January 14, 1948. (4) Associated species: *Lamprocystis roseopersicina* (Kütz) Schröt. and *P. crispa.* (5) Indications of nitrogen rich conditions: *P. crispa;* South Polar skuas [McDonald, 1948; Shumskiy, 1957], snow petrels, and Wilson petrels [Shumskiy, 1957] present in oasis. (6) Remarks: determined by F. Drouet, March 1949; area of Bunger Hills is 400–450 km^2.

Nutt 59, 60, 64, and 66a. (1) Location: Knox Coast, Merritt Island, point AP-11 (66°27′S, 107°10′E). (2) Collector and expedition: D. C. Nutt, U.S. Navy Antarctic Expedition (1947–1948), 'Operation Windmill.' (3) Collection date: January 18, 1948. (4) Associated species: *P. crispa* (all), '*Phormidium tenue*' Gom., predominating (Nutt 61). (5) Indications of nitrogen rich conditions: *P. crispa;* penguins present; pieces of feather with Nutt 59 and 66a; Nutt 66a found on moist gravel and on penguin skeletons; intimately associated with feathers [Nutt, 1948]. (6) Remarks: determined by F. Drouet, March and April 1949.

Nutt 68 and 71. (1) Location: Budd Coast, Holl Island, point AP-13 (66°26′S, 110°29′E). (2) Collector and expedition: D. C. Nutt, U.S. Navy Antarctic Expedition (1947–1948), 'Operation Windmill.' (3) Collection date: January 19, 1948. (4) Associated species: *P. crispa* (Nutt 71), '*Plectonema nostocorum*' Gom. (both), *Nostoc commune* (Nutt 68), and *Calothrix parietina* (Nutt 68). (5) Indications of nitrogen rich conditions: *P. crispa* present (Nutt 69 with bird down intermingled; 70 with feathers, hormidioid growth form; 71; and 72); South Polar skuas and Adélie penguins breeding [Eklund, 1961, Plate III]. (6) Remarks: determined by F. Drouet, April 1949.

Nutt 86 and 87. (1) Location: Ross Island, Cape Royds (77°35′S, 166°09′E), terrestrial. (2) Collector and expedition: D. C. Nutt, U.S. Navy Antarctic Expedition (1947–1948), 'Operation Windmill.' (3) Collection date: January 29, 1948. (4) Associated species: *Nostoc commune.* (5) Indications of nitrogen rich conditions: small Adélie penguin rookery at Cape Royds consisting of about 1500 pairs [Stonehouse, 1965] and South Polar skuas nesting in colonies scattered over the cape [Young, 1963, Figure 1]. (6) Remarks: determined by F. Drouet, April 1949; on soil with crust of *Nostoc.*

Llano 2199a. (1) Location: Victoria Land, McMurdo Sound area, Gneiss Point (77°24′S, 163°44′E). (2) Collector and expedition: G. A. Llano, International Geophysical Year. (3) Collection date: December 3, 1957. (4) Associated species: *Microcoleus paludosus* and *Schizothrix rubella.* (5) Indications of nitrogen rich conditions: skuas breed on Gneiss Point; Llano 2196 (December 3, 1957) collected 'near skua perches.' (6) Remarks: determined by F. Drouet, October 1960 and February 1965.

Llano 2205b and 2206. (1) Location: Ross Island, Cape Evans (77°38′S, 166°24′E). (2) Collector and expedition: G. A. Llano, International Geophysical Year. (3) Collection date: December 6–8, 1957. (4) Associated species: *Prasiola* sp. (Llano 2205b) and *Schizothrix calcicola* (as '*Phormidium valderianum*' ecophene; Llano 2206). (5) Indications of nitrogen rich conditions: *P. crispa* (Llano 2205 and 2207); South Polar skuas breed at Cape Evans [Stonehouse, 1965]. (6) Remarks: determined by G. A. Llano (Llano 2205b) and F. Drouet, September 1962 (Llano 2206).

Llano 2222. (1) Location: Victoria Land, Taylor valley (approximately 77°39′S, 162°52′E). (2) Collector and expedition: G. A. Llano, International Geophysical Year. (3) Collection date: none (probably December 14–18, 1957). (4) Associated species: '*Microcoleus vaginatus* . . . and *Phormidium autumnale* . . . F. Drouet, XI. 1959.' (5) Indications of nitrogen rich conditions: none; east end of Taylor valley opens onto McMurdo Sound and receives marine influences [Jones and Faure, 1969, Figure 25]; skuas breed at east end [Kohn et al., 1971]. (6) Remarks:

determined by F. Drouet; note that both the ecophene and the typical *M. vaginatus* are present.

Leech specimen not numbered. (1) Location: South Shetland Islands, Deception Island, Cathedral Crags (63°00'S, 60°34'W). (2) Collector: R. E. Leech. (3) Collection date: March 10, 1960. (4) Associated species: none. (5) Indications of nitrogen rich conditions: at least four penguin rookeries on Deception Island [U.S. Hydrographic Office, 1965, chart HO 6796] and abundant bird life (R. E. Cameron, personal communication, 1969); apparently, however, there are no rookeries near Cathedral Crags (O. Orheim, personal communication, 1972). *P. crispa* from Cathedral Crags collected March 10, 1960 (Leech, no number). (6) Remarks: determined by F. Drouet, October 1960; '*Microcoleus vaginatus* (Vauch.) Gom., phormidioid.'

Acknowledgments. Much of the work reported herein was done in the Department of Biology, Clark University, with support from the Office of Antarctic Programs, National Science Foundation, grant GA-36. Follow-up work plus further field and laboratory studies were completed at Ohio State University, grant GA-840. Dr. Carroll W. Dodge, University of Vermont, Burlington, kindly identified all lichen specimens sent him; his ready cooperation was crucial to the success of this study. Dr. Stanley W. Greene, University of Birmingham, England, kindly identified the moss specimen, and Dr. Jacques S. Zaneveld, Old Dominion College, Norfolk, Virginia, identified the algae specimens. Mr. Keith A. J. Wise, now of the Auckland Institute and Museum, Auckland, New Zealand, donated many lichen specimens and aided greatly by suggesting collecting sites for lichens. His advice and companionship in the field were invaluable. Messrs. Paul R. Theaker, Ray E. Showman, and Joseph B. Harvey assisted during the second and third field seasons. Mr. Kelvin P. Rennell collected several specimens in West Antarctica and in Victoria Land, and Drs. Yosio Kobayasi, National Museum of Science, Tokyo, and Emanuel D. Rudolph, Ohio State University, donated specimens from Ross Island and Victoria Land. Miss Jen-rong Wang (now Mrs. Yang) isolated several phycobionts. Miss Beverly A. Temple aided in inoculation procedures and generously lent her assistance in many other ways. Mrs. Angharad Holmes and Mr. Henry Adelman provided invaluable guidance in the microkjeldahl analyses. The officers and men of U.S. Naval Development Squadron Six (VX-6) provided excellent logistic support in the field. The support of U.S. Navy Deep Freeze personnel is gratefully acknowledged. Contribution 214 of the Institute of Polar Studies, Ohio State University, Columbus, Ohio.

REFERENCES

Ahmadjian, V.

1958 Antarctic lichen algae. Carolina Tips, *21*(5): 17–18.

1961 Studies on lichenized fungi. Bryologist, *64*(2, 3): 168–179.

1964 Further studies on lichenized fungi. Bryologist, *67*(1): 87–98.

1967 The lichen symbiosis. viii + 152 pp. Blaisdell, Waltham, Mass.

1970 Adaptations of antarctic terrestrial plants. *In* M. W. Holdgate (Ed.), Antarctic ecology. *2:* 801–811. Academic, New York.

Barkman, J. J.

1958 Phytosociology and ecology of cryptogamic epiphytes. 628 pp. Van Gorcum, Assen, The Netherlands.

Bartram, E. B.

1938 The Second Byrd Antarctic Expedition—Botany. 3. Mosses. Ann. Mo. Bot. Gdn, *25*(2): 719–724.

1957 Mosses from the United States Antarctic Service Expedition, 1940–41. Bryologist, *60*(2): 139–143.

Baxter, R. M., and N. E. Gibbons

1962 Observations on the physiology of psychrophilism in a yeast. Can. J. Microbiol., *8*(4): 511–517.

Birt, L. M., and F. J. R. Hird

1956 The uptake of amino acids by carrot slices. Biochem. J., *64*(2): 305–311.

Bose, S.

1944 An iodometric estimation of uric acid in poultry excreta. Poult. Sci., *23*(2): 130–134.

Boyd, W. L., and J. W. Boyd

1962 Presence of *Azotobacter* species in polar regions. J. Bact., *83*(2): 429–430.

Boyd, W. L., J. T. Staley, and J. W. Boyd

1966 Ecology of soil microorganisms of Antarctica. *In* J. C. F. Tedrow (Ed.), Antarctic soils and soil forming processes, Antarctic Res. Ser., *8:* 125–159. AGU, Washington, D. C.

Bryant, H. W.

1945 Biology at East Base, Palmer Peninsula, Antarctica. Proc. Am. Phil. Soc., *89*(1): 256–269.

Burkholder, P. R.

1959 Organic nutrition of some mosses growing in culture. Bryologist, *62*(1): 6–15.

Burton, R. W.

1970 Biology of the Great Skua. *In* M. W. Holdgate (Ed.), Antarctic ecology. *1:* 561–567. Academic, New York.

Cameron, R. E., J. King, and C. N. David

1970 Microbiology, ecology and microclimatology of soil sites in dry valleys of southern Victoria Land. *In* M. W. Holdgate (Ed.), Antarctic ecology. *2:* 702–716. Academic, New York.

Campbell, I. B., and G. G. C. Claridge

1966 A sequence of soils from a penguin rookery, Inexpressible Island, Antarctica. N. Z. Jl Sci., *9*(2): 361–372.

Cavanaugh, G. M. (Ed.)

1956 Formulae and methods of the Marine Biological Laboratory chemical room, 5: 87 pp. Mar. Biol. Lab., Woods Hole, Mass.

Claridge, G. G. C., and I. B. Campbell
1968 Origin of nitrate deposits. Nature, *217*(5127): 428–430.

Clifford, H. T.
1957 New records for antarctic mosses. Aust. J. Sci., *20* (4): 115.

Cochrane, V. W.
1958 Physiology of fungi. xiii + 524 pp. John Wiley, New York.

Darbishire, O. V.
1923 Cryptogams from the Antarctic. J. Bot., Lond., *61* (4): 105–107.

Dawson, E. W., et al.
1965 Balleny Islands reconnaissance expedition. Polar Rec., *12*(79): 431–435.

Dixon, H. N., and W. W. Watts
1918 Mosses. Scient. Rep. Australas. Antarct. Exped., ser. C, *7*(1): 1–9.

Dodge, C. W.
1948 Lichens and lichen parasites. Rep. BANZ Antarct. Res. Exped., ser. B, *7:* 276 pp.
1964 Ecology and geographic distribution of antarctic lichens. *In* R. Carrick, M. W. Holdgate, and J. Prévost (Eds.), Biologie antarctique: 165–171. Hermann, Paris.
1965 Lichens. *In* P. van Oye and J. van Mieghem (Eds.), Biogeography and ecology in Antarctica: 194–200. Junk, The Hague.
1968 Lichenological notes on the flora of the antarctic continent and the subantarctic islands. 7, 8. Nova Hedwigia, *15*(2–4): 285–332.

Drouet, F.
1962 Gomont's ecophenes of the blue-green alga, *Microcoleus vaginatus.* Proc. Acad. Nat. Sci. Philad., *114*(6): 191–205.

Eklund, C. R.
1945 Condensed ornithology report, East Base, Palmer Land. Proc. Am. Phil. Soc., *89*(1): 299–304.
1961 Distribution and life history studies of the South Polar skua. Bird-Banding, *32*(4): 187–223.

Eugster, H. P., and J. Munoz
1966 Ammonium micas: Possible sources of atmospheric ammonia and nitrogen. Science, *151*(3711): 683–686.

Filson, R. B.
1966 The mosses and lichens of Mac. Robertson Land. 169 pp. Antarctic Division, Department of External Affairs, Melbourne.

Fogg, G. E.
1952 The production of extracellular nitrogenous substances by a blue-green alga. Proc. R. Soc., ser. B, *139*(896): 372–397.

Fogg, G. E., and W. D. P. Stewart
1968 *In situ* determinations of biological nitrogen fixation in Antarctica. Bull. Br. Antarct. Surv., 15: 39–46.

Folk, R. L.
1969a Petrography of avian urine. Tex. J. Sci., *21*(2): 117–129.
1969b Spherical urine in birds: Petrography. Science, *166*(3912): 1516–1519.

Follmann, G.
1965 Una asociación nitrófila de líquenes epipétricos de la Antártica Occidental con *Ramalina terebrata* Tayl. et Hook. como especie caracterizante. Publ. 4: 18 pp. Inst. Antártico Chileno, Santiago.

Friedmann, H.
1945 Birds of the United States Antarctic Service Expedition, 1939–1941. Proc. Am. Phil. Soc., *89*(1): 305–313.

Furnari, F., and F. Luciani
1962 Esperienze sulla crescita dei micobionti in *Sarcogyne similis* e *Acarospora fuscata* in coltura pura su vari substrati. Boll. Ist. Bot. Univ. Catania, ser. 3, *3:* 39–47.

Galinou, M.-A.
1954 Sur la mise en évidence de quelques biocatalyseurs chez les lichens. Proc. 8me Bot. Congr., sec. 18: 1–4. Paris.

Gannutz, T. P.
1969 Effects of environmental extremes on lichens. Bull. Soc. Bot. Fr., *1969:* 169–179.

Gates, D. M.
1965 Spectral properties of plants. Appl. Opt., *4*(1): 11–20.

Gibson, G. W.
1962 Geological investigations in southern Victoria Land, Antarctica. 8. Evaporite salts in the Victoria valley region. N. Z. Jl Geol. Geophys., *5*(3): 361–374.

Gimingham, C. H., and R. I. Lewis Smith
1970 Bryophyte and lichen communities in the maritime Antarctic. *In* M. W. Holdgate (Ed.), Antarctic ecology. *2:* 752–785. Academic, New York.

Greene, S. W., et al.
1967 Terrestrial life of Antarctica. *In* V. C. Bushnell (Ed.), Antarctic Map Folio Ser., folio 5: 24 pp., 11 pls. Amer. Geogr. Soc., New York.

Greene, S. W., D. M. Greene, P. D. Brown, and J. M. Pacey
1970 Antarctic moss flora. 1. The genera *Andreaea, Pohlia, Polytrichum, Psilopilum* and *Sarconeurum.* Sci. Rep. Br. Antarct. Surv., *64:* 118 pp.

Gressitt, J. L., C. E. Fearon, and K. Rennell
1964 Antarctic mite populations and negative arthropod surveys. Pacif. Insects, *6*(3): 531–540.

Grønlie, A. M.
1948 The ornithocoprophilous vegetation of the bird-cliffs of Røst in the Lofoten Islands, northern Norway. Nytt Mag. Naturvid., *86:* 117–243.

Gross, M., and V. Ahmadjian
1966 The effects of L-amino acids on the growth of two species of lichen fungi. Svensk Bot. Tidskr., *60*(1): 74–80.

Hakulinen, R.
1962 Ökologische Beobachtungen über die Flechtenflora der Vogelsteine in Sud- und Mittelfinnland. Arch. Soc. Zool. Bot. Fenn. 'Vanamo,' *17*(1): 12–15.

Hale, M. E., Jr.
1958 Vitamin requirements of three lichen fungi. Bull. Torrey Bot. Club, *85*(3): 182–187.

Hirano, M.
1965 Freshwater algae in the antarctic regions. *In* P. van Oye and J. van Mieghem (Eds.), Biogeography and ecology in Antarctica: 127–193. Junk, The Hague.

Holdgate, M. W.
1970 Vegetation. *In* M. W. Holdgate (Ed.), Antarctic ecology. *2:* 729–732. Academic, New York.

Holdgate, M. W., S. E. Allen, and M. J. G. Chambers
1967 A preliminary investigation of the soils of Signy Island, South Orkney Islands. Bull. Br. Antarct. Surv., 12: 53–71.

Holm-Hansen, O.
1963 Algae: Nitrogen fixation by antarctic species. Science, *139*(3539): 1059–1060.
1964 Isolation and culture of terrestrial and freshwater algae of Antarctica. Phycologia, *4*(1): 43–51.

Horikawa, Y., and H. Ando
1961 Mosses of the Ongul Islands collected during the 1957–1960 Japanese Antarctic Research Expedition. Hikobia, *2*(3): 160–178.
1967 The mosses of the Ongul Islands and adjoining coastal areas of the antarctic continent. JARE Scient. Rep., spec. issue 1, 245–262.

Hue, A. M.
1915 Lichens. *In* Deuxième expédition antarctique française (1908–1910). 202 pp. Masson, Paris.

Jackson, T. A., and W. D. Keller
1970 A comparative study of the role of lichens and 'inorganic' processes in the chemical weathering of recent Hawaiian lava flows. Am. J. Sci., *269*(5): 446–466.

Jacobs, M. B., M. J. Gerstein, and W. G. Walter
1957 Dictionary of microbiology. 276 pp. Van Nostrand, New York.

Jennings, D. H.
1963 The absorption of solutes by plant cells. viii + 204 pp. Iowa State University Press, Ames.

Johannesson, J. K., and G. W. Gibson
1962 Nitrate and iodate in antarctic salt deposits. Nature, *194*(4828): 567–568.

Jones, A. S., and R. T. Walker
1964 An organic deposit from the Tottanfjella, Dronning Maud Land. Bull. Br. Antarct. Surv., 3: 21–22.

Jones, L. M., and G. Faure
1969 The isotope composition of strontium and cation concentrations of Lake Vanda and Lake Bonney in Southern Victoria Land, Antarctica. Rep. 4: x + 82 pp. Lab. for Isotope Geol. and Geochem., Inst. of Polar Studies, and Dep. of Geol., Ohio State Univ., Columbus.

Knebel, G.
1935 Monographie der Algenreihe der Prasiolales, insbehondere von *Prasiola crispa*. Hedwigia, *75:* 1–120.

Kobayasi, Y.
1967 *Prasiola crispa* and its allies in the Alaskan Arctic and Antarctica. Bull. Natn. Sci. Mus., Tokyo, *10*(2): 211–222.

Kohn, B. P., V. E. Neall, and C. G. Vucetich
1971 Biological observations from the McMurdo Sound region, Antarctica. Notornis, *18*(1): 52–54.

Kol, E., and E. A. Flint
1968 Algae in green ice from the Balleny Islands, Antarctica. N. Z. J. Bot., *6*(3): 249–261.

Kuprevich, V. F., M. M. Gollerbakh, E. N. Moiseeva, V. P. Savich, and T. A. Shcherbakova
1957 Some data on the biological activity of the ground, soil, and lichens of the eastern Antarctic (in Russian). Dokl. Akad. Nauk SSSR, *126:* 151–154. (Engl. transl., Amer. Inst. of Biol. Sci., Washington, D. C., 1959.)

Lamb, I. M.
1948 Antarctic pyrenocarp lichens. Discovery rep. 25: 30 pp., 4 pls. Cambridge University Press, London.
1968 Antarctic lichens. 2. The genera *Buellia* and *Rinodina*. Sci. Rep. Br. Antarct. Surv., *61:* 129 pp., 26 pls.

Leentvaar, P.
1967 Observations in guanotrophic environments. Hydrobiologia, *29*(3, 4): 441–489.

Letts, E. A.
1913 On the occurrence of the fresh-water alga (*Prasiola crispa*) on contact beds, and its resemblances to the green seaweed (*Ulva latissima*). Jl R. Sanit. Inst., *34*(10): 464–468.

Levine, R., W. Q. Wolfson, and R. Lenel
1947 Concentration and transport of true urate in the plasma of the azotemic chicken. Am. J. Physiol., *151*(1): 186–191.

Lilly, V. G., and H. L. Barnett
1951 Physiology of the fungi. xii + 464 pp. McGraw-Hill, New York.

Lindsay, D. C.
1971 Vegetation of the South Shetland Islands. Bull. Br. Antarct. Surv., 25: 59–83.

Llano, G. A.
1962 The terrestrial life of the Antarctic. Scient. Am., *207*(3): 212–230.

Longton, R. E.
1967 Vegetation in the maritime Antarctic. Phil. Trans. R. Soc., ser. B, *252*(777): 213–235.

Lonsdale, K., and D. J. Sutor
1971 Uric acid dihydrate in bird urine. Science, *172* (3986): 958–959.

Lorius, C., G. Baudin, J. Cittanova, and R. Platzer
1969 Impuretés solubles contenues dans la glace de l'Antarctique. Tellus, *21*(1): 136–148.

Massé, L.
1966a Étude comparée des teneurs en azote total des lichens et de leur substrat: Les espèces 'ornithocoprophiles.'

C. R. Hebd. Séanc. Acad. Sci., Paris, sér. D, *262*(16): 1721–1724.

1966b Étude comparée des teneurs en azote total des lichens et de leur substrat: Les espèces à gonidies Cyanophycées. C. R. Hebd. Séanc. Acad. Sci., Paris, sér. D, *263*(10): 781–784.

Matsuda, T.
1968 Ecological study of the moss community and microorganisms in the vicinity of Syowa Station, Antarctica. JARE Scient. Rep., ser. E, 29: 58 pp.

Mayland, H. F., and T. H. McIntosh
1966 Availability of biologically fixed atmospheric nitrogen-15 to higher plants. Nature, *209*(5021): 421–422.

McCraw, J. D.
1967 Soils of Taylor dry valley, Victoria Land, Antarctica, with notes on soils from other localities in Victoria Land. N. Z. Jl Geol. Geophys., *10*(2): 498–539.

McDonald, E.
1948 Southern cruise by two Navy icebreakers. Proc. U.S. Nav. Inst., *74*(12): 1491–1503.

Moiseeva, E. N.
1961 Biochemical properties of lichens and their practical significance (in Russian with English summary). 82 pp. Akad. Nauk SSSR, Moscow.

Morita, R. Y., and R. D. Haight
1964 Temperature effects of the growth of an obligate psychrophilic marine bacterium. Limnol. Oceanogr., *9*(1): 103–106.

Mueller, G.
1968 Genetic histories of nitrate deposits from Antarctica and Chile. Nature, *219*(5159): 1131–1134.

Nutt, D. C.
1948 Second (1958) U.S. Navy antarctic development project. Arctic, *1*(2): 88–92.

Perkins, J. E.
1945 Biology at Little America III, the west base of the United States Antarctic Service Expedition 1939–41. Proc. Am. Phil. Soc., *89*(1): 270–284.

Petersen, J. B.
1928 The äerial algae of Iceland, 8. *In* L. K. Rosenvinge and E. Warming (Eds.), The botany of Iceland. *2:* part 2, 325–447. Wheldon and Wesley, London.
1935 Studies on the biology and taxonomy of soil algae. Dansk Bot. Ark., *8*(9): 1–180.

Quetsch, M. F., and W. F. Danforth
1964 Kinetics of accumulation of uric acid and xanthine by the yeast, *Candida utilis.* J. Cell. Comp. Physiol., *64*(1): 123–129.

Redón, J.
1969 Nueva asociación de líquenes muscícolas de la Antártica Occidental, con *Sphaerophorus tener* Laur, como especie caracterizante. Bull. 4: 5–11. Inst. Antártico Chileno, Santiago.

Robbins, W. J.
1937 The assimilation by plants of various forms of nitrogen. Am. J. Bot., *24*(5): 243–250.

Rudolph, E. D.
1963 Vegetation of Hallett Station area, Victoria Land, Antarctica. Ecology, *44*(3): 585–586.
1967 Lichen ecology and microclimate studies at Cape Hallett, Antarctica. *In* S. W. Tromp and W. H. Weihe (Eds.), Biometeorology. *2*(2): 900–910. Pergamon, Oxford.

Sáiz, F., and E. R. Hajek
1968 Estudios ecológicos en Isla Robert (Shetland del Sur). 1. Observaciones de temperatura en nidos de petrel gigante *Macronectes giganteus* Gmelin. Publ. 14: 15 pp. Inst. Antártico Chileno, Santiago.

Shields, L. M.
1957 Algal and lichen floras in relation to nitrogen content of certain volcanic and arid range soils. Ecology, *38*(4): 661–663.

Shields, L. M., C. Mitchell, and F. Drouet
1957 Alga- and lichen-stabilized surface crusts as soil nitrogen sources. Am. J. Bot., *44*(6): 489–498.

Shumskiy, P. A.
1957 Glaciological and geomorphological reconnaissance in the Antarctic in 1956. J. Glaciol., *3*(21): 55–61.

Silverman, M. P., and E. F. Munoz
1970 Fungal attack on rock: Solubilization and altered infrared spectra. Science, *169*(3949): 985–987.

Siple, P. A.
1938 The Second Byrd Antarctic Expedition—Botany. 1. Ecology and geographical distribution. Ann. Mo. Bot. Gdn, *25*(2): 467–514.

Siple, P. A., and A. A. Lindsey
1937 Ornithology of the Second Byrd Antarctic Expedition. Auk, *54*(2): 147–159.

Sironval, C.
1947 Expériences sur les stades de développement de la forme filamenteuse en culture de *Funaria hygrometrica* L. Bull. Soc. R. Bot. Belg., *79:* 48–78.

Skottsberg, C. J. F.
1905 Some remarks upon the geographical distribution of vegetation in the colder southern hemisphere. Ymer, *25*(4): 402–427.

Smith, D. C.
1960 Studies in the physiology of lichens. 2. Absorption and utilization of some simple organic nitrogen compounds by *Peltigera polydactyla.* Ann. Bot., *24*(94): 172–185.

Starr, R. C.
1964 The culture collection of algae at Indiana University. Am. J. Bot., *51*(9): 1013–1044.

Stevenson, F. J.
1959 On the presence of fixed ammonium in rocks. Science, *130*(3369): 221–222.

Stewart, W. D. P.
1963 Liberation of extracellular nitrogen by two nitrogen-fixing blue-green algae. Nature, *200*(4910): 1020–1021.
1967 Transfer of biologically fixed nitrogen in a sand dune slack region. Nature, *214*(5088): 603–604.

Stokes, J. L.
1963 General biology and nomenclature of psychrophilic micro-organisms. *In* N. E. Gibbons (Ed.), Recent progress in microbiology. *8:* 187–192. University of Toronto Press, Toronto, Ont.

Stonehouse, B.
1965 Counting antarctic animals. New Scient., *27*(454): 273–276.

Sturkie, P. D.
1964 Avian physiology. 2nd ed., xxvii + 766 pp. Comstock, Ithaca, N. Y.

Swan, L. W.
1963 The aeolian zone in polar and mountain regions (abstract). Bull. Ecol. Soc. Am., *44*(2): 45.
1968 Alpine and aeolian regions of the world. *In* H. E. Wright, Jr., and W. H. Osburn (Eds.), Arctic and alpine environments: 29–54. Indiana University Press, Bloomington.

Syroechkovskii, E. E.
1959 The role of animals in the formation of primary soils in the conditions of the circumpolar region of the earth (in Russian). Bot. Zh. (Trudy̆ Imp. S-Peterb. Obshch. Estest.), *38*(12): 1770–1775.

Ugolini, F. C.
1970 Antarctic soils and their ecology. *In* M. W. Holdgate (Ed.), Antarctic ecology. *2:* 673–692. Academic, New York.

U.S. Board on Geographic Names
1956 Geographic names of Antarctica. Gaz. 14: v + 332 pp. Office of Geogr., Dep. of the Interior, Washington, D. C.

U.S. Hydrographic Office
1960 Sailing directions for Antarctica, including the off-lying islands south of latitude 60°S. H. O. Publ. 27: xiv + 432 pp. Govt. Print. Office, Washington, D. C.
1965 Chart HO 6796. Govt. Print. Office, Washington, D. C.

von Maltzahn, K. E., and I. G. MacQuarrie
1958 Effect of gibberellic acid on the growth of protonemata in *Splachnum ampullaceum* (L.) Hedw. Nature, *181*(4616): 1139–1140.

Ward, M.
1960 Some techniques in the culture of mosses. Bryologist, *63*(4): 213–217.

Weyant, W. S.
1966 The antarctic climate. *In* J. C. F. Tedrow (Ed.), Antarctic soils and soil forming processes, Antarctic Res. Ser., *8:* 47–59. AGU, Washington, D. C.

Wilson, A. T.
1959 Surface of the ocean as a source of air-borne nitrogenous material and other plant nutrients. Nature, *184*(4680): 99–101.

Wilson, A. T., and D. A. House
1965 Chemical composition of South Polar snow. J. Geophys. Res., *70*(22): 5515–5518.

Winsnes, T. S.
1969 What is 'mumiyo' from Antarctica? Årbok 1967: 228–230. Norsk Polarinst., Oslo.

Wise, K. A. J.
1964 New records of Collembola and Acarina in Antarctica. Pacif. Insects, *6*(3): 522–523.

Wise, K. A. J., C. E. Fearon, and O. R. Wilkes
1964 Entomological investigations in Antarctica, 1962–63 season. Pacif. Insects, *6*(3): 541–570.

Young, E. C.
1963 The breeding behaviour of the South Polar skua *Catharacta maccormicki.* Ibis, *105*(2): 203–233.

COMPARATIVE PHYSIOLOGY OF FOUR WEST ANTARCTIC MOSSES

James R. Rastorfer

Department of Botany and Institute of Polar Studies, Ohio State University, Columbus, Ohio 43210

Abstract. *Calliergidium austrostramineum*, *Drepanocladus uncinatus*, *Polytrichum strictum*, and *Pohlia nutans* are major components of well-developed plant communities on the easternmost island of the Corner Islands, Argentine Islands, Antarctica. In all four species the chlorophyll *a*/*b* ratios were about 2:1; however, the total chlorophyll contents based on the grams fresh weight were higher in *D. uncinatus* and *C. austrostramineum* (888–900 μg) than in *P. nutans* and *P. strictum* (313–486 μg). The lipid constituents were 19.0–38.4 mg g^{-1} fresh weight, whereas the soluble carbohydrates were 6.72–13.19 mg g^{-1} fresh weight. The concentration ranges of several macronutrient elements were (in per cent of dry matter): phosphorus, 0.13–0.53; potassium, 0.46–1.25; calcium, 0.11–0.37; magnesium, 0.09–0.12; nitrogen, 1.50–2.95; and sulfur, 0.075–0.195. Among 10 other elements assayed, iron and aluminum were much higher in *P. strictum* and *P. nutans* than in the other two species. The light compensation points (at 10°C) were 0.25 and 0.35 mW cm^{-2} for *C. austrostramineum* and *D. uncinatus*, respectively, but were higher at 0.6 mW cm^{-2} for *P. strictum* and *P. nutans*. However, the light saturation points (at 10°C) were 3.5 mW cm^{-2} for *C. austrostramineum* and *D. uncinatus* and 8 mW cm^{-2} for *P. strictum* and *P. nutans*. The net photosynthetic rates (at 10°C) at light saturation were 588, 450, 345, and 494 μl O_2 evolved hr^{-1} g^{-1} dry weight for the preceding species, respectively. The optimal temperatures for net photosynthesis and respiration were not clearly shown for *C. austrostramineum* and *D. uncinatus* in temperature tests from 5° to 25°C, but optimal temperatures were indicated for *P. strictum* and *P. nutans* in similar tests. The results revealed several physiological differences between the acrocarpous species (*P. strictum* and *P. nutans*) and the pleurocarpous species (*C. austrostramineum* and *D. uncinatus*). Although these observed physiological differences do not provide enough information to elucidate the adaptive relationships of these morphologically dissimilar forms to the antarctic environment, the pigment content and the influences of light intensity and temperature on photosynthesis may be important factors.

The antarctic botanical zone consists of two major floristic regions: (1) the Antarctic Peninsula sector and (2) the rest of the continent with the adjacent islands. The former region (maritime Antarctic), is under oceanic influences, whereas the latter region (continental Antarctic) has a colder and drier climate. As one may expect, the terrestrial floras of the two regions are quite different. In the maritime Antarctic, mosses and lichens are predominant and abundant, but liverworts, algae, and two species of flowering plant also occur. Conversely in the continental Antarctic there are fewer species of mosses and lichens, and they often occur in sparse widely scattered communities. In addition algae are an important floristic component of this region, but only one liverwort has been reported; flowering plants are absent [Greene, 1964; Holdgate, 1964; Llano, 1965; Rudolph, 1965; Steere, 1965].

Since mosses are a major component of the antarctic terrestrial vegetation, one may wonder whether antarctic mosses differ physiologically from those of other regions and if so whether these differences have any apparent survival value. One approach to this problem has involved a 2-year field comparative physiological study of antarctic mosses. The first year's (1967–1968) study was carried out at McMurdo Station, Antarctica. The test plants were *Bryum argenteum* and *Bryum antarcticum*, which were collected from Victoria Land in the continental Antarctic (Figure 1). The results of this work have been reported elsewhere [Rastorfer, 1970]. The present report concerns the second year's (1968–1969) work, which was carried out at Palmer Station. Four species, *Calliergidium austrostramineum* (C. Müll.) Bartr., *Drepanocladus uncinatus* (Hedw.) Warnst., *Polytrichum strictum* Manz. ex Brid., and *Pohlia nutans* (Hedw.) Lindb., were collected from one of the Argentine Islands in the maritime Antarctic (Figure 1). These species were chosen for study because they occurred in the same locality. In addition the

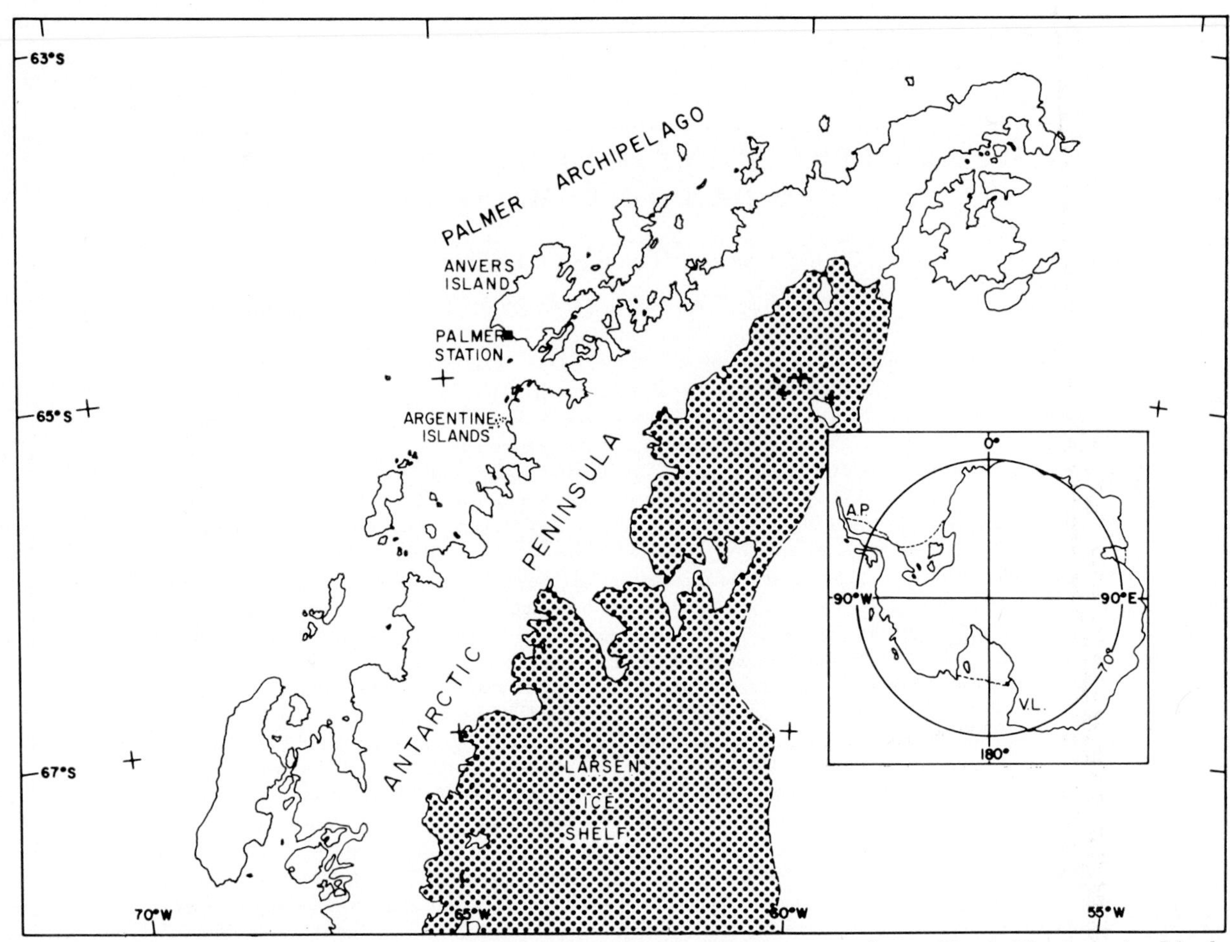

Fig. 1. Portion of Antarctic Peninsula sector of West Antarctica. Palmer Station is on Anvers Island. The Argentine Islands are south of Anvers Island. (A.P., Antarctic Peninsula; V.L., Victoria Land.)

first two species differ in growth form from the last two. Although archegonia and sporophytes were not seen, *C. austrostramineum* and *D. uncinatus* are considered pleurocarpous forms, which grow more or less prostrate and branch freely-forming mats, whereas *P. strictum* and *P. nutans* are considered acrocarpous forms, which grow erect in tufts. In this report the pleurocarpous and acrocarpous forms are compared as to element, chlorophyll, lipid, and carbohydrate concentrations as well as to the effects of different light intensities on photosynthesis and of different temperatures on photosynthesis and respiration. Comparisons with other antarctic plants and with plants of other regions, especially mosses, are also discussed.

PLANT MATERIALS: THEIR SOURCE AND ECOLOGIC NOTATIONS

Source. The four species *(Calliergidium austrostramineum, Drepanocladus uncinatus, Polytrichum strictum,* and *Pohlia nutans)* were collected from one locality on the easternmost island of the Corner Islands, Argentine Islands (Figure 2). The island has a permanent icecap and prominent rock outcrops (Figure 3). These outcrops are free of ice and snow during the summer, especially on their north to west exposures, and they reveal an abundant bryophyte and lichen flora. In addition to exposed cliffs and scattered rock outcrops protruding through the icecap there is an extensively exposed area in the northwest corner of the island that slopes down to sea level (Figure 3). Except for a few scattered patches of snow this area is free of ice and snow during the summer. The bedrock (dacite breccias [Elliot, 1964]) supports a conspicuous lichen flora and well-developed bryophytic communities. The site selected for this study is on a small terrace near the top of the slope (Figure 4). Although the four species occur elsewhere on the island and at other localities in the peninsula region, this site was among the most suitable for comparative physiological studies because all

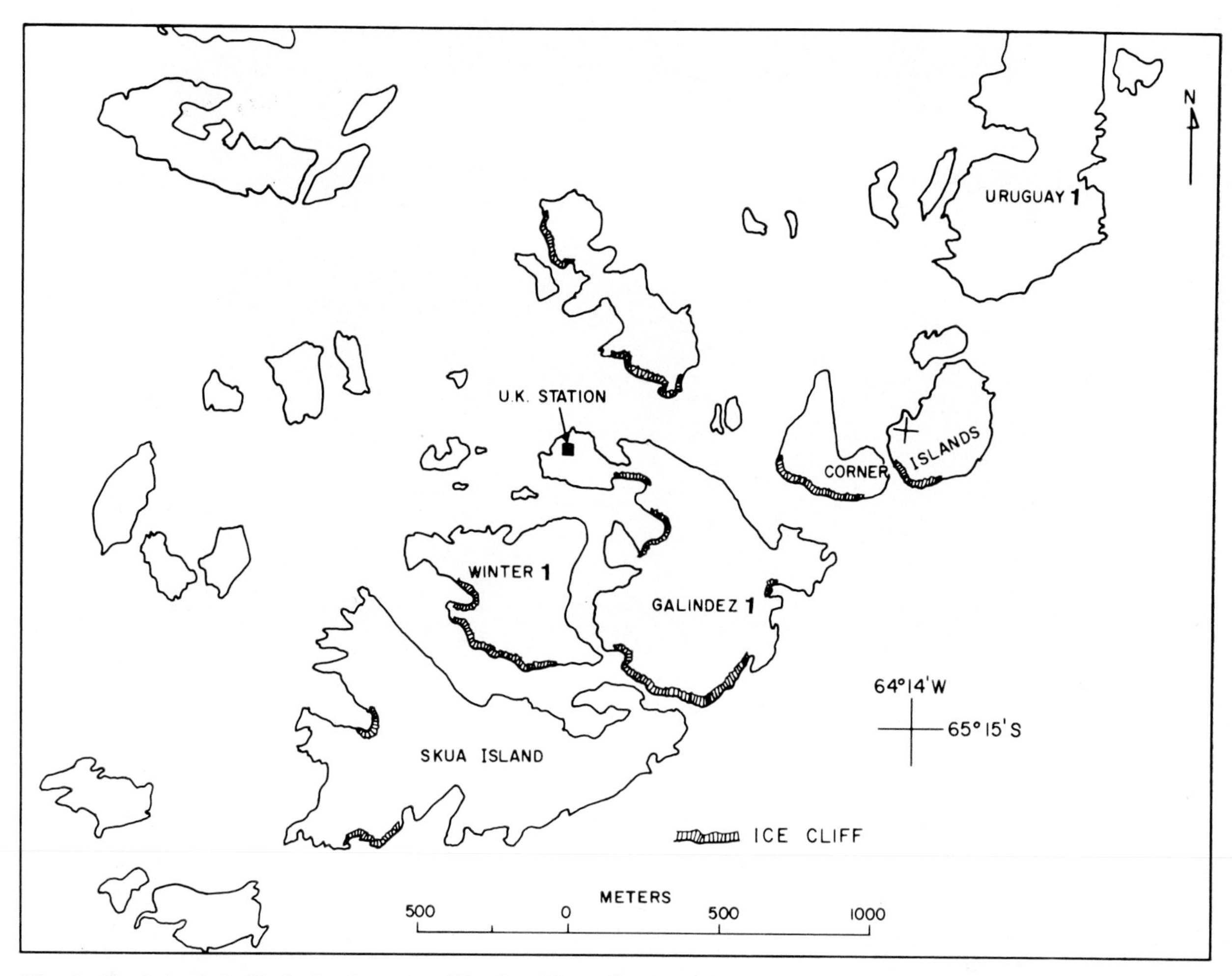

Fig. 2. Portion of archipelagic Argentine Islands. The collection locality (cross) is on the easternmost island of the Corner Islands.

four species were undoubtedly subject to the same macroclimatic influences.

Ecology. No attempt was made to classify the vegetation of the island. However, the particular site at which the plants were collected exhibited distinct bryophytic communities that fit rather well into the physiognomic classification system applicable to the maritime Antarctic. Of the six subformations within the antarctic nonvascular cryptogam tundra formation [Holdgate, 1964; Longton, 1967; Gimingham and Smith, 1970] three are apparent at this site: the moss hummock, the moss turf, and the moss carpet subformations.

The moss hummock subformation consists essentially of *C. austrostramineum,* which occurs in a distinct zone in and adjacent to a small pool (Figure 4a) fed throughout most of the summer from melting snow above the terrace.

The largest bryophytic community is the moss turf subformation, which is dominated by *P. strictum* (Figure 4b). This community consists of contiguous *P. strictum* mounds, some of which are apparently connected by a coalescence of formerly independent mounds. Other mounds are clearly separated by meandering canals probably maintained by melt water drainage. The mounds vary in size, many being $>\frac{1}{2}$ meter in diameter. One mound was 36 cm thick. Only the upper 0.5–1 cm of the *P. strictum* has green leaves, whereas the remainder forms a brown layer consisting of stems and leaves. There appears to be little or no decomposition (Figure 5) except in some cases for a layer 2–3 cm thick at the rock–moss interface that is often quite wet. Interwoven among the stems of *P. strictum* are shoots of *P. nutans* and two leafy liverworts. The relative abundance of these components was roughly esti-

Fig. 3. Aerial photograph of northwest corner of the easternmost island of the Corner Islands. The collection locality is indicated by the cross.

mated by the dry weights of six randomly selected plugs (3 cm in height and 1.8 cm in diameter) from the top portion of the moss turf. On the average, *P. strictum* constituted 95% of the community phytomass, whereas *P. nutans* and the two liverworts collectively constituted only 5% (Table 1). Although these data represent a very small sampling of the community, they do indicate the uniform dominance of *P. strictum*. But the large standard deviation for the weights of the subordinates indicates an uneven

Fig. 4. Moss communities of the collection site: A, *Calliergidium austrostramineum* hummock subformation; B, *Polytrichum strictum* turf subformation; C, *Drepanocladus uncinatus* carpet subformation.

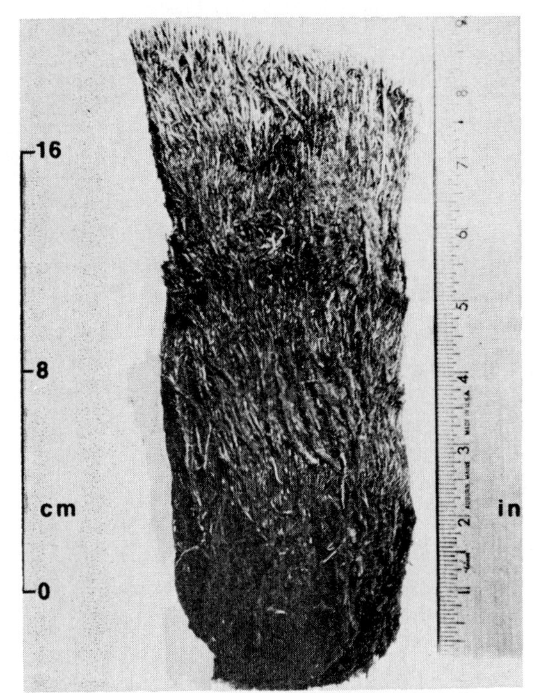

Fig. 5. Portion of *Polytrichum strictum* turf. Only the upper 1 cm of turf is green tissue; the rest is brown stems and leaves.

distribution of these plants within the *P. strictum* turf.

In the moss carpet subformation of this site *D. uncinatus* is the dominant species, *P. nutans* occurring as small tufts within and at the margins of the carpet (Figure 4c). This community differs from the *P. strictum* turf in that an appreciable amount of humic soil has accumulated beneath the upper portion of the carpet (Figure 6). The factors resulting in the differential decomposition between these two moss communities were not ascertained. Water no doubt plays an important role, and *P. strictum* may be more resistant to decay.

The fact that the species were relatively abundant permitted adequate sampling for experimental pur-

TABLE 1. Estimated Relative Abundance of Bryophytes in the *Polytrichum strictum* Turf Subformation

	Polytrichum strictum, mg dry weight	Subordinate Species,* mg dry weight
Sample plug†		
1	833.70	29.44
2	979.96	8.20
3	754.06	25.35
4	810.21	30.98
5	777.68	93.55
6	816.05	77.10
$\overline{X}$	828.61	44.10

The standard deviation and the per cent of total weight are 79.47 and 94.95, respectively, for *Polytrichum strictum* and 33.35 and 5.05, respectively, for the subordinate species.

* *Pohlia nutans* and two leafy liverworts.

† Plugs of turf surface 3 cm in height × 1.8 cm in diameter.

Fig. 6. Portion of *Drepanocladus uncinatus* carpet. Vertical line shows the organic soil that has accumulated between the bedrock BR and the chlorophyllous shoots.

poses without markedly decreasing the populations. Test samples of each species were collected from their respective subformations except for the samples of *P. nutans,* which were collected from the *D. uncinatus* carpet subformation. Specimens were preserved for identification, and voucher specimens were deposited in the U.S. National Herbarium of the Smithsonian Institution, Washington, D. C.

EXPERIMENTAL METHODS

Treatment of materials. On two occasions plants collected in the field were returned to the laboratory at Palmer Station in an ice-cooled insulated box. The plants collected in January were used for chlorophyll and lipid assays. Although samples were prepared and analyzed as quickly as possible, it was necessary to retain them for a few days in the ice-cooled insulated box, which was kept outside near the laboratory. Specimens collected in February were used for carbohydrate assays and for determining photosynthetic and respiratory rates. To maintain the material in reasonably good condition while measurements of photosynthesis and respiration were being made, the specimens for these tests were treated somewhat differently. The upper portions of the colonies of each species were placed in plastic trays and moistened with distilled water, and the trays were covered with a thin sheet of clear plastic. The trays containing the mosses were put in a wooden frame adjacent to the outside wall of the south facing side of the laboratory. The wooden frame was covered with 1.3-cm hardware cloth, which was necessary to keep the trays from blowing away during high winds. It is probable that little change occurred in the plants during the 4 weeks of storage because the prevailing weather conditions were generally poor and consisted of many cloudy days, near freezing temperatures, and occasional snowfalls.

Experimental samples. All the test samples consisted of the upper 2-cm portion of the colonies. The experimental samples of *P. strictum* consisted of the community complex described previously except for the photosynthetic and respiratory tests, in which only individual shoots of this species were used. Prior to the chlorophyll, lipid, and carbohydrate assays, plant samples were thoroughly washed, and fresh weights were determined by centrifuging each sample for 5 min at 2400 rpm in a porous bottomed plastic capsule at a radius of about 6 cm. This method removed surface water without damaging the tissue and gave constant fresh-weight/dry-weight ratios for a given species [Rastorfer and Higinbotham, 1968]. In the photosynthetic and respiratory tests fresh weights and dry weights were determined at the end of each experiment. Dry weights were measured after oven drying the plant samples for 24 hours at 90°C.

Element assays. Plant sample preparations for element assays were adapted from those described by Shacklette [1965b]. However, the initial drying and the air cleaning were eliminated. Instead about 10–15 grams fresh weight of each species was washed with distilled water and distilled–deionized water. After oven drying for 48 hours at 80°C the samples were ground in a Waring blender, and the resulting powders were sealed in screw cap glass vials for shipment to Ohio State University. Portions ranging from 0.5 to 1.0 gram were ashed in a muffled furnace maintained at 550°C for 14 hours after the furnace had reached 550°C from room temperature. Other portions of the powdered samples were assayed for element content at the Department of Agronomy, Ohio Research and Development Center, Wooster, Ohio.

Chlorophyll assays. Plant sample sizes ranged from 0.5 to 1 gram fresh weight. Chlorophylls were extracted by grinding the tissue in cold acetone, filtered through Wattman no. 1 filter paper into a separatory funnel, and then transferred to diethyl ether. The acetone was washed from the ether with distilled–deionized water, and traces of water were removed from the ether with anhydrous Na_2SO_4. Chlorophylls *a* and *b* were determined spectrophotometrically from 50-ml final volumes of the diethyl ether extracts [Koski, 1950; Smith and Benitez, 1955].

Lipid assays. Plant samples (0.8–1.3 grams fresh weight) were homogenized in chloroform-methanol (3:1 vol./vol. [Clark, 1964]) and subsequently centrifuged. The supernatant fluids were filtered into screw cap pyrex bottles to which 0.9% LiCl in H_2O (weight/vol.) was added to be equivalent to one-fifth the volume of each extract. These combined mixtures were shaken and placed overnight in a refrigerator at 4°C. A fluffy white layer of proteolipids [Robinson, 1963] formed at the liquid–liquid interface. It was retained with the organic phase, whereas the aqueous layer was removed and discarded. The lipid extracts were reduced to about 5 ml by vacuum evaporation. These concentrated extracts were transferred to aluminum weighing pans, taken to dryness over $CaCl_2$ in a vacuum desiccator, and after drying weighed to the nearest 0.1 mg.

Carbohydrate assays. The anthrone method was used

TABLE 2. Element Contents and Ash Weights on a Dry Weight Basis

	Calliergidium austrostramineum	*Drepanocladus uncinatus*	*Polytrichum strictum**	*Pohlia nutans*
Element content				
P, %	0.24	0.13	0.44	0.53
K, %	0.48	0.46	0.50	1.25
Ca, %	0.37	0.11	0.23	0.16
Mg, %	0.10	0.09	0.11	0.12
N, %	2.95	2.75	1.50	2.35
S, %	0.190	0.195	0.075	0.150
Mn, ppm	12	<10	23	14
Fe, ppm	162	78	>4000	1157
B, ppm	6	6	32	11
Cu, ppm	5	6	5	9
Zn, ppm	34	21	41	33
Mo, ppm	1.00	0.97	1.03	1.09
Na, %	0.06	0.06	0.03	0.20
Si, %	<0.01	<0.01	0.01	0.02
Al, ppm	167	50	1434	1503
Sr, ppm	38	26	32	28
Ba, ppm	4	<2	6	3
Ash weight,† %	5.32 ± 0.08	3.44 ± 0.08	1.49 ± 0.07	2.56 ± 0.01

* Analysis of *Polytrichum strictum* association, which included *Pohlia nutans* and two leafy liverworts.

† Values are means of two samples plus or minus the standard deviations.

separately in ethanol and water extracts to determine the soluble carbohydrates, which at best can be considered only crude estimates. Plant samples (1.2–2 grams fresh weight) were placed in Erlenmeyer flasks and first extracted for several hours with hot 95% ethanol. After three washes with ethanol the plant residues were extracted again for several hours with hot water. Measured portions from final volumes of 100 ml were used for the anthrone tests. Glucose standards were prepared and run with each spectrophotometric assay [Umbreit et al., 1957].

Measurements of photosynthesis and respiration. Rates of net photosynthesis and respiration were determined by measuring oxygen exchange with differential respirometers. A constant carbon dioxide concentration of 1% (vol./vol.) was maintained within the reaction flasks with a Pardee buffer (diethanolamine). Procedures for preparing this buffer system are found in Umbreit et al. [1957]. Details concerning apparatus, temperature control, light calibration, and so forth are given in an earlier report [Rastorfer, 1970]. Plant test samples consisted of 2-cm terminal portions of gametophytic shoots. Replicate test samples were washed with distilled water and subsequently placed in rectangular reaction flasks that were then retained in a refrigerator at 4°C for 12 hours (overnight) prior to each experiment. The ensuing experimental procedures followed those described by Rastorfer and Higinbotham [1968] and Rastorfer [1970].

RESULTS

Elements and ash weights. The results of element assays and ash weight determinations based on grams dry matter are shown in Table 2. In comparing the four species, one finds the relatively higher amounts of phosphorus, iron, boron, and aluminum obtained for *P. strictum* and *P. nutans* the most striking. Although nitrogen and probably sulfur concentrations should be determined from freshly collected material, the results presented here are useful in comparing the four species, since all the samples were treated and assayed in the same way. Nitrogen and sulfur concentrations were lowest in *P. strictum*

TABLE 3. Chlorophyll, Lipid, and Carbohydrate Contents

	Calliergidium austrostramineum	*Drepanocladus uncinatus*	*Polytrichum strictum**	*Pohlia nutans*
Chlorophyll *a*, $\mu g\ g^{-1}$ fresh weight	617 ± 16	593 ± 33	331 ± 13	213 ± 30
Chlorophyll *b*, $\mu g\ g^{-1}$ fresh weight	283 ± 9	295 ± 16	155 ± 8	100 ± 16
Total *a* and *b*	900	888	486	313
Ratio of *a* to *b*	2.18 ± 0.02	2.02 ± 0.01	2.13 ± 0.03	2.14 ± 0.04
Lipids, mg g^{-1} fresh weight	20.2 ± 2.3	30.4 ± 0.6	19.0 ± 3.8	38.4 ± 11.3
Carbohydrates,† mg g^{-1} fresh weight				
Ethanol extract	5.56 ± 0.29	4.86 ± 0.41	3.70 ± 0.02	8.78 ± 0.35
Aqueous extract	7.15 ± 0.15	1.86 ± 0.05	4.55 ± 1.77	4.41 ± 0.45
Total	12.71	6.72	8.25	13.19

Values are means of two samples plus or minus the standard deviation where indicated.

* Analysis of *Polytrichum strictum* association, which included *Pohlia nutans* and two leafy liverworts.

† Crude estimates based on anthrone tests.

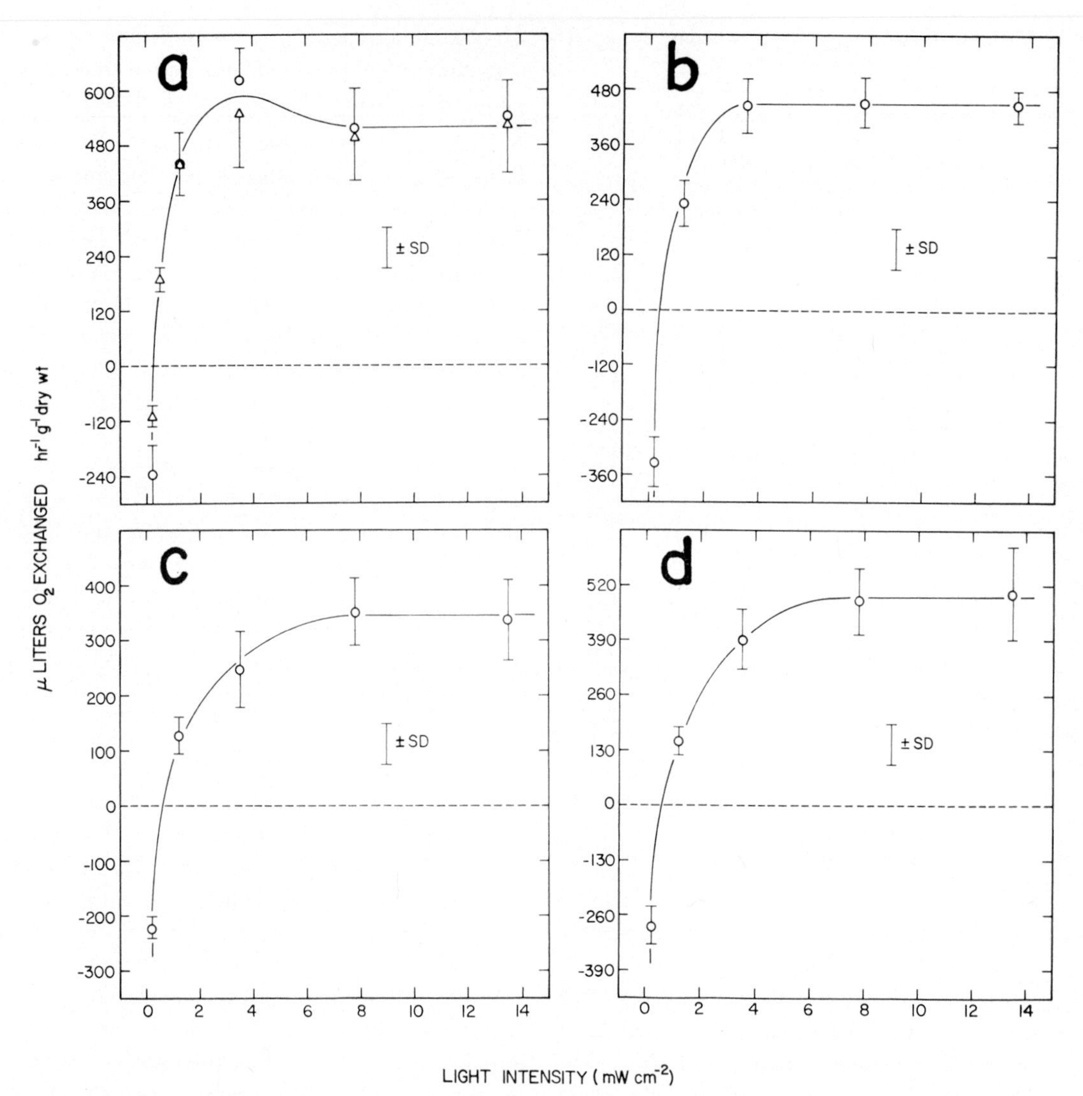

Fig. 7. Effects of light intensity on net photosynthesis at 10°C and 1.0% CO_2. (*a*) *Calliergidium austrostramineum* (circles, $N = 5$; triangles, $N = 4$). (*b*) *Drepanocladus uncinatus*, $N = 4$. (*c*) *Polytrichum strictum*, $N = 5$. (*d*) *Pohlia nutans*, $N = 4$. Points are means of N samples; SD, standard deviation.

in contrast to markedly higher concentrations in the other three species. Percentages of ash weight were surprisingly low for *P. strictum*, comparatively high for *C. austrostramineum*, and intermediate for *D. uncinatus* and *P. nutans*.

Chlorophylls. Data pertaining to the chlorophyll assays are shown in Table 3. Particularly noticeable are the higher total chlorophyll (*a* and *b*) concentrations of *C. austrostramineum* and *D. uncinatus* than of *P. strictum* and *P. nutans*. In spite of the large differences in total chlorophyll concentrations, ratios of chlorophyll *a* to chlorophyll *b* were about the same (approximately 2:1) in all four species.

Lipids. Amounts of extractable lipids in chloroform-methanol were highest in *D. uncinatus* and *P. nutans* (Table 3). The lipid contents of *C. austrostramineum* and *P. strictum* were about the same but were lower than those of the other two species.

Carbohydrates. Table 3 shows the results for extractable carbohydrates in 95% ethanol and water solvents on the basis of glucose equivalence with the anthrone test. Although these values are crude estimates, differences are apparent among the four species. *Drepanocladus uncinatus* and *P. nutans* had larger amounts of carbohydrates in the ethanol extracts than in the aqueous extracts, the amounts differing by approximately a factor of 2. Conversely *C. austrostramineum* and *P. strictum* had larger

TABLE 4. Net Photosynthetic Rates at Various Light Intensities

Light Intensity, mW cm^{-2}	*Calliergidium austrostramineum*, $N = 5$	*Drepanocladus uncinatus*, $N = 4$	*Polytrichum strictum*, $N = 5$	*Pohlia nutans*, $N = 4$
0.18	−47 ± 14	−82 ± 16	−62 ± 3	−64 ± 10
1.17	88 ± 14	57 ± 11	36 ± 9	35 ± 8
3.50	124 ± 13	108 ± 11	70 ± 19	89 ± 17
7.80	102 ± 19	111 ± 10	100 ± 18	109 ± 19
13.5	108 ± 16	109 ± 6	95 ± 20	113 ± 26

Values of net photosynthetic rates are means of N samples plus or minus the standard deviations and are in microliters of oxygen exchanged per hour per gram fresh weight.
Temperature was 10°C, and CO_2 was 1% in air.

amounts of carbohydrates in the aqueous extracts, the differences between the two extracts not being so pronounced. *Calliergidium austrostramineum* and *P. nutans* had essentially the same amounts of total carbohydrates (ethanol and aqueous), whereas *D. uncinatus* and *P. strictum* had somewhat different amounts and less than the other two species.

Effects of light intensity on net photosynthesis. Net photosynthetic rates at 10°C were determined at five different light intensities from 0.18 to 13.5 mW cm^{-2}. The results based on grams dry weight are shown in Figure 7, and those based on grams fresh weight are shown in Table 4. Light compensation points were 0.25 and 0.35 mW cm^{-2} for *C. austrostramineum* (Figure 7a) and *D. uncinatus* (Figure 7b), respectively. But higher light compensation points of 0.6 mW cm^{-2} were obtained for *P. strictum* (Figure 7c) and *P. nutans* (Figure 7d). Furthermore light saturation was reached at 3.5 mW cm^{-2} for *C. austrostramineum* and *D. uncinatus* and at about 8 mW cm^{-2} for *P. strictum* and *P. nutans*.

Calliergidium austrostramineum differed from the other species in having had the highest net photosynthetic rate at 3.5 mW cm^{-2}, which was followed by lower rates with subsequent increases in light intensities (Figure 7a). On the other hand, the other three species showed no further changes in net photosynthesis at light intensities above their saturation points (Figures 7b–d). Whether this response for *C. austrostramineum* was real is uncertain, but the results of two independent experiments (Figure 7a) were similar and thus indicate that there may be a critical light effect for this species around 3.5 mW cm^{-2}.

Maximal net photosynthetic rates varied from a low of 345 μl O_2 evolved hr^{-1} g^{-1} dry weight for *P. strictum* to a high of 588 μl O_2 evolved hr^{-1} g^{-1} dry weight for *C. austrostramineum*. Intermediate rates of 450 and 494 μl O_2 evolved hr^{-1} g^{-1} dry weight were obtained for *D. uncinatus* and *P. nutans*, respectively. Maximal net photosynthetic rates expressed as grams fresh weight show less disparity among the four species than those expressed as grams

TABLE 5. Final Fresh and Dry Weights and Per Cent Dry Matter and Water Contents* of Plant Samples Used in the Photosynthetic and Respiratory Experiments

	Calliergidium austrostramineum	*Drepanocladus uncinatus*	*Polytrichum strictum*	*Pohlia nutans*
Light intensity tests				
Number of samples N	5	4	5	4
Fresh weight, mg	289.0 ± 13.0	372.3 ± 39.8	441.1 ± 61.5	358.0 ± 13.9
Dry weight, mg	57.5 ± 2.9	91.3 ± 10.2	124.8 ± 19.1	80.6 ± 2.6
Dry matter,† %	19.90 ± 0.56	24.52 ± 1.29	28.27 ± 1.12	22.52 ± 0.23
Water content,‡ %	402.3 ± 14.26	308.7 ± 21.88	254.2 ± 14.34	344.2 ± 4.61
Temperature tests				
Number of samples N	5	3	5	3
Fresh weight, mg	318.6 ± 21.4	411.2 ± 10.4	525.8 ± 31.0	478.5 ± 57.7
Dry weight, mg	60.9 ± 4.3	89.3 ± 6.0	149.9 ± 8.8	99.4 ± 7.3
Dry matter,† %	19.13 ± 0.44	21.71 ± 0.93	28.51 ± 0.38	20.85 ± 1.16
Water content,‡ %	423.0 ± 12.0	361.2 ± 20.27	250.8 ± 4.70	380.5 ± 25.92

Values are means of N samples plus or minus the standard deviations.
* Determined at end of respective experiments.
† Equal to the dry-weight/fresh-weight ratio times 100.
‡ Equal to the fresh-weight-minus-dry-weight/dry-weight ratio times 100.

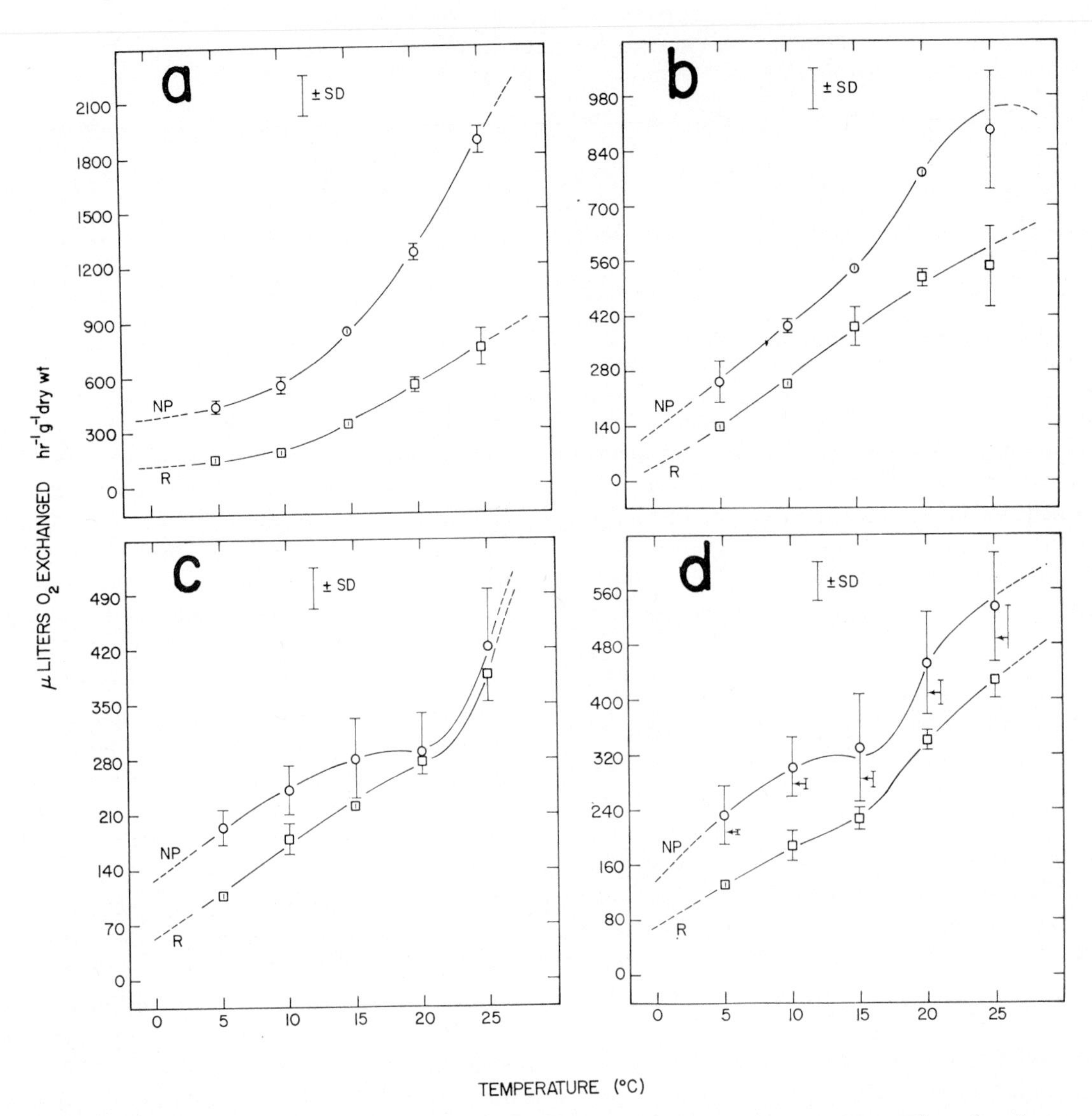

Fig. 8. Effects of temperature on net photosynthesis NP, at a light intensity of 5.27 mW cm^{-2} and on respiration R (1.0% CO_2). (*a*) *Calliergidium austrostramineum*, N = 3–5. (*b*) *Drepanocladus uncinatus*, N = 3. (*c*) *Polytrichum strictum*, N = 4–5. (*d*) *Pohlia nutans*, N = 3 (small arrows and bars indicate the means and the standard deviations of two samples). Points are means of N samples; SD, standard deviation.

dry weight (Table 4, Figures 7a–d). These results can be explained at least in part by the differences in the percentages of dry matter relative to each species (Table 5). *Polytrichum strictum* had the highest percentage of dry matter, and *C. austrostramineum* the lowest. As one may expect, the other two species had intermediate percentages of dry matter.

Effects of temperature on net photosynthesis and respiration. Net photosynthetic and respiratory rates were determined at 5° intervals over a temperature range of 5°–25°C. Net photosynthesis was measured at a light intensity of 5.27 mW cm^{-2}, whereas respiration was measured in the dark. The rates expressed as grams dry weight are shown in Figure 8, and those expressed as grams fresh weight are shown in Table 6.

The net photosynthetic and respiratory rates for *C. austrostramineum* (Figure 8a) and *D. uncinatus* (Figure 8b) increased with increasing temperatures throughout the temperature range tested. No optimal temperature for net photosynthesis is indicated from the curve for *C. austrostramineum;* however, a leveling off from 20° to 25°C is indicated from the net photosynthetic curve for *D. uncinatus*. This temperature range may be interpreted as being close to the optimal temperature for this species. The slopes of the curves for net photosynthesis and respiration differed between the two species. Those for *C. austrostramineum* are initially shallow and thus indicate low Q_{10} at low temperatures. In contrast, those for *D. uncinatus* are initially steeper and thus indicate higher Q_{10} for the same temperatures (5°–15°C).

The net photosynthetic and respiratory responses of *P. strictum* (Figure 8c) and *P. nutans* (Figure 8d) to temperature changes were quite different from those of *C. austrostramineum* and *D. uncinatus*. For *P. strictum* the net photosynthetic rates increased with increasing temperatures from 5° to 15°C, leveled off between 15° and 20°C, but then increased sharply again from 20° to 25°C. For *P. nutans* the responses were similar to those for *P. strictum* except that the net photosynthetic rates initially leveled off between 10° and 15°C and then increased with increasing temperatures. These results indicate that these species may have two or more optimal temperatures for photosynthesis. The respiratory responses to temperature changes were not as pronounced as the net photosynthetic responses; nevertheless deflections in the respiratory curves are noticeable.

Ratios of net photosynthesis to respiration were highest for *C. austrostramineum* but markedly lower for the other three species (Table 6). The ratios were highest at 5°C for all four species and then generally tended to decrease with increasing temperatures.

TABLE 6. Net Photosynthetic and Respiratory Rates and Net Photosynthesis/Respiration Ratios at Various Temperatures

Temperature, °C	*Calliergidium austrostramineum*, $N = 3–5$	*Drepanocladus uncinatus*, $N = 3$	*Polytrichum strictum*, $N = 4–5$	*Pohlia nutans*, $N = 3$
	*Net Photosynthetic Rates**			
5	84 ± 7	55 ± 14	56 ± 7	48 ± 7
10	106 ± 12	86 ± 7	69 ± 9	63 ± 8
15	163 ± 3	117 ± 8	80 ± 15	69 ± 14
20	245 ± 11	170 ± 6	83 ± 13	94 ± 13
25	361 ± 12	192 ± 29	121 ± 21	111 ± 13
	Respiratory Rates			
5	29 ± 6	30 ± 4	30 ± 2	27 ± 2
10	35 ± 2	54 ± 5	51 ± 6	39 ± 3
15	66 ± 7	86 ± 14	63 ± 2	47 ± 5
20	107 ± 10	112 ± 8	79 ± 4	71 ± 2
25	145 ± 23	118 ± 23	111 ± 11	89 ± 4
	Net Photosynthesis/Respiration Ratios			
5	3.05	1.81	1.82	1.76
10	3.01	1.61	1.34	1.61
15	2.48	1.38	1.27	1.46
20	2.30	1.52	1.05	1.32
25	2.54	1.64	1.08	1.24

Values for net photosynthetic and respiratory rates are means of *N* samples plus or minus the standard deviations. Net photosynthetic and respiratory rates are in microliters of oxygen evolved per hour per gram fresh weight and microliters of oxygen uptake per hour per gram fresh weight, respectively.

* Light intensity was 5.27 mW cm^{-2}, and CO_2 was 1% in air.

DISCUSSION AND CONCLUSIONS

Macronutrient elements. Other than hydrogen, oxygen, and carbon the six macronutrient elements considered essential for metabolic processes in plants are phosphorus, potassium, calcium, magnesium, nitrogen, and sulfur. The relative amounts of the latter elements in the mosses included in this study, in several other antarctic mosses, and in one of the two antarctic flowering plants, *Deschampsia antarctica* (a grass), occurring on Signy Island are given in Table 7. In addition the values for *Zea mays* (corn) are given to serve as a general point of reference [Evans and Sorger, 1966] and can be considered to be the internal concentrations adequate for growth in higher plants [Salisbury and Ross, 1969, p. 194]. In comparison to *Zea mays* the antarctic mosses on the average and *Deschampsia antarctica* have smaller amounts of potassium. The other elements occur either in about equal or in greater amounts in the mosses and *Deschampsia antarctica* than in *Zea mays* except for the magnesium content, which is relatively low in the four moss species from the Argentine Islands.

Shacklette [1965b] suggests that north temperate bryophytes have smaller amounts of phosphorus, potassium, calcium, and magnesium than the vascular plants. This interpretation was based on the average contents of various mosses compared to those of vascular plants occurring in the same localities. At present, data are too limited for a similar comparison of antarctic plants, but the data for the Signy Island plants (Table 7) do indicate that there are only minor differences between the average content of five macroelements of some antarctic mosses and that of *Deschampsia antarctica*. However, comparisons of this type, though useful in a general way, must be regarded with some reservations. The seemingly low percentages of some of the macronutrient elements in the four moss species included in this study, e.g., magnesium (Table 2), compared to those in *Zea mays*

TABLE 7. Macronutrient Element Content of Several West Antarctic Mosses, a West Antarctic Flowering Plant, and a Crop Plant in Per Cent of Dry Weight

	Mosses from Argentine Islands*		Mosses from Signy Island†		Flowering Plants	
Element	Mean	Range	Mean	Range	*Deschampsia antarctica*‡	*Zea mays*§
P	0.34	0.13–0.53	0.16	0.07–0.30	0.25	0.20
K	0.67	0.46–1.25	0.46	0.30–0.80	0.40	0.92
Ca	0.22	0.11–0.37	0.64	0.15–1.74	0.36	0.23
Mg	0.11	0.09–0.12	0.41	0.24–0.70	0.44	0.18
N	2.39	1.50–2.95	1.24	0.57–1.66	1.85	1.45
S	0.15	0.08–0.20				0.17

* Values are for surface portions of *Calliergidium austrostramineum, Drepanocladus uncinatus, Polytrichum strictum,* and *Pohlia nutans* and are from Table 3 and the text.

† Values are for surface portions of *Dicranum aciphyllum, Drepanocladus uncinatus, Brachythecium* sp., *Andreaea* sp., *Grimmia* sp., *Tortula* sp., and a mixed *Polytrichum–Dicranum* community and are from Allen et al. [1967, p. 387].

‡ Values are for surface portion and were obtained from plants found on Signy Island [Allen et al., 1967, p. 387].

§ Values are from Evans and Sorger [1966].

cannot be considered mineral deficiencies for these mosses until it has been demonstrated experimentally. Culture studies with a few bryophytes indicate that these elements are required in certain combinations for normal growth; however, their minimal internal concentrations have not been ascertained [Voth and Hamner, 1940; Fulford et al., 1947; Hoffman, 1966]. In addition Tamm [1953] has shown that the differences in the local concentrations of a given element in the Scandinavian forest moss *Hylocomium splendens* are likely to be greater than the differences between comparable samples from different localities. This observation would undoubtedly apply to antarctic moss communities, and therefore the element concentrations reported in this study must be regarded as only approximations.

The concentration of a given element in a moss may depend on the age of the segment assayed. In *Hylocomium splendens* (Sweden and Norway) nitrogen concentrations were highest in young segments but decreased to a consistent level in older segments. Phosphorus and potassium concentrations decreased with each successively older segment, whereas calcium concentrations increased [Tamm, 1953]. Similar patterns between current and older growth were found for potassium and calcium in four epiphytic mosses in Scotland. Nitrogen and phosphorus concentrations, however, were either about the same for both types of growth or were higher in the older segments of some species [Grubb et al., 1969]. Under the antarctic conditions at Signy Island, Allen et al. [1967] found no appreciable difference between the element contents (phosphorus, sodium, calcium, magnesium, carbon, and nitrogen) of a living moss surface and of its underlying brown parent material except for the potassium content which was lower in the underlying brown layer. This observation was applicable only to those taxa that accumulate thick layers of organic matter, such as *Polytrichum, Dicranum, Drepanocladus,* and *Brachythecium.* Lower potassium levels were not detected in the thin organic layers beneath the taxa growing close to rock surfaces (e.g., species of *Andreaea* and *Grimmia*) or subjected to heavy water flushing (e.g., species of *Tortula*).

Studies concerning the macronutrient content of mosses in relation to their sources are quite limited for temperate species. The best known recent works in this field are those of Tamm [1953] and Shacklette [1965b]. Tamm's work is especially important in showing that rain, atmospheric dust, and tree 'leachates' are important factors in supplying major mineral nutrients to a forest carpet moss community dominated by *Hylocomium splendens.* He is somewhat doubtful about the source of nitrogen but suggests that atmospheric ammonia, the ammonia and nitrate in rain, and possibly the nitrogenous compounds in water from tree washings could furnish adequate amounts. Although the environments of forest moss communities and antarctic moss communities are very different, there appear to be some similarities with regard to the sources of certain elements.

To date, the only major studies of the sources and the availability of macronutrient elements to antarctic bryophytic communities are those of Allen et al. [1967], Allen and Northover [1967], and Northover and Allen [1967], which were carried out on Signy Island. Those workers consider the supply of macronutrients to be adequate for plant growth and to be derived from two sources: rocks and the ocean. Rocks are probably the major source of potassium and may also supply some calcium via physicochemical weathering processes. Magnesium is carried in the atmosphere directly from the ocean to the vegetation in precipitation. The ocean indirectly supplies nitrogen and phosphorus from land-breeding marine fauna (seals, penguins, and other sea-feeding birds). Phosphorus probably arrives directly from bird droppings, whereas nitrogenous compounds could be carried in drainage water and ammonia in the atmosphere from breeding sites. The source and distribution of sulfur apparently have not been examined, but the ocean seems to be the most likely source. If these patterns of macronutrient element distribution apply to Signy Island, they should apply to the Argentine Islands as well. The South Polar skua was seen on the island in this study, whereas penguins and seals were not. It seems reasonable that the skua could provide appreciable amounts of organic nutrients and some inorganic nutrients to bryophytic communities. Nevertheless the contribution and the use of mineral nutrients by other organisms cannot be discounted in evaluating the total nutrient balance. Arthropods (collembola, mites, and midges) and nematodes are known to inhabit moss communities in the maritime Antarctic [Strong, 1967; Tilbrook, 1967]. Microorganisms (bacteria, fungi, and protozoa), though fewer here than in areas of warmer climates, are found in moss communities and appear to be decomposers, even though their action is relatively slow [Boyd et al., 1966; Heal et al., 1967]. The extent of the microflora (algae, bacteria, and fungi) in antarctic bryophytic communities and their relationships to nutrient element accumulation and release, especially to nitrogen fixation, require further investigation.

Micronutrient elements. In addition to the macronutrient elements seven other elements (manganese, iron, boron, copper, zinc, chlorine, and molybdenum) are considered essential for plant growth but are required in smaller amounts. Analyses of all these micronutrient elements except chlorine are included in this study (Table 2). When the internal concentrations considered adequate for higher plants [Salisbury and Ross, 1969, p. 194] are used as a basis for comparison, manganese is low in all four species. Boron is low in three species, *P. strictum* being the exception. Iron is slightly low in *D. uncinatus* but excessive in the other species, especially in *P. strictum* and *P. nutans.* The other micronutrient elements are in either adequate or greater amounts than those required for higher plants. However, this comparison is applicable only in a general way, because neither the essentiality nor the minimal concentrations of these elements have been established for bryophytes. In addition the values in Table 2 do not indicate the variations of the element concentrations within the habitats sampled for each species. There is also the possibility of contamination, which is difficult to evaluate. According to Shacklette's [1965b] assessment, the per cent ash weights obtained in this study (all of which are <10%) seem to indicate little or no contamination. Nevertheless the high iron content obtained for *P. strictum* (Table 2) seems excessively large and may be suspect. On the other hand, high iron concentrations have been reported for a number of mosses [Tamm, 1953; Shacklette, 1965b; Touffet, 1966; Paribok et al., 1967].

Other elements. No comments have been made as yet concerning elements that are known to occur in plants but that have not been confirmed as essential for the growth of bryophytes and most higher plants [Evans and Sorger, 1966; Lewin and Reimann, 1969]. Those elements reported in this study are sodium, silicon, aluminum, strontium, and barium (Table 2). These and other elements including the rare earth elements, are reported to occur in a large number of bryophytes. Bryophytes appear likely to contain a greater number and often greater amounts of nonessential elements than vascular plants [Shacklette, 1965a, b, 1967]. Of special note in this report are the relatively high concentrations of aluminum and iron, mentioned earlier, in *P. strictum* and *P. nutans* (Table 2). These elements seem likely to be derived from the rock substratum, which has a high aluminum content (feldspar) and a smaller iron content [Elliot, 1964]. Even though the two pleurocarpous forms (*C. austrostramineum* and *D. uncinatus*) are related to the same substratum, the observed aluminum and iron contents (Table 2) indicate that the two acrocarpous forms were able to accumulate these elements in larger amounts. This observation may be an exception, because there seems to be no definite indication in the literature of specific relationships between growth form and element accumulation. However, the higher aluminum and iron concentrations in *P. strictum* and *P. nutans* may be related to their mode of water absorption and movement. In this respect *P. strictum* and *P. nutans* can be considered endohydric mosses, whereas *C. austrostramineum* and *D. uncinatus* are probably ectohydric mosses [Watson, 1964]. The endohydric condition could be advantageous for water and ion absorption in Antarctica, where there is little precipitation, and in situations where the supply of water and minerals to terrestrial plant communities after the spring snow melt depends on the rock surface movement of melt water from glaciers or persistent snow patches. This interpretation agrees in a general way with the elemental assays (Table 2) of *P. strictum* and *P. nutans* compared to those of the other two species, especially *D. uncinatus*. The higher concentrations of aluminum, iron, and some other elements in *C. austrostramineum* than in *D. uncinatus* are to be expected, since the former species was in a wetter environment (Figure 4).

Chlorophyll. The major plastid pigments of mosses do not differ qualitatively from those of the leaves of vascular plants [Douin, 1956; Freeland, 1957; Goodwin, 1965; Strain, 1966]. Although it seems unlikely that antarctic mosses would differ appreciably in this respect, their pigment composition is too little known to permit an evaluation. The chlorophyll *a*/chlorophyll *b* ratios are generally considered to be about 3 : 1 in the leaves of vascular plants, but this ratio seems generally lower in moss gametophytes. For temperate mosses Wolf [1958] reported chlorophyll *a/b* ratios of 1.40 : 1 for *Polytrichum commune* and 1.45 : 1 for *Dicranum scoparium*. Rastorfer [1962] found higher ratios (2.10 : 1–2.34 : 1) for *Atrichum undulatum*. Mosses in arctic tundra communities had chlorophyll ratios of 0.99 : 1–1.89 : 1, whereas vascular plants in the same community had higher ratios (but <3 : 1) [Tieszen and Johnson, 1968]. Samples of *Bryum antarcticum* freshly collected from the field in Victoria Land, Antarctica, had a chlorophyll ratio of 1.5 : 1 [Rastorfer, 1970]. In contrast all four species included in this study had higher ratios (2.02 : 1–2.18 : 1) (Table 3). Even though the chlorophyll ratios between the pleurocarpous and the acrocarpous forms were about the same, their chlorophyll concentrations were quite different.

The combined chlorophyll *a* and *b* contents of the pleurocarpous forms, *D. uncinatus* and *C. austrostramineum*, were about twice those of the acrocarpous forms, *P. nutans* and *P. strictum* (Table 3). Whether similar trends in the amount of chlorophyll occur among different moss growth forms in temperate regions appears unknown. However, the chlorophyll content of a given moss species may differ between different populations. Lowland forest populations of *Polytrichum juniperinum* in Indiana had 540 $\mu g\ g^{-1}$ fresh weight, whereas alpine populations in Mt. Washington, New Hampshire, had 446 $\mu g\ g^{-1}$ fresh weight. These concentrations seem quite high in comparison to only 120 $\mu g\ g^{-1}$ fresh weight of chlorophyll in *Polytrichum commune* (probably collected in Tennessee) reported by Wolf [1958] but are close to the value reported in this study for *P. strictum*. Although most mosses can be considered evergreens, the amounts of gametophytic chlorophylls likely differ at different times of the year. However, *Atrichum undulatum* gametophytes of southwestern Ohio showed no significant change in chlorophyll content measured over a 13-week period. The concentrations were about 650 $\mu g\ g^{-1}$ dry weight, or roughly 130–162 $\mu g\ g^{-1}$ fresh weight calculated on a 20–25% dry weight basis [Rastorfer, 1962]. On the other hand, Miyata and Hosokawa [1961] found seasonal variations in several epiphytic mosses on *Carpinus tshonoskii* near Itaya-toge, a mountain pass of Seburi-yama, Japan. Maximal chlorophyll concentrations were obtained in October through December (35–70 $\mu g\ g^{-1}$ dry weight).

At present no data on the chlorophyll content of antarctic vascular plants appear to be available, and thus no comparative evaluation between these plants and antarctic mosses can be made. In arctic tundra plant communities near Barrow, Alaska, Tieszen and Johnson [1968] have found that mosses are a significant floristic component in reference to their contribution to chlorophyll per unit area, and in some cases the amount of chlorophyll of the moss synusium may exceed that of the vascular plant groups. Comparative studies of alpine and arctic plant communities have shown that for given species of vascular plants there is less chlorophyll in alpine forms than in arctic forms [Billings and Mooney, 1968]. This difference is attributed to genetic and environmental factors; however, among the environmental factors, the higher light intensities of the alpine environment cause a lower chlorophyll content, whereas the lower light intensities of the arctic environment are less damaging to chlorophyll. To what extent this principle applies to species occurring in the antarctic and southern alpine environments is unknown. On Ross Island the high solar radiation during antarctic summer days is known to reduce the chlorophyll in surface fresh-water phytoplankton [Goldman et al., 1963]. Chlorophyll in the leaves of *Bryum antarcticum* from Victoria Land also appears to be damaged by direct solar radiation [Rastorfer, 1970], but desiccation moisture may cause a lower chlorophyll content as well [Fukushima, 1968].

The effects of light intensity on the chlorophyll content of the mosses included in this study were not ascertained. The experimental results did show that *C. austrostramineum* and *D. uncinatus*, which had the highest chlorophyll contents (Table 3), required less radiation to reach light saturation than the other two species did (Figures 7a–d). However, at light intensities of >7.80 mW cm^{-2} the net photosynthetic rates based on grams fresh weight were about the same in all four species (Table 4). Tieszen and Johnson [1968] found that the cotton grass (*Eriophorum vaginatum*) community had a net production rate just as high as or higher than that of the other three arctic tundra communities, even though it had one of the lowest chlorophyll contents. Direct relationships between chlorophyll concentrations and photosynthesis in relation to net productivity do not appear to be understood clearly, as Tieszen and Johnson [1968] have pointed out.

Lipid and carbohydrate components. Lipids and carbohydrates are two very important classes of organic compounds that contribute to the composition and metabolic processes of each plant cell. Huneck [1969] has briefly reviewed the recent literature pertaining to the organic contents of bryophytes, including lipids and carbohydrates. In general the carbohydrate composition of mosses does not differ substantially from that of higher plants. Except for a few studies pertaining mostly to fatty acids the general lipid composition of mosses remains poorly known. Arachidonic acid, which was formerly thought to be restricted to animals, has unexpectedly been reported in a few mosses and vascular cryptogams, but it seems to be absent in higher vascular plants [Wolf et al., 1966]. However, the physiological significance, if any, of the presence of arachidonic acid in these plants is unknown. No attempt has as yet been made to identify the various lipid and carbohydrate compounds of antarctic mosses.

The principal reserve foods of bryophytes are considered to be starches, lipids, or both. The presence or the absence of starch in a given tissue varies greatly among specific taxa and thus appears to be controlled genetically [Schuster, 1966]. However, the influence of environmental factors coupled with genetic factors on the type(s) of food reserve deposition is not clearly understood. Data for those species occurring under stress conditions (cold and drought) are especially lacking. Larger amounts of lipids than of carbohydrates were found in the growing shoots of all four species examined in this study (Table 3). The lipid contents based on grams fresh weight ranged from 1.9% in *P. strictum* to 3.8% in *P. nutans*, whereas the crude carbohydrate estimates ranged from 0.6% in *D. uncinatus* to 1.3% in *P. nutans*. When they were calculated as the per cent of dry matter of those samples shown in Table 5 (averages), the lipid contents ranged from 6.7% in *P. strictum* to 18% in *P. nutans*, and the carbohydrates from 2.9% in *D. uncinatus* to 6.5% in *C. austrostramineum*. On the basis of the average values of each component there was about 2.6 times more lipid material than soluble carbohydrate material. The greater amounts of lipids cannot as yet be attributed to the physiological responses of these species to the cold antarctic environment. In the temperate moss *Rhodobryum roseum* the stored fatty material is reported to account for about 18% of the dry weight [Watson, 1964]. Since mosses lack the storage organs of higher plants, a relatively higher lipid content would permit the maximum energy reserves in a limited space. Similarly lipids constitute the major food stored in the leaves and the stems of arctic and alpine evergreen shrubs, whereas carbohydrates predominate in the underground storage organs of herbaceous perennials [Hadley and Bliss, 1964; Billings and Mooney, 1968].

Billings and Mooney [1968] report that cold resistance in arctic and alpine vascular plants is closely related to soluble carbohydrates, especially raffinose. This finding concurs with the Heber and Santarius [1967] view that frost hardening requires the production of soluble sugars or other protective substances (which may be sugar derivatives) to protect the sensitive structures, the proteins and the membranes, against the damaging effects of freezing.

Many bryophytes can endure very low temperatures, several tropical species having withstood −16°C [Biebl, 1964]. In Antarctica, mosses endure freezing temperatures for long periods. For example, the temperatures recorded among the leaves of *Polytrichum strictum* (= *P. alpestre* Hoppe) [Greene et al., 1970] were below freezing from April through the middle of November 1965, including a low of −22°C recorded in August [Longton, 1970]. Furthermore, even during the summer, moss colonies usually have to endure daily temperature fluctuations from above to below freezing [Rudolph, 1967; Matsuda, 1968; Longton, 1970]. Longton [1970] reports 25 short-term freeze–thaw cycles for *P. strictum* during February. These observations imply that the antarctic mosses either are permanently cold conditioned (frost resistant) or require no cold hardening, as vascular plants do [Weiser, 1970]. The ability of antarctic mosses to endure frequent and rapid temperature changes and their apparent high lipid to soluble carbohydrate ratio may suggest cold endurance mechanisms different from those proposed for vascular plants; however, this possibility has not been adequately explored.

Photosynthesis and respiration. The light intensity necessary for photosynthetic mechanisms to reach light compensation and light saturation levels differs among various plant species. In this study the effects of light intensity on net photosynthesis were quite different for morphologically dissimilar growth forms. Light compensation and light saturation points were higher by nearly a factor of 2 in the acrocarpous forms, *P. strictum* and *P. nutans,* than in the pleurocarpous forms, *C. austrostramineum* and *D. uncinatus* (Figures 7a–d). However, these observations should not be interpreted as a generalization relating to growth forms and light requirements for mosses occurring elsewhere. The relatively higher chlorophyll contents of *C. austrostramineum* and *D. uncinatus* appear to relate to higher net photosynthetic rates but only at low light intensities. At high light intensities (about 7 mW cm^{-2} or greater) the net photosynthetic rates based on grams fresh weight (Table 4) were about the same for all four species. However, at low light intensities the net photosynthetic rates were higher for *C. austrostramineum* and *D. uncinatus* than for *P. strictum* and *P. nutans,* as the initial slopes (above the compensation points) of the curves in Figures 7a–d and the values in Table 4 indicate. Thus these data indicate that a relatively high combined chlorophyll *a* and *b* content is advantageous only at low light intensities. Similar observations have been reported for various vascular plants [Heath, 1969]. In addition Heath discusses further effects of chlorophyll content and light quality on the photosynthetic mechanism.

Photosynthetic light saturation levels of tundra vascular plants are genetically and environmentally controlled. With regard to the genetic factors alpine populations of *Oxyria digyna* require higher light intensities to reach light saturation than arctic populations do [Billings and Mooney, 1968]. Bazzaz et al. [1970] recently found similar results for alpine and lowland populations of *Polytrichum juniperinum.*

Temperature interacts with light to affect maximal net photosynthetic rates. For example, in tundra seed plants, light saturation increases (or decreases) with increasing (or decreasing) temperatures [Billings and Mooney, 1968]. Rastorfer and Higinbotham [1968] found a higher light saturation value at 20°C than at 4°C for the forest floor moss *Bryum sandbergii* (northern Idaho). Similarly, for the antarctic mosses *Bryum argenteum* and *B. antarcticum* from Victoria Land, light saturation was reached at 4–5 mW cm^{-2} at 5°C but at 8–10 mW cm^{-2} at 15°C [Rastorfer, 1970]. Shifts to lower light saturation (and light compensation) levels with decreasing temperature would be photosynthetically advantageous to antarctic moss communities, since diurnal temperature decreases are often accompanied by decreases in light intensity.

Temperature is an important environmental influence on the rates of photosynthesis and respiration. The temperature range in which photosynthesis and respiration operate most effectively for a particular plant species depends on its genotype and its interrelationships with environmental factors. Besides temperature these factors include light (mentioned earlier), moisture, the availability of metabolic gases, and the environmental history [Billings and Mooney, 1968].

Unfortunately not all these factors as they relate to the effects of temperature on photosynthesis and respiration could be determined on the mosses included in this study. However, net photosynthetic and respiratory rates were measured under controlled conditions at 5° intervals from 5° to 25°C. Since the test samples were collected at the same time and subsequently treated in the same way, prior environmental influences on the observed results may be expected to be minimized. Therefore the temperature curves shown in Figures 8a–d do indicate intrinsic differences among the four species in their net photosynthetic and respiratory responses to temperature. The temperature curves for the two pleurocarpous forms, *C. austrostramineum* and *D. uncinatus,* indicate only relatively high optimal temperatures, whereas those for the acrocarpous forms, *P. strictum* (15°–20°C) and *P. nutans* (10°–15°C), indicate optimums. In addition these data for the acrocarpous forms, especially those for *P. nutans,* may suggest more than one optimal temperature. This result is not entirely unexpected, because multiple optimal temperatures have been reported for other organisms [Oppenheimer and Drost-Hansen, 1960; Ahmadjian, 1970]. Billings and Mooney [1968] suggested that, 'Even though the genetic effect is marked, the effect of environmental regime is greater.' Hence vascular plants grown under low temperatures had their optimal and maximal temperatures for net photosynthesis shifted to lower temperatures than plants grown under higher temperatures. In a similar way Stålfelt [1937] found that mosses in southern Sweden had adapted to the prevailing environmental conditions. He reported that, as the length of the day and the daily temperatures decreased from summer to winter, the optimal temperatures for net photosynthesis also decreased.

Absolute temperature minimums for photosynthesis and respiration in mosses have not been established. Ahmadjian [1970] reports that near −4°C respiration and photosynthesis operate for mosses (apparently of antarctic origin) but that below this temperature photosynthesis ceases while respiration continues. In *Bryum argenteum* oxygen evolution and uptake were measured at −2°C, and the ratios of net photosynthesis to respiration were higher at −2° and 2°C than at 8° and 15°C [Rastorfer, 1970]. Also oxygen evolution exceeded oxygen uptake at −5°C in *Bryum sandbergii* [Rastorfer and Higinbotham, 1968]. However, these data for the *Bryum* spp. are not direct evidence of carbon dioxide assimilation at freezing temperatures. On the other hand, Stålfelt [1937] measured the carbon dioxide exchange of several mosses in atmospheric air and found that the net photosynthetic to respiratory ratios decreased with increasing temperatures (0°–30°C). Of course these observations do not indicate that the absolute temperature minimums for net photosynthesis are lower than those for respiration, but they do suggest that high ratios of net photosynthesis to respiration at low temperatures (0°–15°C) may be expected for antarctic mosses. And for the mosses included in this study the highest ratios were indeed generally obtained at the lowest test temperatures (Table 6). Since mosses do occur and grow in Antarctica, it seems apparent that photosynthesis exceeds respiration at least during part of the growing season. This assumption is supported by Longton's [1970] measurements of the annual net production of *P. strictum* (342 g m^{-2} for Signy Island and 385 and 421 g m^{-2} for the Argentine Islands). In addition these differences in annual net production (dry weight of terminal shoots) were closely correlated with the corresponding summer temperature regimes of the plant communities. His data also show some correlation between the amount of growth of *P. strictum* and the mean daily duration of sunshine.

On the basis of the physiological and ecologic studies of antarctic mosses to date, these organisms do not appear to require a cold environment for survival, in the sense that they would be unable to tolerate more moderate conditions. Nevertheless antarctic mosses do have the capacity to endure freezing temperatures for long periods and to complete their annual growth in a short growing season characterized by frequent freeze–thaw cycles.

The adaptability of these plants may be attributable to phenotypic plasticity. But whether we are dealing with species of broad ecologic amplitudes or with ecotypes remains unknown. To answer this question, further work is required on populations of moss species from diverse climatic environments to ascertain the effects of genetic factors with respect to metabolic processes as influenced by temperature, light, available water, metabolic gases, and other environmental factors that may limit growth.

Acknowledgments. The identifications of the moss species used in this study were made by Dr. Harold E. Robinson, Smithsonian Institution, whose assistance is gratefully acknowledged. Special thanks are due to Mr. John M. Gnau for his competent assistance in Antarctica and to Miss Susan K. Bruns, Mrs. Dorene Rojas Fuller, and Mrs. Kathleen Vian Spangler for their willingness to help in various ways. Many thanks are extended to Dr. Emanuel D. Rudolph, Mr. Geoffrey L. Leister, and Mr. John F. Splettstoesser for reading and correcting the manuscript. This work was supported by National Science Foundation grants GA-1214 and GA-16000, which were awarded to the Ohio State University Research Foundation and administered by the Institute of Polar Studies. Contribution 212 of the Institute of Polar Studies, Ohio State University, Columbus, Ohio.

REFERENCES

Ahmadjian, V.
1970 Adaptations of antarctic terrestrial plants. *In* M. W. Holdgate (Ed.), Antarctic ecology. *2:* 801–811. Academic, New York.

Allen, S. E., and M. J. Northover
1967 Soil types and nutrients on Signy Island. Phil. Trans. R. Soc., ser. B, *252:* 179–185.

Allen, S. E., H. M. Grimshaw, and M. W. Holdgate
1967 Factors affecting the availability of plant nutrients on an antarctic island. J. Ecol., *55:* 381–396.

Bazzaz, F. A., D. J. Paolillo, Jr., and R. H. Jagels
1970 Photosynthesis and respiration of forest and alpine populations of *Polytrichum juniperinum.* Bryologist, *73:* 579–585.

Biebl, R.
1964 Temperaturresistenz tropischer Pflanzen auf Puerto Rico. Protoplasm, *59:* 133–156.

Billings, W. D., and H. A. Mooney
1968 The ecology of arctic and alpine plants. Biol. Rev., *48:* 481–529.

Boyd, W. L., J. T. Staley, and J. W. Boyd
1966 Ecology of soil microorganisms of Antarctica. *In* J. C. F. Tedrow (Ed.), Antarctic soils and soil forming processes, Antarctic Res. Ser., *8:* 125–139. AGU, Washington, D. C.

Clark, J. M., Jr. (Ed.)
1954 Experimental biochemistry. 228 pp. W. H. Freeman, San Francisco.

Douin, R.
1956 Pigments chlorophylliens des Bryophytes. Caroténoïdes des Bryales. C. R. Hebd. Séanc. Acad. Sci., Paris, *243:* 1051–1054.

Elliot, D. H.
1964 The petrology of the Argentine Islands. Sci. Rep. Br. Antarct. Surv., *41:* 1–31.

Evans, H. J., and G. J. Sorger
1966 Role of mineral elements with emphasis on univalent cations. A. Rev. Pl. Physiol., *17:* 47–76.

Freeland, R. O.
1957 Plastid pigments of gametophytes and sporophytes of Musci. Pl. Physiol., Lancaster, *32:* 64–66.

Fukushima, H.
1968 Notes on mosses in Ongul Islands, Antarctica (in Japanese with English summary). Antarctic Rec., *31:* 66–72.

Fulford, M., G. Carroll, and T. Cobbe
1947 The response of *Leucolejeunea clypeata* to variations in the nutrient solution. Bryologist, *50:* 113–146.

Gimingham, C. H., and R. I. L. Smith
1970 Bryophyte and lichen communities in the maritime Antarctic. *In* M. W. Holdgate (Ed.), Antarctic ecology. *2:* 752–785. Academic, New York.

Goldman, C. R., D. T. Mason, and B. J. B. Wood
1963 Light injury and inhibition in antarctic freshwater phytoplankton. Limnol. Oceanogr., *8:* 313–322.

Goodwin, T. W.
1965 Distribution of carotenoides. *In* T. W. Goodwin (Ed.), Chemistry and biochemistry of plant pigments, pp. 127–142. Academic, New York.

Greene, S. W.
1964 Plants of the land. *In* R. Priestley, R. J. Adie, and G. de Q. Robin (Eds.), Antarctic research, pp. 240–253. Butterworths, London.

Greene, S. W., D. M. Greene, P. D. Brown, and J. M. Pacey
1970 Antarctic moss flora. 1. The genera *Andreaea, Pohlia, Polytrichum, Psilopilum* and *Sarconeurum.* Sci. Rep. Br. Antarct. Surv., *64:* 1–118.

Grubb, P. J., O. P. Flint, and S. C. Gregory
1969 Preliminary observations on the mineral nutrition of epiphytic mosses. Trans. Br. Bryol. Soc., *5:* 802–817.

Hadley, E. B., and L. C. Bliss
1964 Energy relationships of alpine plants on Mt. Washington, New Hampshire. Ecol. Monogr., *34:* 331–357.

Heal, O. W., A. D. Bailey, and P. M. Latter
1967 Bacteria, fungi, and protozoa in Signy Island soils compared with those from a temperate moorland. Phil. Trans. R. Soc., ser. B, *252:* 191–197.

Heath, O. V. S.
1969 The physiological aspects of photosynthesis. 310 pp. Stanford University Press, Stanford, Calif.

Heber, U., and K. A. Santarius
1967 Biochemical and physiological aspects of plant frost-resistance. *In* A. S. Troshin (Ed.), The cell and environmental temperature, pp. 27–34. Pergamon, New York.

Hoffman, G. R.
1966 Observations on the mineral nutrition of *Funaria hygrometrica* Hedw. Bryologist, *69:* 182–192.

Holdgate, M. W.
1964 Terrestrial ecology in the maritime Antarctic. *In* R. Carrick, M. Holdgate, and Jean Prévost (Eds.), Biologie antarctique, pp. 181–194. Hermann, Paris.

Huneck, S.
1969 Moosinhaltsstoffe, eine Übersicht. J. Hattori Bot. Lab., *32:* 1–16.

Koski, V. M.
1950 Chlorophyll formation in seedlings of *Zea mays* L. Archs Biochem., *29:* 339–343.

Lewin, J., and B. E. F. Reimann
1969 Silicon and plant growth. A. Rev. Pl. Physiol., *20:* 289–304.

Llano, G. A.
1965 The flora of Antarctica. *In* T. Hatherton (Ed.), Antarctica, pp. 331–350. Methuen, London.

Longton, R. E.
1967 Vegetation in the maritime Antarctic. Phil. Trans. R. Soc., ser. B, *252:* 213–235.
1970 Growth and productivity of the moss *Polytrichum alpestre* Hoppe in antarctic regions. *In* M. W. Holdgate (Ed.), Antarctic ecology. *2:* 818–837. Academic, New York.

Matsuda, T.
1968 Ecological study of the moss community and microorganisms in the vicinity of Syowa Station, Antarctica. JARE Scient. Rep., ser. E, 29: 1–58.

Miyata, I., and T. Hosokawa
1961 Seasonal variations of the photosynthetic efficiency and chlorophyll content of epiphytic mosses. Ecology, *42:* 766–775.

Northover, M. J., and S. E. Allen
1967 Seasonal availability of chemical nutrients on Signy Island. Phil. Trans. R. Soc., ser. B, *252:* 187–189.

Oppenheimer, C. H., and W. Drost-Hansen
1960 A relationship between multiple temperature optima for biological systems and the properties of water. J. Bact., *80:* 21–24.

Paribok, T. A., N. V. Alekseeva-Popova, and B. N. Norin
1967 Content of trace-elements in plants of the forest–tundra zone (in Russian with English summary). Trudȳ Imp. S-Peterb. Obshch. Estest., *52:* 13–23.

Rastorfer, J. R.
1962 Photosynthesis and respiration in moss sporophytes and gametophytes. Phyton, B. Aires, *19:* 169–177.
1970 Effects of light intensity and temperature on photosynthesis and respiration of two east antarctic mosses, *Bryum argenteum* and *Bryum antarcticum.* Bryologist, *73:* 544–556.

Rastorfer, J. R., and N. Higinbotham
1968 Rates of photosynthesis and respiration of the moss *Bryum sandbergii* as influenced by light intensity and temperature. Am. J. Bot., *55:* 1225–1229.

Robinson, T.
1963 The organic constituents of higher plants, their chemistry and interrelationships. 306 pp. Burgess, Minneapolis, Minn.

Rudolph, E. D.
1965 Antarctic lichens and vascular plants: Their significance. Bioscience, *15:* 285–287.
1967 Climate. *In* S. W. Greene et al. (Eds.), Terrestrial life of Antarctica, Antarctic Map Folio Ser., folio 5, pp. 5–7. Amer. Geogr. Soc., New York.

Salisbury, F. B., and C. Ross
1969 Plant physiology. 761 pp. Wadsworth, Belmont, Calif.

Schuster, R. M.
1966 The Hepaticae and Anthocerotae of North America, east of the hundredth meridian. *1:* 802 pp. Columbia University Press, New York.

Shacklette, H. T.
1965a Bryophytes associated with mineral deposits and solutions in Alaska. Bull. U.S. Geol. Surv., 1198-C: 18 pp.
1965b Element content of bryophytes. Bull. U.S. Geol. Surv., 1198-D: 21 pp.
1967 Copper mosses as indicators of metal concentrations. Bull. U.S. Geol. Surv., 1198–G: 18 pp.

Smith, J. H. C., and A. Benitez
1955 Chlorophylls: Analysis in plant materials. *In* K. Paech and M. V. Tracey (Eds.), Modern methods of plant analysis. *4:* 142–196. Springer-Verlag, Berlin.

Stålfelt, M. G.
1937 Der Gasaustausch der Moose. Planta, *27:* 30–60.

Steere, W. C.
1965 Antarctic bryophyta. Bioscience, *15:* 283–285.

Strain, H. H.
1966 Fat-soluble chloroplast pigments: Their identification and distribution in various Australian plants. *In* T. W. Goodwin (Ed.), Biochemistry of chloroplasts. *1:* 387–406. Academic, New York.

Strong, J.
1967 Ecology of terrestrial arthropods at Palmer Station, Antarctic Peninsula. *In* J. L. Gressitt (Ed.), Entomology of Antarctica, Antarctic Res. Ser., *10:* 357–371. AGU, Washington, D. C.

Tamm, C. O.
1953 Growth, yield and nutrition in carpets of a forest moss (*Hylocomium splendens*). Meddn St. Skogsforsk Inst., *43:* 1–140.

Tieszen, L. L., and P. L. Johnson
1968 Pigment structure of some arctic tundra communities. Ecology, *49:* 370–373.

Tilbrook, P. J.
1967 The terrestrial invertebrate fauna of the maritime Antarctic Phil. Trans. R. Soc., ser. B, *252:* 261–278.

Touffet, J.
1966 Sur les teneurs en potassium, sodium et fer des Sphaignes. C. R. Hebd. Séanc. Acad. Sci., Paris, ser. D, *263:* 1086–1088.

Umbreit, W. W., R. H. Burris, and J. F. Stauffer
1957 Manometric techniques. 338 pp. Burgess, Minneapolis, Minn.

Voth, P. D., and K. C. Hamner
1940 Responses of *Marchantia polymorpha* to nutrient supply and photoperiod. Bot. Gaz., *102:* 169–205.

Watson, E. V.
1964 The structure and life of bryophytes. 192 pp. Hutchinson University Library, London.

Weiser, C. J.
1970 Cold resistance and injury in woody plants. Science, *169:* 1269–1278.

Wolf, F. T.
1958 Comparative chlorophyll content of the two generations of bryophytes. Nature, *181:* 579–580.

Wolf, F. T., J. G. Coniglio, and R. B. Bridges
1966 The fatty acids of chloroplasts. *In* T. W. Goodwin (Ed.), Biochemistry of chloroplasts. *1:* 187–194. Academic, New York.

OBSERVATIONS ON THE ORIGIN AND TAXONOMY OF THE ANTARCTIC MOSS FLORA

HAROLD E. ROBINSON

Department of Botany, Smithsonian Institution, Washington, D.C. 20560

Abstract. Recent moss collections from the Antarctic Peninsula are cited, and a listing of the 30 genera and approximately 70 species that have been reported from the continent is given. *Sarconeurum tortelloides* S.W.Greene is transferred to *Tortella*, and the new combination is made for *Platydictya densissima*. The distinctions are given for most of the species. The generic limits of *Pseudodistichium* are discussed, and *P. fuegianum* Roiv. is newly reported from the Antarctic. The possible origin of the moss flora is discussed. Great antiquity of the endemics is doubted, and the frequency of widely ranging species is noted.

Bryophytes are one of the most prominent elements in the limited flora of the Antarctic, and except for two species of vascular plants they are the highest form of plant life on the continent. A number of studies of these plants are available, and a brief summary with a nearly complete listing was provided by Steere [1961], additions being made later by Greene [1968a]. More recently there have been additional works, including some very detailed studies of a few genera and species of mosses by Greene [1968b], Greene et al. [1970], Horikawa and Ando [1963], and Kuc [1969]. More work is to be expected, especially from Stanley Greene, the present specialist in the moss flora. Additional material continues to accumulate, however, from collectors who are specialists in other groups and from those studying physiology and ecology. Such collections, primarily one by Dr. R. M. Schuster in 1969, have led to the observations recorded below.

The moss flora of the Antarctic is less varied than that of any other major area of the earth except the low atolls of Micronesia. As was indicated by Steere [1961], the antarctic flora compares very unfavorably with that of the Arctic, probably because of the much more severe climate of the Antarctic. Most species in the Antarctic are concentrated in the peninsula area, which extends as much as 7° above (northward of) the Antarctic Circle. According to Greene et al. [1970] this flora continues little diminished to about 68°S. Farther south and on the main continental area the moss flora falls off sharply, and <2 dozen species are found. One species, *Sarconeurum glaciale*, has been found as far south as the Dufek Coast near the Queen Maud Mountains, an area nearly 18° south of the Antarctic Circle and within 6° of the South Pole.

The present study has provided certain taxonomic results that are given in the treatment of the genera and the species. However, the following nontaxonomic impressions also seem important. These impressions concern my feeling that the moss flora of the Antarctic has little or none of the relict status evident in regions as close as South Georgia and Tierra del Fuego. I believe that the present flora is composed exclusively of forms that have immigrated or evolved in the last few million years, all of them being capable of reproducing and distributing themselves in the present climate.

As far as the antarctic moss flora is concerned, a general rule applicable elsewhere seems to apply. An individual locality is temporary, and the survival of a species ultimately depends on its ability to reproduce and to colonize new sites. Changes in the Antarctic must be slow, especially in view of the reduced biological competition, but the changes are apparently not slow enough to allow for the indefinite survival of moss colonies. The waxing and the waning of thermal activity and the effects of erosion, both sometimes being catastrophic in nature, must have altered every locality where mosses have been able to grow during the millions of years that the Antarctic has had its present climate, and competition from lichens and the alga *Prasiola* has overtaken many moss specimens that I have seen.

The composition of the antarctic moss flora is certainly indicative of a limited history. Many of the species are wide in their distribution in the world, and some are bipolar in distribution. Such plants could have colonized the continent at any time since its glaciation and perhaps many different times. The limitations that I suggest do not preclude endemic species, but I reject the idea of 1- to 3-million-year holdovers. If such holdovers were possible, there would be many mosses present that are not. Many mosses can survive treatment with liquid air [Lipman, 1936; Morrill, 1950] and long periods of darkness. Additional species of the southern hemisphere, such as *Ditrichum conicum* (Mont.) Mitt. and members of the Chile–New Zealand element, would be able to survive in the Antarctic if those abilities were the only criteria. Many such species may once have occurred on the continent, but they can now be expected to turn up only as fossils in offshore glacial till.

The endemic species of the Antarctic should be more indicative of the floristic history. In my evaluation I eliminate from consideration many species of *Brachythecium*, *Bryum*, *Grimmia*, and *Tortula* that may not be truly distinct from others in or out of the Antarctic, and I omit species of *Andreaea*, *Blindia*, *Calliergidium*, and *Orthotrichum* that often occur at least as far north as the South Orkney Islands or South Georgia. Most of the remaining 'hard core' endemics can be characterized by three rather significant features: (1) All the hard core endemics, such as *Sarconeurum* and *Grimmia antarctici*, have considerable distribution, and none is restricted to a single locality or area. (2) The endemics are capable of reproduction, some by special means. *Sarconeurum glaciale* can reproduce from specialized caducous leaf tips, and the genus *Grimmia* notably can grow from leaf fragments [Keever, 1957]. (3) Many of the endemics differ from their nonendemic relatives primarily by simplified structure, such as the costa of *Sarconeurum* or the enlarged leaf cells of *Ceratodon* or *Brachythecium antarcticum*, characters usually associated with extreme environments in other parts of the world.

The strongest case for a distinctive element in the antarctic flora is the endemic genus *Sarconeurum*. This genus has recently been credited with a second endemic species by Greene [Greene et al., 1970]. The question is simply, How distinct is the genus? My own impression is that, except for the simplified internal structure of the costa, *Sarconeurum glaciale* would be a *Barbula*. The specialized leaf tips of the species are not unique. Such tips occur in a number of species of *Barbula*, including *B. johansenii* Williams of the Arctic and *B. crassicuspis* H.Robinson of Mexico. More significantly such leaf tips are found in *Tortula lithophila* Dus. of southern South America. As I indicate elsewhere, I consider the other species of *Sarconeurum* (*S. tortelloides* Greene) a *Tortella*, and the genus *Tortella* is known to have fragile leaf tips. What has been called *Sarconeurum* impresses me as two rather unrelated Pottiaceous species, each having close relatives outside the Antarctic. It seems indicative of their time of origin that both species differ from their relatives by a character highly adaptive to the present environment of the Antarctic.

The collections cited below are primarily from the 1969 trip of Dr. R. M. Schuster. All the Schuster material is from the 17 localities (Figure 1) that I cite in full here and refer to by general locality and letter in the text:

A. South Shetland Islands, Livingston Island, on and near gravelly irrigated shore, west end of island; 62°40′S, 61°05′W; January 11.

B. Argentine Islands, Graham Coast, insolated ledges and slopes, below snow fields; 65°13′S, 64°13–14′W; January 15–16.

C. Argentine Islands, Graham Coast, richly vegetated slopes, mostly irrigated by snow melt from above; 65°13′S, 64°14′W; January 15–16.

D. Gerlache Strait, Useful Island, limited mossy areas adjoining Chinstrap penquin rookery; 64°43′S, 62°52′W; January 18.

E. Almirante Brown Base, Danco Coast, opposite Bryde Island; 64°54′S, 62°52′W; January 23.

F. South Shetland Islands, Deception Island, south side of Port Foster, above Primero de Mayo (Fumarole) Bay; 62°58–59′S, 60°42′W; January 24.

G. South Shetland Islands, Livingston Island, east side of False Bay; 62°43′S, 60°20′W; January 25.

H. South Shetland Islands, Livingston Island, east side of False Bay, low rocky slopes behind shore; 62°43′S, 60°20′W; January 25.

I. South Shetland Islands, Livingston Island, east side of False Bay, low rocky slopes, north facing; 62°43′S, 60°20′W; January 25.

J. King George Island, Martel Inlet, Admiralty Bay, east side of Keller Peninsula; 20–200 feet; 62°05′S, 58°24′W; January 26.

K. South Shetland Islands, Greenwich Island, northeast cape of Discovery Bay, northeast of Mount Plymouth; 62°27′S, 59°45′W; January 27.

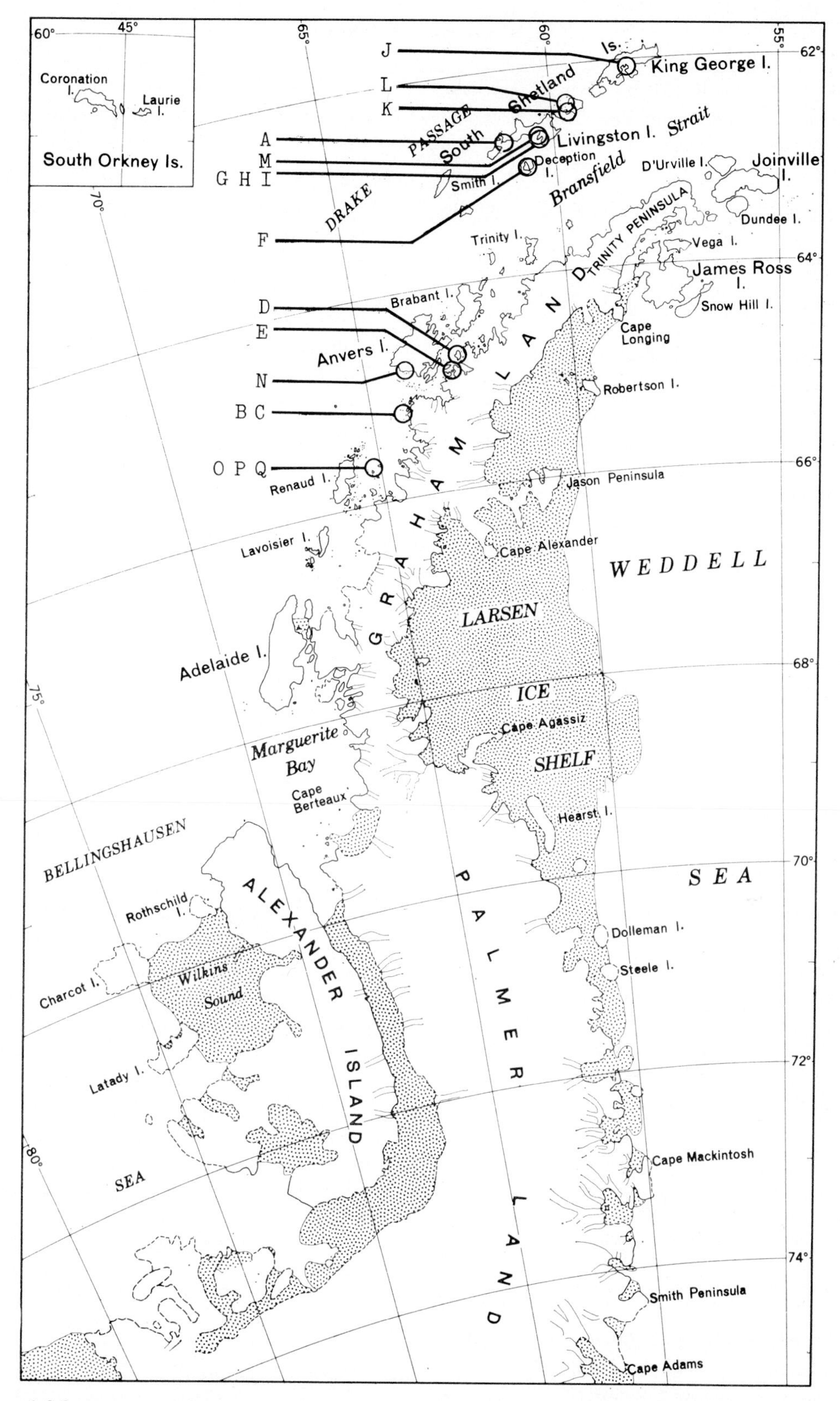

Fig. 1. Sites of Schuster moss collections on Antarctic Peninsula: A, Livingston Island, west end; B–C, Argentine Islands; D, Useful Island; E, Almirante Brown Base; F, Deception Island; G–I, Livingston Island, False Bay; J, King George Island, Martel Inlet, Admiralty Bay; K, Greenwich Island, northeast cape of Discovery Bay; L, Aitcho Islands; M, Livingston Island, Miers Bluff; N, Anvers Island, Norsel Point; and O–Q, Hook Island.

L. South Shetland Islands, Aitcho Islands, English Strait between Robert and Greenwich islands; 62°24′S, 59°45′W; January 27.

M. South Shetland Islands, Livingston Island, west side of Miers Bluff; 62°42′S, 60°25′W; January 28.

N. Anvers Island, Norsel Point, near Palmer Station, coastal low rocky ridge; 64°46′S, 64°05′W; February 3.

O. Graham Coast, Hook Island, on steep mossy irrigated north facing slopes; 65°38′S, 65°10′W; February 4.

P. Graham Coast, Hook Island, on steep mossy irrigated north facing slopes below skua colony; 50–125 feet; 65°38′S, 65°10′W; February 4.

Q. Graham Coast, Hook Island, on steep well-vegetated, north facing irrigated slopes; 40–125 feet; 65°38′S, 65°10′W; February 4.

Details of other collection localities are given with the species. Unless it is otherwise indicated, the specimens cited are in the U.S. National Herbarium. Unless it is indicated (by !), types have not been seen.

TAXONOMIC REVIEW OF ANTARCTIC MOSSES

Andreaeaceae

Genus ***Andreaea***

The antarctic species have been reviewed in two papers [Greene, 1968b; Greene et al., 1970]. The single species *A. depressinervis* is included in the recent treatment of the costate species of *Andreaea* by Schultz-Motel [1970]. All three antarctic species occur widely in the Antarctic Peninsula and northward to the South Orkney and South Sandwich islands or South Georgia. The species are distinguished as follows.

1. Leaves with rather indistinct costa in basal part....... ....***A. depressinervis***
   Leaves unistratose, not costate........................2
2. Most leaves panduriform, basal margins usually crenulate ....***A. gainii***
   Leaves not panduriform, basal margins usually entire... ....***A. regularis***

Andreaea depressinervis Card. Revue Bryol. *27:* 43. 1900. Lectotype: Gerlache Strait, XIIème débarquement, Racovitza 239 (BR).

Collections by R. M. Schuster. Argentine Islands: C. 69-233, 69-270, 69-272, 69-284 part, 69-285b, 69-289 part. Almirante Brown Base: E. 69-352, 69-357, 69-372, 69-387. Graham Coast: G. 69-884 part. South Shetland Islands: L. 69-924; K. 69-967, 69-972, 69-980, 69-1802, 69-1803, 69-1812.

Andreaea gainii Card. C. R. Hebd. Séanc. Acad. Sci., Paris. *153:* 602. 1911. Holotype: Graham Land, Cape Tuxen, Gain 209 (PC).

var. *gainii.*

Collections by R. M. Schuster. Argentine Islands: C. 69-228, 69-270 part, 69-280, 69-289 part. South Shetland Islands: H. 69-540, 69-541, 69-556; L. 69-987; M. 69-814. King George Island: J. 69-949 part, 69-957. Anvers Island: N. 69-846, 69-855, 69-858. Graham Coast: O. 69-513; P. 69-710, 69-718. Q. 69-884 part, 69-895, 69-902.

var. *parallela* (C.Müll.) S.W. Greene. Revue Bryol. Lichen. *36:* 142. 1968.

Andreaea parallela C.Müll. Bot. Jb. *5:* 76. 1883. Holotype: Îles Kerguelen, Bras de la Fonderie (Foundry Branch), Naumann, December 1874 (U).

The variety is maintained for specimens with leaf tips narrower than those of typical *A. gainii* but not as long as those of *A. acuminata* Mitt. No collections have been seen.

Andreaea regularis C.Müll. *In* Neumayer (ed.), Ergeb. Deut. Polar-Exped. Allg. *2:* 286. 1890. Lectotype: Austro-Georgia, Ostseite des Vexirberges, Will 498 (M).

Andreaea pycnotyla Card. Revue Bryol. *27:* 42. 1900. Lectotype: Gerlache Strait, XXème débarquement, Racovitza 270a (BR).

Andreaea pygmaea Card. Revue Bryol. *27:* 43. 1900. Holotype: Gerlache Strait, Xème débarquement, Brabant Island, Racovitza 252d (BR).

Collections by R. M. Schuster. South Shetland Islands: A. 69-040; K. 69-992. Argentine Islands: B. 69-222. Almirante Brown Base: E. 69-355. Graham Coast: P. 69-711.

Other collection on Litchfield Island, 64°46′23″S, 64°04′30″W, Rastorfer and Gnau PAL-156.

Polytrichaceae

The following genus concepts are those in the recent revision of the family by Smith [1971].

Genus ***Polytrichum***

Three species have been reported from the Antarctic. The following key is adapted from the treatment by Greene et al. [1970].

1. Basal cells of leaf sheath short, usually less than five times as long as wide; hair point abrupt, 0.6–1.0 mm long, flexuose . . . . *P. piliferum*
   Basal cells of leaf sheath long and narrow, usually more than five times as long as wide; hair point rather gradual, usually less than 0.5 mm long, rigid 2
2. Clusters of stems compact and densely matted with abundant white rhizoids; margins of lamellae unevenly crenulate with wide deep sinuses between projections . . . . *P. strictum*
   Clusters of stems loose without matted abundant white rhizoids; margins of lamellae rather evenly crenulate with narrow shallow sinuses between projections . . . . *P. juniperinum*

No material of *Polytrichum juniperinum* Hedw. has been seen in this study. The other species are represented as follows:

Polytrichum piliferum Hedw. Species Muscorum Frondosorum. 90. 1801. Holotype: Saxony, Lipsiae ad Bieniz, Schönfeld (G).

Collections by R. M. Schuster. South Shetland Islands: L. 69-927 part, 69-928.

Polytrichum strictum Menz. *ex* Brid. J. Bot. *1800*(2): 286. 1801. Holotype: Newfoundland, Banks, 1765.

Polytrichum alpestre Hopp. Bot. Taschenbuch. *1801:* 198. 1801. Holotype: Untersberg prope Salisburgum, Hoppe.

Collections by R. M. Schuster. Argentine Islands: B. 69-222; C. 69-282, 69-287. Almirante Brown Base: E. 69-353. Anvers Island: N. 69-847, 69-856. Other collections in westernmost island of Corner Islands, 65°14′30″S, 64°13′52″W; Argentine Islands, Rastorfer PAL-78, PAL-79; easternmost island of Corner Islands, 65°14′30″S, 64°14′23″W; Argentine Islands, Rastorfer and Gnau PAL-107, PAL-108.

I consider the species to be very close to *P. juniperinum* and perhaps only an environmental form. Greene et al. [1970] cite *P. strictum* under the authorship of Banks, and the name *P. alpestre* Hoppe is given precedence. Both names were published the same year, but lacking definite information of priority I would retain the familiar name *P. strictum*.

Genus *Polytrichastrum*

The genus has been established by Smith [1971] for nine species of mostly northern distribution. The following single species occurs in the Antarctic.

Polytrichastrum alpinum (Hedw.) G. L. Smith. Mem. N. Y. Bot. Gdn. *21*(3): 37. 1971.

Polytrichum alpinum Hedw. Species Muscorum Frondosorum. 92. 1801. Holotype: Wales, Snowdon Mountain, Dillen (OXF).

Polytrichum brevifolium R. Brown. *In* Perry, J. Voyage Discovery Northwest Passage. 294. 1824. Holotype: Melville Island, Ross.

Collections by R. M. Schuster. South Shetland Islands: A. 69-038; F. 69-912; H. 69-580; I. 69-692 part; K. 69-981 part, 69-998; L. 69-927 part, 69-929, 69-935; M. 69-823 part. King George Island: J. 69-954. Anvers Island: N. 69-848. Graham Coast: O. 69-522, 69-530; P. 69-706a part.

Other collection on Deception Island, Follmann 2871. The distribution of the species is bipolar, and it occurs widely in Europe, Asia, and North America.

Genus *Psilopilum*

A single species, *Psilopilum antarcticum* (C.Müll.) Par., is known from the region. Collection localities include the South Sandwich and South Orkney islands and Deception Island. The species is treated by Greene et al. [1970], but no material has been seen in this study.

Ditrichaceae

Genus *Ceratodon*

The genus is prominent in the Antarctic. The recent treatment of Horikawa and Ando [1963] recognizes four species. I have retained the following three species, and I place the fourth under *Pottia charcotii* Card.

1. Leaf cells 6–9 μ in diameter . . . . *C. purpureus*
   Leaf cells 8–15 μ in diameter 2
2. Leaves 0.8–1.2 mm long, 0.4–0.5 mm wide; margins revolute in median half . . . . *C. grossiretis*
   Leaves 1.8–2.4 mm long, 0.6–0.8 mm wide; margins strongly revolute from near the base . . . . *C. validus*

Ceratodon grossiretis Card. Bull. Herb. Boissier, ser. 2. *6:* 4. 1906. Holotype: Danco Coast, Moss Islands (Île des Mousses), C. Skottsberg, December 1, 1902 (PC).

Collections by R. M. Schuster. Argentine Islands: C. 69-252, 69-270 part. South Shetland Islands: G. 69-412; H. 69-548, 69-557 part, 69-570, 69-589; I. 69-697, 69-698; L. 69-937. Anvers Island: 69-857. Graham Coast: O. 69-514 part, 69-527; P. 69-712; Q. 69-882, 69-890.

Ceratodon purpureus (Hedw.) Brid. Bryol. Universa. *1:* 480. 1826.

Dicranum purpureum Hedw. Species Muscorum Frondosorum. 136. 1801. Type: Europe.

Ceratodon minutifolius Card. C. R. Hebd. Séanc. Acad. Sci., Paris. *153:* 602. 1911. Holotype: Marguerite Bay, Jenny Island, Gain 258a (PC).

Collected by R. M. Schuster. Almirante Brown Base: E. 69-390. South Shetland Islands: H. 69-587. Graham Coast: O. 69-523, 69-525.

Other collections at Molodezhnaya Station, MacNamara, March 1967, December 1967; eastern Thala Hills, MacNamara, March 1967; north Victoria Land, west side of Whitehall Glacier, Wise, November 14, 1964; Deception Island, southeast side near saddle in crater rim, about 250 meters east of Whalers Bay, 62°59.47′S, 60°32.48′W, Waldron and Fehlmann, March 29, 1970. A map of the antarctic distribution is given by Kuc [1969].

Ceratodon validus (Card.) Horikawa & Ando. Hikobia. *3*(4): 278. 1963.

Ceratodon grossiretis var. *validus* Card. Bull. Herb. Boissier, ser. 2. *6:* 14. 1906. Holotype: Danco Coast, Moss Islands (Île des Mousses), C. Skottsberg 446 (PC).

No material of the species has been seen in this study.

Genus *Pseudodistichium*

The present study has resulted in a complete revision of concepts of the genus given in a previous paper [Robinson, 1964]. This revision involves the exclusion of two species previously included in the genus and the recognition of a published species previously overlooked.

The species that has been called *P. atlanticum* Dix. with its synonym *P. taitaoense* Froehl. proves to be a synonym of *Ditrichum conicum* Mont. Portions of the type of *D. conicum* have been seen through the kindness of Mesdames Allorge and Jovet-Ast in Paris, and other specimens have been seen from south central Chile, Tristan da Cunha, and Îles Kerguelen. The species originally constituted the monotypic genus *Aschistodon* Mont., and that rejected name must be used if the species is again separated from *Ditrichum*. Also excluded from *Pseudodistichium* is *Ditrichum buchananii* (R. Brown ter.) Broth. of New Zealand.

The genus *Pseudodistichium* is restricted here to three species having erect, rather sheathing leaf bases, short rostrate opercula, rather short, broad, blunt peristome teeth, and at least a tendency toward enlarged multicellular spores. The genus is distinguished from *Distichium* by the occurrence of the leaves in many ranks and the short rostrate rather than short conical nature of the opercula. The species are distinguished as follows.

1. Peristome teeth not striolate; spores very large, 65–70 μ in diameter .... ***P. brotherusii***
   Peristome teeth striolate; spores moderately enlarged, to 30 μ in diameter 2
2. Capsules narrowed toward the mouth .... ***P. austro-georgicum***
   Capsules relatively narrow and cylindrical .... ***P. fuegianum***

Pseudodistichium austro-georgicum Card. is known from South Georgia, the Falkland Islands, and Tierra del Fuego; *P. brotherusii* (R. Brown ter.) Dix. is from New Zealand; and the following species is known from Tierra del Fuego and the Antarctic.

Pseudodistichium fuegianum Roiv. Suomal. Eläin- ja Kasvit. Seur. van. Julk. *9*(2): 24. 1937. Holotype: Chile, central Tierra del Fuego, Estancia Vicuña, Cerro Pedro Grande, Roivainen, 1928–1929 (H).

Collections by R. M. Schuster. South Shetland Islands: H. 69-536, 69-562, 69-569 with capsule, 69-587; I. 69-691, 69-699a; M. 69-804.

The genus and the species have not previously been reported from the Antarctic, but I suspect that the plants previously reported as *Distichium capillaceum* (Hedw.) B.S.G. from the Antarctic belong here. The range of the latter species includes the South Orkney and South Shetland islands, the Trinity (Louis Philippe) Peninsula, the Danco Coast, and Victoria Land according to Steere [1961], but I have seen no specimens.

Grimmiaceae

Genus *Grimmia*

The genus is well represented in the antarctic region, at least four distinct species being present. Other species that have not been seen and may or may not be distinct are *Grimmia fastigiata* Card., which has been described from Tierra del Fuego and reported from Queen Mary Coast, *G. lawiana* Willis *in* Filson, which has been described from Mac. Robertson Land, and *G. stolonifera* C.Müll., which has been described from Îles Kerguelen and reported from Queen Mary Coast. *Grimmia alpicola* Hedw. has also been reported from the area, but I have not had the same success as others in distinguishing it from *G. apocarpa* Hedw. Of the following species, only *G. antarctici* is endemic. The other three species are common in

Europe and North America. The four species are distinguished as follows.

1. Leaves not keeled; capsules asymmetrical .... ***G. plagiopodia***
   Leaves keeled; capsules symmetrical 2
2. Lower marginal cells of leaves forming distinct hyaline area; capsules exerted .... ***G. donniana***
   Lower marginal cells of leaves not forming sharply distinct hyaline area; capsules immersed 3
3. Leaves narrowly oblong, small and crowded, often in ranks .... ***G. antarctici***
   Leaves usually ovate-lanceolate, 1.5–2.0 mm long, not in distinct ranks .... ***G. apocarpa***

Grimmia antarctici Card. Bull. Herb. Boissier, ser. 2. *6:* 15. 1906. Lectotype: Trinity Peninsula: Cape Kjellman, C. Skottsberg 448 (PC).

var. *antarctici.*

Collection seen in Victoria Land, Prince Olav Mountains, southeast of Shackleton Glacier, north end of small spur on east side of mouth of Barrett Glacier, 84°35′S, 173°50′W, McGregor, summer 1963–1964.

var. *pilifera* Bartr. Ann. Mo. Bot. Gdn. *25:* 722. 1938. Holotype: Marie Byrd Land: Ford Ranges, Mount Rea–Mount Cooper, Siple, Wade, Corey and Stancliff 27 (US).

Collection by R. M. Schuster. South Shetland Islands: H. 69–554.

The species is in the subgenus *Schistidium* and is related to *G. apocarpa.* The small crowded narrowly oblong leaves, which are often in ranks, seem distinctive.

Grimmia apocarpa Hedw. Species Muscorum Frondosorum. 76. 1801. Original material: Lipsiae ad piscinam Lindenthalensem, Gothanam versus Ileburgum, Dresdae ad rivulum Weiseriz ubertim, Hedwig ?

Collections by R. M. Schuster. Argentine Islands: C. 69-271, 69-286. South Shetland Islands: F. 69-920, 69-922; H. 69-542, 69-549, 69-550, 69-557 part, 69-561, 69-591a; M. 69-800. King George Island: J. 69-948, 69-949 part, 69-962. Graham Coast: O. 69-510, 69-512; Q. 69-888.

Grimmia donniana Sm. Engl. Bot. *18:* pl. 1259. 1804. Flora Brit. *3:* 1198. 1804. Holotype: Scotland, on mountain in Angus County, 18 miles north of Forfar, G. Donn (LINN).

Collections by R. M. Schuster. Argentine Islands: C. 69-250. Almirante Brown Base: E. 69-358, 69-386, 69-391, 69-392. South Shetland Islands: G. 69-537, 69-575.

The species is the only *Grimmia* in West Antarctica having exerted capsules. The material seen is frequently fruiting and quite typical of the species. Kuc [1969] has recently studied in detail collections from Bunger Hills in the area of the Knox and Queen Mary coasts. In the process a new form *antarctica* having more abbreviated basal cells is described. Kuc does not mention the various species of *Grimmia* reported from the general area that may represent the same thing. The status of all the taxa from Mac. Robertson Land and Queen Mary Coast needs careful review.

Grimmia plagiopodia Hedw. Species Muscorum Frondosorum. 78. 1801. Holotype: Germany, Saxony, Cl. Flügge (G).

The species is reported from the South Shetland Islands and Marie Byrd Land. Some years ago Dr. Stanley Greene called my attention to the lack of peristome teeth on antarctic material under this name. Dr. Greene's conclusions about the proper disposition of this material are not yet available.

Genus *Rhacomitrium*

Two species of the genus have been reported from the Antarctic. The species are distinguished as follows.

Leaves with tips and upper margins hyaline, very dentate and papillose .... ***R. lanuginosum***
Leaves with little or no hyaline tip, not dentate or papillose .... ***R. crispulum***

Rhacomitrium crispulum (Hook.f. & Wils.) Hook.f. & Wils. Bot. Antarctic Voyage H. M. Discovery. 2. Flora Nova-Zelandiae. *2:* 75. 1854.

Dryptodon crispulus Hook.f. & Wils. London J. Bot. *3:* 544. 1844. Holotype: Campbell Island, J. D. Hooker, 1839–1843.

Rhacomitrium substenocladum Card. Revue Bryol. *38:* 127. 1911. Holotype: Graham Land, Cape Tuxen, Gain 200 (PC).

Collections by R. M. Schuster. South Shetland Islands: H. 69-535, 69-574, 69-576, 69-587.

A more extensive treatment of this and other antarctic species of the genus is intended for my study of the Juan Fernandez moss flora. I do not follow the very broad concept of Clifford [1955].

Rhacomitrium lanuginosum (Hedw.) Brid. has been reported from the South Orkney Islands, but no material has been seen in this study.

SELIGERIACEAE

Genus *Blindia*

One species of the genus has been reported from the Antarctic.

Blindia skottsbergii Card. Bull. Herb. Boissier, ser. 2. *6:* 4. 1906. Holotype: South Georgia, Cumberland Bay, Jason Harbor, C. Skottsberg (S).

Collection by R. M. Schuster. Almirante Brown Base: E. 69–359.

Dicranaceae

Genus *Dicranoweisia*

Two species of the genus have been reported in the Antarctic. The species are distinguished as follows.

Lower cells of leaf lamina broad, oblong to hexagonal, laxly areolate........................ ***D. grimmiacea***

Lower cells of leaf lamina elongate, narrowly rectangular with subincrassate walls.............. ***D. subinclinata***

Dicranoweisia grimmiacea (C.Müll.) Broth. Natürlichen Pflanzenfamilien. *1*(3): 318. 1901.

Blindia grimmiacea C.Müll. *In* Neumayer (ed.), Ergeb. Deut. Polar-Exped. Allg. *2:* 299. 1890. Holotype: South Georgia, 'Brockenthal,' in the Mount Krokisius vicinity of Royal Bay, Will.

Collections by R. M. Schuster. South Shetland Islands: K. 69-979, 69-989, 69-999, 69-1801; M. 69-810, 69-818, 69-819. King George Island: J. 69-946 part. Graham Coast: P. 69-710.

The original descriptions also emphasized the very short setae of *D. grimmiacea* versus the long setae of *D. subinclinata,* but these characters seem to vary.

Dicranoweisia subinclinata (C.Müll.) Broth. Natürlichen Pflanzenfamilien. *1*(3): 318. 1901.

Blindia subinclinata C.Müll. *In* Neumayer (ed.), Ergeb. Deut. Polar-Exped. Allg. *2:* 30. 1890. Lectotype: South Georgia, Ostseite des Vexirberg, Will, February 17, 1883.

Collections by R. M. Schuster. Almirante Brown Base: E. 69-356, 69-367, 69-370. South Shetland Islands: H. 69-533, 69-545, 69-560, 69-599a. King George Island: J. 69-946 part.

Genus *Dicranella*

One species, *Dicranella hookeri* (C.Müll.) Card., has been reported from Ross Island in Victoria Land. No material has been seen in this study.

Genus *Chorisodontium*

Two species of the genus have been reported from the antarctic region [Steere, 1961]. The species are distinguished as follows.

Cells of lower leaf lamina have highly thickened porose walls................................ ***C. aciphyllum***

Cells of lower leaf lamina have thin or nodularly thickened walls.................. ***C. nordenskioeldii***

Only one species has been seen in this study:

Chorisodontium nordenskioeldii (Card.) Roiv. Suomal. Eläin- ja Kasvit. Seur. van. Julk. *9*(2): 42. 1937.

Dicranum nordenskioeldii Card. Bull. Herb. Boissier, ser. 2. *6:* 14. 1906. Holotype: Danco Coast, Moss Islands (Île des Mousses), C. Skottsberg 440 (PC).

Collections by R. M. Schuster. Argentine Islands: C. 69-269, 69-284 part, 69-285a.

Other collection on Litchfield Island: 64°46′23″S, 64°04′30″W, Rastorfer and Gnau PAL-154. My concept of the species is based on the specimens listed above plus a D. Bergström collection from South Georgia. My concept of *C. aciphyllum* (Hook.f. & Wils.) Broth. is based on a Wilkes expedition collection from Bahía Orange, Tierra del Fuego, and some Kunkel collections from Juan Fernandez. These specimens confirm the thicker walled condition of the lower leaf cells emphasized by Cardot [1908] for *C. aciphyllum.* The illustrations of Cardot are misleading, however. The cell walls of *C. aciphyllum* are thicker if anything than they are shown, and they give the whole lower leaf lamina a firmer aspect. The cell walls of all *C. nordenskioeldii* specimens that I have seen have definite nodular thickenings except just above the outer alar cells.

Pottiaceae

Genus *Tortella*

I recognize the following species as a representative of the genus *Tortella.*

Tortella tortelloides (S.W.Greene) H.Robinson, comb. nov.

Sarconeurum tortelloides S.W.Greene. Sci. Rep. Brit. Antarct. Surv. *64:* 38. 1970. Holotype: Alexander Island, Eros Glacier, B. J. Taylor 529 (BIRM).

I am not sure why this species was placed in the genus *Sarconeurum* by Greene. Fragile and even fleshy leaf tips occur in other genera of the Pottiaceae, and the cross section of the costa shows differentiated cells on both sides of the guide cells. The basal cells of the leaf are certainly like *Tortella* as recognized by Greene in his selection of the species name. In *Sarconeurum* the costa has less differentiation throughout and no differentiation in the cells adaxially. The basal cells of *S. glaciale* are completely different and are more like those of *Barbula.*

Genus *Barbula*

The only species reported from Antarctica is the endemic *Barbula byrdii* Bartr. The species is apparently known only from the Ford Ranges of Marie Byrd Land. No material has been seen in this study.

Genus *Didymodon*

One species, *Didymodon gelidus* Card., is reported from Antarctica. The type was from McMurdo Sound in Victoria Land, and the species has since been reported from Deception Island.

Genus *Bryoerythrophyllum*

A specimen of the widely distributed northern species *Bryoerythrophyllum recurvirostre* (Hedw.) Chen from the Bunger Hills in Wilkes Land has served as the basis for a new variety *antarcticum* Sav.-Ljub. & Z. Smirn.

Genus *Pottia*

The two species reported for the Antarctic are distinguished as follows.

Cells of upper leaf lamina not papillose..... ***P. charcotii***
Cells of upper leaf lamina papillose except near margins
.... ***P. heimii***

Pottia charcotii Card. C. R. Hebd. Séanc. Acad. Sci., Paris. *153:* 602. 1911. Holotype: Graham Land, Cape Pérez (cap des Trois-Perez), Gain 272b–273c (PC).

? *Ceratodon antarcticus* Card. Revue Bryol. *27:* 43. 1900. Holotype: Graham Land: Gerlache Strait: Cuverville Island (Île Cavelier de Cuverville), Racovitza 240b (PC).

I have seen no material, and I base the preceding possible synonymy on nearly identical illustrations and descriptions of the two species. If the synonymy can be confirmed, the species name *antarctica* would have priority.

Pottia heimii (Hedw.) Hampe. Flora. *20:* 287. 1837.

Gymnostomum heimii Hedw. Species Muscorum Frondosorum. 32. 1801. Holotype: Germany, near Spandau, Heim.

Collection by R. M. Schuster. South Shetland Islands: M. 69-805.

Genus *Sarconeurum*

The genus with its one species is endemic to the Antarctic. The treatment of the second species recently described by S. W. Greene is found under the heading for the genus *Tortella.*

Sarconeurum glaciale (C.Müll.) Card. & Bryhn. Nat. Antarctic Exped. Natur. Hist. *3* (Musci): 3. 1907.

Didymodon glacialis Hook.f. & Wils. Bot. Antarctic Voyage H.M. Discovery. 1. Flora Antarctica. *2:* 408. 1847 hom. illeg. Holotype: Antarctica, Cockburn Island, 64°S, 57°W, J. D. Hooker 4 (BM).

Leptotrichum glaciale [Hook.f. & Wils.] C.Müll. Synopsis Muscorum Frondosorum. *2:* 611. 1851.

Sarconeurum antarcticum Bryhn, Nyt Mag. Naturvidensk. *40:* 205. 1902. Original material from the Geikie Ridge and Mount Melbourne vicinities of Victoria Land (Terre de Geikie et Terre de Newness), Borchgrevink.

Collections seen in Victoria Land, Ross Island, E. Schofield, December 23, 1963; the Walcott Glacier area near Lake Teardrop, E. Schofield, January 29, 1964; northern Victoria Land, Cape Christie, 72° 18′S, 170°02′E, Strang MCM-10; Ross Island near Scott Base, Rastorfer MCM-21, MCM-22; Observation Hill, 77°51′S, 166°41′E, Locke MCM-36, November–December 1967. The leaf base of *Sarconeurum glaciale* is reminiscent of the type found in *Barbula* and is strikingly different from that found in *S. tortelloides* S.W.Greene. I believe this character to be much more reliable phyletically than the fragile leaf tips. Similar leaf tips have been noted in a number of Trichostomoid and Barbuloid genera in the Pottiaceae and also in a species presently placed in *Tortula,* i.e., *T. lithophila* Dus. of southern South America.

Genus *Tortula*

The genus is prominent in the antarctic moss flora and is in need of a detailed review. The problem is complicated by the need to recognize relationships with various of the innumerable species in adjacent South America. The present study has shown only the two antarctic species listed below. In addition, there are apparently specimens with a smooth hyaline hair tip, including those reported long ago as the widely distributed *T. laevipila* (Brid.) Schwaegr., as well as some more recently referred to *T. fuegiana* (Mitt.) Mitt. and the Falkland Islands species, *T. monoica* Card. An antarctic endemic, *T. heteroneura* Card., according to description is similar to what I call *T. grossiretis,* but many of its leaves never develop a piliferous tip. The cell size of *T. heteroneura* has not been given. The two species that I have seen are distinguished as follows.

Leaf tips sharply acute, without hair tips; cells of upper leaf lamina about 10 μ in diameter.......... ***T. excelsa***

Leaf tips with scabrous hyaline hair tips; cells of upper leaf lamina 15–20 μ in diameter .... ***T. grossiretis***

Tortula excelsa Card. Bull. Herb. Boissier, ser. 2. *6:* 15. 1906. Holotype: South Shetland Islands, Nelson Island, Harmony Cove, C. Skottsberg 447 (PC).

Collections by R. M. Schuster. South Shetland Islands: F. 69-918; M. 68-800, 69-805, 69-808a, 69-809. King George Island: J. 69-940.

Tortula grossiretis Card. Bull. Herb. Boissier, ser. 2. *6:* 6. 1906. Holotype: South Georgia, Cumberland Bay, King Edward Cove (Pot Harbor), C. Skottsberg 299 (PC).

Tortula conferta Bartr. Bryologist. *60:* 140. 1957. Lectotype: Antarctica, Melchior Islands, Omega Island (Lystad Island), Siple 335.12 (US!).

Collections by R. M. Schuster. South Shetland Islands: H. 69-579, 69-585, 69-586; M. 69-805, 69-810. King George Island: J. 69-939, 69-955. Anvers Island: N. 69-843, 69-852, 69-864, 69-867, 69-871.

Other collection on Deception Island, southeast side near saddle in crater rim, about 250 meters east of Whalers Bay, 62°59.47′S, 60°32.48′W, Waldron and Fehlmann, March 29, 1970. I have used the name *Tortula grossiretis* without seeing any authentic material, but the specimens cited do fit the description of that species. A few of the specimens include smaller plants like the type of *T. conferta,* but these plants do not seem specifically distinct. The antarctic distribution of the species has been mapped by Greene [1967] under the name *T. conferta.*

BRYACEAE

Genus *Pohlia*

The antarctic species of the genus have recently been treated by Greene et al. [1970]. The two species known from the Antarctic are both widely distributed in the northern hemisphere. The following key is adapted from Greene [Greene et al., 1970].

Leaf insertion narrow, half or less the width of the leaf; nerve red to reddish-brown; cells of leaf lamina mostly 80–120 μ long; cell walls usually have a sigmoid curve................................. ***P. cruda***

Leaf insertion broad, three-quarters or more of the width of the leaf; nerve dark-green to brownish; cells of leaf lamina mostly 35–70 μ long; cell walls lack a sigmoid curve.............................. ***P. nutans***

Pohlia cruda (Hedw.) Lindb. Musci Scand. 18. 1879.

Mnium crudum Hedw. Species Muscorum Frondosorum. 189. 1801. Original material from Sweden.

Collections by R. M. Schuster. Argentine Islands: B. 69-219; C. 69-247a. Almirante Brown Base: E. 69-363, 69-381. South Shetland Islands: G. 69-408, 69-409; I. 69-694, 69-695; M. 69-829b. Graham Coast: Q. 69-903.

Pohlia nutans (Hedw.) Lindb. Musci Scand. 18. 1879.

Webera nutans Hedw. Species Muscorum Frondosorum. 168. 1801. Original material from Germany (G).

Webera racovitzae Card. Revue Bryol. *27:* 44. 1900. Lectotype: Gerlache Strait, Xème débarquement, Brabant Island, altitude of 350 meters, Racovitza 252a (PC).

Collections by R. M. Schuster. Argentine Islands: B. 69-211, 69-217, 69-226; C. 69-279. Gerlache Strait: D. 69-343. Anvers Island: N. 69-842, 69-860. Graham Coast: O. 69-514 part, 69-515, 69-521, 69-523 part, 69-525 part, 69-529; P. 69-708; Q. 69-872. South Shetland Islands: K. 69-1804, 69-1808.

Other collections on Argentine Islands, easternmost island of Corner Islands, 65°14′30″S, 64°14′23″W, Rastorfer and Gnau PAL-78 part, PAL-110; Avian Island, 67°46′30″S, 68°53′30″W, Rastorfer and Gnau PAL-124.

Genus *Bryum*

There are more species of *Bryum* in the Antarctic than of any other genus of mosses. In addition to the species given below in the key and the discussion, there are three names to be noted. (1) The species *Bryum korotkeviczae* Sav. & Smirn. is an elongate aquatic form from Bunger Hills on the coast of Wilkes Land. No material of this species has been seen, and I can make no evaluation. (2) The species *B. stenotrichum* C.Müll. has been reported from the Antarctic by Bartram [1957] under the name *B. inclinatum* (Brid.) Bland. *(hom. illeg.).* The specimens have not been seen, but from all indications they are actually *B. imperfectum* or *B. inconnexum.* (3) The poorly known species *B. algens* Card. has sometimes been made a synonym of *B. antarcticum* and is usually compared with the latter species. I have been told that *B. algens* is pale green, whereas *B. antarcticum* is always rather brownish. If the original illustration of the species is correct, the distinctions are far more marked. The plate shows excurrent costae and recurved margins, which would relate the species to *B. imperfectum* rather than to *B. antarcticum*.

A very useful discussion of a number of antarctic species of *Bryum* is found in Horikawa and Ando [1961]. A further note on *B. inconnexum* is found in Kuc [1969]. Greene [1967] has mapped part of the antarctic distributions of *B. antarcticum* and *B. argenteum*. The six species recognized here are distinguished as follows.

1. Distal cells of leaf blade hyaline, and plants have whitish appearance.................... ***B. argenteum***
   Leaf blade chlorophyllose throughout or hyaline only at extreme tips.......................................2
2. Costa ends before leaf tip; often subpercurrent.........3
   Costa percurrent to excurrent.........................4
3. Leaf margins plane, denticulate near apex............. ....***B. antarcticum***
   Leaf margins distinctly recurved, essentially entire near apex...............................***B. inconnexum***
4. Leaf margins plane, without distinct border; plants rubescent..............................***B. ongulense***
   Leaf margins usually distinctly recurved, with two to three rows of narrow cells; plants yellowish-green.....5
5. Peristome teeth 55–65 μ wide at base, with hyaline lateral margins; spores 18–29 μ in diameter........ ....***B. imperfectum***
   Peristome teeth only about 40 μ wide at base, without hyaline lateral margins; spores 12–15 μ in diameter ....***B. perangustidens***

I have seen no material of either *B. ongulense* Hor. & Ando from Ongul Island on the Prince Olav Coast or *B. perangustidens* Card. from Petermann Island and Marguerite Bay.

Bryum antarcticum Hook.f. & Wils. Bot. Antarctic Voyage H.M. Discovery. 1. Flora Antarctica. *2:* 414. 1847. Holotype: Cockburn Island, 64°S, 57°W.

Collections seen in Victoria Land, slope of Cape Hallett near Seabee Hook, E. Schofield, December 15, 1963; near Marble Point, 77°26′S, 163°50′E, J. R. Rastorfer MCM-5, MCM-6, MCM-7, MCM-14, MCM-15, MCM-45, MCM-46, November 6 to December 2, 1967, January 19, 1968; near Hobbs Glacier, 77°54′S, 164°21′E, J. R. Rastorfer MCM-38, MCM-42, January 4, 1968; Miers valley, 78°07′S, 164°10′E, J. R. Rastorfer MCM-51, MCM-55, January 21–31, 1968.

I have omitted the names *Webera gerlachei* Card. and *Bryum filicaule* Broth., which are sometimes placed in the synonymy of *B. antarcticum*. Both of these species, *W. gerlachei* from the west coast of the Antarctic Peninsula and *B. filicaule* from the William II Coast, were originally described from outside the range of *B. antarcticum* as recognized by Greene [1967]. Also the description of *Webera gerlachei* indicates a very different plant. I was first inclined to interpret *Bryum antarcticum* much too broadly, and the literature suggests that others have done likewise. The species as recognized here is a small reddish, usually budlike plant, the extreme tips of the leaves being pale and denticulate. This most common form of the species is reminiscent of the Pottiaceous genus *Acaulon*. The range of *Bryum antarcticum* is apparently limited; I know of authentic records or specimens only from the eastern side of the Antarctic Peninsula and from the McMurdo Sound area.

Bryum argenteum Hedw. Species Muscorum Frondosorum. 181. 1801. Original material from Europe.

Bryum amblyolepis Card. Rev. Bryol. *27:* 45. 1900. Holotype: Danco Coast, Gerlache Strait, Beneden Head, Racovitza 229b (PC).

Bryum cephalozioides Card. Bull. Herb. Boissier, ser. 2. *6:* 16. 1906. Holotype: off north Graham Land, Paulet Island, C. Skottsberg 460 (PC).

Bryum siplei Bartr. Ann. Mo. Bot. Gdn. *26:* 723. 1938. Holotype: Edward VII Peninsula, Rockefeller Mountains, Washington Ridge (Mt. Helen Washington), Siple, Wade, Corey, and Stancliff 100 (US).

Collection by R. M. Schuster. South Shetland Islands: A. 69-060.

Other collections in Victoria Land: Protoranker, Marble Point, 77°26′S, 163°50′E, Ugolini 33; Protoranker, The Flatiron, 77°01′S, 162°23′E, Ugolini 36b; Protoranker, Kar Plateau, 76°55′S, 162°28′E, Ugolini 43; Cape Hallett, 72°19′S, 170°18′E, J. R. Rastorfer MCM-9, P. Strang MCM-11; near Marble Point, 77°26′S, 163°50′E, J. R. Rastorfer MCM-13, MCM-47, MCM-48; Spike Cape, 77°18′S, 163°43′E, R. W. Strandtman MCM-17; Cape Roberts, 77°02′S, 163°11′E, R. W. Strandtman MCM-19; Black Island, 78°12′S, 166°25′E, J. R. Rastorfer MCM-24; near Bartley Glacier, 77°32′S, 162°10′E, Wright valley, J. R. Rastorfer MCM-31; Ross Island, near Horseshoe Harbor, 67°36′S, 62°52′E, R. W. Strandtman MCM-32, MCM-34; near Hobbs Glacier, 77°54′S, 164°21′E, J. R. Rastorfer MCM-43; November 19, 1967, to January 19, 1968.

In establishing the preceding concept I have discounted the importance of slender plants having reduced short costate leaves (*B. cephalozioides*) and plants having axillary propagula (*B. siplei*).

Bryum imperfectum Card. Revue Bryol. *27*: 44. 1900. Holotype: Danco Coast, Gerlache Strait, à l'entrée du Lemaire Channel, altitude of 50 meters, Racovitza 268b (PC).

Bryum crateris Dix. Bryologist. *23*: 67. 1920. Lectotype: Deception Island, Robins 451.

Collections by R. M. Schuster. Almirante Brown Base: E. 69-366. South Shetland Islands: G. 69-406; J. 69-947. Anvers Island: N. 69-833.

Other collections on Anvers Island, Biscoe Point, 64°49′S, 63°48′W, Rastorfer and Gnau PAL-83.

Dixon [1920] said his species *B. crateris* differed from *B. imperfectum* in having narrower leaves and a longer peristome. I find variation in both of these characters.

Bryum inconnexum Card. Revue Bryol. *27*: 44. 1900. Original material from Danco Coast, Gerlache Strait, near Cape Anna (Cape Anna Osterrieth), 250b; à l'entrée du Lemaire Channel, altitude of 50 meters, 268a, Racovitza (PC).

Bryum austro-polare Card. Revue Bryol. *27*: 45. 1900. Original material from Danco Coast, Gerlache Strait, near Cape Anna (Cape Anna Osterrieth), 151b, 151c, 205 part; Beneden Head, 233b, 234b, Racovitza (PC).

Webera racovitzae Card. var. *laxiretis* Card. Revue Bryol. *33*: 34. 1906. Gerlache Strait, Wiencke Island *Turquet.*

Collections by R. M. Schuster. South Shetland Islands: G. 69-419, 69-422; M. 69-808, 69-825, 69-829a.

Other collections on Deception Island, Follmann 2870, February 1963; Victoria Land, Protoranker, The Flatiron, 77°01′S, 162°23′E, Ugolini 36a; slope of Cape Hallett near Seabee Hook, E. Schofield, December 15, 1963; Enderby Land, north side of Cape Granat (Mt. Garnet) (2–3 km along coast east of Molodezhnaya Station), MacNamara, January 1968; Ross Island, near Horseshoe Harbor, 67°36′S, 62°12′E, R. W. Strandtman MCM-35; Victoria Land, Cape Roberts, 77°02′S, 163°11′E, Pittard MCM-20; Miers valley, 78°07′S, 164°10′E, J. R. Rastorfer MCM-29, MCM-44, MCM-54; near Hobbs Glacier, 77°54′S, 164°21′E, J. R. Rastorfer MCM-39, MCM-40, MCM-41, December 9, 1967, to January 31, 1968.

Bartramiaceae

Genus *Bartramia*

A number of species have been reported from the Antarctic, but I consider them all to be minor variants of the following species.

Bartramia patens Brid. Muscologia Recentiorum. *2* (3): 134. 1803. Holotype: Strait of Magellan, Commerson.

Bartramia diminutiva C.Müll. Bot. Jb. *5*: 79. 1883. Holotype: Îles Kerguelen, Naumann, 1874–1875.

Bartramia pycnocolea C.Müll. *In* Neumayer (ed.), Ergeb. Deut. Polar-Exped. Allg. *2*: 304. 1890. Holotype: South Georgia, 'Hochthal' über dem oberen 'Whalerthal,' a small ice-free valley at the head of Moltke Harbor, Royal Bay, Will, 1882–1883.

Bartramia subpatens C.Müll. *In* Neumayer (ed.), Ergeb. Deut. Polar-Exped. Allg. *2*: 304. 1890 hom. illeg. Holotype: South Georgia, Whalers Bay, Will, 1882–1883.

Bartramia oreadella C.Müll. *In* Neumayer (ed.), Ergeb. Deut. Polar-Exped. Allg. *2*: 305. 1890. Holotype: South Georgia, 'Whalerthal,' at Moltke Harbor, Royal Bay, Will, 1882–1883.

Collections by R. M. Schuster. Argentine Islands: B. 69-224; C. 69-229, 69-271, 69-274. Almirante Brown Base: E. 69-385, 69-399. South Shetland Islands: G. 69-406a, 69-421; H. 69-532, 69-546, 69-551, 69-581, 69-599a; I. 69-693; M. 69-828. Anvers Island: N. 69-841, 69-844, 69-868.

Genus *Philonotis*

There is one antarctic species endemic to Deception Island that is included in the recent treatment of the genus in Argentina by Matteri [1968].

Philonotis gourdonii Card. C. R. Hebd. Séanc. Acad. Sci., Paris. *153*: 603. 1911. Holotype: Deception Island, Mount Pound (sommet du Mont Pound), altitude of 450 meters, Gourdon 297 (PC).

Collection seen on Deception Island growing in interior of covered 'whale boat' imbedded in beach with warm fumaroles, G. E. Watson, February 18, 1964.

Orthotrichaceae

Genus *Orthotrichum*

Two species have been reported from the Antarctic, both from near Cape Anna in Gerlache Strait. *Orthotrichum antarcticum* Card. is endemic, and *O. rupicola* C.Müll. was described originally from Îles Kerguelen. No specimens have been seen in this study.

Leskeaceae

Genus *Pseudoleskea*

The antarctic species formerly placed in this genus

have been transferred to the genus *Hygroamblystegium*.

Brachytheciaceae

Genus *Brachythecium*

Six species of the genus have been reported from the Antarctic. I consider one of these, *B. turquetii* Card., to be a synonym of *Calliergidium austrostramineum* (C.Müll.) Bartr. A second species, *B. subpilosum* (Hook.f. & Wils.) Jaeg., was described from Isla Hermite, Cape Horn, and has since been reported from an antarctic locality, Cape Kjellman, Trinity Peninsula. I have seen no material of this species. The original description indicates a monoicous plant slightly smaller than *B. rutabulum* (Hedw.) B.S.G. having short cordate-ovate, scarcely plicate leaves with long slender acuminations, margins serrulate in the distal third or half, and the seta rough above. A third species, *B. georgico-glareosum* (C.Müll.) Par., of South Georgia, Tierra del Fuego, Îles Kerguelen, and one site in Antarctica, is known to me only in comparisons with other species. The species is apparently close to *B. antarcticum* but has narrower leaves, recurved margins, and less lax cells. The leaves are broader and have less elongate and less gradual acuminations than those of *B. austro-glareosum*. The lamina is apparently rather deeply plicate. The remaining three antarctic species of the genus are distinguished as follows.

1. Leaf margins slightly to distinctly recurved; alar cells in clusters 10–15 cells high and 5–6 cells wide. . . . . ***B. austro-glareosum***
   Leaf margins mostly plane; alar cells in clusters 8–10 or fewer cells high. 2
2. Median cells mostly 8–12 μ wide; leaves often strongly concave . . . . ***B. antarcticum***
   Median cells mostly 7–9 μ wide; leaves slightly concave . . . . ***B. skottsbergii***

Brachythecium antarcticum Card. Revue Bryol. *27:* 46. 1900. Original material: Danco Coast, Gerlache Strait, Beneden Head, Racovitza 230a part, 232a, 232c, 233c, 234c (PC).

Collections by R. M. Schuster. Argentine Islands: B. 69-215. Gerlache Strait: D. 69-341, 69-345. Almirante Brown Base: E. 69-378. Graham Coast: O. 69-511. South Shetland Islands: M. 69-806 part.

The species has been characterized primarily by the wider cells of the leaf lamina. I have seen some variation, and I suspect a situation similar to that in *B. cirrosum* (Schwaegr.) Schimp. of the northern hemisphere. The latter species, which resembles *B. antarcticum* very closely, has larger leaf cells in the more northern parts of its range.

Brachythecium austro-glareosum (C.Müll.) Par. Index Bryol. 131. 1894.

Hypnum austro-glareosum C.Müll. Bot. Jb. *5:* 82. 1883. Holotype: Îles Kerguelen, Naumann, 1874–1875.

Collections by R. M. Schuster. Argentine Islands: C. 69-246. Almirante Brown Base: E. 69-365, 69-402. Graham Coast: P. 69-705a; Q. 69-889b part.

As in most parts of this study, here I have derived my concepts of the antarctic species of *Brachythecium* without benefit of types or authenticated specimens. The characterizations of *B. austro-glareosum* and *B. georgico-glareosum* have been particularly difficult, and I still question the differences between the two. At present I can confirm only that the Schuster collections from the Antarctic and recent Aubert de La Rüe collections from Îles Kerguelen seem to represent the same species. All the specimens do have the longer less abruptly acuminate leaf tips indicated for *B. austro-glareosum*.

Brachythecium skottsbergii Card. Bull. Herb. Boissier, ser. 2. *6:* 12. 1906. Original material from South Georgia, Cumberland Bay Maiviken (May Harbor), King Edward Cove (Pot Harbor), Jason Harbor, Moraine Fjord, C. Skottsberg (PC).

Collections by R. M. Schuster. South Shetland Islands, Livingston Island: M. 69-801, 69-826, 69-827.

Other collections on Deception Island, Follmann 2872, 2873.

The material I have assigned to *B. skottsbergii* seems to differ from specimens of *B. antarcticum* in that the acuminations on the leaves are longer and more slender and the cluster of alar cells is less exposed on the margin. But these differences, along with the difference in cell size, may prove to be variable.

Amblystegiaceae

Genus *Amblystegium*

One species of the genus, *Amblstegium subvarium* Broth. of Îles Kerguelen, has been reported from Elephant Island in the South Shetland Islands. No material has been seen.

Genus *Hygroamblystegium*

The two species of the genus reported from the Antarctic belong to a group originally described in the genus *Pseudoleskea*. One, *Hygroamblystegium antarcticum* (Card.) Reim., was described from Beneden Head in the Gerlache Strait and later reported from South Georgia. The other species, *H. calochroum* (Card.) Reim., was described from South Georgia and later reported from Cape Pérez on the Graham Coast. The second species supposedly differs by its stiffer subjulaceous branches and narrower more imbricate leaves. No material has been seen in this study.

Genus ***Acrocladium***

The genus is represented in the Antarctic by the following species.

Acrocladium sarmentosum (Wahlenb.) Richs. & Wall. Trans. Br. Bryol. Soc. *1*(4): xxv. 1950.

Hypnum sarmentosum Wahlenb. Flora Lapponica. 380. 1812. Original material in rupibus Lapponiae alpinis, Wahlenberg.

Collections by R. M. Schuster. South Shetland Islands: A. 69-050, 69-059. King George Island: J. 69-947.

The distribution of the species is distinctly bipolar, the species being widely distributed in the northern hemisphere and occurring in southern South America, New Zealand, and Antarctica.

Genus ***Calliergidium***

One species of the genus occurs in the Antarctic.

Calliergidium austro-stramineum (C.Müll.) Bartr. Farlowia. *2*(3): 317. 1946.

Hypnum austro-stramineum C.Müll. *In* Neumayer (ed.), Ergeb. Deut. Polar-Exped. Allg. *2:* 319. 1890. Holotype: South Georgia, Landzunge, Will.

Brachythecium turqueti Card. Revue Bryol. *33:* 34. 1906. Holotype: Booth Island (Île Booth Wandel), Turquet (PC).

Collections by R. M. Schuster. Argentine Islands: B. 69-216 part; C. 69-255. Almirante Brown Base: E. 69-354. South Shetland Islands: G. 69-507. Anvers Island: N. 60-850, 69-851 part. Graham Coast: Q. 69-889a.

Other collections on westernmost of Corner Islands, 65°14′30″S, 64°13′52″W, Argentine Islands, Rastorfer PAL-80; easternmost island of Corner Islands, 65°14′30″S, 64°14′23″W, Rastorfer and Gnau PAL-106; Litchfield Island, 64°46′23″S, 64°04′30″W, Rastorfer and Gnau PAL-157.

I have not seen types, but I can see no basis for separating *Brachythecium turqueti* Card. from *Calliergidium austro-stramineum*.

Genus ***Drepanocladus***

One species of the genus is found in the Antarctic, where it is very abundant.

Drepanocladus uncinatus (Hedw.) Warnst. Beih. Bot. Zbl. *13:* 402, 417. 1903.

Hypnum uncinatum Hedw. Species Muscorum Frondosorum. 289. 1801. Original material from locis montosis Saxoniae metalliferae, montis piniferi Franconiae, alpinis Austriae, Salisburgi, et in Suecia.

Collections by R. M. Schuster. Argentine Islands: B. 69-216 part, 69-225; C. 69-283. Gerlache Strait: D. 69-342. Almirante Brown Base: E. 69-364, 69-382, 69-393. South Shetland Islands: G. 69-411; H. 69-543, 69-568, 69-582; I. 69-692 part, 69-699; K. 69-981 part, 69-982, 69-985; L. 69-931, 69-933, 69-936; M. 69-823 part. King George Island: J. 69-950, 69-953, 69-964. Anvers Island: N. 69-869. Graham Coast: Q. 69-873, 69-889a part, 69-889b part.

Other collections on Avian Island, Tom Berg, January 20, 1963; King George Island, Potter Cove, 62°13.9′S, 58°40′W, Schmitt, USS *Staten Island* expedition, 1962–1963; Livingston Island, False Bay, 62°42′S, 60°22′W, Schmitt, 1962–1963; Anvers Island, Biscoe Point, 64°49′S, 63°48′W, Rastorfer and Gnau PAL-82, PAL-84; easternmost island of Corner Islands, 65°14′30″S, 64°14′23″W, Argentine Islands, Rastorfer and Gnau PAL-109; Avian Island, 67°46′30″S, 68°53′30″W, Rastorfer and Gnau PAL-125; Litchfield Island, 64°46′23″S, 64°04′30″W, Rastorfer and Gnau PAL-155.

Most antarctic specimens of the species are a somewhat reduced form, the leaves having little or no longitudinal plication.

Plagiotheciaceae

Genus ***Plagiothecium***

Only one species of the genus, *Plagiothecium simonovi* Sav. & Smirn., has been reported from the Antarctic. No material of this endemic species has been seen in this study.

Hypnaceae

Genus ***Platydictya***

One species of the genus is known from the Antarctic.

Platydictya densissima (Card.) H.Robinson, comb. nov.

Amblystegium densissimum Card. Revue Bryol. *27:* 46. 1900. Holotype: near Cape Anna, Racovitza 205a (PC).

Collections by R. M. Schuster. Almirante Brown Base: E. 69-395. South Shetland Islands: M. 69-806 part.

All other species of the genus are in the northern hemisphere. The antarctic species was originally compared with *P. jungmannioides* (Brid.) Crum and *P. subtile* (Hedw.) Crum, but the antarctic plants were considered denser and more julaceous and to have shorter more concave leaves.

Genus ***Hypnum***

One species occurs in the Antarctic. The species is known otherwise only from the northern hemisphere.

Hypnum revolutum (Mitt.) Lindb. Oefversigt Kongl. Vetensk. Akad. Foerhandlingar. *23:* 542. 1867.

Stereodon revolutus Mitt. J. Linn. Soc. Bot. Suppl. *1:* 97. 1859. Holotype: in Tibet. occid. reg. alp., in summo montis 'Hera La,' altitude of 18,700 ped., H. Strachey.

Collections by R. M. Schuster. King George Island: J. 69-941, 69-956.

Acknowledgments. I wish to acknowledge the cooperation shown during the study by the New York Botanical Garden and its director, Dr. W. C. Steere. I also appreciate the loan obtained from the Laboratoire de Cryptogamie, Muséum National d'Histoire Naturelle, Paris, through the kindness of Mesdames Allorge and Jovet-Ast.

REFERENCES

Bartram, E. B.
1957 Mosses from the United States antarctic service expedition, 1940–41. Bryologist, *60:* 139–143.

Cardot, J.
1908 La flore bryologique des terres magellaniques, de la Géorgie du Sud et de l'Antarctide. Wiss. Ergebn. Schwed. Südpolarexped., *4*(8) : 1–298, pls. 1–11.

Clifford, H. T.
1955 On the distribution of *Rhacomitrium crispulum* (H.f. & W.) H.f. & W. Bryologist, *58:* 330–334.

Dixon, H. N.
1920 Contributions to antarctic bryology. Bryologist, *23:* 65–71.

Greene, S. W.
1967 Bryophyte distribution. *In* V. Bushnell (Ed.), Terrestrial life in Antarctica, Antarctic Map Folio Ser., folio 5, pp. 11–13. Amer. Geogr. Soc., New York.
1968a Studies in antarctic bryology. 1. A basic check list for mosses. Revue Bryol. Lichen., *36*(1–2) : 132–138.
1968b Studies in antarctic bryology. 2. *Andreaea, Neuroloma.* Revue Bryol. Lichen., *36*(1–2) : 139–146.

Greene, S. W., D. M. Greene, P. D. Brown, and J. M. Pacey
1970 Antarctic moss flora. 1. The genera *Andreaea, Pohlia, Polytrichum, Psilopilum,* and *Sarconeurum.* Sci. Rep. Br. Antarct. Surv., *64:* 1–118.

Horikawa, Y., and H. Ando
1961 Mosses of the Ongul Islands collected during the 1957–1960 Japanese antarctic research expedition. Hikobia, *2*(3) : 160–178.
1963 A review of the antarctic species of *Ceratodon* described by Cardot. Hikobia, *3*(4) : 275–280.

Keever, C.
1957 Establishment of *Grimmia laevigata* on bare granite. Ecology, *38*(3) : 422–429.

Kuc, M.
1969 Some mosses from an antarctic oasis. Revue Bryol. Lichen., *36*(3–4) : 655–672.

Lipman, C. B.
1936 The tolerance of liquid air temperatures by dry moss protonema. Bull. Torrey Bot. Club, *63:* 515–518.

Matteri, C. M.
1968 Las especies de '*Philonotis* (Bartramiaceae)' del sur de Argentina. Revta Mus. Argent. Cienc. Nat. Berbardino Rivadavia Inst. Nac. Invest. Cienc. Nat., *3*(4) : 185–234.

Morrill, J. B.
1950 Mosses in liquid air. Bryologist, *53:* 163–164.

Robinson, H.
1964 Five bryophytes of interest from Chile. Bryologist, *67:* 53–55.

Schultz-Motel, W.
1970 Monographie der Laubmoosgattung *Andreaea.* 1. Die costaten Arten. Willdenowia, 6: 25–110.

Smith, G. L.
1971 A conspectus of the genera of Polytrichaceae. Mem. N. Y. Bot. Gdn, *21*(3) : 1–83.

Steere, W. C.
1961 A preliminary review of the bryophytes of Antarctica. Science in Antarctica. 1. The life sciences in Antarctica. Nat. Acad. Sci. Publ. 839: 20–33. Washington, D. C.

NEW BASIDIOMYCETE FROM THE ANTARCTIC

ROLF SINGER

Department of Botany, Field Museum of Natural History, Chicago, Illinois 60605

Abstract. The southernmost fungus yet collected fructifying is a specimen described here as new, *Gerronema schusteri* Sing. n. sp., named for its collector, R. M. Schuster, and considered to be closely related to *Gerronema fibula* (Bull. ex Fr.) Sing.

During his collection trip to the Antarctic in 1969 Dr. R. M. Schuster observed and prepared a specimen of basidiomycetes encountered on the south coast of Anvers Island; thus it is the southernmost fungus yet observed. It is very closely related to *Gerronema fibula,* but thanks to a colored slide accompanying the collection it can be stated that the fungus in question is not merely a form of *Gerronema fibula* but a new species that is described below.

Gerronema schusteri Sing. spec. nov. Pileo aurantiaco-flavo, sicco grisascente, ad marginem magis aurantiaco; subglabro, convexo, centro depresso, 4–6 mm lato. Lamellis aurantiacis, subdistantibus vel distantibus, mediocriter latis, profunde decurrentibus, aciebus integris, haud intervenosis. Stipite lamellis concolori, subglabro, haud insititio, ±5 × 1 mm. Contexto pallido. Sporis 5.5–6 × 2.5–2.8 μ, hyalinis, oblongis vel cylindraceis; basidiis tetrasporis; cystidiis 15–43 × 3.5–9.5 μ, obtusis, ventricoso-ampullaceis, integris. Superficie pilei et stipitis dermatocystidiis numerosis ornatis. Hyphis fibulatis, inamyloideis. Ad terram inter muscos, haud lichenisata ad basin stipitis, gregatim ad terram. Anvers Island, Antarctica, R. M. Schuster 69-3031 (National Fungus Collections, Beltsville, Maryland), Typus.

Pileus orange yellow when fresh, remaining more orange on the margin in most dried specimens, tending to be paler in the center, otherwise becoming somewhat grayish in the herbarium ('camels hair,' 'racquet,' according to Maerz and Paul [1930]), appearing glabrous but under a good lens finely pubescent from the dermatocystidia, convex with depressed center, 4–6 mm broad. Lamellae lighter colored than the surface of the pileus, dried orange between 11-B-10 and 'carrot r' of Maerz and Paul [1930], distant or subdistant, not intervenose, with entire edge, medium broad, deeply decurrent. Stipe concolorous with the lamellae, macroscopically glabrous but under a lens finely pubescent like the pileus, somewhat hollow in the apical portion, less so below; in the lowest part, strongly white myceloid and not insititious, with white mycelium, equal or tapering somewhat downward, about 5 × 1 mm. Context pallid, thin but moderately so.

Spores (4.5)–5.5–6 × 2.5–2.8 μ, oblong to cylindric, hyaline, smooth, inamyloid. Hymenium: basidia 20–23 × 4–5 μ four spored. Cystidia on edges and on sides of lamellae, ventricose ampullaceous, broadest in the middle or near the base (neck portion 1.7–4 μ in diameter) with obtuse tip with clamped base, hyaline. Hyphae of the trama filamentous and multiseptate, generally rather broad, not gelatinized, hyaline, with clamp connections, inamyloid. Hymenophoral trama regular about 150 μ broad near pileus trama, thinning to about 25 μ near edge, more or less interwoven but axillarly arranged, hymenopodium 5–20 μ broad, consisting of more subparallel to parallel hyphae, with firm walls (wall ⩽0.5 μ thick); hyphal cells narrower here than in the mediostratum, where they are 4–25 μ broad (some 27 × 15 μ; 35 × 25 μ; others narrower), slightly divergent in a narrow zone near the hymenium (Clitocybe subtype of hymenophoral trama). Cortical layers: epicutis of the pileus made up of erect, rather dense, and very numerous but neither palisadically crowded nor hymeniformly arranged dermatocystidia; these rising mostly without septum from the broad hyphal cells of the cutislike hypodermium, ventricose to almost subulate. Most of them distinctly capitate to subcapitate, with thin to firm not distinctly thickened wall, simple and entire, with hyaline to yellowish (in KOH) cell sap, 25–64 × 6–15 μ. Constriction below the apex 2.5–4 μ across, capitulum 3–7 μ in diameter, obtuse. Similar dermatocystidia on the surface of the stipe.

Growing on earth among mosses (not *Sphagnum*) not associated with lichenized tissue, with ample

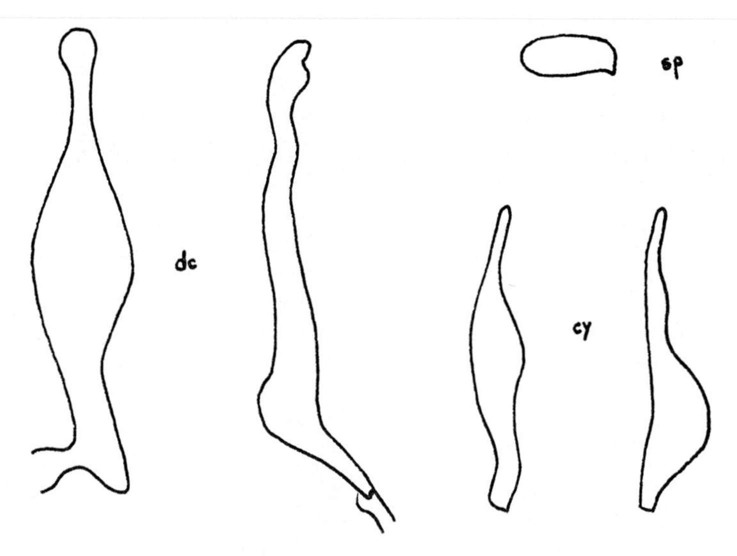

Fig. 1. *Gerronema schusteri* Sing (dc, dermatocystidia, 1000×; cy, hymenial cystidia, 1000×; sp, spores, 2000×).

white mycelium over the moss thalli, gregarious. Collected by R. M. Schuster (69-3031), February 1969, south shore of Anvers Island west of Palmer Station, Norsel Point. Type collection deposited at National Fungus Collections, Plant Industry Station, Beltsville, Maryland.

This species is so close to *G. fibula* that it was thought at first to be an antarctic race of that species. It belongs in the same section and subsection as *G. fibula.* However, the habit of this species is remarkably and constantly different from that of the typical *G. fibula,* the stipe being shorter, the colors slightly different, and the lamellae more consistently distant or subdistant.

In the material I have collected in Tierra del Fuego, the southernmost point at which *G. fibula* has been found, the habit and the colors were quite typical. Two forms could be distinguished: the short stemmed (but still longer stemmed than *G. schusteri*) form was growing among *Polytrichum* on and around logs and had both cystidia and dermatocystidia capitate or subcapitate; the long stemmed form was found among mosses on open pasture, the hymenial and the dermatocystidia also being mostly capitate or subcapitate and the lamellae white and distant. Both of these forms are obviously somewhat different from *G. schusteri.* In the oakwood zone of Colombia a form with ampullaceous noncapitate hymenial and dermatocystidia was found, the latter being smaller than those in *G. schusteri* and the former much more scattered on the sides of the lamellae and often incrusted on the apex. Most specimens of *G. fibula* in North America studied by A. H. Smith have equally subcapitate hymenial and dermatocystidia, some noncapitate ones being intermixed. Similarly subcapitate hymenial cystidia have also been observed in the Faeroe Islands by Møller. One may therefore conclude that the slight dimorphism of the cystidia in the antarctic species is an additional character that may be helpful in differentiating the new species from forms of *G. fibula.*

According to the observations by Gams [1962] *G. fibula,* like other representatives of the same genus, frequently seems to be associated with a basidiolichen. This association would be expected under the specific conditions of Antarctica. However, *G. schusteri,* at least the type collection, was not found to be associated with basidiolichens. On the other hand, the fact that Heikkilä and Kallio [1966] have found *G. fibula* nonlichenized in Finland should be taken into consideration.

REFERENCES

Gams, H.
1962 Die Halbflechten *Botrydina* and *Coriscium* als Basidiolichenen. Öst. Bot. Z., *109:* 376–380.

Heikkilä, H., and P. Kallio
1966 On the problem of subarctic badidiolichens, 1. Annls Univ. Turku, ser. A, *36*(2): 48–74.

Maerz, A., and M. R. Paul
1930 A dictionary of color. McGraw-Hill, New York.

ORNITHOGENIC SOILS OF ANTARCTICA

F. C. Ugolini

College of Forest Resources, University of Washington, Seattle, Washington 98105

Abstract. Because of the intense cold and the paucity of liquid water, Antarctica supports a very meager terrestrial flora. In contrast Antarctica supports a rich oceanic life. Penguin rookeries occurring on ice-free areas bordering the sea represent the most extensive source of organic matter for the terrestrial ecosystem. Ornithogenic soils consist of a well-defined layer of guano resting sharply on mineral soil. Analysis of the guano showed a high nitrogen content resulting from the high protein diet of the penguin. The phosphorus content was also high and mainly inorganic. The high salinity of the guano is due to the bordering sea, the penguin nasal excretions, and the aridity of the climate. Differential thermal analysis of guano, algal, and moss materials showed distinct differences in the thermoreactivity of the organic components. X-ray analysis revealed few differences in the clays of the mineral horizon underlying the guano and the soils uncontaminated by guano. The sharp separation of the guano layer and the underlying mineral soil indicates little frost mixing. The recognition of specific characteristics of soils associated with guano deposits may be important in locating and identifying relict or buried ornithogenic soils; dating the guano of these soils may have a bearing on the glacial history of the continent.

The continent of Antarctica supports such a meager and primitive flora that botanically it may be considered a desert. Intense cold and persistent drought are factors limiting the establishment of plants. In contrast the oceanic life is very abundant, as the guano deposits that occur in the ice-free areas bordering the sea and seasonally occupied by Adélie penguin colonies (*Pygoscelis adeliae*) indicate.

The ornithogenic soils of Antarctica, which were so named by Syroechkovsky [1959], were later discussed by others [Ugolini, 1965; Tedrow and Ugolini, 1966; Allen and Northover, 1967; Holdgate et al., 1967; Ugolini, 1969; Allen and Heal, 1970]. Campbell and Claridge [1966] and McCraw [1967] also investigated these soils; Campbell and Claridge [1969] refer to them as avian soils. Ornithogenic soils consist of a layer of guano resting sharply on unconsolidated coarse sand. The surfaces of these soils are made of indurated guano crust and scattered mounds of pebbles used for nest building. The surface conditions vary considerably according to the age and topography of the site.

Information relative to soil conditions in continental Antarctica has been provided by a number of investigators [Jensen, 1916; Blakemore and Swindale, 1958; Glazovskaia, 1958; McCraw, 1960, 1967; Ugolini, 1963, 1967a, b, 1970, also unpublished data, 1964; Claridge, 1965; Claridge and Campbell, 1968; Campbell and Claridge, 1967, 1969; Ugolini and Bull, 1965; Ugolini and Grier, 1969; Ugolini and Bockheim, 1970; Everett and Behling, 1968; Cameron et al., 1969, 1970a, b]. These investigations have been conducted on the most widespread soils, i.e., those lacking an organic cover. Few details are available for the soils of the rookeries. The purpose of this paper is to report the results of studies conducted on the ornithogenic soils as they occur in the Adélie penguin rookeries of Cape Royds, Ross Island. Ornithogenic soils derived from the guano of skuas, snow petrels, and other birds nesting in coastal colonies are not discussed here.

AREA OF STUDY

Cape Royds (77°33′S, 166°09′E) is situated on the west coast of Ross Island, Antarctica, about 35 km north of the main American base, McMurdo Station (Figure 1). The cape is bordered by the Ross Sea, which is normally free of ice in the vicinity of the cape from November through the rest of the austral summer. A polynya that forms between Cape Royds and Cape Bird almost every boreal winter or spring makes it possible for colonies of Adélie penguins to exist at Cape Bird and Cape Royds [Stonehouse, 1967]. The Adélie penguin rookery, the southernmost on the continent, is located on the west side of the cape and occupies an area of approximately 2 ha. Stonehouse [1963] reported a population of 1250

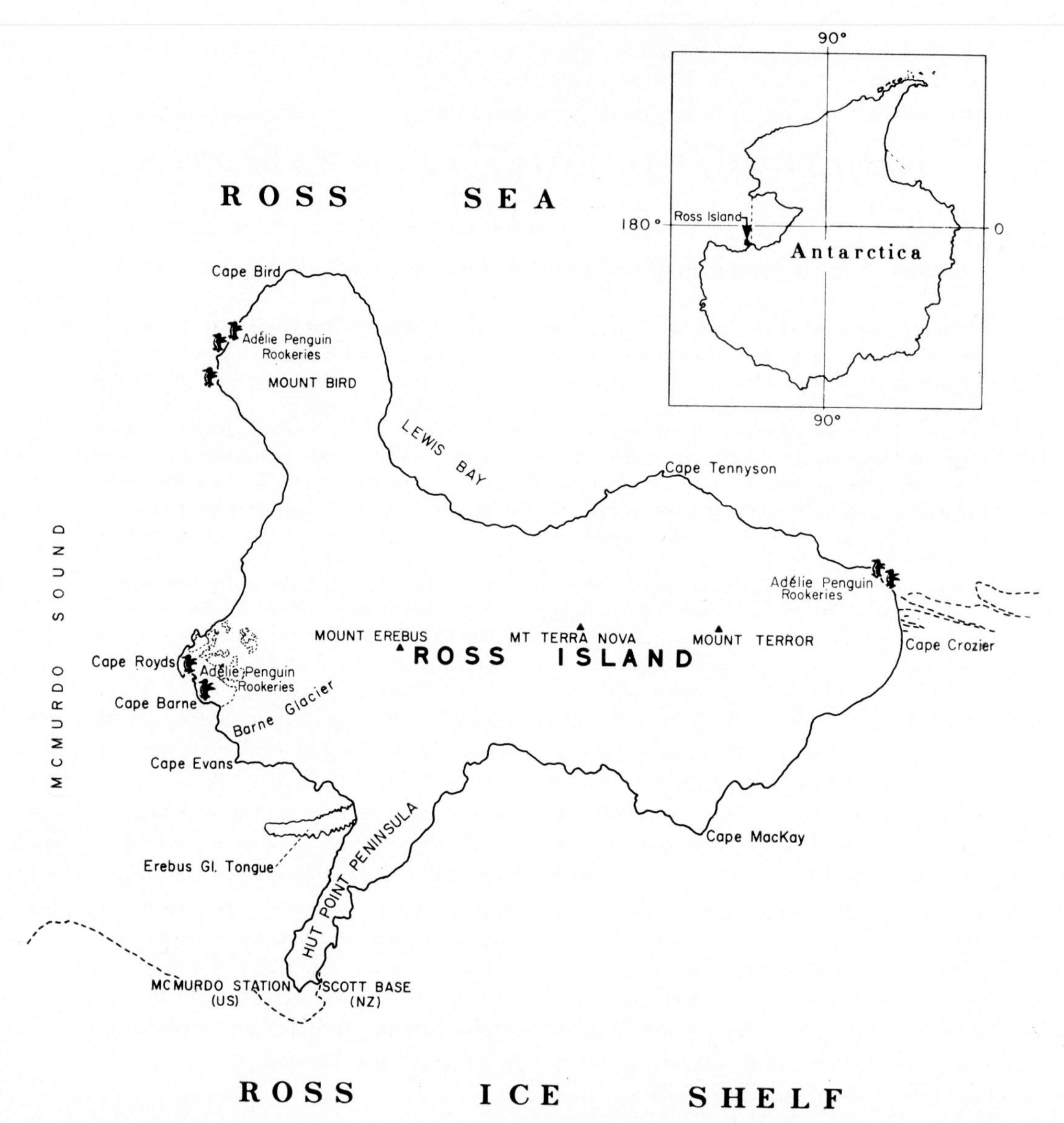

Fig. 1. Ross Island, Antarctica. The dotted lines inland of Cape Royds and Cape Barne (at left of figure) show the extent of ice-free areas.

penguin pairs in 1961 and indicated a continuous decline since 1957.

CLIMATE

Antarctica has the most severe climate on this planet, a climate characterized by very low temperatures, scant precipitation, and very strong winds. At McMurdo Station 35 km south of Cape Royds the mean annual air temperature is −17.6°C. The mean annual precipitation is 11.9 cm of water equivalent; however, blowing snow mixed with falling snow makes accurate measurements of precipitation difficult. The mean relative humidity is 57%. The mean temperature for the warmest months (December through February) is −5.6°C and for the coldest months (June through August) is −25.9°C. The minimum and the maximum temperatures are −50.6° and 5.6°C, respectively [U.S. Navy Weather Research Facility, 1961; Rudolph, 1966].

Climatologic data collected by the British antarctic expedition of 1907–1909, which wintered at Cape Royds, indicate that December and January are the warmest months of the year; however, the mean temperatures for those 2 months never rose above 0. The maximum temperature recorded during the 2 years (1907–1909) was −3°C (in January), and the minimum was −28°C (in July). The snowfall from February 1908 to February 1909 was estimated to be about 24.1 cm of water equivalent. No rain was recorded. The winds were either gently northerly with speeds below 20 km/hr or from the south-southeast or southeast. The latter were increasingly intense during blizzards [Shackleton, 1909]. Intermittent measurements of soil temperatures obtained by W. Boyd [Boyd and Boyd, 1963] and by the author indicate a progressive warming of the soil surface (0–2 cm) from −2.0°C in mid-November 1961 to 6.0°C in mid-December 1961. Positive temperatures persist into January. The soil moisture fluctuates greatly during the season and locally because of snow melting, sea sprays, and penguin activity.

GEOLOGY

Cape Royds may consist of three kenyte lava flows, which in turn are covered by volcanic agglomerate and lithic tuff [Treves, 1962]. The glassy kenyte of the surface flow consists of large rhombohedral feldspar phenocrysts (oligoclase) immersed in a matrix composed of glass containing plagioclase, aegerine-augite, apatite, ilmenite, and magnetite. The volcanic agglomerate and breccia are made of fragments of kenyte and basalt in a matrix of tuff. The lithic tuff consists of fragments of basalt, broken plagioclase, and pyroxene crystals in a glassy matrix. Erratics of igneous, sedimentary, and metamorphic rock cover the area. The bedrock is covered in places by morainic deposits [Treves, 1962].

The present rugged and rough surface of the cape consists of kenyte. In the area occupied by the rookery the microrelief is modified by guano deposits and the mounds of pebbles gathered by the Adélie penguins for use as nests.

SOILS AND SOIL-FORMING PROCESSES

The soils of continental Antarctica are restricted by the limited extent of ice-free areas and are forming under continuous low temperatures, limited amounts of water, and reduced biological activity. The pedogenic processes acting on these soils are similar to those of desert regions. Only in a few areas where a sparse cover of mosses or a mat of algal material is present are the soils covered with an organic layer. Another exception is the areas bordering the sea and occupied by penguin colonies, where guano has accumulated.

In general the soils of continental Antarctica are devoid of an organic cover and have been called ahumic soils by Tedrow and Ugolini [1966] and frigic soils by Claridge [1965] and McCraw [1967]. Because chemical weathering is low, the clay content is meager, and coarse textures prevail. The limited amount of leaching results in an accumulation of soluble salts and carbonates. The oxidation of iron-bearing minerals and the iron coatings of the soil particles impart yellowish-red colorations to the soils. These soils appear morphologically rather featureless and show little horizonation. The surface is protected by a desert pavement and in places displays salt efflorescences. A zone of salt accumulation may be present below the surface. Indurated zones weakly cemented by carbonates or gypsum are sometimes found. Soils along the coasts of Enderby Land, East Antarctica, appear more developed than any other soils of the continent. MacNamara [1969] reports a red ahumisol showing considerable horizon differentiation and clay cutans. The soils of maritime or oceanic Antarctica are more diverse and more developed than the soils of continental Antarctica [Allen and Heal, 1970].

The soils of Antarctica are underlain by permafrost. Because of the extremely dry conditions even when temperatures are well below 0, soil material is not solidly frozen in places; therefore not all the permafrost is ice cemented.

Weathering in Antarctica is dominated by physical processes [Blakemore and Swindale, 1958; Kelly and Zumberge, 1961]; however, evidences of chemical alteration have been described by Glazovskaia [1958], Claridge [1965], Linkletter [1971], and the author (unpublished data, 1964).

BIOTIC FACTORS

The terrestrial vegetation of Antarctica is notoriously impoverished and dominated by cryptogams such as mosses, lichens, algae, and fungi. Cape Royds does not escape this condition. The crusts of lichens, the cushions of mosses, and the algae are found at the cape [Schofield and Rudolph, 1969]; however, because of saline conditions, trampling by birds, and the possible presence of antibiotics in the guano [Sieburth, 1959], plant establishment in the rookery is retarded [Rudolph, 1966]. (Preliminary studies by the author and R. E. Cameron (personal communications, 1971) showed considerable microbial growth

Fig. 2. Ornithogenic soil profile. The crust of desiccation is visible at the surface; the darkened zone above the marker is the guano layer, Cape Royds, Antarctica.

in media inoculated with guano from Cape Royds. This finding demonstrates that antibiotics were not limiting biological activity.) Springtails (*Gomphiocephalus hodgsoni*) and trombidiform mites were collected at the cape [Gressitt, 1965]. Members of the *Azotobacter* genus, a heterotrophic nitrogen fixer, and lactate-reducing bacteria were reported by Boyd and Boyd [1963] in the penguin rookery. Accumulated guano deposits in the coastal rookeries represent the most abundant source of organic matter in the continental antarctic terrestrial ecosystem. Being ocean-feeding species, penguins and other birds supply the impoverished land with organic matter from an extremely rich sea. According to Hutchinson [1950] guano accumulates when a highly productive sea is associated with an arid environment; Cape Royds and the surrounding Ross Sea meet these prerequisites. The guano of Cape Royds consists of Adélie penguin excreta mixed with feathers, decaying crustacean food, egg shells, and penguin carcasses. Colonies of skuas (*Catharacta maccormicki*) preying on the penguins also contribute to a lesser extent to the accumulation of organic debris. Syroechkovsky [1959] estimated in the Mirnyy area (East Antarctica) that a population of 1800 penguins per square kilometer removes 556 tons of marine organisms from the ocean during the summer. At Cape Crozier, Ross Island (Figure 1), where another Adélie penguin rookery is located, large sections of the area are literally covered with dead penguins only slightly decomposed. Decomposition by heterotrophic microorganisms evidently remains below the rate of guano accumulation. Even more dramatic are the mummified bodies of dead seals found in inland areas; these bodies are perfectly preserved except for wind abrasion.

The population size and the areas occupied by the penguin colonies may vary with time. Although the reasons for these natural fluctuations are unknown (B. Stonehouse, personal communication, 1970), it is evident from soil studies that areas once colonized were abandoned and covered by drifting sand. Buried surfaces of old rookeries were exposed during excavations made in the vicinity of Cape Royds. Near Blacksand Beach north of the cape a detailed map

TABLE 1. Profile 1 (Figure 2)

Horizon	Depth, cm	Description
011	13–11	Crust of desiccation that can be removed and pealed off easily. Surface of the crust contains salt efflorescences. Smoothly polished pebbles are imbedded in the crust. Crust includes organic debris, feathers, and fragments of egg shells and bones. Upper part of the crust contains numerous pores.
012	11–0	Light yellowish-brown (10YR6/4) horizon that consists of fine fluffy organic debris well homogenized with feathers and small bones; strong odor of ammonia. Mineral matter is scarce and consists of polished pebbles and angular grains of coarse sand that appear bleached. Abrupt and wavy boundary; temperature of this layer was −2°C.
C	0–15	Dark gray (N3/0) sandy mineral horizon, single grain structured, loose, with strong odor of ammonia, but with no visible organic material. Rests on an ice-cemented layer; temperature of this layer was −1.0°C.

Another soil profile was opened at about 15 meters above sea level in the vicinity of Blacksand Beach, Cape Royds, where a rookery found abandoned in 1910 is now completely covered by aeolian sand (Table 2).

TABLE 2. Profile 2

Horizon	Depth, cm	Description
1	0–8	Very dark gray (N3/0) coarse sand layer, loose and gritty; dry; wavy and abrupt boundary.
2	8–13	Light yellowish-brown (10YR6/4); pebbly; faint odor of ammonia; pebbles in this layer resemble the ones used at present by the penguins to build their nests; wavy and abrupt boundary; thickness of 5–15 cm; dry.
3	13–35	Very dark gray (N3/0); coarse sand with angular pebbles; loose; dry; rests directly on an ice-cemented layer.
4	35	Ice-cemented layer.

made during the British antarctic expedition of 1910–1913 shows a completely exposed abandoned rookery. When the area was visited in November 1961, no surface evidences of a rookery could be found. A continuous, pebbly, light-yellowish layer was found after the excavation of a soil profile. This layer betrayed the old surface of the abandoned rookery, which is now under 8 cm of coarse sand.

According to Harrington [1960] the thickness of the Adélie penguin guano deposits provides a rough measure of the relative ages of the rookeries. After studying a sequence of soils from an Adélie penguin rookery at Inexpressible Island, Ross Sea, Antarctica, Campbell and Claridge [1966] found that, after an area is abandoned by the penguins, there is a gradual decomposition and disappearance of the organic matter. Because the destruction of organic matter under antarctic conditions is considered to be a slow process, these authors suggested a considerable time interval for the degradation of the guano layer.

FIELD INVESTIGATION AND SOIL DESCRIPTION

Most of the field investigation was carried out in the Cape Royds area; however, a number of ornithogenic soil profiles were also examined in the rookery of Cape Crozier, Ross Island. Field observations and chemical parameters in the Cape Royds area revealed a sequence of surficial conditions related to the length of occupancy by the Adélie penguins [Ugolini, 1969]. Sites presently occupied show a number of nests consisting of irregularly spaced pebble mounds. Sites no longer used for nesting were still covered by guano but were devoid of pebble mounds. The guano at these sites had acquired a hard crust produced by desiccation and penguin traffic. Still older sites were completely drifted over by aeolian sand. Descriptions of two ornithogenic soil profiles from sites at Cape Royds not presently occupied by penguins are given in Tables 1 (Figure 2) and 2. Samples were collected on November 8, 1961, when the air temperature at 5:00 P.M. was −10°C.

METHODS OF ANALYSIS

Total carbon was determined by chromic acid reduction [Purvis and Higson, 1939], some samples being rechecked by the method suggested by Jackson [1958] by which organic carbon is determined as CO_2. Total nitrogen including nitrate was estimated by the improved Kjeldahl method [Association of Official Agricultural Chemists, 1955]. Loss on ignition was calculated after the samples were ignited at 700°C. A high-temperature pressure vacuum furnace was used for differential thermal analysis [Lodding and Hammel, 1959]. Dynamic air at 10 oz/in.^{2} was used to favor the combustion of the organic materials. A modified sandwich technique was used [Ugolini et al., 1963]. To allow a semiquantitative comparison of the different thermocurves, samples were corrected for percentage of carbon.

Because of the high water-holding capacity of guano

TABLE 3. Chemical Analyses of the 1:7 Soil–Water Extract of an Ornithogenic Soil, Cape Royds, Antarctica

Horizon	Depth, cm	Electrical conductivity, Mmhos/cm	*p*H	Cations, meq/l Na^+	K^+	Ca^{++} and Mg^{++}	Anions Cl^-, meq/l	$SO_4^=$, meq/l	Total $CO_3^=$, %
011	13–11	18.0	7.2	105.3	38.6	9.9	30.5	29.3	0.7
012	11–0	11.0	7.3	161.0	8.7	0.9	27.9	55.0	0.3
C	0–15	3.3	7.1	13.0	5.1	0.3	8.3	6.4	0.0

and the limited supply of samples, a rather large (1 : 7) soil to water ratio had to be used for obtaining a water extract. The *p*H and the electrical conductivity of the extract were determined. The colored water extract was digested in acids to oxidize suspended and soluble organic material prior to elemental analysis. Such treatment was not necessary for the soil outside the rookery. Chlorides were determined by titration with silver nitrate [Richards, 1954], whereas sulfates were determined gravimetrically [Richards, 1954]. Calcium and magnesium were analyzed by an atomic absorption spectrophotometer and by EDTA (ethylenediaminetetraacetic acid) titration, carbonate by hot hydrochloric acid [Richards, 1954], and sodium and potassium by flame photometer. Total phosphorus was determined colorimetrically [Jackson, 1958] after the samples were wet digested. Organic phosphorus was evaluated by the method of Metha et al. [1954]. Clays were obtained from the mineral horizon underlying the guano layer and from two horizons of an ahumic soil profile near the rookery uncontaminated by penguin droppings. Clays were X-rayed after 'cleaning' [Aguilera and Jackson, 1953] with a Siemens Crystalloflex 4 diffractometer.

RESULTS

A partial chemical analysis of the water extract of the ornithogenic soil showed a very high soluble salt content (Table 3). Among the cations sodium is the most abundant and is generally concentrated in the guano layer. Potassium is high at the surface but decreases drastically with depth. Calcium and magnesium are relatively low and most abundant at the surface. Chloride and sulfate are abundant in the guano layer and considerably less abundant in the mineral horizon below it. Carbonates are low and restricted to the guano. The sums of the cations and the anions are not equal; this discrepancy will be discussed below.

The water extract of the ahumic soil collected near the rookery (Table 4) appears less saline and less enriched in sodium. The combined calcium and magnesium content is slightly higher than the sodium content. Among the anions the chloride content is almost 6 times the sulfate content. The total anion and cation concentrations in this soil are nearly equal. The soil reaction in both ornithogenic and ahumic soils is neutral.

The carbon/nitrogen ratio of penguin guano is very narrow (Table 5). Bird guano is the richest in nitrogen among natural organic materials, and penguin guano appears to be no exception. Phosphorus, which averages about 5% in recent Peruvian guano [Jacob and v. Uexküll, 1960], is 6.3% in Cape Royds guano (Table 5). The guano at Cape Royds is not made entirely of organic components, 50% consisting of mineral matter. Recent penguin droppings are richer in nitrogen than the guano itself (Table 6).

Differential thermocurves of the penguin guano (Figure 3) show the recurrence of three exothermic peaks at approximately 330°, 460°, and 530°C and two endothermic peaks at 50° and 75°C. The mineral horizon underlying the guano layer shows a pattern

TABLE 4. Chemical Analyses of the 1:7 Soil–Water Extract of an Ahumic Soil, Cape Royds, Antarctica

Horizon	Depth, cm	Electrical Conductivity, Mmhos/cm	*p*H	Cations, meq/l Na^+	K^+	Ca^{++} and Mg^{++}	Anions Cl^-, meq/l	$SO_4^=$, meq/l	Total $CO_3^=$, %
1	5–10	2.3	7.2	11.3	2.59	14.3	17.6	3.1	not determined

TABLE 5. Partial Chemical Analysis of an Ornithogenic Soil Profile, Cape Royds, Antarctica

Horizon	Depth, cm	C	N	C:N Ratio	P Total	P Organic	P Inorganic	Ca^{++}	Mg^{++}	H_2O*	Loss on Ignition
011	13–11	21.8	11.6	1.8	6.3	0.8	5.5	3.1	1.4	19.6	52.5
012	11–0	17.1	8.5	2.0	5.9	1.3	4.6	3.0	1.3	14.2	40.8
C	0–15	0.4	0.4	1.0	0.5	...	...	...	...	0.5	1.7

All values except the depth and the C:N ratio are in per cent. Three center dots indicate that no value was determined.
* 105°C.

similar to that of the guano itself, but the peaks are more subdued. Fresh penguin droppings show mainly a two exothermic peak system: peaks at 345° and 550°C and a small peak at 425°C. The endothermic peak occurs at 45°C.

The buried guano layer of the old rookery has a pronounced exothermic reaction at about 290°C and a reduced one at 410°C. The endothermic reactions occur at 45° and 98°C. Combustion curves for moss exhibit a three exothermic peak system (Figure 4). The peaks occur at about 330°, 420°, and 540°C. Thermograms of the algal material (Figure 4) show two exothermic reactions in the 280° and 500°–520°C regions. The endothermic reactions occur at 40°–50°C and 100°C.

Diffractograms of clay collected from the mineral horizon underlying the guano layer (at 0–15 cm) show a 15.5-Å peak in addition to peaks for a micaceous and a kaolinitic mineral, quartz, and feldspars (Figure 5). The diffractogram of clay of the ahumic soil horizon uncontaminated by penguin droppings and covered by 15 cm of mineral soil shows an assemblage of minerals similar to the mineral horizon underlying the guano except for a 14.2-Å peak (Figure 6).

DISCUSSION

The close proximity of the sea and the high percentage of soluble salts in penguin excreta are responsible for the high conductivity values given in Table 3. Like other birds that gather food from the sea, penguins are especially equipped to secrete hypertonic solutions through the nasal glands [Douglas, 1964]. The arid environment also contributes to the retention in the soil of soluble salts and soluble products of weathering. Deposits of mirabilite are present at Cape Barne near Cape Royds; other accumulations of salts in the ice-free areas of the McMurdo Sound region are common [David and Priestley, 1914; Rivard and Péwé, 1962; Gibson, 1962; Nichols, 1963; Tedrow and Ugolini, 1966]. Campbell and Claridge [1966] also report high soluble salts in soils from an antarctic rookery.

The discrepancy between the sums of the cations and the anions in the ornithogenic soil (Table 4) may be attributed to the presence of other anions such as phosphate, urate, and iodate, which were not

TABLE 6. Partial Chemical Analysis of Recent Penguin Droppings, Cape Royds, Antarctica

Depth, cm	C	N	C:N Ratio	H_2O*	Loss on Ignition
0–2	21.2	17.3	1.2	4.8	62.6

All values except the depth and the C:N ratio are in per cent.
* 105°C.

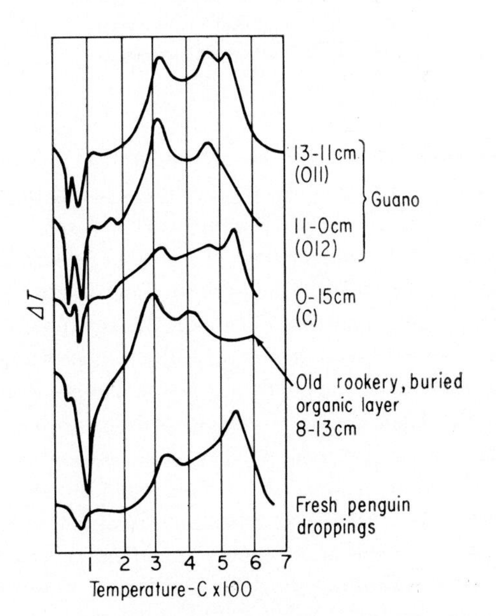

Fig. 3. Differential thermal curves of the ornithogenic soil profile, buried organic layer (old rookery), and fresh penguin droppings, Cape Royds, Antarctica. Analyses were made in dynamic air.

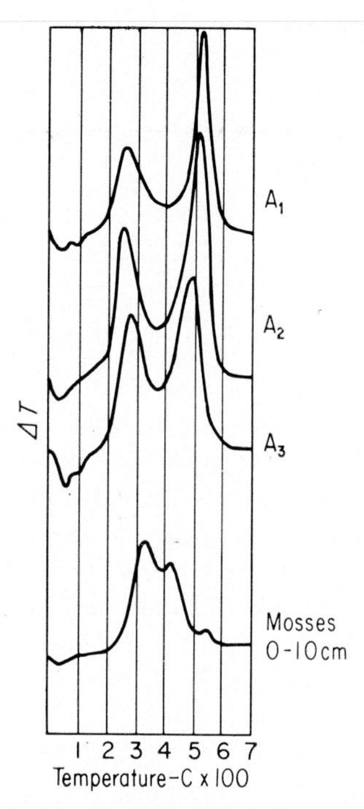

Fig. 4. Differential thermal curves of algal material and mosses, Cape Royds, Antarctica. Analyses were made in dynamic air: A_1, fresh algal material from present shore, 0–3 cm; A_2, algal material from old shore 30 cm above present shore, 0–30 cm; A_3, algal material from older shore 60 cm above present shore, 0–2.5 cm.

determined. The paper chromatographic analysis of the guano performed by the author showed that uric acid is present. Carboxylic radicals of amino acids, proteins, and compounds of protein degradation, which can react with the sodium ion in circulation, probably also contribute to the discrepancy. Boyd et al. [1966] also report a high concentration of vitamins and amino acids in a soil sample collected from the Cape Royds rookery. These amino acids, together with proteins, polypeptides, and other polymers of colloidal size, found their way into the water extract; any sodium absorbed or combined with the polymers and the amino acids would be released and eventually quantitatively determined after digestion of the water extract. The good agreement between the sums of the cations and the anions and between these sums and the electrical conductivity value in the ahumic soil uncontaminated by penguin debris (Table 4) tends to support the contention that guano constituents acting as anions are responsible for increasing the sodium ion retention and concentration.

The field observations and analyses in Tables 3 and 4 point out the sharp segregation between the guano and the underlying mineral horizon. This segregation clearly demonstrates the absence of mixing between these two layers and implies that frost action or other disturbing agents are not effective in modifying or distorting the existing stratigraphy.

The high nitrogen content of the guano results from the protein rich diet of penguins. Campbell and Claridge [1966] also reported a high percentage of nitrogen for an antarctic soil containing penguin guano. Allen and Northover [1967] also found high concentrations of nitrogen in soils from penguin rookeries as well as in soils from seal wallow grounds at Signy Island. Analyses performed on other guanos also agree with the Cape Royds data [Wheeler, 1913; Voorhees, 1917; Hutchinson, 1950; Jacob and v. Uexküll, 1960]. The nitrogen in guano is essentially in the form of uric acid and ammonia, there being lesser amounts of purines, kerotin, and nitrate [Hutchinson, 1950]. As we previously mentioned, uric acid was detected in the Cape Royds guano; however, according to Boyd and Boyd [1963] nitrates were absent; these authors found large amounts of ammonia instead. The total nitrogen content of recent droppings (Table 6) was higher than the nitrogen content of the guano in profile 1 (Table 1). This finding suggests that decomposition occurs and that some nitrogen is lost in the process. Hutchinson [1950] reported that the decomposition of uric acid results in the production of water-soluble ammonium salts that may eventually be leached out; other products of uric acid decomposition include carbon dioxide, ammonia, and oxalic acid. The persistence of ammonia in the guano of the Cape Royds rookery must be due to the low magnitude of leaching.

Most of the phosphorus in the guano at Cape Royds is inorganic. It is derived from droppings, penguin bones, and egg shells [Hutchinson, 1950]. The contribution of phosphorus from mineral soil is practically nil; the ahumic soil in the vicinity of the rookery uncontaminated by excreta has only 0.1% total phosphorus. Allen and Northover [1967] obtained similar data from Signy Island when they compared the extractable phosphorus of soils contaminated and uncontaminated with guano. The decomposition of Peruvian guano under moist and warm conditions results in a depletion of nitrogen and a relative accumulation of phosphorus; consequently the nitrogen/

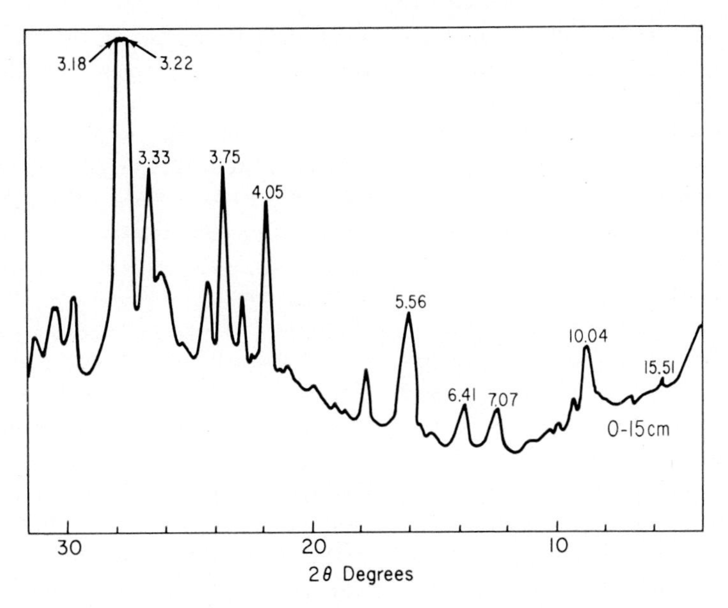

Fig. 5. X-ray diffractogram of oriented magnesium-saturated clay (<0.002 mm) obtained from an ornithogenic soil profile below the guano layer, Cape Royds, Antarctica ($CuK_{\alpha_1,\alpha_2} = 1.54$).

phosphorus ratio becomes narrow with respect to time. The nitrogen/phosphorus ratio of surficial guano at Cape Royds is similar to the ratios of modern Peruvian guano [Hutchinson, 1950]. Despite similar nitrogen/phosphorus ratios the antarctic guano and the modern Peruvian guano may not be the same age. The reasons for this may include differences in the species of the birds, the type of food, and the environmental factors. To characterize the organic components of the ornithogenic soil and antarctic plant material further, samples of guano, mosses, and algae were analyzed by the differential thermal method (Figures 3 and 4). The low-temperature endothermic reactions at 35° and 70°C in the penguin guano may

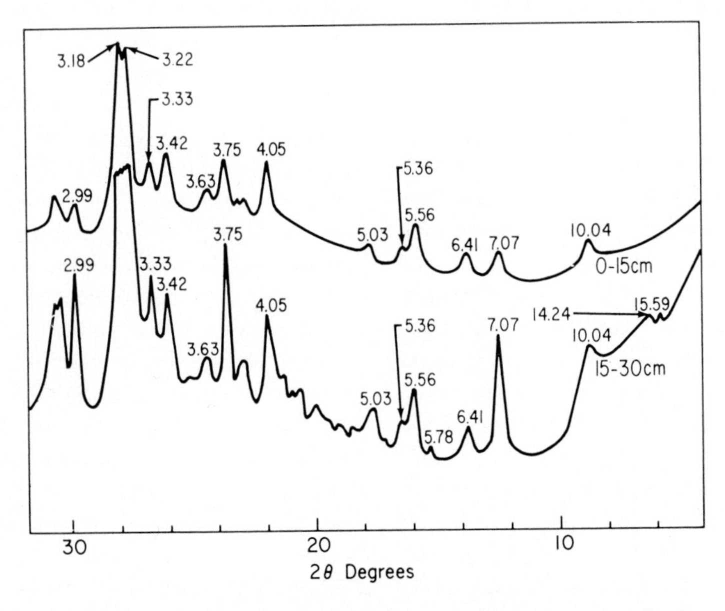

Fig. 6. X-ray diffractogram of oriented magnesium-saturated clay (<0.002 mm) obtained from an ahumic soil profile, Cape Royds, Antarctica ($CuK_{\alpha_1,\alpha_2} = 1.54$).

be ascribed to the loss of water by salts at different stages of hydration. The interpretation of the exothermic reactions is difficult, owing to the complex composition of the guano and to a lack of information on the thermoreactivity of penguin droppings. Although some of the peaks in the differential thermograms of guano may approximate those of guanine (530°C), uric acid, and oxalic acid (330°C) [Stefanovits, 1957; Chesters et al., 1959], a direct equivalence at this time seems rather uncertain.

On the other hand, a very close resemblance is apparent between the thermogram of the mosses and those reported by Mitchell [1960] for *Sphagnum* peat with an intermediate degree of humification. Differential thermal analysis of soil organic matter has not yet progressed to the point at which specific compounds can be positively identified, but valuable information can be obtained by comparing and contrasting the curves of different substances. In this context it appears that the three prevalent sources of soil organic matter in the ice-free areas of Victoria Land (i.e., guano, mosses, and algae) could be distinguished from each other. These results should be considered with some caution, since only a limited number of samples were analyzed.

The local and well-delineated distribution of guano deposits provides a unique opportunity to study the effect of organic matter on mineral soils and to compare these studies with barren areas topographically, climatologically, and apparently geologically similar. Clays collected from the mineral horizon underlying the guano layer and from the subsurface horizon of a nearby ahumic soil profile devoid of guano or other organic matter were X-rayed for possible differences (Figures 5 and 6). As the diffractograms revealed, the clay constituents of the mineral soil below the surficial guano layer are similar to those of the horizon below the surficial mineral layer except that the latter layer has a 14-Å peak. From this preliminary experiment the guano seems to have little effect on the alteration of clay minerals. This single analysis is undoubtedly insufficient to define the role of guano in mineral alteration. One of the limitations is the lack of information about the mineralogy of the sand and silt fractions of two mineral horizons. It is possible that in spite of their proximity the ahumic and ornithogenic profiles have not been derived from the same parent material. Campbell and Claridge [1966] also investigated the effect of guano on mineral alteration and concluded that guano had little impact on the mineralogy of the soil. The only prolonged effect of penguin habitation on the soil recognized by these authors was the increased number of pebbles from nests.

Ornithogenic soils, which are found in coastal areas of continental Antarctica and are also widespread along the western coast of the Antarctica Peninsula, the islands of the Scotia Ridge, and the subantarctic islands, are an important and unique segment of antarctic pedology. These soils display an exceptional abundance of organic material in contrast with the rest of the soils of continental Antarctica. They also play an important role in the entire antarctic terrestrial ecosystem. The redistribution of nutrients from the guano through leaching and wind transport helps to supply phosphorus and nitrogen to areas poor in these elements [Allen and Northover, 1967].

The recognition of ornithogenic soils derived from snow petrel debris either exposed or buried above the moraines and erratics of the last ice expansion and the dating of the lowest layers of guano would be useful for documenting the last major ice retreat, as Hollin [1970] has suggested. Furthermore penguin occupation of areas above present sea level may be important for documenting sea level changes, which are in turn related to the glacial history of Antarctica.

CONCLUSIONS

Field and analytical studies of the ornithogenic soils were undertaken to provide information about these little known soils of Antarctica. These soils, which are found in the penguin rookeries, display an accumulation of organic matter unknown to the rest of the soils of continental Antarctica.

Analyses of an ornithogenic soil profile show a high nitrogen content resulting from the high protein diet of the penguins. The phosphorus content is also rather high and mainly inorganic. The salinity of the soil has been attributed to the vicinity of the ocean, the arid environment, and the salt-enriched excreta of the Adélie penguins. Of considerable interest is the sharp segregation between the guano layer and the underlying mineral horizon, which indicates that frost churning is not effective in disturbing the soil. Differential thermal analysis of the three main sources of organic matter found in Victoria Land shows distinctive curves for guano, algae, and mosses. The interpretation of these thermocurves still remains uncertain. However, if the thermal analysis could distinguish among these organic sources, a very valuable tool would be available for identifying buried organic layers and paleosurfaces.

Studies initiated to evaluate the impact of guano on the mineral substrate were very preliminary and probably insufficient for a definite conclusion. These limited data show that guano has little influence on mineral alteration; additional work is required.

Acknowledgments. The project was financially supported by National Science Foundation grants G-17212, G-23787, and GA-74. Gratitude is expressed to U.S. Navy Task Force 43 for logistic support in the field. The author is thankful to Dr. Janetschek, University of Innsbruck, for his congenial companionship and to Dr. J. C. F. Tedrow for inspiring this project. Thanks are extended to Mr. M. Perdue and to my wife, Dr. Siino-Ugolini, for some analyses. The author is also grateful to Dr. B. Stonehouse for his assistance in describing the Cape Royds rookery. Part of this study was conducted while the author was at Rutgers University.

REFERENCES

Aguilera, N. H., and M. L. Jackson
1953 Iron oxide removed from soils and clays. Proc. Soil Sci. Soc. Am., *17:* 359–364.

Allen, S. E., and O. W. Heal
1970 Soils of the maritime antarctic zone. Antarct. Ecol., *2:* 693–696.

Allen, S. E., and M. J. Northover
1967 Soil types and nutrients on Signy Island. Phil. Trans. R. Soc., ser. B, *252:* 179–185.

Association of Official Agricultural Chemists
1955 Official methods of analysis. 7th ed., 910 pp. Washington, D. C.

Blakemore, L. C., and L. D. Swindale
1958 Chemistry and clay mineralogy of a soil sample from Antarctica. Nature, *182:* 47–48.

Boyd, W. L., and J. W. Boyd
1963 Soil microorganisms of the McMurdo Sound area, Antarctica. Appl. Microbiol., *11:* 116–121.

Boyd, W. L., J. T. Staley, and J. W. Boyd
1966 Ecology of soil microorganisms of Antarctica. *In* J. C. F. Tedrow (Ed.), Antarctic soils and soil forming processes, Antarctic Res. Ser., *8:* 125–159. AGU, Washington, D. C.

Cameron, R. E., D. D. Lawson, and C. Pazaree
1969 Thermoluminescence characterization of antarctic dry valley soils (abstract). *In* Agronomy abstracts, p. 106. American Society of Agronomy, Madison, Wis.

Cameron, R. E., J. King, and C. N. David
1970a Microbiology, ecology and microclimatology of soil sites in dry valleys of southern Victoria Land, Antarctica. Antarct. Ecol., *2:* 702–716.
1970b Soil microbial ecology of Wheeler valley, Antarctica. Soil Sci., *109:* 110–120.

Campbell, I. B., and G. G. C. Claridge
1966 A sequence of soils from a penguin rookery, Inexpressible Island, Antarctica. N. Z. Jl Sci., *9:* 361–372.
1967 Site and soil differences in the Brown Hills region of the Darwin Glacier, Antarctica. N. Z. Jl Sci., *10:* 563–577.
1969 A classification of frigic soils—The zonal soils of the antarctic continent. Soil Sci., *107:* 75–85.

Chesters, G., O. N. Allen, and O. J. Attoe
1959 Differential thermograms of selected organic acids and derivatives. Proc. Soil Sci. Soc. Am., *23:* 454–457.

Claridge, G. G. C.
1965 The clay mineralogy and chemistry of some soils from the Ross Dependency, Antarctica. N. Z. Jl Geol. Geophys., *8:* 186–220.

Claridge, G. G. C., and I. B. Campbell
1968 Soils of the Shackleton glacier region, Queen Maud Range, Antarctica. N. Z. Jl Sci., *11:* 171–218.

David, T. W. E., and R. E. Priestley
1914 Reports on the scientific investigations, glaciology, physiography, stratigraphy, and tectonic geology of southern Victoria Land, British antarctic expedition, 1907–1909. *1:* 208–217. Heinemann, London.

Douglas, D. S.
1964 Extra-renal salt excretion in the Adélie penguin chick. *In* Biologie antarctique, pp. 503–512. Hermann, Paris.

Everett, K. R., and R. E. Behling
1968 Chemical and physical characteristics of Meserve Glacier morainal soils, Wright valley, Antarctica: An index of relative age? *In* A. J. Gow, C. Keeler, C. C. Langway, and W. F. Weeks (Eds.), Proceedings of the International Symposium on Antarctic Glaciological Explorations, pp. 459–460. Hanover, N. H.

Gibson, G. W.
1962 Geological investigations in southern Victoria Land, Antarctica. 8. Evaporite salts in the Victoria valley region. N. Z. Jl Geol. Geophys., *5:* 361–374.

Glazovskaia, M. A.
1958 Weathering and primary soil formations in Antarctica (in Russian). Sci. Pap. Inst. 1: 63–76. Fac. of Geogr., Moscow Univ., Moscow.

Gressitt, J. L.
1965 Entomological field research in Antarctica. Bioscience, *15:* 271–274.

Harrington, H. J.
1960 Adélie penguin rookeries in the Ross Sea region. Notornis, *9:* 33–39.

Holdgate, M. W., S. E. Allen, and M. J. G. Chambers
1967 A preliminary investigation of the soils of Signy Island, South Orkney Islands. Sci. Rep. Br. Antarct. Surv., *12:* 53–71.

Hollin, J. T.
1970 Antarctic glaciology, glacial history and ecology. Antarct. Ecol., *1:* 15–19.

Hutchinson, G. E.
1950 Survey of existing knowledge of biogeochemistry. 3. The biogeochemistry of vertebrate excretion. Bull. Am. Mus. Nat. Hist., *96:* 554 pp.

Jackson, M. L.
1958 Soil chemical analysis. 498 pp. Prentice-Hall, Englewood Cliffs, N. J.

Jacob, A., and H. v. Uexküll
1960 Fertilizer use. Translated from German by C. L. Whittles, 617 pp. Verlagsgesellschaft für Ackerban, Hanover.

Jensen, H. I.
1916 Report on scientific investigations. Report on antarctic soils, British antarctic expedition, 1907–1909. *2:* 89–92. Heinemann, London.

Kelly, W. C., and J. H. Zumberge
1961 Weathering of a quartz diorite at Marble Point, McMurdo Sound, Antarctica. J. Geol., *69:* 433–446.

Linkletter, G. O., Jr.
1971 Weathering and soil formation in antarctic dry valleys. Ph.D. thesis, 122 pp. Univ. of Wash., Seattle.

Lodding, W., and L. Hammel
1959 High-temperature pressure-vacuum furnace. Rev. Scient. Instrum., *30:* 885–886.

MacNamara, E. E.
1969 Biological research opportunities at the Soviet antarctic station Molodezhnaya. Antarct. J. U.S., *4:* 8–12.

McCraw, J. D.
1960 Soils of the Ross Dependency, Antarctica, A preliminary note. N. Z. Soc. Soil Sci. Proc., *4:* 30–35.
1967 Soils of Taylor dry valley, Victoria Land, Antarctica, with notes on soils from other locations in Victoria Land. N. Z. Jl Geol. Geophys., *10:* 498–539.

Metha, N. C., J. O. Legg, C. A. I. Goring, and C. A. Black
1954 Determination of organic phosphorus in soils. 1. Extraction method. Proc. Soil. Sci. Soc. Am., *18:* 443–448.

Mitchell, B. D.
1960 The differential thermal analysis of humic substances and related materials. Scient. Proc. R. Dubl. Soc., ser. A, *1:* 105–114.

Nichols, R. L.
1963 Geological features demonstrating aridity of McMurdo Sound area, Antarctica. Am. J. Sci., *261:* 20–31.

Purvis, E. R., and G. E. Higson
1939 Determining organic carbon in soils. Ind. Engng Chem. Analyt. Edn, *11:* 19.

Richards, L. A.
1954 Diagnosis and improvement of saline and alkali soils. U.S. Department of Agriculture handbook 60: 160 pp. Agricultural Research Service, Washington, D. C.

Rivard, N. R., and T. L. Péwé
1962 Origin and distribution of mirabilite, McMurdo Sound region, Antarctica (abstract). *In* Abstracts for 1961, p. 119. Geological Society of America, New York.

Rudolph, E. D.
1966 Terrestrial vegetation of Antarctica: Past and present studies. *In* J. C. F. Tedrow (Ed.), Antarctic soils and soil forming processes, Antarctic Res. Ser., *8:* 109–124. AGU, Washington, D. C.

Schofield, E., and E. D. Rudolph
1969 Factors influencing the distribution of antarctic terrestrial plants. Antarct. J. U.S., *4:* 112–113.

Shackleton, E. H.
1909 Heart of the Antarctic. *1:* 110–193. Heinemann, London.

Sieburth, J. M.
1959 Gastrointestinal microflora of antarctic birds. J. Bact., *77:* 521–531.

Stefanovits, P.
1957 Investigation of humus substances on the basis of DTA curves (in Hungarian). Agrokém. Talajt., *6:* 129–136.

Stonehouse, B.
1963 Observations on Adélie penguins *(Pygoscelis adeliae)* at Cape Royds, Antarctica. *In* Proceedings of the 13th International Ornithological Congress, pp. 766–779. American Ornithologists' Union, Baton Rouge, La.
1967 Occurrence and effects of open water in McMurdo Sound, Antarctica, during winter and early spring. Polar Rec., *13:* 775–778.

Syroechkovsky, E. E.
1959 The role of animals in the formation of primary soils under the conditions of circumpolar regions of the earth (Antarctica). Zool. Zh. Ukr., *38:* 1770–1775.

Tedrow, J. C. F., and F. C. Ugolini
1966 Antarctic soils. *In* J. C. F. Tedrow (Ed.), Antarctic soils and soil forming processes, Antarctic Res. Ser., *8:* 161–177. AGU, Washington, D. C.

Treves, S. B.
1962 The geology of Cape Evans and Cape Royds, Ross Island, Antarctica. *In* H. Wexler, M. J. Rubin, and J. E. Caskey, Jr. (Eds.), Antarctic research, Geophys. Monogr. Ser., *7:* 40–46. AGU, Washington, D. C.

Ugolini, F. C.
1963 Soil investigations in the lower Wright valley, Antarctica. NAS-NRC Publ. 1287: 55–61. Nat. Acad. of Sci., Washington, D. C.
1965 Ornithogenic soils of Antarctica (abstract). *In* Agronomy abstracts, p. 109. American Society of Agronomy, Madison, Wis.
1967a Soils of Mount Erebus, Antarctica. N. Z. Jl Geol. Geophys., *10:* 431–442.
1967b Soils. *In* V. C. Bushnell (Ed.), Terrestrial life of Antarctica, Antarctic Map Folio Ser., folio 5, pp. 3–15. Amer. Geogr. Soc., New York.
1969 A sequence of soils from a penguin rookery, Cape Royds, Antarctica (abstract). *In* Program and abstracts, p. 2. Western Society of Soil Science, Pullman, Wash.
1970 Antarctic soils and their ecology. Antarct. Ecol., *2:* 673–692.

Ugolini, F. C., and J. Bockheim
1970 Biological weathering in Antarctica. Antarct. J. U.S., *5:* 122.

Ugolini, F. C., and C. Bull
1965 Soil development and glacial events in Antarctica. Quaternaria, *7:* 251–269.

Ugolini, F. C., and C. C. Grier
1969 Biological weathering in Antarctica. Antarct. J. U.S., *4:* 110–111.

Ugolini, F. C., J. C. F. Tedrow, and C. L. Grant
1963 Soils of the northern Brooks Range, Alaska. 2. Soils derived from black shale. Soil Sci., *95:* 115–123.

U.S. Navy Weather Research Facility
1961 Climatology of McMurdo Sound. Rep. 16-1261-052: 67 pp. Norfolk, Va.

Voorhees, E. B.
1917 Fertilizers. 365 pp. Macmillan, New York.

Wheeler, H. J.
1913 Manures and fertilizers. 389 pp. Macmillan, New York.

MICROBIAL AND ECOLOGIC INVESTIGATIONS IN VICTORIA VALLEY, SOUTHERN VICTORIA LAND, ANTARCTICA

ROY E. CAMERON

Bioscience Section, Jet Propulsion Laboratory, California Institute of Technology, Pasadena, California 91103

Abstract. Soil microbial and ecologic studies were undertaken at 11 sites in Victoria valley (77°23′S, 162°00′E), where 13 soil samples were collected from the surface to the depth of the ice-cemented permafrost. Microclimatic measurements pertinent to the soil microbial ecosystem were made at one of the sites northwest of Lake Vida for approximately 1 week during the mid-austral summer. In situ environmental measurements were made for soil and/or air temperature, relative humidity, wind profile, evaporation rate, and solar, global, and net exchange radiation flux. Soils were analyzed or characterized for physical, physicochemical, and chemical properties, including texture, bulk density, porosity, hardness, predominant minerals, abundance of chemical elements, moisture content and constants, color and Munsell notation, reflectivity, *p*H, *E*h, *r*H, electrical conductivity, osmotic pressure, loss on ignition, cation exchange capacity, buffer capacity, '*N*' value, cations and anions (Na^+, Ca^{++}, Mg^{++}, K^+, NH_4^+, Zn^{++}, Fe^{+3}, Al^{+3}, Cl^-, NO_3^-, NO_2^-, HCO_3^-, BO_3^{-3}, $SO_4^=$, and PO_4^{-3}), and total and organic carbon and nitrogen and their ratios. The distribution, abundance, and kinds of microorganisms were determined, including aerobic, microaerophilic, anaerobic, heterotrophic and chemoautotrophic, psychrophilic, mesophilic, and thermophilic bacteria, yeasts, molds, algae, and protozoa. Microbiology was correlated insofar as possible with microclimate, soil, and other features of the sites and surrounding terrain of the dry valley environment. The solclime (hydrothermal regime), characterized by favorable temperatures and the quantity, the quality, and the duration of available moisture, was important in the survival and the growth of microorganisms during the diurnal freeze–thaw cycles. Some soils were too dry or saline and did not contain any detectable microorganisms. No streptomycetes, molds, coliforms, or thermophiles were detected. Only one site (near Lake Vida) contained algae and as many as 10^5 bacteria per gram of soil. All microorganisms grew at +2°C as well as at +20°C. Soil diphtheroids (*Corynebacterium* spp. and *Arthrobacter* spp.) were the most prominent microorganisms. Investigations supported the observations of an ecologic sequence in microbiota along a gradient of unfavorable to favorable environmental factors. The simple soil microbial ecosystems in the cold, barren, arid, wind-swept antarctic dry valleys are valuable in understanding and comparing soil microbial ecosystems in hotter arid areas. The harsh antarctic environment also serves as a useful model for the detection of extraterrestrial life, for example, on Mars.

Soil microbial, ecologic, and microclimatic investigations were undertaken in the antarctic cold desert, especially in the dry valleys of southern Victoria Land during the austral summer of 1966–1967. The main purpose of these investigations was to determine the presence (or absence), the abundance, the distribution, and the kinds of microorganisms in the dry valleys in relation to environmental factors, including the microclimate and soil habitat. The microbial ecology in the dry valleys is not complexed by extant or decaying higher animals and plants or by a large influx of the air-borne microorganisms frequently present in temperate and tropical deserts [Benoit and Cameron, 1967; Cameron, 1967, 1969b]. However, on a limited environmental scale, the dry valleys show variations in the degree and the extent of their glaciation and the nature of their morainic deposits, orientation, age, exposure, slope, moisture, drainage patterns, weathering and degree of soil development, elevation, latitude, and local climate. The dry valleys therefore provide gradients for investigating microbial ecology in a truly natural harsh environment. The environment of the dry valleys is approached only by comparable areas in high mountains and basins and in the high Arctic [Boyd, 1967; Cameron, 1969b; Chen, 1963; Davies, 1961a, b; Fang, 1963; Fristrup, 1952, 1953; Gerasimov and Zimina, 1968; Gromov, 1957; Krasil'nikov, 1956; Kriss, 1940; Lo, 1963; Mikhaylov, 1962; Mishustin and Mirzoeva, 1964; Murzayve, 1967; Sushkina, 1960; Swan, 1961, 1968; Tedrow, 1968a; Tedrow and Brown, 1962].

Additional objectives for investigating antarctic soil

Fig. 1. Victoria valley, view toward McMurdo Sound showing sample sites 626, 537-539a, 540-543, 634, and 574-576. U.S. Navy photograph by VX-6.

microbial ecosystems were to obtain environmental or microclimatic data for selected sites in exposed areas of soil in the Antarctic, to characterize and analyze soils for their indigenous properties, and to obtain microbial isolants for the design and the testing of extraterrestrial life detection experiments before the unmanned search for possible life on Mars [Cameron, 1963, 1966, 1969b; Horowitz et al., 1969; Hubbard et al., 1968, 1970]. This last purpose does not mean that Antarctica possesses a Martian environment, but it is a valuable study area because it is the only naturally harsh terrestrial region approaching Martian conditions in a number of ways, including its low magnetic field, comparatively high short-wave ultraviolet and other biologically detrimental radiation, desiccative and abrasive winds, low relative humidity, low incidence and duration of solar radiation and correspondingly low temperatures (−50° to +20°C for Victoria valley versus about −70° to +20°C for Mars), lack of liquid precipitation, sublimation of surface moisture, diurnal freeze–thaw cycles, presence of permafrost and subsurface ice layers, biologically detrimental concentrations of moisture absorbent salts, low moisture activity, unfavorable exposure, bare and eroded soil, and possession of a few populations of surface or subsurface microorganisms as the only possible life forms. In some areas of the Antarctic the environmental factors are so limiting to life that no microorganisms have been detected by either cultural or radiorespirometric methods [Benoit and Cameron, 1967; Benoit and Hall, 1970; Boyd, 1967; Boyd and Boyd, 1963a; Boyd et al., 1966; Cameron, 1969b, c, 1971, 1972; Cameron et al., 1968, 1969b, c, 1970b; Horowitz et al., 1969, 1972; Hubbard et al., 1968, 1970; Meyer et al., 1963].

Mars is expected to be an extreme desert, but even the Antarctic has some areas so arid that there is insufficient moisture available for frost action processes. However, where frost action is apparent, polygon and permafrost studies have been useful in postulating a Martian permafrost [Wade and de Wys, 1968]; and the presence of deliquescent salts [Cameron, 1971, 1972; Meyer et al., 1962; Tedrow et al., 1963] may provide an explanation for the phenomenon of frost heaving in some antarctic soils similar to that postulated for a Martian soil [Anderson et al., 1967]. Evidence of extraterrestrial frost heaving would be important not only for the possible presence of moisture and salts and their implications for both present and historical Martian physiography but also for the existence of an important environment for microorganisms.

Antarctica is useful for establishing a model of Martian ecology, in terms of both operative environmental factors and the low number, variety, activity, and survival of microorganisms. The soil microbial ecology in the arid areas of Antarctica is simple, as that of Mars is also expected to be, and, despite the comparatively recent inroads by man into the antarctic continent, can still be considered a relatively virgin territory, not extensively contaminated by man's few habitats and activities there. The similarities and dissimilarities between the antarctic continent and Mars have also been indicated for testing equipment, facilities, logistics, and interpersonal relationships prior to manned space exploration and the establishment of extraterrestrial bases [Johnson, 1966; Johnson and Smith, 1970; Lindsey, 1966; Matthews and Madden, 1968; Stuhlinger, 1969]. The experience gained on Russian occupied antarctic bases is considered applicable to the establishment of lunar bases [Cameron, 1969b].

PHYSIOGRAPHY OF VICTORIA VALLEY

Within southern Victoria Land the Victoria valley system occupies approximately 17×10^4 km^2 of soil and rock that are essentially ice free. The constituent valleys of the system (counterclockwise in Figure 1) are Victoria, Barwick, Balham, and McKelvey. Victoria valley is located at 77°23′S, 162°00′E, has a maximum width of 10 km and a length of approximately 32 km, and forms a broad arc at an elevation of 390–550 meters. It is delineated in the north by Saint Johns Range and in the south by the eastern extent of the Olympus and Insel ranges. These ranges rise abruptly to 950 meters above the valley floor and are generally characterized by smooth glaciated walls, cirque glaciers, hanging valleys with accordant floors, and loose slopes of talus and scree. The Barwick and McKelvey valleys converge at the eastern end of the Insel Range, where they join Victoria valley. Victoria valley is limited in the east by the relatively stagnant Victoria Lower Glacier (77°19′S, 162°40′E), which extends into the valley as a tongue of the Wilson Piedmont Glacier, which faces McMurdo Sound. The Packard Glacier (77°21′S, 162°10′E) extends its snout slightly west of Victoria Lower Glacier. It is the most active of the four large glaciers in the Victoria valley system [Calkin, 1963].

Victoria valley is limited in the northwest by Victoria Upper Glacier, an alpine glacier that is approximately 20 km from Victoria Lower Glacier and is located at 77°17′S, 161°33′E. It is apparently stagnant and retreating [Webb and McKelvey, 1959]. During the midsummer weeks it provides an intermittent glacial melt water stream that flows southeast through a series of small lakes before it emerges through an extensive silt delta into Lake Vida (77°23′S, 161°54′E) at a surface elevation of 390 meters, the lowest point in the valley. Melt water streams derived from Victoria Lower Glacier flow westward into the lake, and smaller glaciers in the eastern Olympus and Saint Johns ranges can also contribute melt water inflow, but not during every summer. Other minor influxes of moisture into the frozen lake are from the melting of local snowbanks and the snow accumulation on the lake surface. Lake Vida, which is approximately 3.5 km long and 1 km wide, has no outflow. This lack of outflow is typical of other dry valley lakes, including Bonney, Fryxell, Vanda, and Vashka. Lake Vida is permanently ice covered, except for some melting around its periphery during midsummer (Figure 2), and is probably frozen to its base [Calkin and Bull, 1967]. It shows a history of recession, and, although the shore line is delineated principally by pebbly beaches, there are also frozen silty and sandy cliffs up to 3 meters high.

In addition to glacial lake features the geology, glacial geology, and geomorphology have been studied extensively [Allen, 1962; Allen and Gibson, 1962; Berg and Black, 1966; Black and Berg, 1963a; Bull et al., 1962a, b; Calkin, 1963, 1964; Calkin and Cailleux, 1962; Dort, 1967, 1970; Gibson, 1962; Gunn and Warren, 1962; Harrington, 1965; Nichols, 1964, 1966; Ugolini and Bull, 1965; Warren, 1965; Webb and McKelvey, 1959]. It is quite important from the ecologic viewpoint to understand geologic features, because the life forms in the dry valleys, except those in aqueous environments, not only are soil microbial inhabitants but, where the soil is nonexistent or poorly developed, also are geomicroorganisms and lithophytes and have played a minor role in soil formation.

In the dry valleys a basement complex of rocks, gneisses, schists, marbles, and metagreywackes was intruded by masses of granite. In Victoria valley the exposed basement is primarily uniform sheets of quartz–orthoclase granite, the younger and mostly horizontal dolerite sills being separate. These sills are the most conspicuous geologic feature [Webb and McKelvey, 1959]. (Dolerite is not included among the

Figs. 2–5. Victoria valley, Antarctica. Fig. 2, view along the northwestern shore of Lake Vida during midsummer peripheral melt. Fig. 3, dark doleritic material over Beacon sediments, Olympus Range; east end of Lake Vida and sand wedge polygons are in foreground. Fig. 4, dune area in eastern end of valley with view toward McKelvey valley (site 634); Olympus Range is at left, and Insel Range is at right. Fig. 5, sand dune in east end of valley undergoing active saltation (site 634).

more than 175 minerals reported for Antarctica [Stewart, 1964] because it is a common antarctic rock type and the name is given to those varieties of diorite and gabbro too fine grained for the nature of the dark colored mineral to be determined by megascopic examination. The 'usual' dolerite is approximately 40% plagioclase (typically An_{60-75}), 12% quartz–feldspar micropegmatite, and the remainder accessory minerals of biotite, hornblende, and iron ores [Warren, 1965].) The Beacon sediments resting on the basement are sandstones, silt stones, and shales, and there are also some sparse, generally pulverized coal and fossils. Some time after the deposition of the sediments, great sills and sheets of dolerite magma solidified to form the prominent dark, hard, jointed, mostly horizontal rock that is the most conspicuous geologic feature (Figure 3). This doleritic material is not readily eroded by glacial action in the antarctic environment [Clark, 1965], but it contributes to morainic deposits, desert pavement, and polygon and soil formation.

Four glaciations have occurred in Victoria valley, as well as in Wright valley [Bull et al., 1962a], and the records of several major glaciations can be observed in the Victoria valley system [Calkin, 1963, 1964]. The Insel glaciation, the first major flow of ice, moved eastward from the inland plateau through the valleys to the coast and left behind a drift of some lake salts, very silty till, and erratic pebbles and cobbles on mesas 300–600 meters above the valley floors [Calkin, 1964]. Deep melt water channels were cut during the recessional phase of the Insel glaciation, and the shapes of the major valleys, including that of Victoria valley, have not changed significantly since that time. The Victoria glaciation was the second major influx of ice and was marked primarily by strong invasions from local ice fields and the coastal area. The Victoria glaciation, which began <30,000 years ago, has included three successive episodes: the Bull, Vida, and Packard drifts [Calkin, 1964]. These drifts have contributed to the topography of moraines and outwash fans in Victoria valley. Till from the Bull drift can be observed east of the Insel Range, where the lower McKelvey valley merges into Victoria valley southwest of Lake Vida, as well as on the approach to the southeast shore of the lake. The Vida drift, the most vigorous of the three, left thick, sandy ground and end moraine deposits, including outwash fans in the marginal and eastern dune areas of the valley (Figure 4). The Packard drift episode, which is still active, has also left a very sandy till that occurs largely as ground moraine representing a slow, regular glacial retreat in both the Victoria Upper and the Victoria Lower glacial areas [Calkin, 1964]. As the result of glaciation the walls of the valley show remnants of glacial benches at approximately 800 meters [Bull et al., 1962a].

Generally good agreement has been found between the age of the major glacial deposits and the proportion of fine materials in the Victoria valley system [Calkin, 1963]. For example, the Vida drift at Lake Vida is more than 9700 ± 350 years old [Calkin, 1963] on the basis of the dating of buried algal peat. The nature and the extent of frost crack polygons have been useful for dating, and they show that the age of exposure of the dry valleys is 3000–10,000 years B.P. [Berg and Black, 1966]. On the basis of chemical data the flow of major glaciers may be judged to have ceased about 50,000 years ago [Wilson, 1964]. Some areas may be more than 60,000 years old, as determined by the chloride content of lakes, e.g., that of Lake Vanda in Wright valley [Wilson, 1970]. The deglaciation in all the areas investigated in Victoria Land is not older than the late or classical Wisconsin glacial stage (40,000–60,000 years ago) [Black and Berg, 1963b]. However, alpine glaciers in the dry valleys were much more advanced in the 3000–6000 years immediately prior to 3000 years B.P., and the period from 3000 to 1200 years B.P. shows alpine glacial recession; more recently (1200 years ago) a glacial advance took place, and only 100 years ago a minor advance occurred [Wilson, 1970]. At present there are a colder climate and reduced precipitation, which have resulted in glacial starvation and drastically reduced activity [Dort, 1970]. However, on the basis of concurrent northern hemisphere activity, it is quite likely that the climate is undergoing a warming trend, that precipitation will increase, and that glacial growth and activity will increase correspondingly [Dort, 1970].

A number of factors can serve as criteria for indicating older dry valleys or older exposed areas within a dry valley. These factors include the nature and the extent of exposed moraines, especially older drift, the absence of frost crack polygons, an increased depth to the ice-cemented permafrost, the gradual hardening of the soil with depth, an increase in the concentration and the formation of salt crystals in soils, salt efflorescence around rocks, the presence of desert pavement with varnish, cavernous weathering and the formation of ventifacts, the reduction of boulders to ground level, and the presence of poorly developed, arid soils of a more mature yellowish-brown color.

Figs. 6–9. Victoria valley, Antarctica. Fig. 6, stone wedge polygons and flagstone pavement at southwest end of Lake Vida with view toward Victoria Upper Glacier. Fig. 7, desert pavement in sand wedge polygon (site 537-539a). Fig. 8, extensive desert pavement on sloping glacial end moraine (site 540-543) with view toward Victoria Lower Glacier. Fig. 9, closeup of desert pavement (site 537-539a) ; the minute surface irregularities provide favorable microenvironments for microorganisms.

Fig. 10. Closeup of dark polished dolerite-derived pavement (site 540-543).

On the basis of the preceding factors, some of the areas in Victoria valley can be judged to approach the age of the Bull drift episode, which is dated at 30,000 to <35,000 years (Figure 1, near site 540-543) [Calkin, 1963].

The valley floor is thickly mantled with morainic deposits occupying more than 30 km² of the valley floor. The moraines have been modified by the action of melt water streams and their drainage channels, and lower Victoria valley has been reworked into extensive flood plain areas. The morainic forms are generally poorly developed, ground moraine being the most extensive and nearly covering the valley except for small areas near Lake Vida [Allen and Gibson, 1962]. A poorly developed lateral moraine occurs in upper Victoria valley. The moraines contain boulders of porphyritic granite and gneiss with some Beacon sandstone. The general distribution and structure of the rock types are a continuation of that previously described for Wright valley [McKelvey and Webb, 1962].

In the eastern part of the valley between Lake Vida and Victoria Lower Glacier, there is a large procession of sand dunes, sheets, and mantles, including easterly-facing (windward sloped) barchans (Figure 1, near sites 634 and 574-576). One isolated barchan is approximately 3 km wide and 25 meters high [Allen and Gibson, 1962]. This dune area is the largest aeolian deposit in the dry valleys, and the dunes are advancing westward. Saltation is an active process (Figure 5), and dune movement is as much as 5 cm/day during the warmer parts of the summer [Calkin, 1963]. A permafrost layer can be found below the surface of the dunes that will impede major movement unless the permafrost is uncovered, is exposed to solar radiation, and subsequently thaws.

One of the most significant features in large areas of the dry valleys is the frost crack polygons. The surficial characteristics of polygons and their associated wedges indicate not only their age, material characteristics, and glacial fluctuations [Black and Berg, 1963a, 1964, 1965] but also their hydrothermal environment [Black and Berg, 1963b]. They are absent in most of the older morainic deposits and, if they are present, are composed of finer textured materials and have a low moisture content. Polygons with sand wedges (Figure 3) and stony debris (Figure 6) indicate that moisture is evaporating and sublimating from the soil; sand layers also indicate a low moisture content and the recession and the reduction of ice in permafrost [Berg and Black, 1966; Black and Berg, 1965]. The most well developed polygons in Victoria valley are those at the fronts of glaciers, where moisture is ample [Calkin, 1963]. In less moist areas the polygons are broadly concave and are separated by sand-filled troughs, but in coarse moraine they are smaller and have deeper peripheral troughs. The highest double raised-rim polygons in Victoria Land are in Victoria valley 4 km west of Victoria Lower Glacier, south of the sand dunes and Packard Glacier. The rim heights of these polygons are up to 1.5 meters above the polygonal center and 3–10 meters across the 0.5- to 1.5-meter-deep troughs from crest to crest, there being active sand-filled wedges beneath the troughs. The overall polygon diameters are 20–30 meters [Berg and Black, 1966]. Little or no attention has been given to

Fig. 11. Boulder-strewn ventifacts in general area of site 626 with view toward Victoria Upper Glacier and Saint Johns Range.

these important and conspicuous features of the dry valleys as habitats for organisms.

Other surface features in the valley include well-developed desert pavements (Figures 7–10) strewn with ventifacts (Figures 8, 11, and 12); cavernous weathering is evident in lower Victoria valley in the erratic boulders occurring on end moraines [Calkin and Cailleux, 1962]. The nature and the extent of cavernous weathering, as well as the soil formation, can be useful in interpreting glacial chronology [Calkin, 1963; Linkletter, 1970; Ugolini and Bockheim, 1970; Ugolini and Grier, 1969]. A flagstone pavement caused by frost heave is found in alluvial deposits at the southwest end of Lake Vida (Figure 6). Many of the talus slopes are now stable because of more arid conditions, although there are a few small, recently active mudflows [Calkin, 1963].

Although no abiotic weathering may be considered to occur in the Antarctic [Polynov, 1953], the active physical processes of wind erosion and temperature changes have contributed to the disintegration and some decomposition of surficial deposits (Figure 11). Some minor chemical weathering of hydration and oxidation may be active in rock weathering as well, as is shown by the concentrations of salts and white salt crusts, indicative of salinity in arid soils [McGeorge, 1940], the oxidation and staining of rock and soil materials, and the occurrence of fine desert varnish on dolerite pebbles, cobbles, and boulders [Calkin, 1963]. Minor traces of salt evaporites are widespread on rock surfaces [Allen and Gibson, 1962; Gibson, 1962], especially in drier areas (Figure 13). A pictorial description of the significant surface features in the dry valleys has been given by McCraw [1967a]. Many of the desert surface features described for Victoria valley and other dry valleys of southern Victoria

Fig. 12. Sloping boulder-strewn glacial end moraine at center of Victoria valley with view from site 626 toward Lake Vida and Victoria Lower Glacier.

Fig. 13. White salt evaporites ringing partially imbedded surface stone (site 540-543).

Land can be compared with similar desert features and phenomena pictorially described for temperate and tropical desert regions [Cameron et al., 1966a; Stone, 1967] as well as for North Polar areas [Davies, 1961a, b; Davies and Krinsley, 1961; Davies et al., 1963; Fristrup, 1952, 1953; Tedrow, 1966a, b, 1968a, b]. The aeolian characteristics of the dry valleys, especially the sand dunes of Victoria valley, are of interest for the Mars landing by the Viking lander in 1975 (E. Morris, personal communication, 1971).

Climatic factors exert a selective influence on terrestrial biota over a wide area of the Antarctic, and these factors have been mentioned by a number of investigators [Ahmadjian, 1970; Boyd et al., 1966; Dodge, 1965; Dodge and Baker, 1938; Gannutz, 1969, 1970; Greene, 1965; Greene et al., 1967; Greene and Longton, 1970; Janetschek, 1967, 1970; Lamb, 1970; Llano, 1962, 1965, 1970; Rudolph, 1965, 1966, 1970; Schofield and Rudolph, 1969; Siple, 1938; Ugolini, 1970]. The influence of geographical and topographic features on the weather and climate of the dry valleys has been noted previously, particularly with reference to the katabatic winds that sweep off the antarctic plateau and the heat transfer and radiation balance between the atmosphere and exposed surfaces [Bull, 1966; Calkin, 1963; Janetschek, 1967; Rubin, 1965, 1970; Rubin and Weyant, 1965; Rudolph, 1966; Rusin, 1964; Shaw, 1960; Solopov, 1967; Tauber, 1960; Ugolini, 1963; Weyant, 1966].

The katabatic winds produce special weather features in the valleys that are characterized by a greater proportion of clear skies, winds of steady direction and speed, low temperatures, and reduced atmospheric humidity [Tauber, 1960]. In general, the character and intensity of the katabatic winds depend largely on the time of year and the direction and intensity of solar heat. In periods of less solar heat, i.e., in transitional seasons and in the winter, the wind blows with a steady constancy and with little or no diurnal variation in speed, and the wind speed is twice that in areas subject to katabatic air movement [Tauber, 1960]. Glacial hollows and valleys greatly affect the formation of the local climate [Rusin, 1964]. The relief and the underlying surface function as wind regulators and influence the thermophysical properties of the valleys [Rusin, 1964; Solopov, 1967]. Other main causes for wind variations are the strong thermal gradients between the open water near the coast and the inland ice and the synoptic pressure gradient [Shaw, 1960].

The wind speed in the dry valleys also depends on the surface temperature distribution, variations resulting from strong insolation and radiation [Shaw, 1960]. The low albedo of dark rock is largely responsible for the increase in the local air temperature in the summer, when incident radiation is received on the surface; this condition in turn reduces the relative humidity and further increases ablation [Calkin, 1964]. Waves of thermal radiation ascending from the valley floor can be observed and photographed during nonwindy periods [Cameron, 1969a]. An obviously important macroclimatic and microclimatic factor is the strong positive radiation balance, which is responsible not only for the increased size of the ice-free area but also for the survival and adaptation of biota. During the frequent windy periods, organisms are subject to wind chill, scour, abrasion, and burial. Wind is still the most active agent responsible for both the deposition and the erosion of materials in the Victoria valley system [Calkin, 1963]. Wind, temperature, relative humidity, and blowing particles of snow and/or sand interact to determine the survival and the adaptation of surface organisms.

Figs. 14–17. Victoria valley, Antarctica. Fig. 14, field station and orientation of instruments at northwest end of Lake Vida with view toward Victoria Upper Glacier. Fig. 15, instruments for in situ environmental measurements near site 537-539a with view toward Victoria Lower Glacier. Fig. 16, excavated site 537-539a; permafrost melt occurred at the bottom of the exposed pit. Fig. 17, closeup of profile at site 537-539a showing faint stratification from stream deposition.

The strongest winds in the Victoria valley system are katabatic and blow from the southwest quadrant. They have probably blown for thousands of years from this quadrant, as the cirque erosion shows, and they predominate during the winter. However, easterly winds from McMurdo Sound are both the strongest and the most predominant winds in the eastern part of the Victoria valley system during the summer [Calkin, 1963]. These winds carry more moisture and are colder than the westerly winds, which are heated adiabatically in their rapid descent from the interior ice plateau. The easterly winds also help determine the survival of organisms, particularly on the surface. During observations made at both ends of Lake Vida during the austral summer of 1961–1962, the wind nearly always blew from the east at a mean speed of 10 knots (11.5 mph) [Bull, 1966].

Some prior data have been published for temperature measurements made in Victoria valley. During three summers of observations at stations west and east of Lake Vida from 1959 to 1962, the mean air temperatures ranged from −23.7°C in October to 1.3°C in early January to −7.4°C during the first 2 weeks in February [Bull, 1966]. A minimum thermometer left in a screen west of Lake Vida showed −62°C for the winter of 1960 [Bull, 1966]. An overall average mean temperature for the Victoria valley system is approximately −20°C [Wilson, 1970], which is 4°C higher than that for comparable periods at McMurdo Station; the relative humidity for the Victoria valley system is also lower [Bull, 1966; Péwé, 1960]. Another estimate of the annual mean surface air temperature for the dry valley region is −25° to −35°C [Rubin, 1970]. During summer any snow falling in the valley system usually does not remain on the floor for more than a day. In our study snow accumulation rarely exceeded 6–10 cm in Victoria valley during the summer field season [Calkin, 1963] and did not remain for more than a few hours in the dunes area (A. Allen et al., unpublished data, 1963). A winter meteorologic station was maintained in Wright valley during 1969, but no data are yet available.

MATERIALS AND METHODS

For our investigations in Victoria valley, a field station was established near the northwest end of Lake Vida (Figure 14, site 537-539a) for approximately 8 days (December 27, 1966, to January 4, 1967) during the austral summer. This station served as a base camp for traverses into various areas in the valley and for measurements of environmental parameters (Figure 15). These parameters included the following variables for air and/or soil: temperature, relative humidity, dew, wind direction and velocity, solar radiation flux, environmental radiation flux (total sun, sky, and terrain), net thermal exchange, evaporation rate, visible light intensity, barometric pressure, and extent and nature of cloud cover. These measurements were recorded continuously by machine or every 3 hours by hand, except for the maximum and minimum thermometers, which were observed and reset once a day at 1200 hours (local time). The instruments used for the preceding measurements included Yellow Springs telethermometers, Taylor maximum–minimum air and soil thermometers, Taylor standard 3-cup vane anemometers, a Belfort hygrothermograph, a Hygrodynamics electric hygrometer with type TH LiCl sensors, a Lambrecht recording evaporation gage, a Greentube snow gage, a Wallin-Palhemus dew duration recorder, a Michelson bimetallic pyrheliograph (actinometer), a Bellani spherical distillation pyranometer, a Gier and Dunkle net exchange radiometer, a Photovolt foot-candle meter, and an American Paulin aneroid barometer. Clouds were visually observed and compared with U.S. Department of Commerce cloud photographs [Cantzlaar, 1964].

A few of the preceding environmental factors were also measured at the time of collection (appendix) at selected sites (site 540-543 in Figures 8, 10, 13, and 18; site 574-576 in Figures 19 and 22; site 634 in Figures 4, 5, 20, and 23; and site 626 in Figures 11, 12, and 21) in addition to the main site (site 537-539a in Figures 14–17). Observations were made at each site, and descriptions were entered on a soil sampling information sheet [Cameron et al., 1966b]. This sheet included information not only on point measurements for microclimatic factors but also on elevation, general topography, location, position, slope, surface microrelief, and structure and development of the soil profile and its parent material (appendix). Each area, site, soil profile, and other conspicuous feature, e.g., desert pavement (Figures 9 and 10), salt evaporite (Figure 13), algal crust (Figure 24), contamination source (Figures 25 and 26), was photographed to aid in defining and characterizing the habitat or even in explaining the absence of organisms. In desert areas a photographic record is not only desirable but often necessary because of the monotony of the terrain and the lack of significant unique features. It is also the easiest and most certain method for returning to a site if further work

Figs. 18–21. Victoria valley, Antarctica. Fig. 18, rolling bouldery glacial debris lobe (site 540-543 near fractured boulder); view toward west end of Lake Vida and glacial melt water stream from Saint Johns Range. Fig. 19, undifferentiated sandy profile to depth of ice-cemented permafrost (site 574-576). Fig. 20, extensive dune mantles at site 634 with views toward Olympus Range (left), McKelvey valley (center), and Insel Range (right). Fig. 21, rough, uneven, and pebbly site (626) of mixed morainic debris including decomposed sandstone; sample collected for use by the NASA Lunar Receiving Laboratory, Manned Spacecraft Center, Houston, Texas.

must be performed, and several of the sites were actually revisited in January 1971.

At each of the sample sites, soils were collected from the surface to the depth of the hard, ice-cemented permafrost by aseptic techniques developed for sampling, handling, and processing of desert soils [Cameron, 1967; Cameron et al., 1966b]; two of the samples (634 and 626) were collected in duplicate from the surface 5 cm of soil in association with Dr. James Turnock, Deputy Director of the Apollo program, for use at the NASA Lunar Receiving Laboratory, Manned Spacecraft Center, Houston, Texas. Care was taken at each site to clean and sterilize (usually by propane torch) all collecting instruments before and after each sample was obtained from the surface or from lower depths of the soil pit. This sterilization was done to reduce microbiological as well as chemical contamination during the sampling sequence. A portion of each sample from Victoria valley was placed in a sterilized glass jar, and the bulk of the sample was put into standard plastic-lined canvas soil sample sacks or 'whirl-pak' bags. Another aliquot of each sample was taken with a soil tin sampler for subsequent determination of soil moisture, porosity, and bulk density. After the 1966–1967 season, soil samples for trace organic analyses were collected in nonplasticized bags and other 'inert' containers after it was discovered

Fig. 23. Slightly rippled dune site (634); sample collected for use by the NASA Lunar Receiving Laboratory, Manned Spacecraft Center, Houston, Texas.

Fig. 22. Coarse sandy dune site (574-576).

that the plastic-lined bags were contaminated by phthalic acid esters [Bauman et al., 1967].

After collection, all samples were kept out of direct sunlight in the field and were later kept frozen at −25° to −30°C during transport and storage at McMurdo Station, the Jet Propulsion Laboratory at the California Institute of Technology (JPL), and Virginia Polytechnic Institute (VPI). At JPL part of each bulk sample was processed in a cold isolation booth (antarctic simulator) [Cameron and Conrow, 1968]. For most analytical purposes each sample was homogenized by passing it through a sterilized stainless steel ≤2-mm sieve (no. 10 mesh) unless the particles or aggregates were already ≤2 mm in diameter, the size range for soil analyses. For physical, physicochemical, and chemical analyses, a fraction of each sample was air dried (or vacuum dried if the soil contained hygroscopic salts), and a subsequent fraction was powdered by mortar and pestle when necessary. For most microbiological determinations a frozen portion was processed at below freezing temperatures (commonly at −30°C) and retained in storage at

Fig. 24. Closeup of algal surface crusts at edge of melt water pools (see Figure 2).

−5° and −30°C for culture and radiorespirometric studies, respectively, usually at the in situ moisture content.

Soil physical, physicochemical, chemical, thermoluminescent, and mineralogic analyses were performed on sieved or powdered portions of the unfractionated, or whole, soil. These properties were determined by standard procedures [Chapman and Pratt, 1961; Jackson, 1958; Richards, 1954; Soil Survey Staff, 1951, 1960] or by other methods developed especially for desert soils or similar geologic materials from harsh environments [Cameron, 1969a, 1970; Cameron and Blank, 1963; Cameron et al., 1966b]. Because cold desert soils are unique in some characteristics and similar to temperate and tropical desert soils in others, it is important to consider the methods used to determine the soil physical and chemical properties not only for interpretation of results but also for comparison with the soil data of other regions. These methods included the following:

1. Mechanical analyses for soil separates were performed by the hydrometer method modified for saline or alkali desert soils; textures were subsequently determined by reference to the U.S. Department of Agriculture textural triangle (Table 1).

2. The bulk density and the porosity were calculated from samples carefully taken with the soil tin sampler, excluding surface pavement and hard frozen soil (Table 1).

3. The in situ hardness was obtained with a calibrated, spring-loaded soil penetrometer (Table 1).

4. The predominant minerals were obtained on the unfractionated, powdered sample by X-ray diffraction

Fig. 25. Remains of old food cache from International Geophysical Year study. Field station tent at site 537-539a.

Fig. 26. Opened food cans and weathered trail ration box (see Figure 25). Wind-blown materials from this cache continually blew across and into site 537-539a. Five years later these materials were still in place and were still contributing to site contamination.

and interpretation, which were substantiated by 'total' elemental and chemical ion analyses (Table 1).

5. The soil color and Munsell notation for hue, value, and chroma were observed on soils at the in situ moisture content, when the soil was air dry and when it was moist (field capacity or approximately 0.5 bar) (Table 1).

6. The reflectivity (albedo) values were determined on the same samples used for the soil color and moisture tests by using a Photovolt reflectivity meter with a tristimulus green filter. The method of zero suppression had reference standards of 4.0 and 74.5% reflectivity (Table 2).

7. The moisture content for in situ samples was

TABLE 1. Soil Physical Characteristics: Separates, Texture, Bulk Density, Porosity, Hardness and Minerals, Victoria Valley, Antarctica

Soil Sample	Sample Depth, cm	Sand, 2.00–0.02 mm, %	Silt, 0.02–0.002 mm, %	Clay, ≤2 μ, %	Texture	Bulk Density, g/cm³	Porosity Volume, %	Hardness, kg/cm²	Predominant Minerals
537	0–2	97	3	0	sand	1.76	33.7	<0.25	quartz; labradorite; augite (?)
538	2–15	97	2	1	sand	1.81	31.6	0.5–2.5	
539	15–25	96	1	3	sand	1.53	42.4	>4.5	
539a	25–30	98	2	0	sand	frozen	frozen	>4.5	
540	0–2	88	11	1	sand	1.62	38.9	<0.25	
541	2–15	73	21	6	sandy loam	1.45	45.2	0.5–1.0	quartz; labradorite; albite; microcline salts of Na, Ca, Mg, Cl, NO_3, SO_4; dolerite
542	15–25	68	26	6	sandy loam	1.49	43.6	0.75–1.25	
543	25–30	71	23	6	sandy loam	1.38	48.1	>4.5	
574	0–2	100	0	0	sand	1.94	26.9	<0.25	
575	2–15	100	0	0	sand	1.92	27.4	2 to >3.5	quartz; labradorite; diopside
576	15 ~ 25	100	0	0	sand	1.70	35.9	>4.5	
626	0–5	99	1	0	sand	1.46	44.9	<0.25–0.5	quartz; labradorite; dolerite; diopside; anorthite
634	0–5	100	0	0	sand	1.83	31.0	<0.25	quartz; labradorite; dolerite; diopside

TABLE 2. Soil Physical Characteristics: Moisture, Color, and Reflectivity, Victoria Valley, Antarctica

Soil Sample	Soil Depth, cm	Moisture Constants, weight % H_2O: in situ	Maximum H_2O Holding Capacity	0.1 bar	0.5 bar	15 bars	Hygroscopic Coefficient at 98% RH	Oven Dry at 105°C ± 5°	Soil Color* and Munsell Notation: in situ	Air Dry†	Moist†	Reflectivity, %: Air Dry†	Moist†	Air Dry/Moist Ratio
537	0–2	0.24	21.5	8.8	5.7	0.90	0.22	0.11	5Y 7/3 py	10YR 6/3 pb	10YR 4/2 dgb	24.0	8.5	2.8
538	2–15	0.28	20.2	8.3	5.6	0.77	0.65	0.08	2.5Y 6/2 lbg	10YR 6/3 pb	10YR 5/3 b	26.5	11.0	2.7
539	15–25	1.16	20.0	9.0	7.1	2.2	1.37	0.38	5Y 7/3 py	10YR 6/3 pb	10YR 4/4 dyb	26.5	7.0	3.8
539a	25–30	19.2	21.3	5.0	2.9	1.27	0.25	0.10	5Y 4/2 og	10YR 6/2 lbg	10YR 4/4 dyb	26.5	8.0	3.3
540	0–2	1.83	19.9	11.8	8.8	3.3	14.8	0.75	2.5Y 6/4 lyb	2.5Y 6/2 lbg	2.5Y 4/4 ob	27.5	8.0	3.4
541	2–15	3.5	24.6	15.0	10.1	5.1	59.8	1.40	5Y 7/3 py	2.5Y 5/4 lob	2.5Y 4/4 ob	25.5	6.0	4.3
542	15–25	4.4	28.8	15.3	11.8	6.0	42.0	1.28	5Y 6/2 log	2.5Y 4/2 dgb	2.5Y 4/2 dgb	15.0	4.0	3.8
543	25–30	4.9	36.8	16.2	11.7	6.7	24.6	1.05	5Y 5/2 og	2.5Y 5/2 gb	2.5Y 4/2 dgb	14.5	3.0	4.8
574	0–2	0.35	15.6	9.6	0.86	0.49	0.22	0.17	2.5Y 6/2 lbg	2.5Y 5/2 gb	2.5Y 4/2 dgb	27.5	6.5	4.2
575	2–15	0.20	15.6	11.7	0.98	0.59	0.22	0.17	2.5Y 6/2 lbg	2.5Y 5/2 gb	2.5Y 4/4 ob	24.0	10.0	2.4
576	15 ~ 25	1.70	16.0	10.1	0.88	0.45	0.21	0.09	5Y 5/4 o	2.5Y 5/2 gb	2.5Y 4/2 dgb	18.5	8.5	2.2
626	0–5	2.1	21.9	10.0	2.3	0.72	0.58	0.64	10YR 6/4 lyb	10YR 6/3 pb	2.5Y 5/4 lob	20.0	7.5	2.6
634	0–5	0.14	20.6	10.2	6.9	0.54	0.19	0.30	10YR 6/1 g	2.5Y 6/2 gb	2.5Y 4/2 dgb	18.5	7.0	2.6

* The colors are abbreviated as follows: py, pale yellow; g, gray; og, olive gray; log, light olive gray; lbg, light brownish gray; o, olive; b, brown; pb, pale brown; gb, grayish brown; ob, olive brown; lyb, light yellowish brown; lob, light olive brown; dgb, dark grayish brown; dyb, dark yellowish brown.

†Processed soil, ⩽2 mm.

TABLE 3. Soil Physicochemical Characteristics, Victoria Valley, Antarctica

		*p*H		*E*h + mv[a]		*r*H[b]		Electrical Conductivity × 10^{-6} mhos/cm at 25°C							
Soil Sample	Soil Depth, cm	Saturated Paste	1:5 Extract	Saturated Paste	1:5 Extract	Saturated Paste	1:5 Extract	Saturated Paste	1:5 Extract	Paste 1:5 Extract	Osmotic Pressure,[c] atm	Loss on Ignition,[d] % weight	Cation Exchange Capacity, me/100 g	Buffer Capacity, me/100 g	*N* Value[e]
537	0–2	8.9	7.2	130	80	22.3	17.2	200	88	2.3	0.03	0.74	4	1	−0.20
538	2–15	8.5	6.9	160	100	22.5	17.2	900	190	4.7	0.07	0.67	3	1	−17.2
539	15–25	7.8	6.6	180	130	21.8	17.7	2,700	490	5.5	0.18	0.60	5	1	−1.68
539a	25–30	8.2	6.8	130	120	20.9	17.7	300	112	2.7	0.04	0.39	4	1	+11.9
540	0–2	8.0	6.6	180	150	22.2	18.4	14,500	3,900	3.7	1.40	2.31	4	1	−0.46
541	2–15	7.9	6.7	200	150	22.7	18.6	47,000	14,000	3.4	5.04	4.35	8	2	+1.38
542	15–25	7.5	6.7	200	140	21.9	18.2	28,000	12,200	2.3	4.39	4.94	10	3	+1.37
543	25–30	7.7	6.9	180	120	21.6	17.9	21,000	7,000	3.0	2.52	3.82	11	5	+0.90
574	0–2	9.0	7.3	75	150	20.6	19.8	910	49	18.6	0.02	0.76	0	0.3	−16.7
575	2–15	9.0	7.2	70	150	20.4	19.6	520	80	6.5	0.03	0.81	0	0.2	−31.6
576	15 ∼ 25	8.8	7.0	100	150	21.0	19.2	450	40	11.3	0.01	0.82	0	0.2	−5.9
626	0–5	7.0	6.4	210	190	21.2	19.0	700	680	1.0	0.24	0.70	1.5	2.6	+0.52
634	0–5	7.4	6.6	150	140	19.9	18.0	95	49	1.9	0.02	0.09	1	1.1	−23.8

[a] Does not include cell K (+244 mv).
[b] $rH = (Eh/0.029) + 2\ pH$.
[c] Extract electrical conductivity times 360.
[d] 30 min at 600°C.
[e] The *N* value equals the per cent of H_2O field-2 divided by the per cent of clay plus 3 OM.

determined by oven drying to 105° ± 5°C. Additional moisture constants were determined by standard methods for maximum water-holding capacity, for hygroscopic coefficient, and at 0.1, 0.5, and 15 bars (Table 2).

8. After 1 hour of equilibration the *p*H and *E*h (oxidation reduction potential) were determined potentiometrically on both the saturated soil pastes and the 1:5 soil–water extracts; the *r*H milieu was calculated (Table 3).

9. The electrical conductivity was determined on both the saturated pastes and the 1:5 soil–water extracts; osmotic pressure values were based on extract data (Table 3).

10. The weight loss of ignitable solids was determined by heating at 600°C for 30 min (Table 3).

11. The cation exchange capacity was determined by means of the modified barium chloride–triethanolamine procedure (Table 3).

12. The buffer capacity was determined titrametrically to the methyl orange or phenolphthalein end points (Table 3).

13. The *N* values for in situ subsidence characteristics were calculated from values for moisture, fines, and empirical organic matter content (Table 3).

14. The abundance of major chemical elements was determined primarily by the cathode layer arc method, separate analyses being used for total carbon,

TABLE 4. Abundance of Major and Minor Chemical Elements, Victoria Valley, Antarctica

		Element Content, % weight												
Soil Sample	Sample Depth, cm	O	Si	Al	Ca	Fe	K	Mg	Na	Ti	S	Cl	H	C
537	0–2	50.4	27.0	9.5	4.3	2.7	<1.0	2.5	2.2	0.71	0.142	0.0025	0.197	0.165
538	2–15	50.9	28.0	7.6	4.0	1.6	2.4	2.0	2.5	0.40	0.140	0.0115	0.129	0.134
539	15–25	50.5	28.0	8.0	3.9	1.8	2.5	1.6	0.41	0.41	0.092	0.0500	0.194	0.068
539a	25–30	50.2	24.0	11.0	3.3	2.4	3.3	2.3	2.5	0.51	0.173	0.0070	0.142	0.051
540	0–2	49.8	23.0	9.7	5.4	3.1	1.9	2.6	2.4	0.66	0.397	0.4000	0.279	0.149
541	2–15	48.1	27.0	7.4	4.9	2.8	1.5	2.4	2.4	0.39	0.453	1.90	0.270	0.192
542	15–25	49.8	25.0	9.0	4.1	2.4	1.7	2.2	2.5	0.53	0.239	1.60	0.311	0.287
543	25–30	50.3	26.0	8.5	4.1	3.0	1.3	2.2	2.2	0.55	0.151	0.750	0.271	0.371
574	0–2	51.2	26.0	9.5	3.7	3.0	1.4	2.3	1.7	0.62	0.155	0.0025	0.167	0.111
575	2–15	50.9	24.0	9.2	4.1	2.8	2.4	3.1	2.2	0.70	0.136	0.0025	0.115	0.096
576	15 ∼ 25	47.2	19.0	12.0	3.8	2.3	8.8	2.5	3.3	0.56	0.217	0.0015	0.096	0.086
626	0–5	45.9	26.0	8.9	3.7	3.3	3.0	5.6	2.3	0.55	0.136	0.0105	0.103	0.264
634	0–5	50.0	27.0	6.8	2.7	2.1	1.6	7.4	1.6	0.46	0.066	0.0020	0.066	0.059

TABLE 5. Abundance of Trace Chemical Elements, Victoria Valley, Antarctica

Soil Sample	Sample Depth, cm	Element Content, % weight												
		Mn	N	Sr	Cr	Zr	Va	Ni	Ga	Cu	Co	P	B	Zn
537	0–2	0.084	0.001	0.028	0.021	0.017	0.013	0.0095	0.0053	0.0049	0.0085	0.002	0.0001	0.00001
538	2–15	0.061	0.001	0.031	0.026	0.014	0.012	0.0065	0.0075	0.0055	0.0054	0.002	0.0001	0.00002
539	15–25	0.063	0.007	0.028	0.025	0.012	0.012	0.0092	0.0029	0.0041	0.0044	0.004	0.0002	0.000025
539a	25–30	0.057	0.002	0.014	0.036	0.011	0.014	0.013	0.0048	0.0041	0.0084	0.002	0.0001	0.000005
540	0–2	0.068	0.034	0.013	0.021	0.013	0.015	0.011	0.0066	0.0056	0.0059	0.002	0.0003	0.000025
541	2–15	0.056	0.161	0.014	0.0088	0.013	0.0098	0.0063	0.0069	0.0046	0.0029	0.002	0.0006	0.00001
542	15–25	0.062	0.163	0.017	0.014	0.021	0.010	0.0080	0.0074	0.0065	0.0068	0.002	0.0003	0.00002
543	25–30	0.068	0.125	0.016	0.012	0.019	0.010	0.0072	0.0058	0.0064	0.0038	0.002	0.0002	0.00003
574	0–2	0.068	<0.0005	0.011	0.029	0.0085	0.013	0.0091	0.0054	0.0029	0.0062	0.005	0.0004	<0.00001
575	2–15	0.082	<0.0005	0.027	0.033	0.012	0.018	0.011	0.0067	0.0038	0.0100	0.005	0.0003	<0.00001
576	15 ~ 25	0.046	0.001	0.024	0.034	0.015	0.015	0.0087	0.0072	0.0039	0.0065	0.005	0.0002	<0.00001
626	0–5	0.063	0.003	0.018	0.023	0.013	0.012	0.011	0.0045	0.0049	0.0071	0.048	0.0003	0.00004
634	0–5	0.059	0.004	0.023	0.045	0.0053	0.015	0.015	0.0029	0.0031	0.0085	0.011	<0.0001	<0.00001

TABLE 6. Water-Soluble ('Available') Soil Cations and Anions, Victoria Valley, Antarctica*

Soil Sample	Soil Depth, cm	Ion Content, ppm in 1:5 Soil–Water Extract														
		Na^+	Ca^{++}	Mg^{++}	K^+	NH_4^+	Zn^{++}	Fe^{+++}	Al^{+++}	Cl^-	NO_3^-	$SO_4^=$	HCO_3^-	$PO_4^{\equiv}$	BO_3^{-3}	NO_2^-
537	0–2	8	8	1	1	0.2	0.02	0.04	0	5	2	8	24	0.2	0.2	0.02
538	2–15	22	9	2	1	0.2	0.04	<0.01	0	23	6	22	21	0.3	0.2	0.03
539	15–25	48	24	5	2	0.1	0.05	0.03	0	100	25	20	18	0.2	0.3	0.01
539a	25–30	14	2	6	0	0.1	0.01	0.03	0	14	4	14	15	0.4	0.2	0.02
540	0–2	640	185	31	2	0.2	0.05	0.01	0	800	185	670	18	0.2	0.6	0.02
541	2–15	2450	375	200	5	0.2	0.02	0.01	0	3800	960	825	21	0.1	1.2	0.02
542	15–25	2200	275	250	5	<0.1	0.04	0.01	0	3200	940	1200	18	0.2	0.6	0.02
543	25–30	960	140	150	3	0.1	0.06	0.02	0	1500	620	870	24	0.2	0.3	0.03
574	0–2	6	2	0.5	1	2.0	<0.1	0.3	0.3	5	0	2	21	0.1	0.7	0.02
575	2–15	8	3	1	1	2.2	<0.1	0.3	0.3	5	0	6	31	0.1	0.5	0.01
576	15 ~ 25	6	1	0.3	1	1.3	<0.1	0.1	0.3	3	0	3	18	<0.01	0.4	0.02
626	0–5	25	100	3	1.2	0.6	0.08	0.02	0	21	5	270	24	0.1	0.5	<0.01
634	0–5	5	0.5	0.3	0.6	0.4	0.01	0.02	0	4	1	4	6	0.06	<0.1	0.01

* No extractable manganese, cobalt, copper, molybdenum, hydroxide, or carbonate was obtained.

TABLE 7. Soil Carbon, Nitrogen, and Hydrogen Values, Victoria Valley, Antarctica

Soil Sample	Soil Depth, cm	Total Organic Matter,* % weight	Total Carbon, % weight	Total Nitrogen, % weight	Organic Carbon, % weight	Organic Nitrogen, % weight	Carbonate Carbon, % weight	Total Hydrogen, % weight	Organic C:N Ratio	Total C:N Ratio	Total N:H Ratio	Total H:C Ratio
537	0–2	0.029	0.165	0.001	0.017	0.002	0.134	0.197	8.5	165	0.01	1.2
538	2–15	0.026	0.134	0.001	0.015	0.003	0.105	0.129	5.0	134	0.01	1.0
539	15–25	0.072	0.068	0.007	0.042	0.004	0.051	0.194	11.0	9.7	0.04	2.9
539a	25–30	0.048	0.051	0.002	0.028	0.004	0.024	0.142	7.0	26	0.01	2.6
540	0–2	0.093	0.149	0.034	0.054	0.0005	0.090	0.279	108	4.4	0.12	1.9
541	2–15	0.121	0.192	0.161	0.070	<0.0005	0.110	0.270	>140	1.2	0.60	1.4
542	15–25	0.193	0.287	0.163	0.112	0.001	0.168	0.311	112	1.8	0.52	1.1
543	25–30	0.355	0.371	0.125	0.206	0.002	0.145	0.271	103	3.0	0.46	0.7
574	0–2	0.033	0.111	<0.0005	0.019	0.002	0.098	0.167	9.5	>222	<0.003	1.5
575	2–15	0.019	0.096	<0.0005	0.011	0.001	0.080	0.115	11.0	>192	<0.004	1.2
576	15 ~ 25	0.017	0.086	0.001	0.010	0.001	0.073	0.096	10.0	86	0.01	1.1
626	0–5	0.064	0.264	0.003	0.037	0.002	0.196	0.103	18.5	88	0.03	2.6
634	0–5	0.026	0.059	0.004	0.015	0.003	0.037	0.066	5.0	14.7	0.06	0.9

* Organic carbon times 1.724.

TABLE 8. Cultural Conditions for Microbial Abundances and Distribution in Antarctic Soils

Culture Medium	Microbial Test	*p*H	*E*h + mv (Uncompensated)	Electrical Conductivity × 10^{-6} mhos/cm at 25°C	Incubation Temperature, °C	Incubation Period, average days	First Observed Growth, average days
Actinomycete isolation agar	aerobic bacteria and streptomycetes	7.8	124	3,680	20	21	14
Burk's 'ion' agar	nitrogen-fixing and nonchromogenic bacteria	6.9	172	2,800	20	21	14
Desoxycholate agar	coliform bacteria	7.4	131	6,400	37	14	7
di Menna's dextrose–neopeptone agar	yeasts and molds	4.5	160	880	5, 20	42, 21	21, 14
Fluid thioglycollate	microaerophiles, aerobic and anaerobic bacteria	7.1	5	7,300	20	56	16
Lactose broth	lactose fermenters	6.8	155	2,180	20	42	6
Nitrate reduction broth	nitrate reducers	7.1	158	3,020	20	42	7
Rose bengal agar	fungi (molds and yeasts)	5.7	208	2,320	5, 20	28	10
Salt–organic agar (STV); simulated Taylor valley extract, neopeptone, and yeast extract	aerobic and chromogenic bacteria	7.1	153	2,240	5, 20	42, 30	21, 14
Starkey's solution, modified	sulfate reducers (anaerobic bacteria)	6.3	110	560	20	42	12
Thornton's salt medium, 'Gro-Lux' lights (~200 ft-c)	algae and protozoa	7.8	158	2,670	20	180	14
Trypticase soy agar	aerobic bacteria, streptomycetes and fungi; anaerobic bacteria in CO_2 or Gas Pak	7.5	140	5,680	5, 20, 45, 55; 20 or room temperature	42, 30, 5, 5; 49	21, 14, 2, 2; 40
Trypticase soy broth	bacteria (broad spectrum), halotolerants	7.4	142	12,800	20	42	4
Van Delden's sulfate reduction agar	sulfate reducers (microaerophiles) and nonchromogenic bacteria	5.8	176	5,120	20	21	14

hydrogen, nitrogen, phosphorus, and sulfur. Total carbon, hydrogen, and nitrogen were determined by ignition in Coleman carbon–hydrogen and nitrogen analyzers modified for small samples; phosphorus was determined by perchloric acid extraction and colorimetric determination of vanadomolybdophosphoric; and sulfur was determined by sodium peroxide–sodium carbonate fusion followed by barium precipitation. Oxygen was calculated by difference (Tables 4, 5, and 7).

15. Anions and water-soluble, or 'available,' cations (Na^+, Ca^{++}, Mg^{++}, K^+, NH_4^+, Zn^{++}, Fe^{+3}, Al^{+3}, Cl^-, NO_3^-, NO_2^-, $SO_4^=$, BO_3^{-3}, HCO_3^-, and PO_4^{-3}) were obtained with a slightly acidified (CH_3COOH) 1:5 soil–water extract and followed by colorimetry, flame photometry, or atomic absorption spectrometry (Table 6).

16. Organic and inorganic (carbonate) carbon were determined microgravimetrically as evolved CO_2 after chromic acid digestion (Table 7).

17. Organic nitrogen was determined by the micro-Kjeldahl method (Table 7).

The thermoluminescence of soils was determined by means of approximately 10-mg samples passed through a 100-mm sieve (no. 100 mesh). Natural thermoluminescence, as well as thermoluminescence following irradiation with 50 krad of a Co_{60} source, was determined. The relative peak heights of light intensity

TABLE 9. Microbial Abundance, 'Standard Determinations,' Victoria Valley, Antarctica

Soil Sample	Sample Depth, cm	Aerobic Bacteria per Gram of Soil				Microaerophiles Positive at Highest Dilution per Gram of Soil at 20°C[c]	Anaerobes per Gram of Soil at Room Temperature[d]	Fungi, Molds, Yeasts per Gram of Soil at +2°C and +20°C[e]	Algae Positives at Highest Dilution per Gram of Soil at +20°C[f]	Protozoa Positives at Highest Dilution per Gram of Soil[f]
		+2°C[a]	+20°C[a]	+2°C[b]	+20°C[b]					
537	0–2	2.8×10^4	8×10^3	600	900	10^4	0	0	10^3	10
538	2–15	2.6×10^3	1.2×10^3	275	20	10^4	0	0	0	0
539	15–25	5.6×10^4	1.8×10^4	3.1×10^3	7.3×10^3	10^4	1	0	10	0
539a	25–30	2.7×10^3	2.5×10^3	2.2×10^3	975	10^3(1) *	0	0	10	0
540	0–2	<10	55	5	<10	10^3(1) *	0	0	0	0
541	2–15	30	40	<10	<10	10	0	0	0	0
542	15–25	0	0	0	0	0	0	0	0	0
543	25–30	5	<10	<10	<10	10	0	0	0	0
574	0–2	730	150	5	20	10^3(1) *	0	<10†	0	0
575	2–15	55	400	10	55	10^3(3) *	0	5†	0	0
576	~25	35	150	155	210	100(2) *	0	0	0	0
626	0–5	<10	<10	<10	<10	100	0	0	0	0
634	0–5	<45	<10	<10	<10	10	0	0	0	0

[a] Trypticase soy agar (TSA).
[b] Salts (simulated Taylor valley), yeast extract, and neopeptone (STV agar).
[c] Fluid thioglycollate.
[d] TSA and STV agar in CO_2.
[e] Rose bengal agar and di Menna's dextrose–neopeptone agar.
[f] Thornton's salt medium (without organics) and Sylvania 'Gro-Lux' fluorescent lights at about 250 ft-c.
* Colony count where discernible.
† Yeasts.

were plotted as glow curves versus temperature when the soils were heated from ambient to approximately 350°C. The details of instrumentation and the methods for analyzing soils have been given previously [Lawson et al., 1970], including analyses of antarctic soils [Cameron et al., 1969d].

The abundance and the distribution of general and specific groups of microorganisms and their presence or absence in a sample were determined by the direct inoculation of several grams of soil onto agar plates in the field. After sufficient incubation time and the subsequent development of microbial colonies, these preliminary results helped determine the need for further dilutions and the selection of culture media. It was not always feasible in the field to keep agar plates from freezing, but pocket hand warmers were helpful.

In the McMurdo biology laboratory and subsequently at our respective stateside laboratories the abundance and the groupings of microorganisms were determined by the direct inoculation of 1–10 grams of soil onto agar plates or by the pretempering of the spread plates to the desired temperatures for low temperature incubations and by dilution tube culture methods. These methods were designed and tested for the investigation and study of low abundances of cold desert soil microorganisms [Benoit and Hall, 1970; Benoit and Cameron, 1967; Cameron, 1967, 1970; Cameron and Benoit, 1970; Cameron et al., 1970b, c; Hall, 1968; Hall and Benoit, 1968]. Agar pour plate methods were found to be unsatisfactory for this study; the agar pour plate method for determinations of antarctic dry valley microorganisms showed a drastic reduction and selection of microorganisms compared with the spread plate method [Cameron, 1967; C. N. David and J. King, personal communication, 1968]. The cultural conditions used to determine microbial abundance and distribution in antarctic soils have been summarized in Table 8.

Besides the bacteria, streptomycetes, molds, and yeasts (Tables 9 and 10), protozoa and algae were also determined in dilution tube cultures (Table 9). However, some soil crusts likely to contain algae were incubated in moist chambers, and any subsequent growth was examined as described previously [Cam-

TABLE 10. Microbial Abundance, 'Supplementary Determinations,' Victoria Valley, Antarctica

Soil Sample	Sample Depth, cm	Aerobic Bacteria per Gram of Soil: Positive at Highest Dilution at +20°C[a], Set 1	Set 2	+20°C[b]	+45° and +55°C[c]	Lactose Fermenters Positive at Highest Dilution per Gram of Soil at +20°C[d]	Nitrate Reducers Positive at Highest Dilution per Gram of Soil at +20°C[e]	Nitrogen Fixers per Gram of Soil at +20°C[f]	Sulfate Reducers per Gram of Soil at +20°C[g]	Coliforms per Gram of Soil at +20° and +37°C[h]	Incubation Required for First Observed Growth, days[i]
537	0–2	10^4	10^4	7×10^3	0	10^3	10^4	400	700	0	7
538	2–15	10^3	10^4	<10	0	10	100	10	285	0	9
539	15–25	10^5	10^4	1.9×10^5	0	100	100	5.3×10^3	2.8×10^4	0	4
539a	25–30	10^3	10^4	50	0	100	100	25	5	0	2
540	0–2	100	10^3	<10	0	10	10	<10	<10	0	3
541	2–15	100	10	160	0	100	10	<10	160	0	7
542	15–25	0	0	0	0	0	0	0	0	0	5
543	25–30	100	10	<10	0	10	<10	<10	<10	0	7
574	0–2	10^3	100	125	0	100	10^3	5	125	0	7
575	2–15	10^3	100	4.7×10^3	0	10^3	10^3	25	470	0	7
576	~25	10^3	10	600	0	10	10^3	35	600	0	7
626	0–5	<10		<10	0	10	100	<10	<10	0	14
634	0–5	10		<10	0	<10	<10	<10	<10	0	21

[a] Trypticase soy broth (TSB) at 0.1% concentration and full strength.
[b] Actinomycete isolation agar (no streptomycetes present).
[c] Trypticase soy agar.
[d] Lactose broth.
[e] Nitrate reduction broth.
[f] Burk's N-free ion agar.
[g] Van Delden's sulfate reduction 'ion agar.'
[h] Desoxycholate agar.
[i] Any culture medium.

eron and Devaney, 1970]. One of the methods designed for extraterrestrial life detection was of considerable value in substantiating microbial abundances [Cameron et al., 1969c; Horowitz et al., 1969; Hubbard et al., 1968, 1970]. Some of the Victoria valley soils were subsequently tested for microbial activity by this method, which used the metabolic assimilation of $^{14}CO_2$ into organic compounds (Table 11). The final characterization of the bacterial isolants was performed by K. Byers Kemper and W. Bollen (personal communication, 1968), and that of the yeasts by M. di Menna (personal communication, 1968).

ENVIRONMENTAL FACTORS

Temperature. Environmental measurements of air and/or soil were made at site 537-539a (Figures 2 and 7). The aerial measurements were not made above 1 meter because this elevation extends above the microclimatic zone for antarctic organisms. In fact, in a consideration of the life forms and their distribution in the dry valleys, the microclimatic and microenvironmental zone can for all practical purposes be considered to extend only a few centimeters above the soil surface and a few additional centimeters below the surface to the boundary layer of the ice-cemented permafrost layer. If the soil microbial ecosystem is active, it is usually at the immediate air–soil interface and extends down through the soil profile as far as heat fluxes can interact with moisture to provide a favorable solclime.

The temperatures of the air and the soil for the period of measurement are shown in Figures 27–30. The daily variations recorded continuously with a hygrothermograph are shown in Figure 28. The diurnal variations are not as pronounced (Figure 28) as the variations of the manual readings plotted every third hour (Figure 27). The values obtained with thermistors (Figure 27) are much more accurate, and the response is also more rapid than that obtained with the hygrothermograph (Figure 28), which was more susceptible to wind currents. As both Figures 27 and 28 show, during a significant portion of each day during midsummer the temperature is above freezing, but there is a diurnal cycle whereby the temperatures of both the air and the soil are below freezing for a given 24-hour period. The average high temperature

TABLE 11. Metabolic Soil $^{14}CO_2$ Production, Victoria Valley, Antarctica*

Soil Sample	Sample Depth, cm	CO_2 Evolved,† counts/min Un-treated	Steri-lized	Net	Estimated Microbial Count per Gram of Soil
537	0–2	1008	109	899	10^4
540	0–2	157	37	120	10^2–10^3
541	2–15	56	62	−6	10–10^2
542	15–25	32	36	−4	<1
543	25–30	28	61	−33	<10–10
574	0–2	323	121	238	10^3

* Substrate mixtures contained 144 ng (nanograms) of ^{14}U C glucose (5×10^4 counts/min) plus 20 ng of ^{14}U C amino acids (3.9×10^4 counts/min) in 0.3 ml of water.

† CO_2 per hour per 300 mg of soil.

TABLE 12. Air and Soil Temperatures in Degrees Celsius in Victoria Valley, December 28, 1966, to January 3, 1967

Station	Maxi-mum	Average Daily Maxi-mum	Mini-mum	Average Daily Mini-mum	Average Total
Air					
1 meter	6.0	4.6	−4.0	−1.3	1.8
30 cm	8.0	5.8	−3.0	−1.2	2.1
15 cm	10.0	7.0	−4.5	−0.9	2.6
Surface					
Sun	19.0	16.6	−4.0	−1.3	5.8
Shade	9.5	6.2	−2.3	−0.4	2.5
Soil					
5 cm	6.0	5.3	−1.8	−0.6	2.1
10 cm	4.3	2.7	−2.0	−0.9	1.2
15 cm	0.8	−0.5	−4.0	−3.2	−2.1
20 cm	−2.0	−2.8	−5.0	−4.3	−3.6
30 cm	−6.0	−7.3	−8.0	−7.9	−7.6

for air determined by the hygrothermograph was 42°F (6°C), and the average low temperature was 29°F (−1.5°C). The values determined by thermistors for the same period are shown in Table 12. The highs and lows determined by soil and air maximum–minimum thermometers for the same period are shown in Figures 29 and 30. The highest air temperature (at 15 cm) was 10°C, and the lowest only −4.5°C. The highest soil temperatures were recorded for the open soil surface exposed to the sun (16.6°C), but the lowest soil temperature almost reached that of the air (−4°C). The lowest values were usually recorded after the sun had been obscured by a mountain in the south at 0025 hours. After 5 min of shading the soil surface temperature would drop 1.5°C after a 2°C drop in the air temperature at 15 cm. Two and a half hours later the air temperature would rise to 0°C, and the soil surface in the sun to 2°C. This phenomenon could be significant for the growth of psychrophils (provided moisture was also available).

As other investigators [Janetschek, 1967; Rudolph, 1966] have shown, there can be pronounced differences between soil surface and air temperatures at relatively the same time or after a short lag period. Although there was a <1°C average difference between the temperature of the air at 1 meter and that at 15 cm, the average soil surface temperature in the

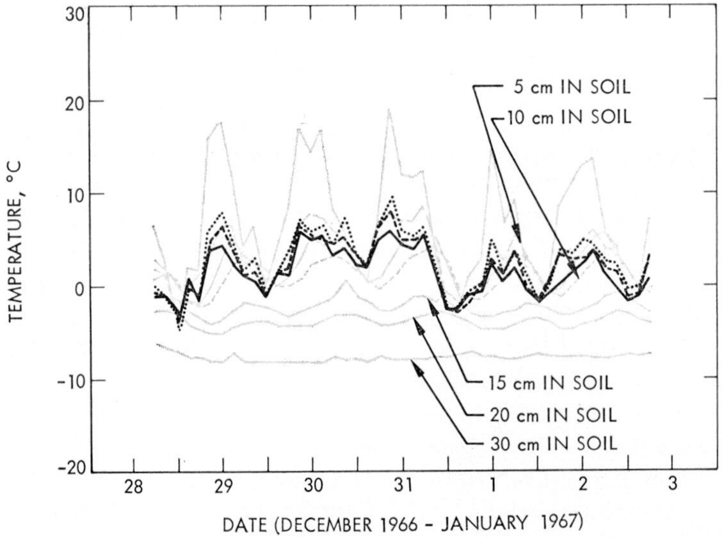

Fig. 27. Temperature of air and soil, Victoria valley, Antarctica. (Solid, dashed, and dotted black lines represent 1 meter, 30 cm, and 15 cm above the surface, respectively. Solid and dashed gray lines represent soil surface in sun and shade, respectively.)

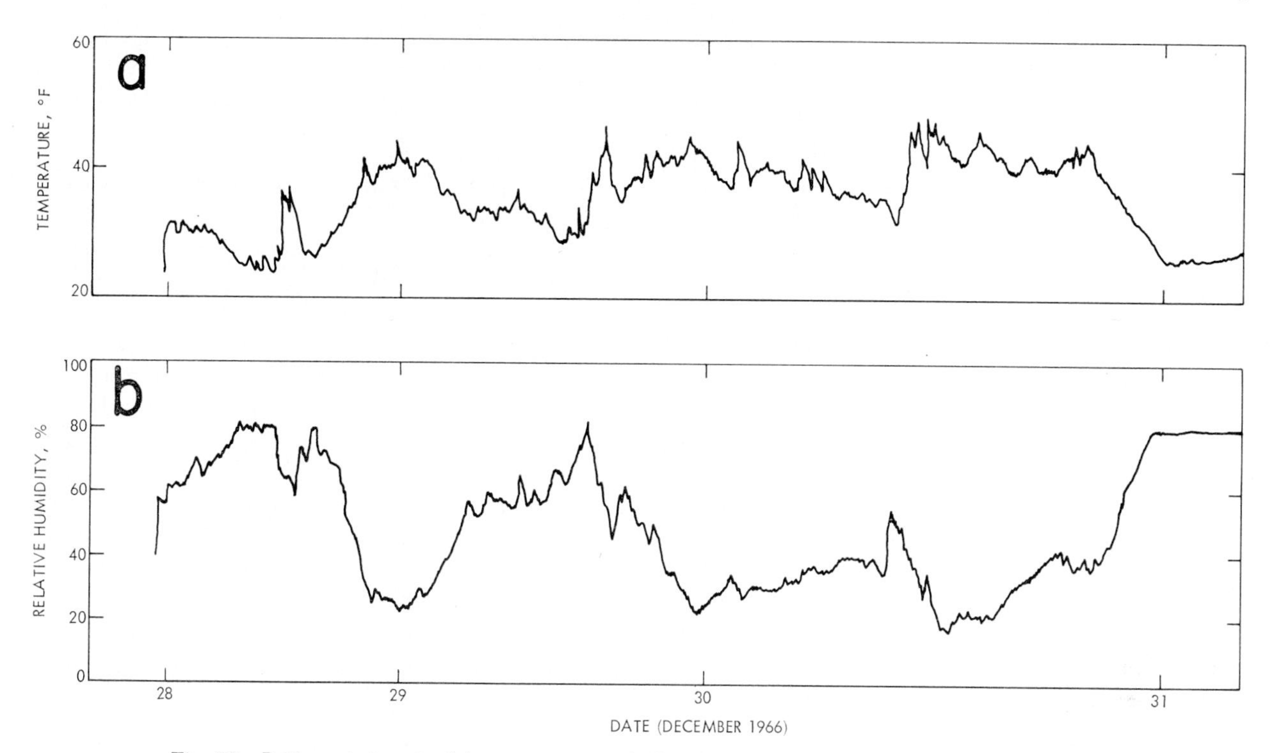

Fig. 28. Daily variations in (*a*) temperature and (*b*) relative humidity, Victoria valley, Antarctica.

sun was 5.8°C and more than 3°C warmer than the air temperature. However, there is a very narrow range of soil warming, and the average temperature at 5 cm in the soil was 2.1°C, which was the same as the air temperature at 6 cm. There was a gradual decrease in soil temperature with soil depth to the ice-cemented permafrost boundary layer at 30 cm, where the temperature was −7.6°C, and the soil temperature remained below freezing and showed very little variation during our study. The soil temperature at a depth of 15 cm was infrequently above freezing, but the average soil temperature at a depth of 10 cm was just above freezing (1.2°C). As Weyant [1966] indicated, the active layer of the soil is above the permafrost, i.e., the ice-cemented layer. The depth from the soil surface to the ice-cemented layer is a function of the mean annual surface temperature, the annual temperature wave at the surface, and the heat transmission characteristics of the soil into the upper layers [Weyant, 1966]. The thawing of the soil in the Bunger Hills begins in the first part of November and reaches its highest value by the end of January; freezing begins again during the second half of January [Solopov, 1967]. The same phenomena appear to hold true for the dry valleys of Victoria Land as the data of Berg and Black [1966] and Ugolini [1970] show.

The temperature measurements show one essential component of the solclime (hydrothermal regime) for microorganisms in the Victoria valley soils. The relatively narrow range of soil temperatures and the higher temperatures of unshaded soil surfaces frequently exposed to the sun show that temperature is an important consideration for the growth and the reproduction of microorganisms. Unfortunately the soil surface is also more subject to ultraviolet irradiation, desiccation by wind, and erosion by blowing sand and ice particles and has the least moisture. Freezing and thawing are also detrimental to the survival as well as to the growth and reproduction of various groups of microorganisms [Cameron and Blank, 1967; Cameron et al., 1969a, 1970a; Farrell and Rose, 1967; Goos et al., 1967; Holm-Hansen, 1963a; Mazur, 1961; Young et al., 1963, 1967]. The death rate of *Escherchia coli* is extremely rapid in the Antarctic [Boyd and Boyd, 1963b]; the repeated freezing and thawing of *E. coli* in a liquid medium result in a linear decrease in the number of surviving cells as a function of the number of freeze–thaw cycles [McLaren and Skujins, 1964]. There is a gradual decrease in viability for mesophilic spore-forming bacilli in soil subjected to controlled diurnal freezing and thawing [Hawrylewicz et al., 1965; Hagen and Hawrylewicz, 1969]. Freeze–thaw phe-

TABLE 13. Air Relative Humidity in Per Cent for Victoria Valley, December 28, 1966, to January 3, 1967

Station, cm in air	Maximum	Average Daily Maximum	Minimum	Average Daily Minimum	Average Total
100	63	48.4	12.4	20.8	32.4
30	64	49.1	11.6	20.7	31.7
15	65	49.1	11.6	20.7	32.5

TABLE 14. Soil Relative Humidity in Per Cent for Victoria Valley, December 28, 1966, to January 3, 1967

Station, cm in soil	Maximum	Average Daily Maximum	Minimum	Average Daily Maximum	Average Total
Surface	63	50.0	9.4	19.0	31.3
5	47.4	43.8	35.0	37.5	40.1
15	63	62.5	40.5	57.3	61.7
30	92.5	90.6	77.5	88.3	90.4

nomena are undoubtedly largely responsible for reduction or elimination of coliform and spore-forming *Bacillus* species in antarctic soils, although the moisture content and activity, the in situ association of bacteria with soil particles, and the cryotolerance of the bacteria must also be considered.

In laboratory experiments with isolants as well as with soil incubated on agar plates and in soil–water solutions, the growth of microorganisms has not been rapid (at either +2° to +5°C or +20° to +25°C) compared to that found in similar tests for temperate desert soils. Bacterial cells markedly lose their viability on rapid chilling, but Ca^{++}, Mn^{+++}, and especially Mg^{++} are effective in protecting the cells from the effects of cold shock [Sato and Takahashi, 1969], and the presence of these ions in antarctic soils is certainly an important factor for the survival of bacteria. The magnitude of the chilling through a definite temperature zone was found to be more important in effecting cold shock in bacteria [Sato and Takahashi, 1969]. Soils stored at sub-zero temperatures subsequently have an increased release of soil nutrients [Allen and Grimshaw, 1960], and this effect could be an important concurrent factor in the survival of antarctic microorganisms. Although few, if any, antarctic soil bacteria are obligate psychrophils [Straka and Stokes, 1960], it is significant that bacteria grown at 10°C are relatively insensitive to cold shock [Farrell and Rose, 1968]. Freezing can markedly affect the number of viable fungi in arable soils [Mack, 1963], but some antarctic yeasts are obligate psychrophils [Bab'eva and Golubev, 1969; di Menna, 1966; Goto et al., 1969; Sinclair and Stokes, 1965]. Some antarctic microorganisms can increase in numbers when they are incubated at a constant temperature above freezing with sufficient moisture [Cameron and Conrow, 1969b; Cameron and Merek, 1971].

Relative humidity. Relative humidity (RH) measurements taken every third hour for air and soil are plotted in Figure 31. Measurements obtained with the hygrothermograph are shown in Figure 28. As the temperature measurements with the hygrothermograph indicate, it was not the most reliable instrument, LiCl sensors being much more accurate. However, the humidity graph obtained from the hygrothermograph does show a continual measurement that indicates an approximate semiquantitative short-term response of humidity changes to concurrent temperature changes.

The average high air RH obtained with the hygrothermograph was 76%, and the average low was 32%. Air measurements taken with the LiCl sensors were practically the same at any point in time regardless of whether the measurements were made at 1 meter, 30 cm, or 15 cm above the soil surface. The lowest daily averages were obtained for January 1 (27.0% RH) and January 2 (24.9% RH); these values were influenced by the lower total wind and the lower humidity of westerly winds. The higher temperatures and the lower humidities of the westerly winds are a consequence of the lack of ice in the western sector of the Victoria valley system [Bull, 1966]. The RH values during our study are shown in Table 13.

Soil RH showed a much greater range of values, and the expected higher RH values were encountered closer to the ice-cemented permafrost. The RH increased with soil depth from the surface to 30 cm, from an average of 31.3 to an average of 90.4% RH (Table 14). Although the immediate surface of the soil showed the greatest range of RH and closely followed the air curves, these values were considerably influenced by the method of measurement. It was later determined that LiCl sensors needed a longer equilibration period when they were placed on or in the soil than when they were placed in the air. Subsequent measurements in other valleys showed a greater damping of the vapor flux intensity into and out of the soil surface and less direct similarity to variations in the air RH above the soil [Cameron, 1969b].

Relative humidity measurements are not usually made for the Antarctic and, if they are made, should be interpreted cautiously because of errors due mainly to the cold temperatures [Rudolph, 1967] and to the lack of proper sensors (e.g., LiCl electrolytic cells).

However, LiCl sensors were used at Cape Hallett [Rudolph, 1966], and the RH was usually <20% at midday and <50% later in the day. During a 6-week period of summer observations in the vicinity of Lake Vanda, westerly winds had extremes of RH between 5 and 60%, the average being 45% at a mean temperature of 2.2°C, but easterly winds had a mean RH of 68% at −0.5°C [Bull, 1966]. One-half mile from Wright Lower Glacier, the RH varied between 25 and 70% during the summer [Ugolini, 1963]. Errors in the hygrothermograph at Lake Vida limited the quantitative measurements made at that site [Bull, 1966], although dew points were calculated. No dew (frost) was obtained during our study, in which a Wallin-Palhemus dew duration recorder was used. The dew point was at all times below the actual temperature at the site. This finding is consistent with data obtained by Rudolph [1966] at a more favorable site (Cape Hallett), where the dew point was also seldom reached.

In the Antarctic away from the coast, RH values can be expected that are similar to or even less than those for temperate or tropical deserts, such as those in central Asia, and that are 15–20% lower than those in the Arctic, because the drainage winds prevailing over the ice undergo adiabatic descent and drop below the saturation point of the water vapor [Rusin, 1964]. In general the RH in temperate arid regions ranges between 20 and 40% [Fairbridge, 1967]. As a rule the RH in dry valleys in East Antarctica increases with height during the summer; this relationship is attributed to the entrance of water vapor into the atmosphere from below, and it also results in a drier RH over the nonglacial surfaces [Solopov, 1967]. Although the lowest RH values were observed during the winter, relatively dry values (between 30 and 40% RH) were observed in summer winds from almost all directions in the dry valleys [Rusin, 1964] and were similar to our measurements in Victoria valley. Differences between both RH and temperature are equalized during the transition seasons in ice-free valleys and glaciers [Solopov, 1967], and this trend probably occurs in the dry valleys of Victoria Land as well.

As was expected, the lowest air RH values were obtained when air temperatures were the highest (Figures 27 and 30). Both the RH and the temperature curves became less pronounced within the soil and with increasing depth, although there could be substantial vapor pressure gradients if the moist surface soils were warmed. Whether there are moisture movements in the soil as the result of a thermal gradient has been discussed previously [Ugolini, 1963], and in situ studies with $Na^{36}Cl$ have subsequently shown that ionic transport can indeed occur in continuous liquid phase water films under the arid conditions of the ice-free areas of southern Victoria Land [Ugolini and Grier, 1969]. However, the presence of salts at the soil surface and within the profile must also be considered [Ugolini, 1963], especially for the availability of moisture for microbial activities. There were $CaCl_2 \cdot 6\ H_2O$ crystals (antarcticite) [Torii and Ossaka, 1965] in Don Juan pond in the south fork of Wright valley and also in Balham valley; the RH of these areas must be <45% [Wilson, 1970].

As Wilson [1970] indicated, lichens should not be expected in dry valleys except between areas of 80% RH and the snow line, and this expectation should generally hold true for the over-all dry valley systems except where there is a favorable microenvironment, including exposure, moisture supply and availability, and drainage. Although lichens have shown a sufficient ability to absorb water vapor from the air to carry on measurable rates of photosynthesis [Ahmadjian, 1970], the Victoria valley RH value is too low for this purpose. The microorganisms previously reported from the extremely saline Don Juan pond in Wright valley [Meyer et al., 1962] are not actually present in the pond per se and apparently occur only in the periphery of the pond near the stream inlet [Hall, 1968]. The latter observation is consistent with the kinds and the concentrations of salts (predominantly $CaCl_2 \cdot 6\ H_2O$) [Torii and Ossaka, 1965], the freezing point of $-57° \pm 2°C$ [Tedrow et al., 1963], and the expected low a_w of 0.45 [Wilson, 1970].

Unfortunately, as temperatures decrease, the salt tolerance of antarctic soil microorganisms is also lowered [Hall, 1968]. The $CaCl_2 \cdot 6\ H_2O$, for example, shows an increase of 32–40% RH as the temperature is lowered from 20° to 5°C [Winston and Bates, 1960], a favorable temperature range for the growth of antarctic bacteria, but not a favorable humidity range. Note also that the freezing point of bacteria cells is similar to or slightly lower than the freezing point of the culture medium [Christian and Ingram, 1959], and this relationship does not aid the survival of bacteria in antarctic soils or in other soils when the soil freezes and liquid water is unavailable. These factors either singly or in combination would present an unfavorable habitat for known terrestrial microorganisms, but they are intriguing from the viewpoint of a possible extraterrestrial biology [Cameron et al., 1970d].

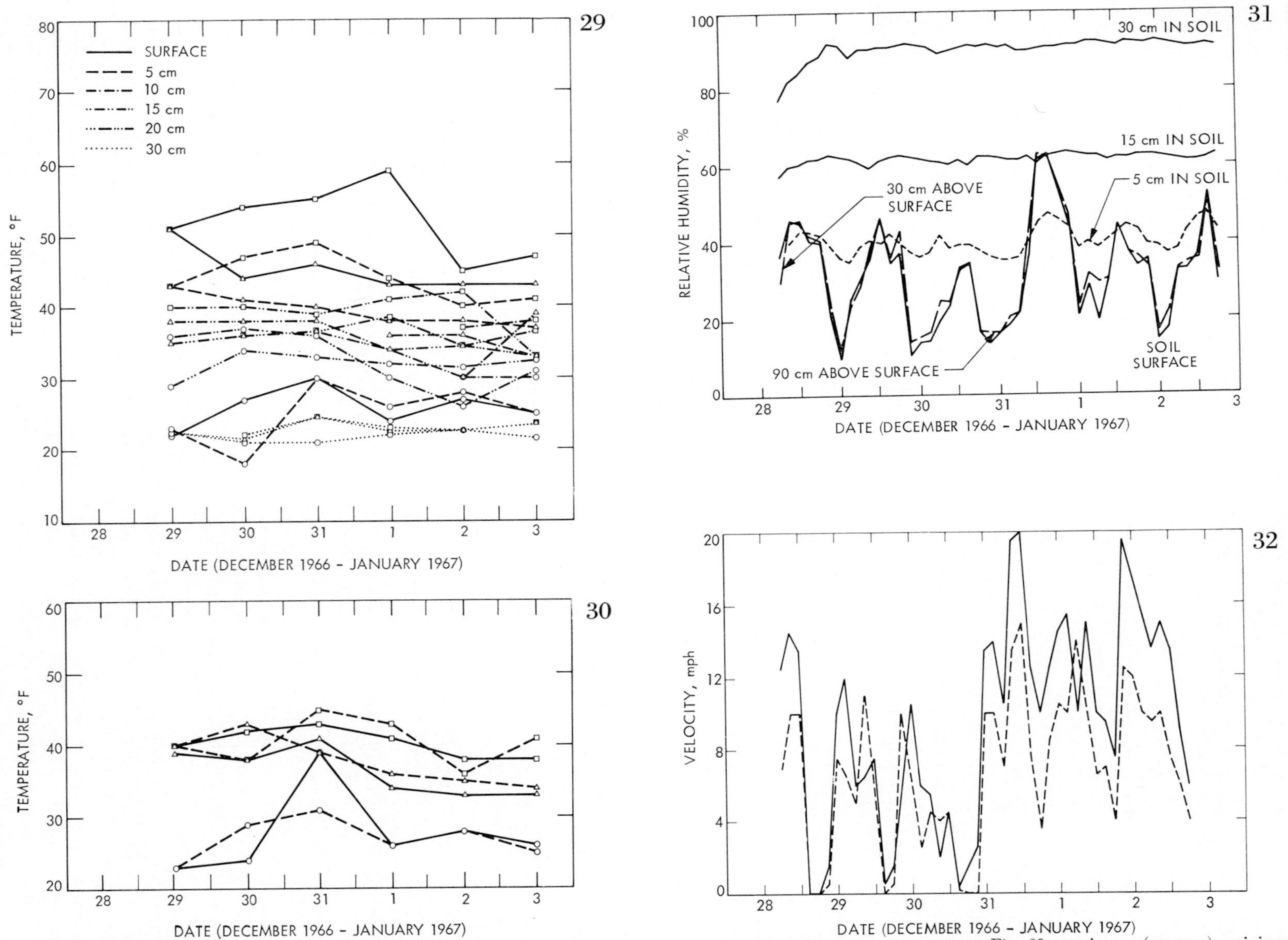

Figs. 29–32. Victoria valley, Antarctica. Fig. 29, maximum (squares), minimum (circles), and actual (triangles) temperatures. Fig. 30, maximum (squares), minimum (circles), and actual (triangles) air temperatures (the solid line represents 1 meter above the surface, and the broken line 30 cm above the surface). Fig. 31, relative humidity of air and soil. Fig. 32, wind velocity (the solid line represents 1 meter above the surface, and the broken line 30 cm above the surface).

The activities, the growth, and the reproduction of microorganisms in antarctic soils under the conditions of diurnal and fluctuating relative humidities and temperatures are not known. When the surface humidities are low, the temperatures are relatively high (Figures 27 and 31). Also, as the depth of the soil increases and the ice-cemented permafrost approaches, the temperatures decrease, and the relative humidities increase (the RH approaching 100%). However, the subsurface microfloras do not appear to be active and do not grow as rapidly as the surface microfloras when they are first isolated on sprinkle or spread plate dilutions and incubated at a high humidity and at temperatures of +2° to +20°C.

In dry temperate soils, microorganisms are primarily active at very high RH values in the range of the permanent wilting point (99% RH), but mycoflora activities reportedly do not occur below 75% RH, probably owing to distortion and disordering of the DNA molecule [Chen and Griffin, 1966]. In our study, xerophytic fungi and actinomycetes in semiarid soils did not show further development below 80–85% of maximum hygroscopicity except on larger sized soil particles; bacteria were not active at maximum hygroscopicity (98% RH) [Novogrudsky, 1946]. Studies of the moisture relations of food spoilage microorganisms, including a number of yeasts, molds, and bacteria that also occur in soils, showed that the lower limit of growth approached a water activity a_w of 0.90 [Scott, 1957]. The freezing point of the soil at this a_w is −10°C. As a group, bacteria could grow in liquid media at as low an a_w as 0.75 (saturated sodium chloride), and some yeasts and molds could grow slowly at an a_w of 0.62, but not on solids at the same a_w [Scott, 1957]. The growth of obligate halophiles may be optimum in a 4–5 *M* NaCl concentration at 0.80–0.85 a_w [Scott, 1961]. In tests of the survival of halotolerant, facultative, anaerobic bacilli and cocci in a variety of soils from harsh terrestrial environments, the limiting a_w for bacilli and cocci have been found to be 0.86 and 0.84, respectively [Hagen and Hawrylewicz, 1968a]. The kind of soil and the native (geographical) location (especially desert, high altitude, and tundra) appeared to correlate more with bacterial survival than minimum a_w did [Hagen and Hawrylewicz, 1968b].

The low RH of antarctic air could be an important factor in the survival of air-borne bacteria. Bacteria in aerosols can be more resistant to ultraviolet light (2800–3200 Å) at high than at low RH, and there is an abrupt decline in cell resistance between 55 and 65% regardless of wave length [Webb, 1963]. This critical RH level corresponds to that at which bound water is removed from nucleic acid bases. There is little or no ultraviolet or nonirradiated damage to aerosolized bacteria above 70% RH, but there is rapid death of cells exposed to ultraviolet at approximately 65% RH and of nonirradiated cells at between 65 and 45% RH [Webb et al., 1964]. In a nonultraviolet-irradiated environment at temperatures of 18°–21°C, gram positive coccoid bacteria may have a greater resistance to a wide range of humidities (12–98% RH) but gram negative bacilli may be completely destroyed at an RH of <40% [Vlodavets, 1962]. The spores of four *Bacillus* species stored for 2–6 years in air at an a_w of 0.20–0.50 and in a vacuum showed a marked decrease in viability [Marshall et al., 1963], and *Bacillus* species are absent or are not prevalent in antarctic soils. Although dormant spores of *Bacillus subtilus* are resistant to cold shock, they lose viability almost completely on germination [Sato and Takahashi, 1968], and this property may be another factor in the low abundance or absence of *Bacillus* species in most antarctic soils.

More micrococci, which form colored colonies on agar plates, are found in the surface of antarctic soils and in the aerial plankton than in the subsurface soils. Their predominance and survival may be related to greater resistance to high ultraviolet irradiation and low humidities, and the same factors must apply to the diphtheroids, which constitute the predominant group of bacteria in antarctic soils. The soil coryneform bacteria may constitute more than half the total bacterial colonies determined by plate counts in a variety of soils including those near the Arctic Circle [Clark, 1967]. The diphtheroids or nocardoids are aerobic, are gram positive to gram variable pleomorphic cocci rods, and are not easily taxonomically distinguishable [Barksdale, 1970; Davis and Newton, 1969; Jones and Sneath, 1970; Sukapure et al., 1970]. An important factor in their survival in antarctic soils is undoubtedly the presence of 'antidesiccant' and 'antifreeze'; i.e., a significant number of strains of *Corynebacterium*, *Nocardia*, and *Mycobacterium*, as well as three species of *Arthrobacter* [Shaw, 1970], contain lipidic substances in their cells [Sukapure et al., 1970]. Lipids are also formed by certain yeasts, including *Hansenula*, *Rhodotorula*, *Candida*, and other species [Imshenetskii, 1969], and, since some antarctic yeasts are obligate psychrophils, this property may aid in their survival. Whether antarctic soil bacteria and other microorganisms can conduct their activities,

TABLE 15. Wind Velocity in Miles per Hour for Victoria Valley, December 28, 1966, to January 3, 1967

Station	Maximum	Average Maximum	Minimum	Average Minimum	Average Total
1 meter above surface	20	16.6	0.0	3.3	9.5
30 cm above surface	15	12.4	0.0	1.5	6.8

grow, and reproduce at low RH, under ultraviolet irradiation, and at other concurrent phenomena of low temperature and high salinity has not been investigated.

Wind velocity. Wind velocity (Figure 32) showed diurnal maximums and minimums that occurred concurrently with other variations in the environment. Calm periods occurred most frequently in early morning, and intensity increased toward noon and afternoon. There was some decrease in the intensity of wind velocity with proximity to the soil surface. A frost crack polygon pressure ridge rose 30 cm above the location of the lower wind scope and resulted in some eddy currents and a reduction in wind velocity. A few measurements were made at the soil–air interface, and these measurements did not usually exceed 5 mph; however, the characteristics of the surface, i.e., desert pavement and lack of fines, indicated that the immediate surface wind velocity has previously reached high velocities. Regardless of the wind velocity the soil surface at the soil–air interface was desiccated, as we determined by macroscopic observation, 'feel' of materials, and measurement of moisture content. For the period of our report the wind velocities were as shown in Table 15.

During our study there were some periods of obviously higher wind velocities, including occasional peak gusts of ≤35 mph, but these gusts were not recorded unless they occurred during every third hour, when measurements were taken. The total miles of wind were not measured, but wind gradient curves would have been evident, as they have been at other microclimatic stations where wind profiles have been measured [Whitman and Wolters, 1967].

The wind was predominantly from the east and was from other directions <10% of the time (usually from the west with lower humidity). This finding is consistent with observations made earlier by geologists at Lake Vida [Bull, 1966], where the wind was noted nearly always to blow from the east, although measurements were taken only at 0800 and 2000 hours. Westerly winds occurred only before 2100 hours on January 2, before 0300 hours on January 3, and again at 0600 hours on January 3. In the Antarctic, as in many other regions, there is an increase in wind speed with height, and the average monthly wind speeds at the surface are reportedly higher in winter than in summer [Sergeyeva and Sokolov, 1966]; however, the latter characteristic needs to be substantiated for the dry valleys.

Annual wind speeds for McMurdo Station based on 6 years of measurements are about 12 knots (14 mph) [Wilson, 1968], and there is not too much monthly variation. The wind speed frequency versus direction for McMurdo Station during January is also high for east winds and next highest for east by northeast winds [Wilson, 1968]; the same conditions would apply for our site at Victoria valley.

There are good correlations between our measurements for high wind velocity, high thermal influx and outflux, high light intensity, absence of cloud cover, high evaporation rate, and high relative humidities of easterly winds, as a comparison of other figures for environmental measurements shows. Although periods of high wind would not be the optimum time for microbial activities, other environmental conditions would be favorable. Wind and degree of slope are more important than temperature in determining the ablation of coastal glaciers [Loewe, 1967a], and the same relationship may also be true in determining the desiccation of exposed soil surfaces. Wind in combination with a topographic feature could either promote or inhibit the growth of lichens at Cape Royds, Cape Crozier, the Kar plateau, or Miers valley [Schofield and Rudolph, 1969]. In temperate arid regions, light winds and strong surface heating prevail, promote convection over advection, and thereby limit the areal extent of possible microclimatic modification [Fritschen and Nixon, 1967]. A niche that is protected from the wind but at which other favorable environmental conditions would be allowed to prevail would have an advantage for the growth and the survival of microorganisms in Victoria valley as well as in other dry valleys.

Barometric pressure. The only environmental measurement performed during our study that did not show rather distinct diurnal changes was that of the barometric pressure. Average diurnal variations are not to be expected at high latitudes [Ichiye, 1967]. However, significant changes in actual barometric pressure occurred after two successive 3-day intervals (Figure 33). For the period of our report the highest barometric pressure reading was 28.53 in. Hg (966.14

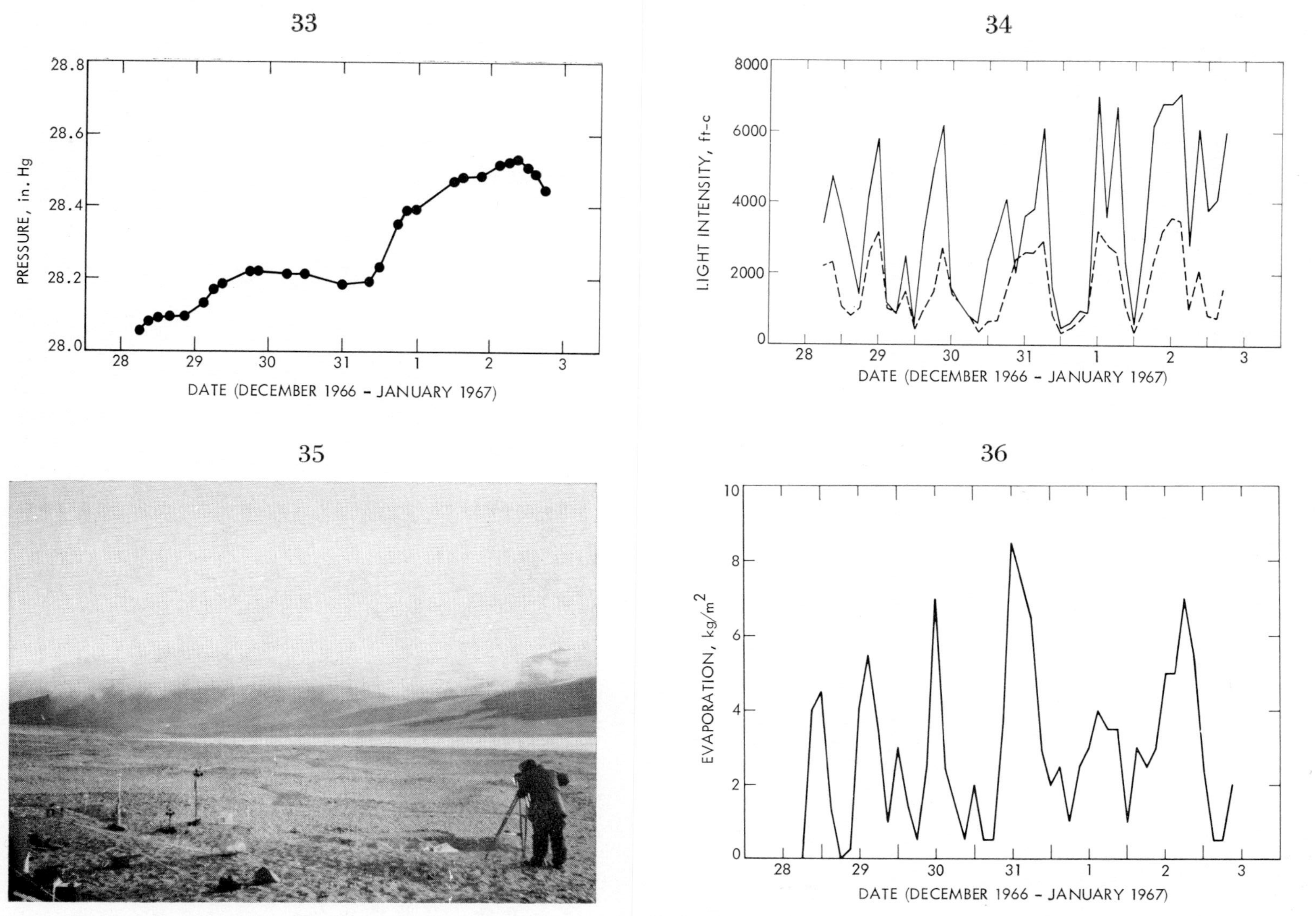

Figs. 33–36. Victoria valley, Antarctica. Fig. 33, barometric pressure (uncompensated) showing two rises in air pressure. Fig. 34, light intensity (the solid line represents the sun, and the broken line the soil surface). Fig. 35, low westward-moving cloud cover and snow flurries over the Olympus Range. Fig. 36, evaporation rate.

TABLE 16. Light Intensity in Foot Candles for Victoria Valley, December 28, 1966, to January 3, 1967

Station	Maximum	Average Maximum	Minimum	Average Minimum	Average Total
Perpendicular to sun	7100	6440	480	830	3370
Soil surface	3600	3120	340	500	1610

mb), the average high reading was 28.31 in. Hg (958.69 mb), the lowest reading was 28.06 in. Hg (950.22 mb), and the average low reading was 28.24 in. Hg (956.32 mb). The total average reading was 28.30 in. Hg (948.19 mb). These values are slightly lower than those reported for McMurdo Station. Pressure can be expected to increase toward the end of summer, and pressure variations can be expected to proceed with a component moving from east to west (as well as from north to south in the vicinity of the Ross Sea) and with lag periods of ≤ 60 hours for pressure variations moving opposite to the east–west variations [Loewe, 1967b].

Barometric pressure is only one of five distinct factors influencing the aeration of the soil [Bouyoucos and McCool, 1924]. Other factors include partial pressures and diffusion of gases, temperature fluxes, meteoric precipitation or other moisture fluxes, and wind. Temperature changes can have a very large influence on soil aeration, but a comparison of daily variations in air barometric pressure with latitude shows a significant damping of the harmonics of the sinesoidal pressure curve with increases in latitude [Bouyoucos and McCool, 1924]. After a change in barometric pressure a lag phase would be expected before a change in aeration would occur with soil depth in the antarctic dry valleys, and the rate of change Δt would decrease with the depth to the permafrost, an increase in the hardness of the soil, or an increase in the moisture content of the soil. Measurements of gas concentrations in McKelvey valley, Taylor valley, and Conrow Glacier (in Asgard Range) did not show any significant changes in the gas concentrations of O_2, N_2, CO_2, and A with soil depth or between the soil and the air above the soil [Cameron and Conrow, 1969b], although significant changes in the ratios of O_2 to N_2 have been found with increases in latitude along the Transantarctic Mountains from 77°23′S to 87°22′S (R. E. Cameron, unreported results, 1966–1970).

Light intensity. The light intensity at the site is indicated in Figure 34, which shows diurnal variations comparable to those of the other environmental phenomena, including solar radiation, relative humidity, and evaporation rate. Two of the most important factors causing variations in the light received are the humidity of the atmosphere and the amount of cloudy weather [Shirley, 1935], and the results of these phenomena are shown as irregularities in what would otherwise be a regular harmonic pattern. The light intensities are shown in Table 16.

Since full sunlight with zenith sun produces a luminance of 10,000–14,000 ft-c on a horizontal surface [Platt and Griffiths, 1964], it is evident that during midsummer a significant intensity of light is received at the soil surface for photosynthesis. The actual light intensity required for the survival of plants is actually quite low and is only 1–5% of the total [Shirley, 1935].

Although it is not readily apparent from Table 16, there were times when the light intensity values measured perpendicular to the sun were the same as those measured at the soil surface, primarily because of the nature and the extent of the cloud cover. There were also a few occasions when because of complete cloud cover the diffusion of light rays resulted in values slightly higher for the soil surface than for points perpendicular to the direction of the hidden sun. This cloudy period would also be the time of greater reception of ultraviolet and blue wave lengths than of infrared and red wave lengths [Shirley, 1935]. Regardless of cloud cover the lowest light intensities were frequently obtained after 2400 hours, when the sun was at its lowest angle, and between 0025 and 0050 hours, when the sun was also behind Mt. Orestes in the Olympus Range south of our camp.

Light was not necessary for the growth of most of the soil microorganisms in our sites in Victoria valley, except for a very low abundance of microscopic blue-green algae in the few surface centimeters of site 537-539a. The bacteria were essentially heterotrophes. High light intensities are in fact detrimental to the growth of antarctic algae, lichens, and mosses, and better growth and photosynthesis are obtained in dim light [Cameron, 1972; Fogg and Horne, 1970; Gannutz, 1967; Rastorfer, 1970]. The growth of algae is usually light saturated at 4000 lux (~370 ft-c) [Fogg and Horne, 1970].

Where plants do not receive direct sunlight, e.g., on steep poleward slopes, sky light is relied on and is only about 17% as intense as full direct sunlight on a level surface [Daubenmire, 1959]. High light intensities promote rapid transpiration, which would be detrimental in the cold arid antarctic environment;

soil moisture conditions are also more favorable in partial shade rather than in full sunlight [Daubenmire, 1959], provided, of course, temperature conditions are also favorable. However, as was shown for antarctic mosses, an adjustment to lower light compensation and light saturation occurs when the temperature decreases [Rastorfer, 1970]. Algae may grow heterotrophically in dim light if glucose is available [Van Baalen et al., 1971], and growth may be doubled at temperatures between 10° and 28°C [Fogg and Belcher, 1961]. The same phenomenon can be expected for antarctic algae [Fogg and Horne, 1970]. High light intensities during midsummer are modified at the soil surface by the low angle of receivable incident radiation, the minute irregularities in the soil surface (which provide shading) (Figures 9, 10, and 24), and the protection by translucent and transparent surface materials, such as white quartz, marble, and chalcedony, which constitute a 'microgreenhouse' [Durrell, 1956] for adherent subsurface algae and associated microorganisms. The effect of the spectral quality of light on antarctic plants is not known, but it could be expected to differ from one species to another [Daubenmire, 1959].

The light in polar regions is richer in red than normal daylight [Shirley, 1945]; this property could be important for plants. Red algae generally have lower light requirements than other algae, and the noticeable occurrence of red pigments (phycoerythrin) in some antarctic blue-green algae may be partially due to the reduction of light intensity in the fall, although freeze–thaw, salt, pH–Eh changes, desiccation, aging, and other phenomena cannot be ruled out. However, high light intensities and correspondingly large amounts of usually red carotenoid pigments, such as those in snow algae, may be expected to aid in energy conversion. Red cell pigments have been induced in snow algae by increasing the light intensity [Curl and Sutton, 1967].

The effect of light on algal nitrogen assimilation, fixation, and reduction is not well understood, although both NO_3^- and NO_2^- reduction occur at high light intensities [Fogg, 1962], and there may be a correlation between decreasing light intensities in the fall and NO_3^- build-up from algal activity. Low temperatures and high light intensities have been found to decrease nitrogen assimilation in blue-green algae, and the optimum conditions for photosynthesis apparently are also optimal for nitrogen fixation [Fogg, 1962]. For plants in general, unless enough light is received for rapid photosynthesis, nitrogen is accumulated but is not used effectively. Because NO_3^- is the end product of the oxidation of nitrogen compounds for photosynthetic organisms [Olson, 1970], the algae undoubtedly play an important role in soil nitrogen accumulation in the Antarctic, and this accumulation is largely dependent on light as well as on temperature and moisture.

Clouds and precipitation. Cloud observations were made concurrently with other environmental measurements. The extent and the duration of cloud cover were closely correlated with light intensity, heat fluxes, and other associative measurements. When clouds were present, they were varieties of cumulus, stratus, and occasionally cirrus formations, and they had a high frequency distribution of altostratus, altocumulus, cirrostratus, and stratocumulus. However, nearly all types of clouds were observed at one time or another during our study, and their extent ranged from nil or slight haze to complete storm cloud cover, some freshly fallen snow being deposited on the walls of the valley, usually above the 450- to 500-meter level, as in Wright valley [Ragotzkie and Likens, 1964]. Most of the low cloud cover was stratus, was variable in the direction of movement (commonly between east and south, as was observed previously at Lake Vashka [Bull, 1966]), and moved throughout the length of the valley. The ice-free dry valley areas are usually less cloudy than the coastal areas of McMurdo Sound. Cloud cover at McMurdo Station is greatest during late summer and autumn, altostratus and altocumulus being the most frequent types, and it is almost always above 600 meters [Wilson, 1968]. The mean cloudiness of the ice-free areas, such as Victoria valley, is probably about 10% less than that in the vicinity of McMurdo Station and Scott Base [Bull, 1966].

There were no snow showers over the lake or at our site during the study, although it has been noted that snowfall on the dry valley floors is greater to the east [Bull, 1966]. We also observed more storms and snowfall in the east (Figure 35) as well as in McKelvey and Barwick valleys, but they did not reach Victoria valley.

The precipitation at McMurdo Station varies between 8 and 22 cm, the average annual water equivalent being 15 cm [Rubin and Weyant, 1965], and it is probably lower for the dry valleys. Snowfall appears to be slightly heavier during the summer in southern Victoria Land [Gunn and Warren, 1962]. At present, snowfall in the Antarctic is very light, the glaciers are starving, and there will not be more snowfall until there is a higher temperature regime [Dort, 1970].

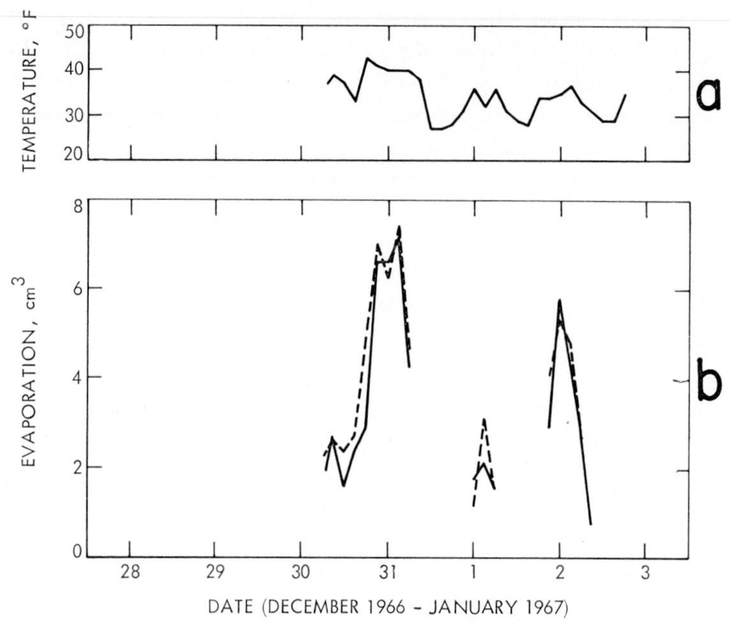

Fig. 37. Victoria valley, Antarctica. (*a*) Temperature at station 1 evaporimeter. (*b*) Evaporation rate (corrected Piché values; the solid line represents station 1, and the broken line station 2).

The peak intensities of soil surface temperature (Figure 27) were broken by clouds and also resulted in reduced maximums of soil subsurface temperatures. Any microorganism dependent on an extended diurnal optimum of temperatures above freezing would consequently be caught by rapid and sometimes precipitous changes in temperature and related phenomena detrimental not only to its growth and reproductive phases but also to its very survival. However, antarctic algae, lichens, and mosses can respire and photosynthesize at temperatures slightly below freezing [Becker, 1970; Gannutz, 1970; Rastorfer, 1970]. None of these plants was present at our site in Victoria valley, and only the algae were macroscopically evident on the shores of Lake Vida.

Although the nature, the extent, and the distribution of clouds were not important microclimatic phenomena per se, they nevertheless would have important effects on the microbiota, especially in terms of interrupted increments of not only heat but also the quality and the intensity of light. An increase in cloudiness would further increase the proportion of red and infrared rays [Shirley, 1945], which are already in a higher proportion in the Antarctic because of latitude as well as altitude.

Evaporation rate. The evaporation rate was determined by a continuous open pan recording evaporimeter and a Piché (wetted paper type) atmometer. Evaporation rates were plotted every third hour (Figures 36 and 37) for comparison with other environmental data plotted at similar time intervals. Determining the evaporation rate by Piché atmometers was not feasible during periods of freeze, as is indicated in the temperature curve in Figure 37; the fact that the atmometers broke from the pressure of expanding ice and had to be replaced is one of the disadvantages of these instruments [Platt and Griffiths, 1964]. The average evaporation rates are given in Table 17. The Piché values given in Table 17 are only for periods when temperatures were above freezing. Although measurements by two different methods are difficult to correlate, a comparison of Figures 36 and 37 shows a similarity in peak height intensity for periods of

TABLE 17. Evaporation Rate* for Victoria Valley, December 28, 1966, to January 3, 1967

Station	Maximum	Average Maximum	Minimum	Average Minimum	Average Total
Piché atmometer	7.3	4.3	1.2	1.5	3.6
Recording evaporimeter	8.5	5.4	0	0.5	2.9

* The Piché atmometer values are in cubic centimeters of water per day, including a correction factor of $\frac{1}{2}\pi(R^2 - r^2)$, where R and r are the radii of the paper and the tube, respectively. The recording evaporimeter values are in kilograms per square meter of water per day.

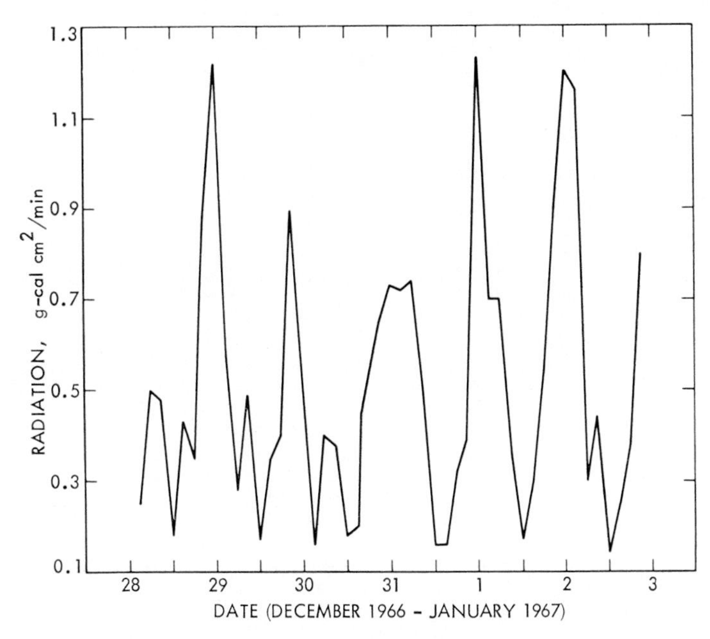

Fig. 38. Total solar, sky, and reflected radiation at normal incidence, Victoria valley, Antarctica. (The solid line represents 100 cm above the surface, and the broken line 35 cm above the surface.)

maximum radiation during midday on December 31, January 1, and January 2. However, the Piché readings show an interruption in peak height intensity between December 30 and January 1 not indicated by the open pan instrument, and they also show more correlation with the total radiation determined by the Bellani pyranometers during that same time period (Figure 38). The Piché atmometer tubes were not shielded or stationary and were therefore more influenced by surface wind and outflows of thermal radiation from the soil than the open pan evaporimeter. However, both measurements gave diurnal curves similar to those obtained for other microclimatic environmental measurements.

Very little evaporation data are available for the Antarctic; mean estimates of annual evaporation for longitudinal belts are 20 cm between 60° and 70°, 4 cm between 70° and 80°, and 1 cm between 80° and 90°S [Loewe, 1957]. Evaporation may approach 12 cm/yr in the dry valleys, as in the Arctic [Swinbank, 1967], more than 4–8 cm on the coast and Antarctic Peninsula, as measured for the summer at Point Barrow [Mather and Thornwaite, 1958], but less than that of hot arid areas such as Death valley, which has the highest evaporation rate in the United States (approximately 17 cm/yr) [Greene and Sellers, 1964]. Even estimates of evaporation from lake and ice surfaces are rare in the dry valley areas, but they are reportedly between 0.19 cm/day for Lake Vanda during midsummer [Ragotzkie and Likens, 1964] and 0.05 cm/day for newly fallen snow on Ferrar Glacier 50 km south of Wright valley [Gunn and Warren, 1962].

In the dry valleys of East Antarctica, evaporation was measured for a firn basin having temperatures similar to that of the dry valley; the daily evaporation amounted to 2–3 mm and thus would require about 250–360 g-cal of solar radiation and one-third this amount of net radiation [Rusin, 1964]. Our data showed evaporation values not nearly as high as those for the dry valley, the daily average being 2.9 kg/m^2.

It is obvious that evaporation exceeds precipitation and that the net precipitation/evaporation balance is negative; otherwise the dry valleys would not be ice free [Wilson, 1970]. If snow falls on the surface, it does not last more than a day or two at most, as Ragotzkie and Likens [1964] also observed; and it rapidly disappears by sublimation after the cloud cover is broken and the radiation balance becomes more positive. These phenomena were observed in McKelvey valley prior to our study in Victoria valley and subsequently in Conrow Glacier and Wheeler valleys after storms that deposited approximately 15 cm of snow. During the final sublimation of the snow, there was very little subsequent melt into the surface of the soil. This melt was to a depth of only approxi-

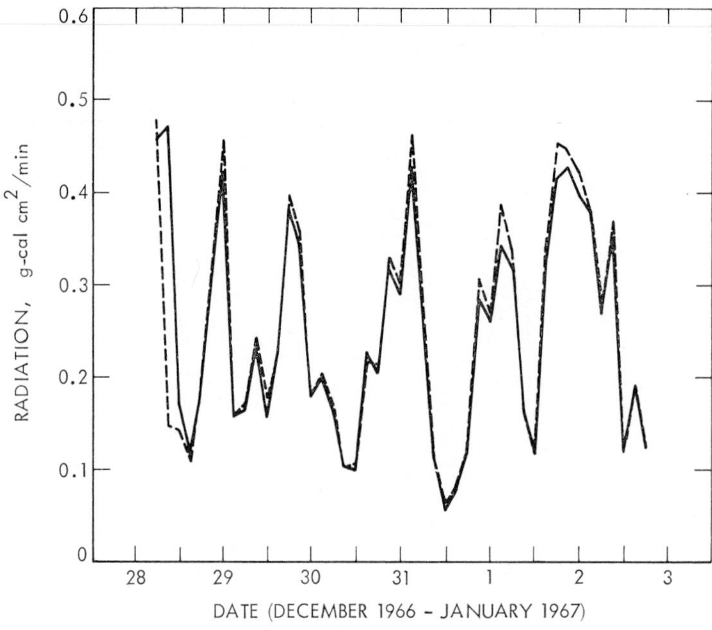

Fig. 39. Solar radiation at normal incidence, Victoria valley, Antarctica.

mately 1 cm, and subsequent thermal influx into the soil surface soon dried out the soil. This short-lived condition of available soil moisture, which is also subject to intermittent and diurnal freeze–thaw, is obviously not favorable to the survival of soil microorganisms, especially possibly germinating spores of *Bacillus* species [Sato and Takahashi, 1968]. The over-all conditions in the dry valleys indicate a predominance of evaporation over condensation or precipitation for most of the year, there being a net annual transfer of water vapor into the air [Janetschek, 1967]. Evaporation in hot deserts may exceed precipitation by 7–50 times [Smith, 1966], and these values are probably approached in the dry valleys. The highest rates of evaporation would of course be expected during the summer, concurrent with the strong positive radiation balance. There may or may not be any obvious relationship between the direct measurement of moisture losses from evaporimeters and the loss of moisture from the upper layers of soil [Sutton, 1953]; it is also difficult to correlate moisture loss with other environmental data, including temperature. In fact, in bare soils in temperate regions there may be no direct relationship between soil temperature and net radiation [Hanks et al., 1961]. Net radiation over bare soil may even decrease, and evaporation will continue at the potential rate [van Bavel, 1967]. Nevertheless, our data (Figures 27 and 38–40) indicate concurrent changes in radiation and surface temperature, and, without an in situ measurement of evaporation, they roughly indicate the potential evaporation rate and intensity. Factors influencing evaporation include saturation vapor pressure, temperatures of both the air and the evaporating surface, wind speed, precipitation, radiation, barometric pressure, soil moisture profile, exposure, nature and extent of the evaporating surface, and other variables [Swinbank, 1967]. The most important factors affecting evaporation are solar radiation, air temperature, humidity, and wind, but solar radiation is more important than air temperature in promoting evaporation [Hordon, 1967]. Evaporation is not easy to measure; nor are the results readily correlated with other environmental conditions and measurements. The effects of the rate, the intensity, and the intermittent and diurnal variations of evaporation on the abundance, the distribution, and the kinds of activities of microorganisms are generally unknown. Rapid evaporation, which occurs in arid areas with high vapor pressure gradients between the air and the soil surface, would be detrimental to microorganisms needing available moisture for their activities. Although a high and rapid rate of evaporation from a moist soil may not be detrimental while moisture is available, the approach toward a relatively static equilibrium condition of dry soil is not favorable, even though the evaporation rate from the soil is decreased. When the soil is dry, evaporation, as well as microorganisms, is more subject to the heat flux of the soil [de Vries, 1958]. A dry and frozen soil has a lower specific

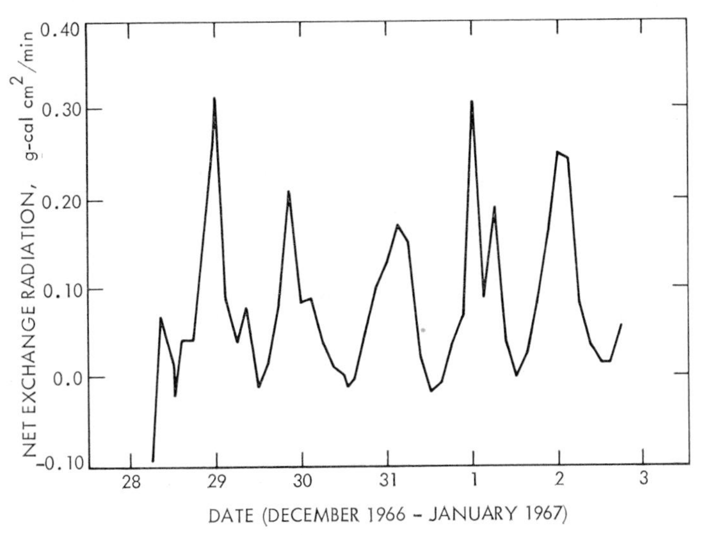

Fig. 40. Net thermal exchange radiation, Victoria valley, Antarctica.

heat and a lower thermal conductivity than a wet and nonfrozen soil, and a sandy soil has a much lower specific heat and thermal conductivity than a finer textured soil [Sutton, 1953]. Evaporation from the surface of saturated sand is slightly greater than that from a free water surface, and evaporation from a loamy soil is about 90% of that from a free water surface [Berry et al., 1945]. In a study of a hot desert area (i.e., Arizona during midsummer), both the amount and the distribution of evaporation over bare wet soil were similar to those values over open water [van Bavel et al., 1963]. The depth to the water table (or the 'ice table') also determines the amount of evaporation from a bare soil [Berry et al., 1945]. Evaporation from the surface is one of the principal components of the water and heat balance and, in association with other environmental factors, imposes important limiting conditions on the microbiota in the barren and generally dry, cold, shallow, sandy soils of the Antarctic. Microenvironmental factors that would reduce or modify the rate and intensity of evaporation in the microbial habitat while they maintained favorable conditions of soil moisture and temperature would be obvious advantages for microorganisms in the Antarctic as well as in hot deserts.

Radiation. Radiation measurements were performed with a bimetallic Eppley (or Michelson) recording pyrheliometer (pyranometer) for solar radiation on 1 cm^2 of horizontal surface (Figure 39), with Bellani spherical distillation pyranometers for total or diffuse (global) sun, sky, and reflected radiation on 1 cm^2 of spherical surface (Figure 38), and with a Gier and Dunkle ventilated net thermal exchange radiometer for the net flux of downward and upward total (solar, terrestrial surface, and atmospheric) radiation through a horizontal surface (Figure 38). Both the recording and the spherical pyranometers measure incident short-wave radiation (0.3^{-3} μ) because the radiation is received through a glass-covered surface. The radiation is less with the Bellani instrument (Figure 38) than with the Eppley instrument (Figure 39) because 'half of the spherical surface is not exposed to the sun and sky radiation and receives only the weak reflected radiation (except in the case of fresh fallen snow)' (F. Prohaska, personal communication, 1968). The net thermal exchange radiometer measures the total radiation budget in both short and long wave lengths (3–50 μ), but, because the 'net long wave radiation in polar zones is rather small, it is reasonable that the Bellani values are more similar to the net radiation, than the global radiation' (F. Prohaska, personal communication, 1968). A comparison of Figures 38 and 40 bears out this statement, and there are only small differences in measured values, which were generally slightly higher for the Bellani measurements (Figure 38). Because the Bellani instrument 'reacts not only on the sun and sky, but also on the reflected radiation, [it] is the most appropriate instrument for biological research on the total short wave radiation input on a body above the earth surface' (F. Prohaska, personal communication, 1968). There is some question as to the accuracy of all the instruments in measuring local radiation effects such as reflection from valley walls (W. D. Sellers, personal communication, 1968), and there is some concern about the 'mere chance caused

TABLE 18. Radiation Values for Victoria Valley, December 28, 1966, to January 3, 1967

Type of Radiation Measured	Maximum	Average Maximum	Minimum	Average Minimum	Average Total
Solar	1.23	1.00	0.14	0.17	0.50
Total*	0.43	0.40	0.05	0.10	0.26
Net	0.32	0.25	−0.10	−0.02	0.08

All values are in gram calorie square centimeters per minute.
* Values are averages of two measurements.

perhaps by very special local radiation conditions' resulting in similar readings for the Bellani and net thermal exchange radiometers (E. Flach, personal communication, 1968). Of the three instruments used for radiation measurements, the Bellani pyranometer is considered to have the most reasonable degree of accuracy [Hariharan, 1961]. All the radiometers measured the exchange of energy between the receiving surface and its immediate surroundings within the limitations, the orientation and setting, and the design of the instruments. Additional details on the operation, general use, accuracy, sensitivity, response time, advantages, and disadvantages of various radiation instruments have been presented by the World Meteorological Organization [1965]. The radiation values obtained with the three kinds of instruments during our study are summarized in Table 18.

The figures in Table 18 indicate that, during periods of maximum receivable radiation, the average maximum solar radiation was 1.00 g-cal cm^2/min and that the average total was 0.50 g-cal cm^2/min; however, these values may be as much as 20% in error on the basis of calculated values taking into account latitude and absorption and scattering of radiation by the atmosphere (W. D. Sellers, personal communication, 1968).

As our data indicate (Figure 39), the minimum solar radiation was 0.14 g-cal cm^2/min; thus our readings could be in error by approximately that amount. The extreme antarctic temperatures are known to produce readings 14% too high [Flowers and Helfert, 1966], but reflected sunlight from the valley walls, the lake, the glacial surfaces, and the relatively high albedo of light colored rocks around the lake and at the site could slightly elevate the measurements.

Note also that the intensity of direct solar radiation in the Antarctic is 7% higher than that at the same solar heights and latitudes in the Arctic because of the location of the earth at perigee in addition to the high atmospheric transparency and high albedo of the snow, which result in a magnitude of direct solar radiation approaching the solar constant (1.87 cal/cm^2/min) [Rusin, 1965]. Even in hot deserts the low humidities allow $\leqslant$90% of the solar insolation to penetrate the atmosphere and heat the ground [Smith, 1966]. The transparency coefficients for the Antarctic range between 0.77 and 0.90, the highest values being at higher altitudes and latitudes during spring and fall; the minimum values are obtained during the summer [Rusin, 1964]. The mean direct solar radiation under clear skies for various Soviet bases during the summer ranges between 1.58 and 1.71 g-cal cm^2/min [Rusin, 1964]. More than 300 g-cal of daily net gain have been measured during January for the rock and soil surface of Oazis Station in the Bunger Hills [Weyant, 1966].

In view of the relatively large time lag of the recording pyrheliometer [Platt and Griffiths, 1964], greater precision can be expected on dull or bright consistent days than on variable days [Holt and Youngberg, 1971]. The net thermal exchange radiometer, however, gave more rapid responses than the recording pyrheliometer, but measurements could not be made continuously because of a lack of generator power and the necessity for fuel conservation. The net thermal values were comparable to those obtained at Scott Base on Ross Island for December through January [Thompson and MacDonald, 1962] but were slightly lower, probably because of the higher albedo of the surface materials at site 537-539a. The values are one-half to one-third lower than the peak intensities obtained during midsummer in a hot desert (e.g., in Arizona) [van Bavel et al., 1963].

The average maximum values for the Bellani instruments were 0.40 g-cal cm^2/min, and the net thermal values were only 0.25 g-cal cm^2/min. For the period of measurement the average radiation obtained with the Bellani instruments was about 50% of the average solar radiation, and the net thermal exchange was only about 15%. Some negative net thermal exchange values were recorded when incident solar radiation was intercepted by Mt. Orestes, and more heat flowed out from the soil surface than was received. As Figure 27 indicates, there were definite periods of sub-zero temperatures. Rapid and immediate responses (within a few seconds) were obtained by the net thermal radiometer whenever solar radiation was intercepted by clouds as well as by the mountain, and the radiation flux could change drastically. When the sun was obscured by a mountain or was below the horizon, the amount of cloud cover was the major factor governing net radiation; cloud cover functions essentially as a black body, and the net outflow of radiation will be diminished during overcast conditions; or, if the temperature of the cloud base is higher than that of the ground, a net incoming balance can occur [Thompson and MacDonald, 1962].

The ozone and CO_2 concentrations must also be considered in transforming solar energy to heat [Rubin and Weyant, 1965]. Increases in both the CO_2 and the ozone concentrations can increase the amount of radiation at the surface [Conaway and van Bavel, 1967]. Atmospheric ozone over the Antarctic increases in November, concurrent with an influx of air from lower latitudes, an increase in pressure, the sinking motion of air, and an increase of ozone in the stratosphere [Rubin, 1970]. Carbon dioxide may also increase noticeably [Cameron and Conrow, 1969b] and is gradually increasing over the Antarctic [Rubin, 1970]. The carbon dioxide concentration was measured southwest of Lake Vida on January 17, 1971, and found to be 0.05% (M. Frech, personal communication, 1971). This value was higher than that usually reported for the Antarctic [Rubin, 1970] and was probably due to the growth of algal vegetation in the lake in addition to a higher concentration of CO_2 in easterly air blowing across the open and photosynthetically active Ross Sea beyond Wilson Piedmont Glacier [Cameron and Conrow, 1969b].

The over-all radiation balance of Victoria valley, as well as that of the other dry valleys in southern Victoria Land, is apparently positive [Bull, 1966]. The annual radiation figures obtained for Scott Base [Thompson and MacDonald, 1962] indicate an annual net balance of radiation in the dry valleys [Wilson, 1970], and there may be a gain of as much as 29,000 cal cm^2/yr in the dry valleys [Bull, 1966] and 26,700–37,600 cal cm^2/yr in the slightly warmer dry valleys of East Antarctica [Rusin, 1964].

The thermal regime of the soil and the heat transfer are complicated by the nature and the extent of the stony surface, which acts as a mulch to reduce evaporation [Rusin, 1964], the color and the albedo of soil and rocks, the soil moisture status, and the inhomogeneities of soil structure, including stone and salt inclusions. Any rapid change in an environmental parameter, especially a rapid and precipitous change in solar and net thermal radiation, can damage the soil microorganisms, especially those in the few surface millimeters. Absorption of both short-wave and long wave radiation occurs in a soil layer about 1 mm thick [Van Wijk and de Vries, 1963]. Although the biota are somewhat protected by soil and/or rocks, their small size in the dry valleys couples them to the physical and chemical environment, as the large convection coefficient indicates [Gates, 1968], and makes them highly vulnerable to environmental changes. As workers have indicated for extraterrestrial environments [Lederberg and Sagan, 1962] as well as for harsh terrestrial environments [Cameron, 1963; Cameron and Blank, 1966], favorable microenvironments are also highly desirable habitats for antarctic microorganisms. In addition, inherent protective mechanisms (such as lipids and mucilagenous membranes, which guard against adverse environmental conditions of high evaporation potential, freezing, and desiccation) and the ability to reproduce rapidly are obvious advantages for the survival and the growth of soil microorganisms in the cold and arid Antarctic.

SOIL HABITAT

The most extensive areas of cold desert soils on the antarctic continent are principally within the limits of a 4800-km^2 arid zone in southern Victoria Land [Gunn and Warren, 1962]. This area ranges in elevation from sea level to approximately 2000 meters. Not only are the extent and the distribution of soils limited by the ice-free area, but their development is also restricted by an unfavorable solclime of low temperatures and scarce available (liquid) water, there being a consequent slow rate of weathering and a low 'biotic pressure' that has been limited in turn by unfavorable environmental factors of the harsh antarctic climate. In general the soils of southern Victoria Land indicate their arid nature and are primarily weakly or poorly developed, poorly drained sandy materials of shallow depth that are underlain by permafrost, as a number of investigators have indicated [Berg and Black, 1966; Calkin, 1963, 1964; Claridge, 1965; Jensen, 1916; McCraw, 1960, 1967a, b; Tedrow and Ugolini, 1966; Ugolini, 1963, 1970; Ugolini and Bull, 1965; Ugolini and Perdue, 1968].

The formation and the development of soils in these areas approximate the chronologic sequence of glacial and geomorphologic events described previously [Ugolini and Bull, 1965]. They show differences not only because of the prior and extant hydrothermal–solclime regime but also because of the differences in time of exposure and rate of weathering, the influence of glacial types and deposition of various parent materials, the variations in degree of slope, the extent of exposure and drainage of a particular site, and the intensity of local past and present climatic features [Berg and Black, 1966; Cameron, 1971; Cameron and Conrow, 1969b; Campbell and Claridge, 1967, 1969; Claridge, 1965; Janetschek, 1967; McCraw, 1960, 1967a, b; Tedrow and Ugolini, 1966; Ugolini, 1963, 1970; Ugolini and Bull, 1965]. With few exceptions the soils are sandy or gravelly materials, although a small percentage of fines can be present, especially in older surface soils or near the ice hard permafrost boundary. The identifications and characteristics of the clays of Taylor and Barwick valleys have been given previously [Claridge, 1965]. The older dry valley soils are commonly overlain by a desert pavement showing varnish or polish and surface colors of brown or yellowish brown [Cameron, 1969b, 1971; Campbell and Claridge, 1969; Tedrow and Ugolini, 1966; Ugolini, 1970].

Until our present studies, soils had not been investigated in Victoria valley, although one profile had been studied near Lake Vashka in Barwick valley (indicated as Victoria valley by Claridge [1965]), and four profiles had been studied in McKelvey valley earlier in the season [Cameron, 1967]. Some of the soil characteristics indicated in this report have not been measured or described previously for antarctic soils, especially not in such detail for a given site.

Descriptive information is given for each area and site in Table 1. Only five sites were investigated thoroughly in Victoria valley, but these sites by no means include all the representative areas in the valleys remaining to be investigated (e.g., adjacent lake shores and beaches, drainage channels, a variety of frost crack polygons, local salt-encrusted areas, slopes of overhanging valleys, and areas closer to glaciers). However, small quantities of soil samples were collected from many diverse areas within the valley for the VPI studies underway to determine the content and the kinds of bacteria, actinomycetes, fungi, and algae.

Some of the physical characteristics of the soil are given in Table 2. Except for site 540-543 the soils were exceptionally sandy regardless of depth. However, site 540-543 contained a significant proportion of silt-sized fines, probably related to the degree and the nature of the weathering of the predominant minerals in morainic materials of the Bull drift. In addition to quartz, labradorite and secondarily dolerite were the predominant minerals. It has been noted previously that the principal minerals in Victoria valley are feldspar (usually occurring as labradorite) and pyroxenes (augite being the most common) as well as amphibole, apatite, biotite, iron ore, and chlorite [Webb and McKelvey, 1959]. Labradorite An_{60} constitutes $\leq 60\%$ of the rock in the valley [Webb and McKelvey, 1959], but quartz is predominant in the dunes. Typical antarctic soil samples that we have collected show quartz and plagioclase as the major minerals, there being trace or minor constituents of diopside, olivine, and chlorite [Gardner, 1970]. The total mineralogic analysis performed for our soils does not show clay minerals unless they are present in large abundance, i.e., in a proportion of $>10\%$. More than 2–5% clays may be high for antarctic cold desert soils [Claridge, 1965]. However, a total analysis does help to indicate the degree of soil development and the stratification and source of parent materials [Barshad, 1964]. Prolonged weathering of minerals in the antarctic climate contributes to the formation of clay minerals, such as montmorillonite, under a high base status [Claridge, 1965]. For microorganisms it is important to consider that the types of clay minerals determine the extent of nitrification, and montmorillonite exhibits the highest degree of oxidation [Alexander, 1965]. The dominant clay mineral types of aridisols (desert soils) are mica, vermiculite, interstratified layer silicates, and chlorite [Jackson, 1964]. The lack of clays is important for antarctic bacteria.

Bulk density values are also generally indicative of coarse textured materials and are highest for the dune sands. In general, bulk density values decrease with depth, and porosity values increase inversely to the level of the hard ice-cemented permafrost. The degree of soil hardness, though related to texture, structure, bulk density, and porosity, was more influenced by the moisture–temperature regime and the depth of the soil. The location of the soil with regard to the relatively noncohesive immediate air–soil interface, the desert pavement, or the proximity to the hard ice-cemented permafrost was an important factor. Hardness exceeded 4.5 kg/cm^2 close to the permafrost and the subsurface freeze–thaw line; it was also influenced by the presence and the concentrations of salts. Only in McKelvey valley, in a few other dry valley areas, and farther south (e.g., Coalsack Bluff moraine), did we encounter increasing soil hardness instead of the usual distinct ice-cemented subsurface boundary layer similar to a frozen water table or a very dry soil with a much greater depth to the permafrost. A significant depth of dry soil has also been encountered by others in parts of Taylor (Nussbaum Riegel area) and Wright valleys [Ugolini, 1970]. The lack of a distinctive subsurface ice pan can be considered to be an indication of other areas or sites exposed for a longer period.

41

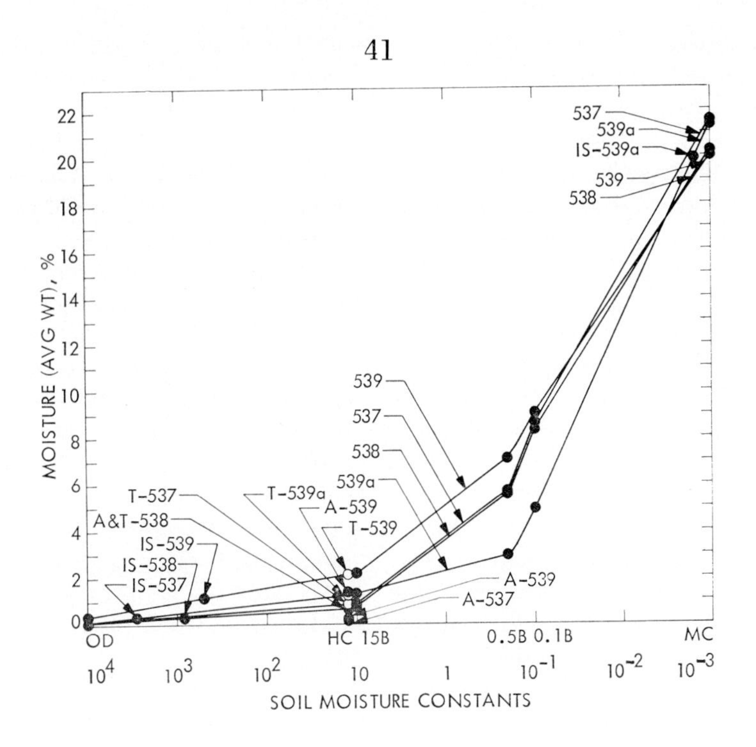

42

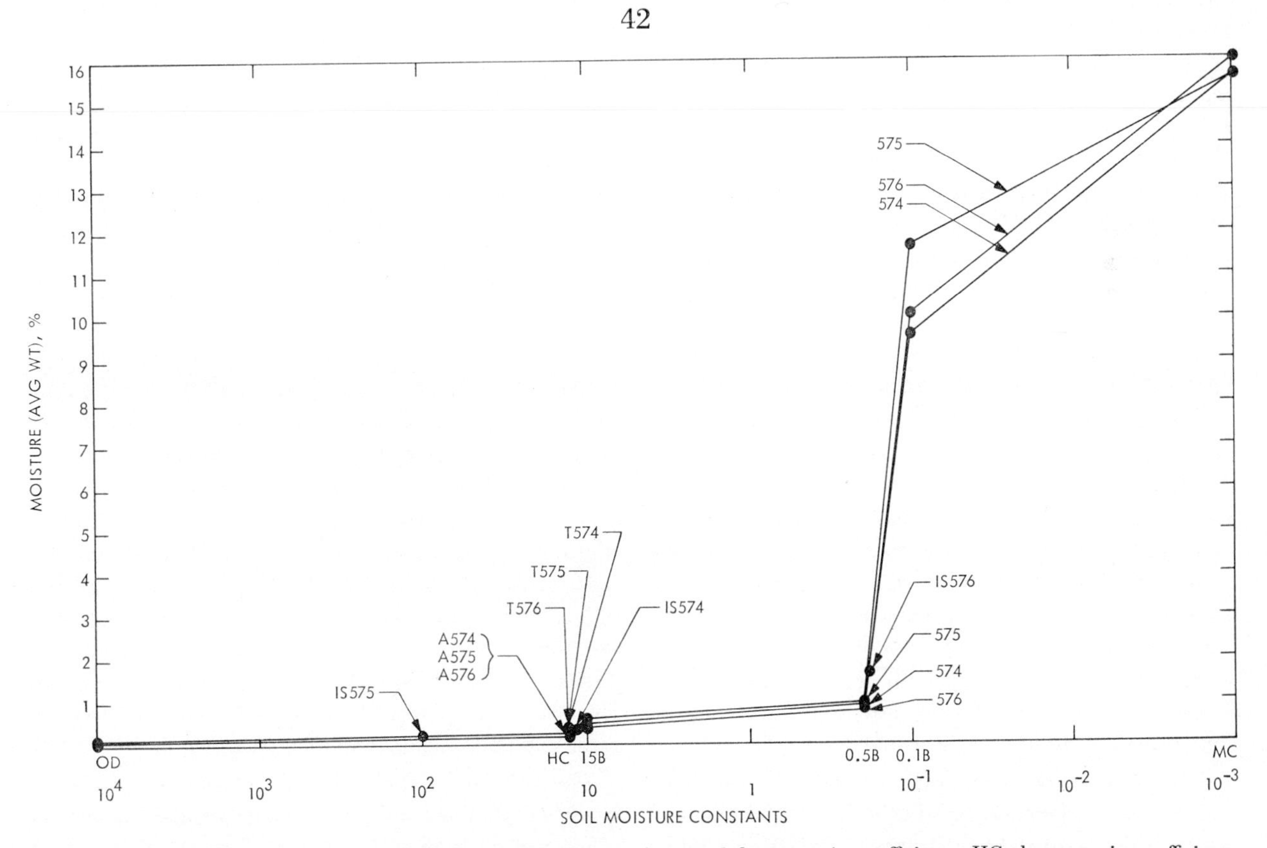

Figs. 41–42. Victoria valley, Antarctica, soil moisture curves: A, actual hygroscopic coefficient; HC, hygroscopic coefficient; IS, in situ moisture; OD, oven dry moisture; T, theoretical hygroscopic coefficient; MC, moisture capacity; 15B (for example), bar moisture. Fig. 41, samples 537, 538, 539, and 539a. Fig. 42, samples 574, 575, and 576.

Other physical characteristics of our soils are given in Table 2. The in situ moisture content of the soils tested showed a relatively narrow range except at the permafrost boundary layer (i.e., sample 539a). Moisture curves for three of the five sites are given in Figures 41–43. These figures illustrate three types of curves for 'ahumic' cold desert soils. The sandy wind-mixed alluvial materials at site 537-539a (Figure 41) showed little variation with depth except at the 0.5-bar moisture level. Dune site 574-576 (Figure 42) showed an abrupt displacement in the moisture stresses of finer textured materials and salts. The ability of the soil at this site to absorb moisture with increasing depth also indicates an active or dynamic condition of equilibrium rather than a static one, as not only the steepness of the curves but also the inclusion of the displacement of theoretical hygroscopic coefficient values shows. Changes in the slopes of curves for these soils are most closely correlated with increases in organic matter (organic carbon) (Table 7) plus silt and clay content and corresponding increases in cation exchange capacity and buffer capacity (Table 3). For arable temperate zone soils the available water capacity of a soil has been shown to be negatively correlated with the percentage of silt and organic matter [Slater and Williams, 1969]. Temperate arid soils have an optimum moisture content of 4% for fine sand and 10–12% for sandy loams [McGeorge, 1940]. For arable soils it has also been demonstrated that the distribution of pore sizes can be calculated directly from the moisture curves [Slater, 1969], and this correlation is approximately true for our antarctic samples, although it does not apply to ice hard permafrost soils, which would correspond to 'shrinking' temperate zone soils in this respect [Slater, 1969].

The large displacement of the actual hygroscopic coefficient from the theoretical hygroscopic coefficient for site 540-543 indicated the influence of hygroscopic salts on the reduction of moisture activity at this site. A similar phenomenon was found for a McKelvey valley site, and thus the moisture ranking of the soil as a microbial habitat in comparison with drier sites was altered [Cameron and Conrow, 1969a]. Salty temperate desert soils also show this phenomenon [Cameron, 1966], but highly leached sandy soils do not [Cameron, 1970].

In situ moisture values were plotted on each moisture curve for comparison with the range of 'available' water, which approaches 10^2 bars as an extreme lower limit. As we expected, soil moisture contents and corresponding relative humidities (Figure 31) increase with soil depth and with proximity to the hard ice-cemented permafrost, as nearly all our antarctic samples have shown [Cameron and Conrow, 1969a]. For some samples obtained within the ice table, such as those at Wheeler valley [Cameron et al., 1970c], the percentage of moisture could exceed that of soil (e.g., in areas of morainic deposits over ice or where there was an influx of subsurface moisture from glacial melt). The in situ moisture content of the Victoria valley soils showed a relatively narrow range except for sample 539a, which was obtained from the permafrost boundary layer. In general the driest antarctic cold desert soils have moisture contents approaching or similar to the driest desert soils from temperate or tropical desert regions [Cameron, 1969a; Cameron and Conrow, 1969b].

Soil colors and Munsell notations were determined under three conditions: (1) in situ and in the laboratory on processed (2) moist and (3) air dry samples with particles or aggregates of <2 mm. Neither the in situ method nor the laboratory methods are satisfactory because of the quality and the relatively large amount of red wave length illuminating the soil in the field and because of the alteration of the soil structure (including mottles) and the natural arrangement of particles through laboratory processing, except for loose cohesionless sands. Not only soil color but also Munsell notation for hue, chroma, and value showed shifts for a particular soil sample. Obvious shifts usually occurred in value and chroma but not in hue when the soils were moistened and compared with determinations of the same samples at in situ and air dry moisture.

Most soils in the dry valleys (e.g., those of Taylor valley) show little change in color for a given profile unless a zone of calcium carbonate or sulfate is encountered [McCraw, 1967b]. Soils developed on older moraines, such as those in Wright valley, are usually a strong brown, and soils developed on the youngest moraines are gray to pale yellow and lower in free iron oxides [Ugolini, 1970]. In this regard our site 537-539a was the youngest site and corresponded to the glacial chronology and the influence of stream deposits, and sites 540-543 and 626 were the oldest. However, the interpretation of soil color depends to some extent on the method of determination.

The soil colors and Munsell notation were similar to those obtained for soils from warmer desert regions and were an indication not only of depositional factors, mineral and inorganic composition, and salinity but also of degrees of weathering, erosion, and

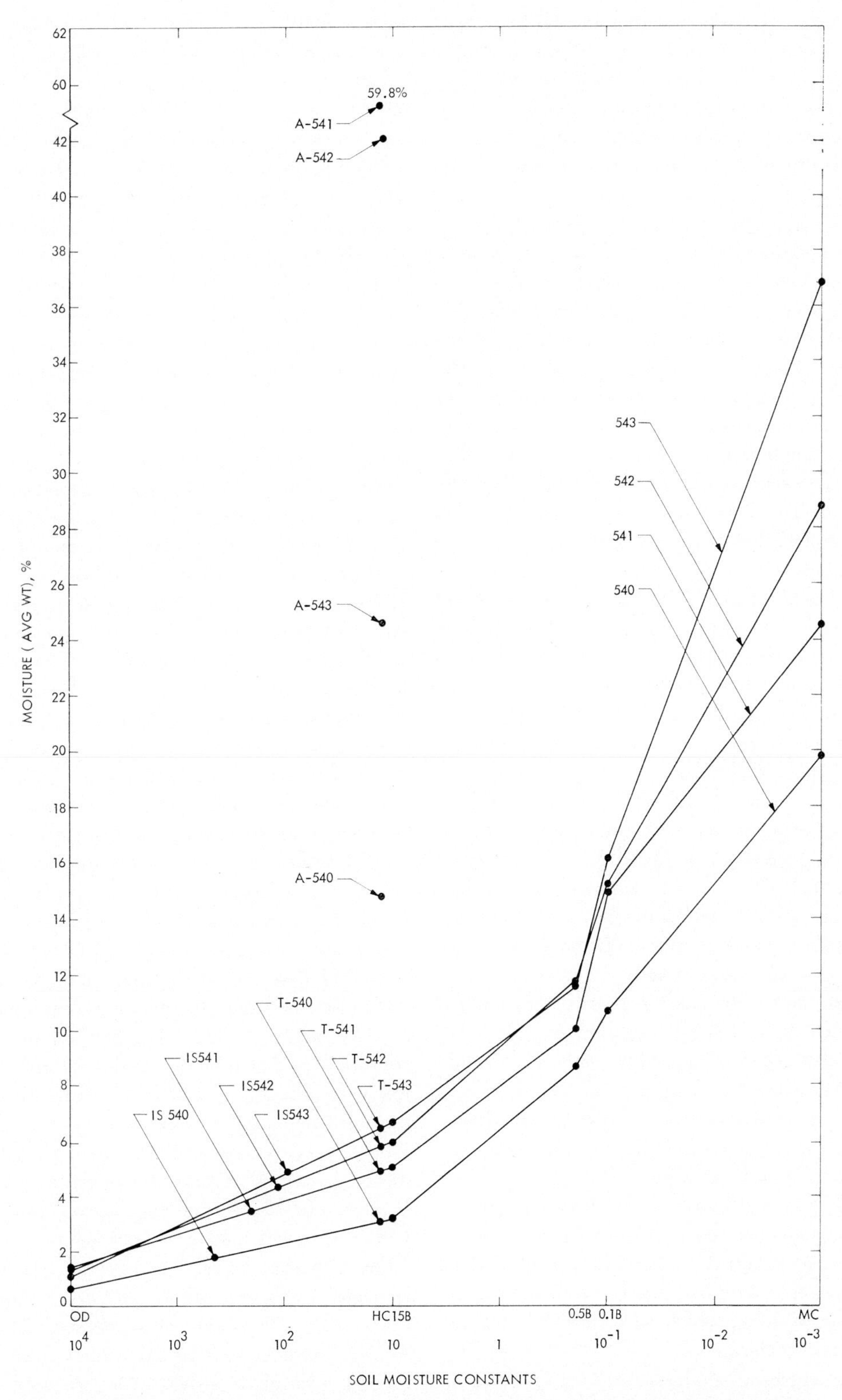

Fig. 43. Soil moisture curves for samples 540, 541, 542, and 543: A, actual hygroscopic coefficient; HC, hygroscopic coefficient; IS, in situ moisture; OD, oven dry moisture; T, theoretical hygroscopic coefficient; MC, moisture capacity; 15B (for example), bar moisture.

oxidation state. The darker colored surface soils were an obvious advantage for microorganisms with regard to warming, but this advantage could be offset by sublimation and evaporation of soil moisture and the resultant increase in aridity, unless there was a renewable moisture supply. The over-all solclime could be unfavorable.

Many of the same factors that determine soil color and Munsell notation also apply to reflectivity characteristics. Our reflectivity values were obtained for both the processed air dry and the moist samples. For sites 537-539a and 574-576 similar reflectivity values with depth of soil profile indicated that the soils had been subject to mixing. However, for site 540-543 the reflectivity value decreased with depth regardless of whether the sample was air dry or moist and thus indicated a more stable older soil. Only the reflectivity of exposed surface soils is important for heat fluxes of antarctic soil profiles. The reflectivity of overlying desert pavement and varnished rocks is usually more important because of the close spacing of pavement compared to the extent of exposed soil, which frequently has been removed by wind. Reflectivity values have been determined for a number of antarctic dry valley soils and are similar to the values and the air dry/moist ratios obtained for a number of California desert soils [Cameron, 1962].

Reflectivity, as well as color and Munsell notation, was measured for a number of surface rocks collected from Victoria, McKelvey, Balham, King, and David valleys. Although colors, hues, chromas, and values for many rocks were similar compared with those for soils, the determinations were usually lower for most dark reddish brown, dark grayish brown, dark gray, or black rocks whether the air was dry or moist. Light brownish gray, light reddish brown, or pale olive rocks gave reflectivities comparable to those of sandy soils of similar colors. However, white colored rocks gave reflectivity values of $\leqslant 75\%$ (dry) and $\leqslant 60\%$ (moist), and the lightest colored soils were yellowish brown and gave reflectivity values of 35% (dry) and 7% (moist). Although it was not determined directly, the albedo of light colored granitic surface rocks in Wright valley is assumed to be 20% [Bull, 1966], and this value approximates the average air dry reflectivity value of 23%, which was obtained for Victoria valley soils (Table 2).

Soil physicochemical characteristics are shown in Table 3. All the *p*H values of saturated pastes were $\geqslant 7.0$ (mostly 7.5–8.5), values typical of many antarctic soils [Cameron, 1971], but the extract values were usually <7.0. For highly leached, acidic, volcanic soils the opposite relationships have been found [Cameron, 1970]. The values of the Victoria valley samples generally indicated slightly to strongly alkaline or saline–alkaline soils and were similar to those of many other cold and hot desert soils [Cameron, 1969a, 1971]. In alkaline soils, *p*H is principally influenced by exchangeable Ca^{++} alone, by Ca^{++} and an excess of $Ca(HCO_3)_2$ together, or by exchangeable Na^+ [Seatz and Peterson, 1964], which are also important factors in antarctic cold desert soils.

The *E*h (oxidation reduction potential) values obtained for the Victoria valley samples indicate moderately oxidized and aerated soils [Andreason, 1952]. The values are similar to those for a high mountain range [Cameron, 1970]. Because of dilution factors the *E*h values of water extracts were lower than those for saturated pastes, followed the same pattern as that for *p*H values, and further substantiated the arid, unleached, and low compacted condition of the soils. However, the *E*h measured in suspensions of arable soils has been found to be far lower than that measured in situ, which is a condition of aeration and determined primarily by oxygen [Quispel, 1946], mineral oxidation state, and other soil factors. The oxygen content of dry valley soils (e.g., that of McKelvey valley soil) is essentially the same as that of the air above the soil surface. Some variation in *E*h with soil depth can be related not only to aeration but also to soil moisture content and soil horizon characteristics. The widest variations in *E*h of soils are caused by climate, activities of microorganisms, and influxes and outfluxes of the ecosystem [McKenzie et al., 1960]. For cold desert soils of the antarctic climatic zone the most important factors determining *E*h potential are the aeration and oxygen content, the oxidized condition of soil minerals and salts, the resultant *p*H, the variations in soil structure, and the increase in moisture and compaction with approach to the hard ice-cemented permafrost. Elevated *E*h values may indicate increased nitrate concentrations [Bailey and Beauchamp, 1971], and this assumption is consistent with our results, as a comparison of the values in Tables 3 and 6 shows. The resultant *r*H milieu, which is based on factors of *p*H and *E*h, showed a narrow range and also indicated moderately oxidized soils. All the extracts showed *r*H values lower than those for the saturated pastes. The *r*H values are lower than those for sandy soils of temperate deserts (values were determined at the compensated *E*h value of cell K equal to +244 mv) [Cameron, 1970] but intermediate between those of normal and inundated arable soils [Quispel, 1946]. The *r*H milieu would not be favorable for anaerobes.

Electrical conductivity values showed a range typical of cold deserts [Cameron, 1971]. The higher values obtained for the saturated pastes compared with those for the extracts again demonstrated the unleached nature of the soils and the presence of reserve salts that could be solubilized with time if liquid moisture were available. The electrical conductivity values were also typical of those of the temperate desert soils of California, but site 540-543 had high values similar to those encountered in the high nitrate soils of Chile [Cameron, 1969a]. Paste extract ratios, except that of dune site 574-576, were slightly lower than those of the temperate desert soils of California (approximately 8.5) and again indicated the arid unleached condition of antarctic cold desert soils.

Extract electrical conductivity values of $>250 \times 10^{-6}$ mhos/cm are high enough to be osmotically detrimental to organisms under the normal conditions of a favorable moisture–temperature regime [Richards, 1954], and such values have been reported for a number of antarctic soils [Cameron, 1971]. Values of $>2000 \times 10^{-6}$ mhos/cm, such as those for site 540-543, indicate a very high proportion of total soluble salts [Metson, 1961] and constitute a very high salinity hazard in temperate arid areas [Smith et al., 1964]. Such a site would provide an environment for halotolerant or osmophilic organisms [Hall, 1968; Hall and Benoit, 1969]. Most of the antarctic soils that have yielded culturable algae have electrical conductivity values of $<250 \times 10^{-6}$ mhos/cm [Cameron, 1972]. It has been noted previously that the osmotic pressure of antarctic soils is increased because of aridity and desiccation [Boyd, 1967]. Although the *p*H–*E*h milieu of antarctic soils is within the geothermal environment of gypsum–halite and calcite–dolomite evaporites [Baas Becking et al., 1960] and should not be limiting for osmophiles or halophiles, this advantage is offset by unfavorable temperature relationships. A temperature of $>25°$C is optimal for the growth of microorganisms in solutions of high salt concentrations [Ingram, 1957].

The weight loss of soils after ignition was an indication of soil moisture stress determined above oven dry values ($105° \pm 5°$C) and related to the percentage of fines, the loss of moisture of hydration from salts, and to a lesser degree the loss of organic matter. As we expected, the combustion values were highest for site 540-543, which had finer textured, salty soils that were slightly higher in organic matter content. Values were correspondingly lower for sand samples. The total weight loss on ignition of antarctic cold desert soils was similar to that of temperate desert soils of comparable texture, concentrations and kinds of salts, and organic matter content [Cameron, 1969b].

The cation exchange capacity of the Victoria valley soils was dominated by the monovalent and divalent ions listed in Table 6. Approximate relationships can be shown between the cation exchange capacity, the total extractable ions, and the electrical conductivity [Metson, 1961], between the percentages of clay and organic matter, and between the *p*H of the soil and the method of determination [Fiskell, 1970; Yuan et al., 1967; Pratt, 1961]. Care must be taken to remove excess salt in the determination, or the values will be too high [Rich, 1962]. Recent studies on a chronosequence of soils have shown that clays contribute very little to the cation exchange capacity of sandy soils and that organic matter contributes the most [Syers et al., 1970]. Consequently, in view of the preceding factors, a very low cation exchange capacity can be expected for coarse textured soils, including those of the Antarctic, which have only small amounts of fines and organic matter. As we expected, relatively unweathered antarctic cold desert soils have a very low cation exchange capacity [Cameron, 1971], although weathered shale, slate, and coal have a higher cation exchange capacity because the 'sand' is a stable aggregate of claylike substances [Tedrow, 1966a, b]. No measurable cation exchange capacity was obtained for the dune sands (site 754-756), and a higher, though still moderately low, cation exchange capacity was obtained only in subsurface rocks (site 540-543). The presence of Ca^{++} undoubtedly contributed to the weakly cohesive structure of the subsurface soils in this profile, but the presence of Mg^{++} and Na^{+} offset it. For comparison, cation exchange capacity values of 4.0–6.0 me/100 g have been measured for powdered pumice, kaolin, silicious sand, and limonite [Cameron, 1970], 20–21 me/100 g for 20 California soils [Kelley, 1948], and 3.5–23 me/100 g for 33 California desert soils (R. E. Cameron, unreported results, 1961–1963). Although ion exchange may be considered the most important of all the soil processes [Wiklander, 1964], it is not as important in antarctic cold desert soils unless a more favorable temperature–moisture regime is operative.

The buffer capacity of the Victoria valley soils is quite low. It correlates well with cation exchange capacity values, but it also includes the kinds and amounts of water-soluble salts. In general, antarctic soils do not show a high buffer capacity [Cameron, 1971]; this property is also indicated by the interrelated and dependent factors of a low clay content and a coarse texture, a low colloidal organic matter content, a *p*H at or above neutral, and the presence of available carbonates and phosphates. As we expected, older soils that have experienced more weathering (e.g., site 540-543) have a higher buffer capacity than other sites. Antarctic soils with the highest buffer capacity generally have not provided optimum conditions for microorganisms, despite a favorable moisture content [Cameron, 1971]. Nevertheless, soils with discernible algal organic matter can have a high buffer capacity, and they provide a more favorable and evolving environment for populations and communities of microorganisms. For temperate areas, 33 California desert soils had a wide range of buffer capacities (1.5–94 me/100 g), the average buffer capacity being 21.4 me/100 g (R. E. Cameron, unreported results, 1961–1963).

Tables 4 and 5 show the abundance of chemical elements in the Victoria valley. A spectrochemical analysis is a convenient means of obtaining an overview of the inorganic composition of the soil [Mitchell, 1957]. Although the total content of these elements is not nearly as important as that of water-soluble ions, the total abundance does indicate the mineralogy, degree of soil formation, and potential of the soil to supply nutrients after further weathering. In general the elemental composition of the Victoria valley soils is comparable to that of temperate desert soils, such as those of the Mojave, Colorado, Great Basin, and upper Sonoran deserts [Cameron, 1970]. As we expected, some reduction in the abundance of oxygen, silicon, iron, and aluminum could be shown, as for other desert soils, when there were increases in the abundance of other elements, such as calcium, magnesium, and sodium from marine sources. The usual sequence of abundance of chemical elements in typical soils is $O > Si > Al > Fe > C > Ca \geqslant K > Na \geqslant Mg$ [Vinogradov, 1959]. However, in desert soils, including that of the Antarctic, this sequence is $O > Si > Al \geqslant Fe \geqslant Ca > K > Na \geqslant Mg$ [Cameron, 1970]. For Victoria valley this sequence is $O > Si > Al > Ca > Fe \geqslant K \geqslant Mg \geqslant Na$. The carbon present in typical soils is entirely displaced from the series for desert soils because of the low biotic potential and turnover in arid areas.

For typical soils, the calcium, magnesium, and sodium contents are 1.37, 0.63, and 0.63%, respectively [Vinogradov, 1959]. As a comparison of values for these elements in Table 4 shows, values are higher in the Victoria valley soils, and thus not only the mineralogy and salt composition of the soils but also the over-all aridity and unleached character of the soil substratum are indicated. A fourth element (potassium) is also found in a high concentration in some arid soils of the southwestern United States, but it was in an abnormally high concentration in only one sample in Victoria valley: subsurface dune sample 576 had 8.8% potassium. The distribution of potassium in soils follows a definite geomorphic pattern and is related to weathering conditions of potassium feldspars and micas [Jackson, 1964]. The concentration of phosphorus, another essential biophilic element, is usually between 0.02 and 0.5% in most mineral soils but is only about one-tenth this value in Victoria valley soils. Approximately half of the inorganic phosphorus occurs as $Ca_3(PO_4)_2$ in alkaline and calcareous mineral soils, and the other half occurs in combination with soil organic matter [Jackson, 1964], which is extremely low in Victoria valley and other dry valley soils.

The titanium content was slightly higher than that in most soils [Vinogradov, 1959], although much higher values would be expected for highly leached soils, such as laterites [Jackson, 1964]. In typical soils the sulfur content is as high as the phosphorus content [Vinogradov, 1959], but in Victoria valley soils, as in other arid soils (aridisols, gray and red deserts, and sierozems), the sulfur content was much higher, and in site 540-543 there was as much titanium as sulfur. The content of some of the trace elements (cobalt, copper, and water-soluble boron) was slightly higher than that of most soils [Vinogradov, 1959]. The trace element composition is usually higher in the surface than in the subsurface of arable soils because of surface enrichment from plant residues [Mitchell, 1964], especially that of the biophilic elements [Jackson, 1964]. The elemental composition of the nonsalty Victoria valley soils also showed similarities to that of high altitude California soils [Cameron, 1970]; this finding would correspond to the existence of similar weathering processes and erosion at high elevations as well as at high latitudes in the Antarctic. The low concentrations of the major biophilic elements (e.g., carbon, hydrogen, oxygen, nitrogen, and phosphorus) indicate the low biotic pressure of the antarctic dry valleys.

The principal cations in cold desert soils are Ca^{++}, Mg^{++}, Na^{+}, and K^{+} [Boyd et al., 1966; Claridge,

1965; Claridge and Campbell, 1968; Tedrow and Ugolini, 1966; Ugolini, 1970], the Ca^{++} and Na^{+} usually dominating the exchange complex. This finding was the same for all the sites in Victoria valley (Table 6). Soils of weathered doleritic debris can contribute a large amount of calcium, sodium, and magnesium salts, as in the vicinity of Lake Vashka [Claridge, 1965]. Sodium is also contributed to antarctic soils through the weathering of sedimentary rocks and precipitation from fogs or clouds that have previously passed over marine areas. High concentrations of Ca^{++}, Na^{+}, and Mg^{++} salts, especially $CaCl_2$, which occurs in some antarctic soils, can depress the soil freezing point and lower the activity of the water available to organisms. The accumulation of $CaCO_3$ on the undersurface of stones (Figure 13) is a common feature not only of the arid antarctic environment but also of extremely arid salty areas in the temperate zone, such as the northern Chilean Atacama Desert (R. E. Cameron, unpublished results, 1966–1967).

The descending order of extractable cations was $Na^{+} > Ca^{++} > Mg^{++} > K^{+} > NH_4^{+} > Zn^{++} > Fe^{+3} > Al^{+3}$. (Boron is a nonmetal with a valence of +3 in all its compounds and is usually reported as B or B^{+3}. However, water-soluble boron can be found in significant concentrations in neutral and alkaline soils under arid conditions, and in its water-soluble or available state it should be reported as BO_3^{-3} (S. Manatt, personal communication, 1971).) Not much NH_4^{+} is usually detected unless there are bird-derived (ornithogenic) soils [Boyd and Boyd, 1963b; Ugolini, 1967] or decaying plant organic matter, such as algae and fossiliferous materials. Small concentrations of Zn^{++} and Fe^{+3} were found at all sites. Practically no water-soluble Al^{+3} was found, but it is not normally expected in saline desert soils. Oxides of iron and aluminum as well as of titanium do not markedly increase unless soil minerals have undergone pedogeochemical transformations [Jackson, 1964].

The principal anions in the sites at Victoria valley were Cl^{-}, $SO_4^{=}$, NO_3^{-}, and HCO_3^{-}, there being lower concentrations of PO_4^{-3}, BO_3^{-3} and NO_2^{-}. Only one site (540-543) contained appreciable quantities of water-soluble Cl^{-}, NO_3^{-}, and $SO_4^{=}$. Except for NO_3^{-}, which can occur in high concentrations in soils of temperate arid areas such as the Chilean Atacama Desert [Cameron, 1969b], the concentrations of anions were comparable to those in other arid areas. An NO_3^{-} concentration of 20 ppm is considered a good supply for temperate arid soils [McGeorge, 1940]. Nitrate losses are increased through volatilization when arable soils are considerably heated, are acidic, or contain much exchangeable Al^{+3} [Thomas and Kissel, 1970], but these conditions are rarely encountered in antarctic cold desert soils. Nitrite is rarely found in nature [Alexander, 1961] and rarely accumulates in typical soils even under conditions of rapid nitrification [Stevenson, 1964], but it can accumulate in frozen antarctic soils [Cameron, 1971]. Detectable quantities of NO_2^{-} were found in soils at both site 537-539a and site 540-543 when the frozen samples were analyzed immediately after they were taken from the freezer instead of after they had been air dried at room temperature. For site 537-539a the NO_2^{-} concentrations were 8, 6, 4, and 1 ppm at increasing soil depths. For site 540-543 these values were 4, 14, 32, and 1 ppm.

Although boron is an essential plant nutrient, notably in nitrogen fixation, the concentration of BO_3^{-3} has reached toxic levels in some cold desert soils [Cameron et al., 1969b]; however, the boron concentration has reached much higher levels (>4 ppm) in temperate arid areas and can be tolerated by some higher plants [Haas, 1944; Moghe and Mathur, 1966] as well as by microorganisms, including bacteria, streptomycetes, and algae (R. E. Cameron, unpublished results, 1961–1966). Only plants that can tolerate boron levels of >3.0 ppm are considered 'tolerant' in temperate arid areas [Richards et al., 1952; Smith et al., 1964], but the fact that some temperate zone microorganisms can grow at higher levels (R. E. Cameron, unpublished results, 1961–1966) probably indicates a longer period of adaptation. For most plants the limits of boron availability are rather narrow [Mitchell, 1964]. Relatively unweathered soils derived from alluvium, limestone, shale, and glacial drift are high in boron [Whetstone et al., 1942]. The boron content has also been found to increase in finer textured soils, and it will also increase with an increase in electrical conductivity [Jain and Saxena, 1970]. Soils of temperate arid and semiarid areas may contain 10–40 ppm or as much as 130 ppm [Whetstone et al., 1942]. The available boron content can increase with an increase in *p*H (i.e., from *p*H 7.3 to *p*H 10.3), but there is a strong correlation between decreasing boron availability and increasing carbonate content [Moghe and Mathur, 1966; Reeve et al., 1948]; the same factors should apply to antarctic cold desert soils. Therefore boron toxicity in antarctic soils is not entirely dependent on the total concentration of this nutrient.

Nitrate as well as NO_2^- accumulations indicate extreme aridity. Reduction of soil moisture from the wilting point (~99% RH) to the air dry state can result in a marked increase in NO_3^- [Munro and MacKay, 1964]; this factor could be important in the Antarctic. Nitrite accumulation is associated with alkaline *p*H values (>7.7 *p*H in temperate desert soils) [Martin et al., 1942], and as little as 15 ppm may be toxic to higher plants [Jones and Hedlin, 1970]. Nitrite losses are larger in the presence of high NH_4^+ levels [Jones and Hedlin, 1970], and, although some NH_4^+ was found at all sites, these concentrations were not high enough to account for any significant loss of nitrogen. However, site 574-575 did have the highest concentrations of NH_4^+ and no detectable NO_3^- and was a dune subjected to movement, exposure of permafrost, and subsequent warming by solar radiation, which could have resulted in denitrification. In Canadian soils, freezing and thawing were not found to affect the NO_3^- content [Hinman, 1970]. Low temperature storage of temperate zone soils at 1° to −3°C prevents significant changes in both mineralizable sulfur and nitrogen [Chaudhry and Cornfield, 1971].

A completely satisfactory explanation of the origin of nitrate deposits, whether in the Antarctic or in Chile, has not yet been fully substantiated. The atmospheric formation of nitrates in the Aurora Australis has been indicated on the bases of precipitation of South Polar snow [Wilson and House, 1965] and of soils derived from various parent materials [Claridge and Campbell, 1968]. Although subsequent evaporation and sublimation of snow deposits can account for some accumulation of soil nitrates, the biological origin should also be considered. In the Antarctic, nitrates are found in combination with sodium, calcium, and magnesium sulfates [Claridge and Campbell, 1968]. The nitrates may be present with chlorides in low or undetectable concentrations [Claridge and Campbell, 1968], but the opposite has also been found [Cameron, 1971]. Nitrification can result in an increase in Ca^{++} and Mg^{++} in addition to NO_3^- [Larsen and Widdowson, 1968]. The temperatures on the antarctic continent are too low to allow for significant leaching of salts, but high concentrations of salts, especially NaCl, could probably be tolerated by denitrifying bacteria [Ingram, 1957].

For photosynthetic organisms, nitrate is the end product of the oxidation of nitrogen compounds [Olson, 1970], and these factors, plus the complete mineralization following decomposition or the lack of denitrification, could account for the accumulation of nitrate. It is also important that a substantial proportion of our desert bacterial isolants, including our antarctic isolants (R. M. Johnson, personal communication, 1969–1972), will grow on nitrogen-free media [Johnson, 1970] and could also contribute to the formation and accumulation of nitrate. Sulfate-reducing bacteria, which may be widely distributed in the Antarctic [Barghoorn and Nichols, 1961], also have the ability to fix nitrogen [Riederer-Henderson and Wilson, 1970]. The algae may include a nitrogen-fixing species, or they may at least contribute to heterotrophic ammonification and otherwise enter into the microbial food chain of the dry valleys [Boyd et al., 1966; Janetschek, 1967]. Algae have also been present on the continent in the past, as the stromatolite deposits show (W. Breed, unpublished data, 1970). Sulfate-reducing bacteria are found in the Antarctic in association with decaying organic matter of algal origin [Barghoorn and Nichols, 1961].

In addition to nitrates, nitrites, and borates, sulfides may also accumulate in frozen soils and present a toxicity problem. Analyses of soils 537–539 showed concentrations of 2, 3, and 44 ppm, respectively (nitrite and sulfide analyses by Lois L. Taylor, Jet Propulsion Laboratory Polymer Research Section).

Regardless of concentrations, kinds, and balances of chemical ions in soils the greatest single factor restricting growth at low temperatures, whether for macroorganisms or microorganisms, is undoubtedly the low absorption of water [Richards, 1952]. This assertion would also apply to antarctic soils. If nutrients are deficient, the addition of needed nutrients may improve growth at low temperatures [Nielsen and Humphries, 1966]. If phosphate is deficient even for temperate desert soils [McGeorge, 1940], climatic conditions should be considered in interpreting nutrient analyses (such as those of PO_4^{-3} and its relation to the field) [Sutton, 1969]. Soil nutrients (e.g., nitrogen, phosphorus, sulfur, and calcium in organic form) are released when the temperature favors microbial decomposition [Nielsen and Humphries, 1966].

The organic carbon and nitrogen content and the total organic matter content of the soils were quite low and in general were in the same range as that obtained for other dry valley soils [Cameron, 1971]; they were also similar to the lowest values obtained for soils from harsher temperate desert regions of the western United States, the Egyptian Sahara, and the Negev, Patagonian, and Atacama deserts [Cameron, 1969b]. Higher values are reported for scrub deserts such as the upper Sonora and the arid and Saharan areas of Morocco and central Australia (R. E. Cameron, unreported results). Organic carbon does not usually make up more than 0.05% of most antarctic soils, and only site 540-543 showed this much organic carbon, although site 537-539a contained algae, which can lead to higher values of both organic carbon and nitrogen [Cameron, 1972]. These low values correspond to criteria recently established for antarctic cold desert zonal (frigic) soils [Campbell and Claridge, 1969].

The small amount of organic matter present in some antarctic soils is anthracite coal [Bauman et al., 1970; Horowitz et al., 1969]. Farther south (e.g., at Mt. Howe, 87°21′S) some of this carbon is graphitic (R. E. Cameron, unreported results, 1970–1971). Graphitic carbon may be oxidized by microorganisms in certain soils [Shneour, 1966], but it may be necessary to sulfomethylate or sulfonate coal before it can be used by organisms [Cairns and Moschopedis, 1971]. Ammoniated coal does not appear to show any chemical change from soil microorganisms [Berkowitz et al., 1970]. However, some of the organic matter may have been derived from algal material laid down in the past history of Victoria valley; it is presently being formed during part of the summer when conditions are favorable for growth. *Nostoc commune* may be considered the primary synthesizer of organic matter in the dry valleys [Boyd et al., 1966], although autotrophic bacteria and photosynthesizers (i.e., lichens and mosses), as well as algae, are considered to be primary producers [Janetschek, 1967]. Windborne organic matter from the sea, wind-blown debris from McMurdo Station and Scott Base, penguin rookeries on Ross Island, skua nesting sites, a few mummified seals and penguins within Victoria, Taylor, and Wright valleys, and field party debris are other possible sources of organic matter. East of site 537-539a (Figures 25 and 26) organic materials were continually being torn off or displaced and blown into the area from a New Zealand food cache left from the International Geophysical Year investigations. In January 1971 this area was revisited, and the same conditions were operative.

The total carbon and nitrogen in some soils (e.g., site 537-539a) closely approximated the values obtained for organic carbon and nitrogen. Slight discrepancies can be noted in some cases where there is a difference in the method of analysis. However, most of the carbonate carbon occurred at relatively the same low concentration as the organic carbon for a given soil sample. Total nitrogen (e.g., that for site 540-543) was increased over total organic nitrogen by the amount of nitrate nitrogen in the sample. Total hydrogen was more influenced by the presence of clays resulting from the weathering and powdering of rocks by the mechanical action of glacial movement than by the presence of soil organic matter. Mechanical translocation of clay within the soil profile is considered consistent with the increase in the trace element content [Mitchell, 1964], but this pattern is not indicated for our samples either in terms of fine textured materials, trace metals, or hydrogen.

The ratios of organic carbon/nitrogen, total carbon/nitrogen, nitrogen/hydrogen, and hydrogen/carbon are also shown in Table 7. Except for the ratio of organic carbon/nitrogen, ratios are rarely determined or reported for soils [Cameron, 1970]. The organic carbon/nitrogen ratios, with the exception of that of site 540-543, are considered 'narrow,' i.e., ≤ 10. The organic carbon/nitrogen ratio for microorganisms is usually between 4:1 and 9:1 [Buckman and Brady, 1960]. The narrowing of the carbon/nitrogen ratio may be due to fixed NH_4^+ [Stevenson, 1959], but that relationship is not true for these soils, as shown by the NH_4^+ concentrations (Table 6). The organic carbon/nitrogen ratio for antarctic cold desert soils is similar to that for arid temperate zone soils [Cameron, 1971], and the narrow ratios obtained for all samples except that for site 540-543 correspond to the values reported for subsurface soils of more favorable climatic zones [Buckman and Brady, 1960]. A critical ratio of organic carbon/nitrogen between 20:1 and 25:1 indicates that nitrogen mineralization would occur in all soils except that of site 540-543, whereas the wider ratios occurring at site 540-543 would favor almost complete immobilization of the available (leachable) nitrogen as microbial protoplasm [Stevenson, 1964], provided there were a favorable solclime, nutrient balance, and so forth. For arable soils both carbon and nitrogen are mineralized after incubation and drying–rewetting cycles and at higher temperatures, there being a concurrent widening in the carbon/nitrogen ratio above 8 [Agarwal et al., 1971]. It is not known if the same results could be obtained for antarctic soils.

Although the organic carbon/nitrogen ratios are relatively narrow for four of the five Victoria valley sites, the total carbon/nitrogen ratios showed almost the reverse relationship. The total carbon/nitrogen ratio includes organic as well as inorganic forms of

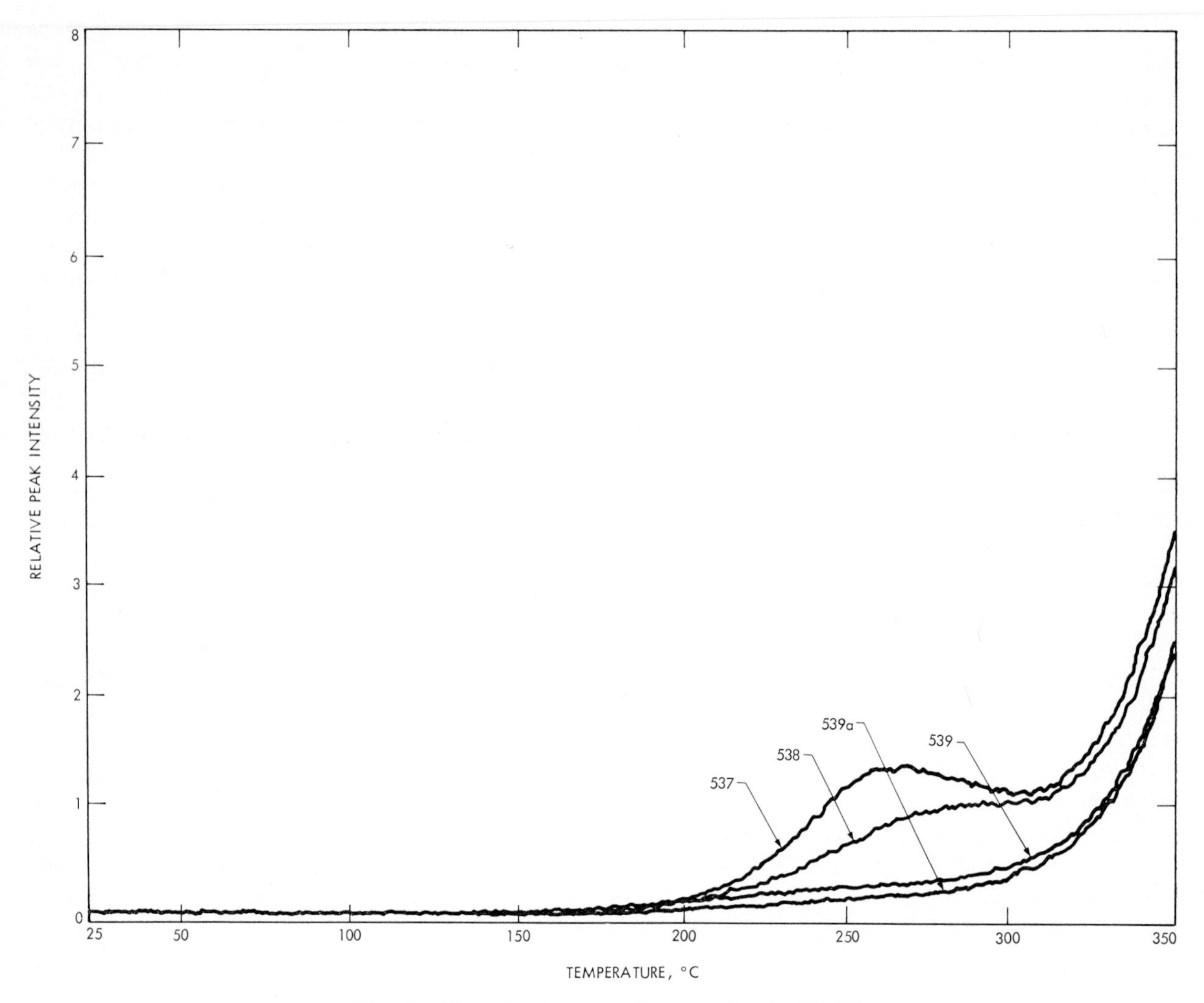

Fig. 44. Thermoluminescence glow curve for site 537-539a.

the two elements, especially NO_3^-. Site 540-543 had exceptionally narrow ratios for all samples, and sites 537-539a and 574-576 had relatively large ratios, especially at the surface, and some narrowing of the ratios with increasing soil depth.

The higher percentages of total nitrogen are also reflected in very narrow ratios for all sites except site 540-543, where the values are 10 to >100 times those of the other sites. However, the relative abundance of total carbon and hydrogen had a narrow range and thus showed that there probably was very little leaching, transport, or deposition of these elements for a given site and soil depth. For the Antarctic, the build-up of nitrates is indicated not only by the relatively wide organic carbon/nitrogen ratios but also by the correspondingly narrow total carbon/nitrogen ratios and higher total nitrogen/hydrogen ratios. These ratios substantiate other antarctic conditions for nitrate accumulation, including aridity, oxidized soils, low temperatures, lack of organic matter and leaching, and absence of denitrifying bacteria.

Thermoluminescence curves are shown for three of the sites (537-539a, Figure 44; 540-543, Figure 45; and 574-576, Figure 46). These curves are distinct for the three different sites (Figures 44–46 and appendix). They have some similar as well as some different soil properties (Tables 1–8); spectrochemical analyses for elements (Tables 4 and 5) do not show any apparent correlation with thermoluminescence; nor do those for predominant minerals, which were the same for each site except for some salts at site 540-543.

The influence of site exposure, slope, heat and moisture, age and derivation of materials, nature of salts,

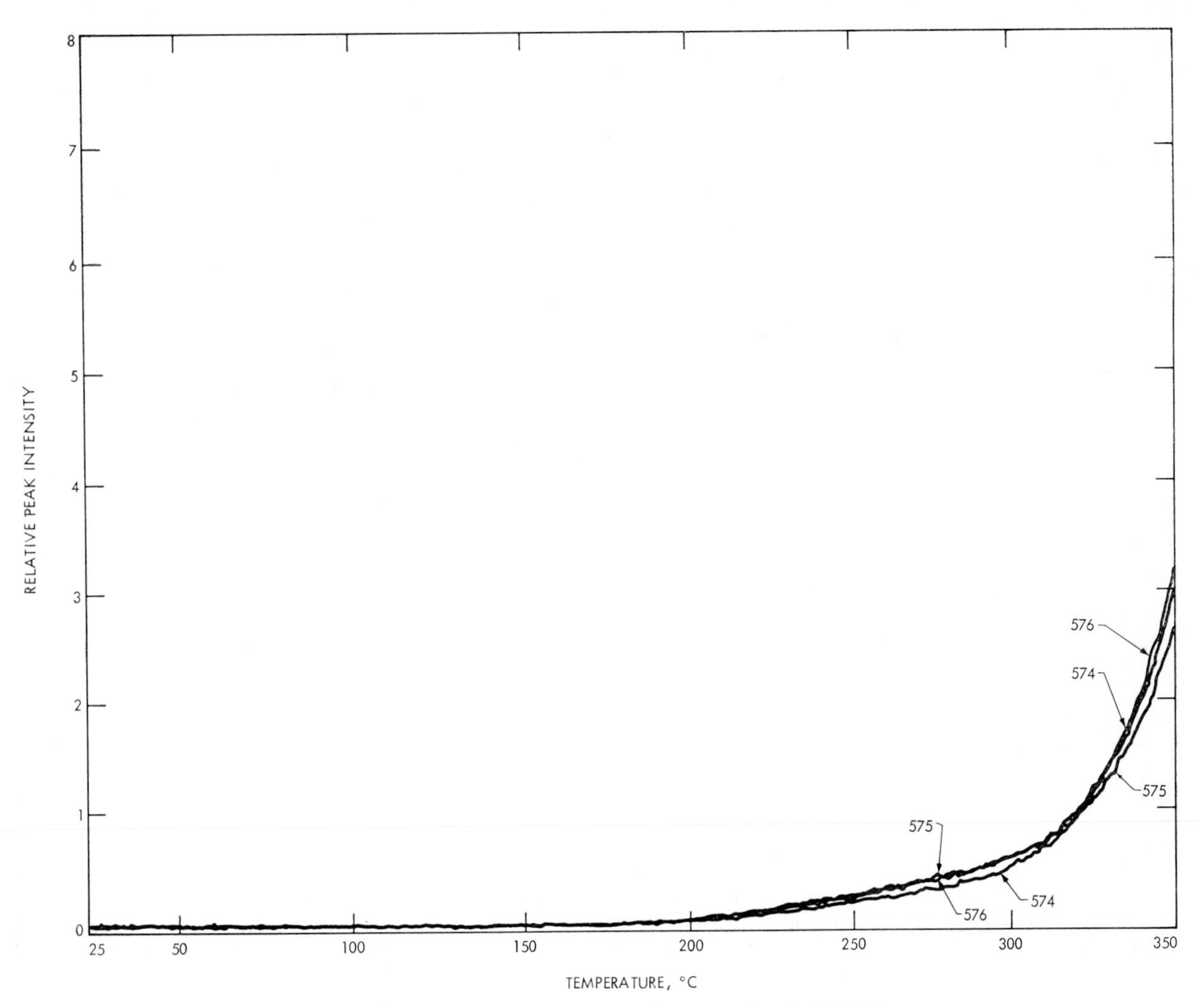

Fig. 45. Thermoluminescence glow curve for site 574-576.

and possibly soil texture and structure is indicated for the sites. For the dune sand (Figure 22) the thermoluminescence curves are replicates of each other (Figure 45). This result is to be expected if one considers the unstable nature of dunes and their continual movement, saltation effects, and turnover of structureless, single grained materials of essentially uniform composition. As a 'soil,' a dune would be of very recent origin and show little or no profile development because of continual turnover. The moisture curves were also similar for the three samples of soil from this site (Figure 42).

Relatively stable and older materials showing the influence of fluvioglacial action, stream deposition, and surface heating are plotted for the thermoluminescence of soils obtained from site 537-539a (Figure 44). The two subsurface samples (539 and 539a) from depths of 15–30 cm show very little modification from heat influxes.

In contrast to those for site 537-539a the curves for site 540-543 (Figure 46) show considerably elevated peaks but also a noticeable decrease with soil depth. Site 540-543 is older and more mature, not only in terms of deposition and derivation of materials from the Bull drift but also in terms of the surface-polished and salt-encrusted stones and soil profile characteristics (i.e., brownish color, mottles, macro salt crystal inclusions, presence of fines, and lack of distinct ice-cemented permafrost boundary layer at ~30 cm). A plot of soil moisture constants, especially of the hygroscopic coefficient (Figure 43), indicates the influence of hygroscopic salts at this site. The salt concentrations for all three samples are unusually high, as the electrical conductivity and osmotic pressure values

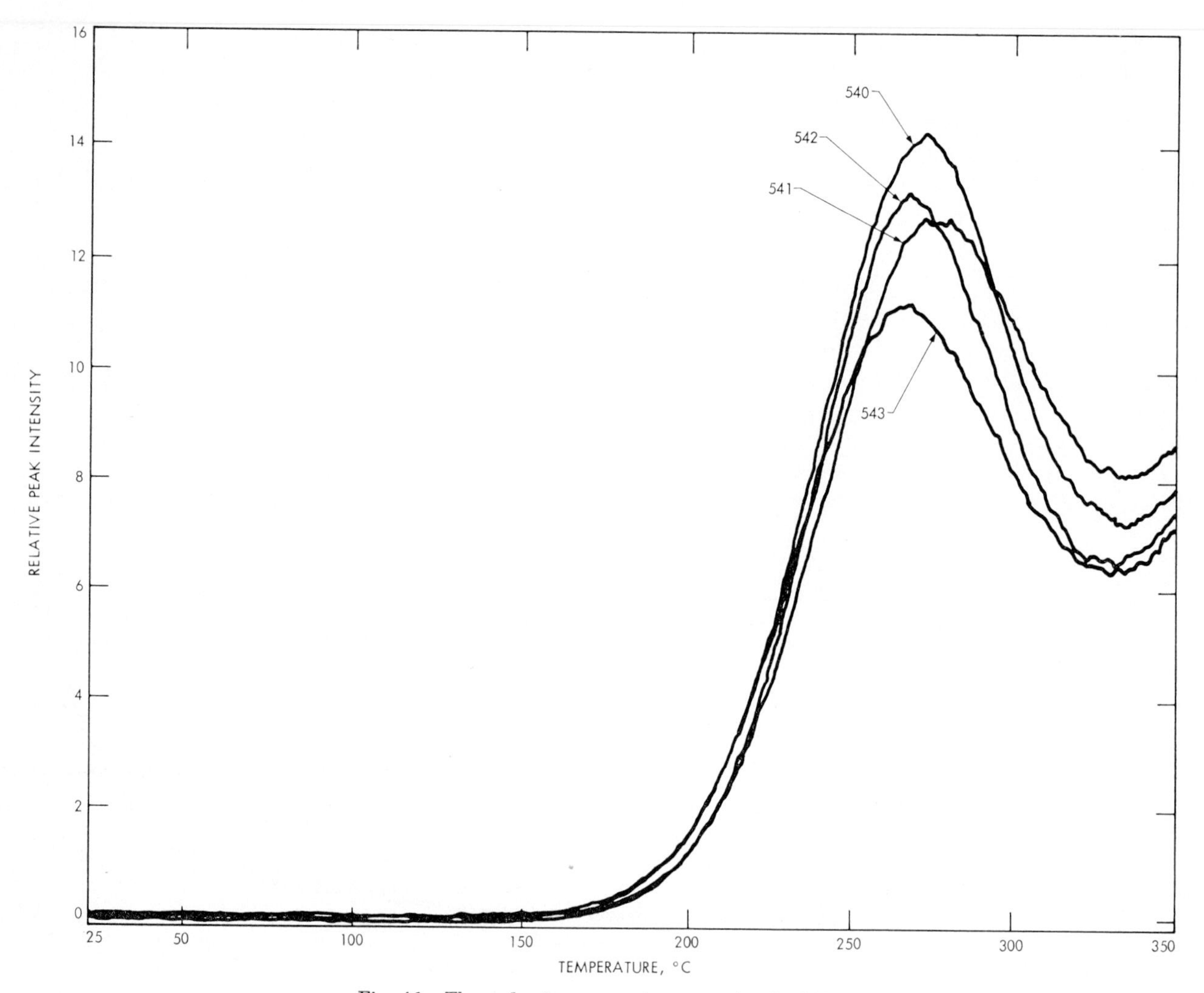

Fig. 46. Thermoluminescence glow curve for site 540-543.

(Table 3) and the concentrations of available ions, Na^+, Ca^{++}, Mg^{++}, Cl^-, NO_3^-, and $SO_4^=$ (Table 6), which influenced the slope and intensity of the thermoluminescence curves, show. The shift and slightly broad peak of the thermoluminescence curve for sample 541 and the slight elevation of the curve for the deeper sample 543 over that of sample 541 are not readily understood but are probably related to heating fluxes and the influence of moisture movement. It may be more than coincidental that sample 543 showed no culturable microorganisms. Site 540-542 obviously received more heat (and other solar radiation), as its northern exposure and slope show. It would be a favorable habitat for microorganisms in terms of thermal regime but an unfavorable one in terms of moisture content as well as moisture activity because of the concentrations and the kinds of salts.

Although it is a relatively new tool for characterizing soils, thermoluminescence has been used to study geologic age, microclimate, mineral composition and radiation of rocks [Daniels, 1968; Ronca, 1964; Zeller and Pearn, 1960; Zeller, 1968], and gamma ray irradiation of soils [Nishita and Hamilton, 1970]. Some thermoluminescence data have been presented previously for a traverse in the Matterhorn Glacier–Taylor valleys area to show the influences of favorable versus unfavorable microclimates, soils, and habitats for microorganisms [Cameron et al., 1969; Cameron, 1971]. Additional data have been obtained to show not only that there are distinct differences for individual sites and soils but also that there is a consistent pattern for the orientation, the microclimate, and the soils of a given valley and that there appears to be a good correlation with decreased microbial abundance and increased thermoluminescence from Wheeler to Matterhorn Glacier–Taylor to Pearse valleys (J. D.

Ingham and D. D. Lawson, personal communication, 1971). On the basis of preliminary analyses for the dry valleys, thermoluminescence should prove to be a useful survey method, either singly or in combination with a few on-site soil tests (i.e., those of moisture and salts) to determine the presence, the absence, or the likelihood of microorganisms at a given site. It could also be useful in extraterrestrial studies, and the analysis of a few milligrams of soil by thermoluminescence or in combination with X-ray diffraction, differential thermal analysis, and infrared spectrophotometry could yield a significant amount of information relevant to the geologic history, microclimate, chemical and physical soil properties, and favorable or unfavorable attributes of a site on the basis of the known terrestrial microbial life forms and their ecology.

MICROBIOLOGY

Microbial abundance. The abundance of groups of microorganisms was determined with the culture media listed in Table 8. The *p*H, the *E*h, and the electrical conductivities of the various media are given for comparison with the same characteristics of sites and soils in Victoria valley (Table 3). Several of the media had *p*H values lower than the *p*H values of the soils, the *p*H of di Menna's agar being only 4.5; however, this medium has proved excellent for primary isolation and cultivation of both yeasts and molds. Fluid thioglycollate medium had a lower *E*h than the soils or other media, but the *E*h is an average value, and the tubed medium provides an excellent range of conditions, from an aerobic interface at the surface of the medium through a microaerophilic stratum to anaerobic conditions at the bottom of the tube. At lower dilutions bacteria usually grew at the surface of the tube or secondarily in the subsurface, either as discrete globular colonies or as shafts extending down into the medium. Except for trypticase soy broth, most of the culture media were within the electrical conductivity of most of the nonsalty soils. The trypticase soy broth had a much higher electrical conductivity because it contained 5% NaCl, and it should have been a more favorable medium for any possible halotolerant or salt-adapted microorganisms (e.g., those at site 540-543).

Incubation temperatures for the cultures are also given, including the average length of incubation and the first observed growth. In most cases either no growth was observed beyond the indicated average incubation period (Table 10) or it was difficult or impractical to maintain the quality of the media beyond the indicated dates; i.e., media dehydration and crystal growth interfered with the validity of the tests. Soil sprinkled on plates as well as low dilutions (1:5 or 1:10) were used in attempts to determine possible low abundances of microorganisms. This method approximated soil enrichment and capillary techniques.

Soil microbiological properties for bacteria, fungi, algae, and protozoa are given in Tables 9 and 10. The 'standard' determinations used for the cultural detection and enumeration of microorganisms are given in Table 9; 'supplementary' tests performed to confirm information or to obtain further information on physiological groups of microorganisms are given in Table 10. As these tables indicate, the total number of microorganisms cultured at 20°C was between 0 (undetectable) and 10^5 per gram of soil. The total of samples incubated at +2°C was approximately the same as that at 20°C. No obligate psychrophilic bacteria were obtained. Only one soil (sample 539) had as many as 10^5 microorganisms per gram of soil. This sample was from a site in a more favorable ecologic niche than the other sites, and it also contained a low abundance of algae. The number of bacteria at this site was similar to that of comparable ecologic niches in Wheeler valley [Cameron et al., 1970c] but less than that for nearby and less favorable Bull Pass and McKelvey valley [Cameron, 1971]. There are also a number of unfavorable localities within a given dry valley where the bacterial counts may be lower than those for site 537-539a [Benoit and Hall, 1970; Boyd et al., 1966; Cameron et al., 1968; Cameron, 1971].

The abundance of aerobic bacteria is lower than or similar to that found in typical temperate or hot desert soils lacking macrovegetation [Cameron, 1969a] and less than that reported for arctic (polar) deserts [Jensen, 1951; Mishustin and Mirzoeva, 1964], which were more favorable habitats because a significant number of molds were also reported. Only one anaerobic colony was isolated from any of our samples, and it was a facultative *Bacillus* species. Obligate anaerobes should not be expected in the least favorable or driest niches in the dry valleys [Cameron, 1969a]. Only one other of our dry valley sites has shown the presence of potentially anaerobic bacteria, and these bacteria were in an ice-cemented subsurface sample in Wheeler valley [Cameron et al., 1970c].

In general, similar abundances of bacteria were obtained, frequently within 1 log unit, whether the

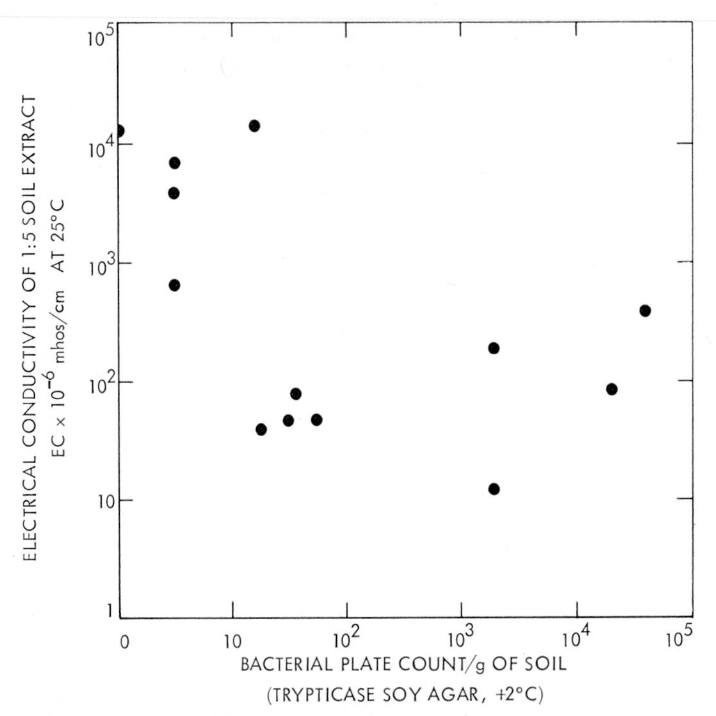

Fig. 47. Relationship between salt concentration and soil bacterial abundance, Victoria valley, Antarctica.

culture medium was trypticase soy agar, similated Taylor valley salts–organic agar, trypticase soy broth, or fluid thioglycollate. Actinomycete agar was also a reasonably good medium, but the results were more variable with lactose broth, Burk's agar, and Van Delden's agar. The bacterial abundances obtained with Burk's agar do not necessarily indicate strict nitrogen fixers, even though ion agar and deionized redistilled water were used, because some samples had an obvious carry-over of NO_3^-, there being trace amounts of NO_2^-, NH_4^+, and organic nitrogen in soil dilutions (Tables 6 and 7). The abundance of sulfate reducers also does not indicate obligate forms because the agar plates were not incubated in an anaerobic atmosphere, although it has been postulated that sulfate reducers may be widely distributed in the Antarctic [Barghoorn and Nichols, 1961]. Either several samples did not show any culturable microorganisms, or the abundances were approximately $\leqslant$10 per gram of soil (e.g., sites 540-543, 626, and 634). The failure to obtain culturable microorganisms could have been due to the methods employed, but a radiorespirometric method (Table 11) (J. Hubbard, personal communication, 1970) essentially verifies the abundance of microorganisms by culture media. However, better correlations were obtained for abundances of microorganisms in surface soils than were obtained for those in subsurface soils, and the fact that a greater response per culturable organism was obtained in the surface soils [Hubbard et al., 1970] may indicate greater activity, readier response, lower age, or greater abundance of microorganisms in the surface soils. As we indicated previously, both edaphic and microclimatic factors are limiting at these sites, especially in terms of the solclime (hydrothermal regime), and are complicated by salt factors, including possible toxicity in the presence of moisture. There was a general decline in the abundance of bacteria with increasing salt concentration (Figure 47), and most of our bacterial isolants either could not grow or did not grow well in added NaCl at concentrations of >2%; however, a few others could grow in concentrations of 7 and 10% added NaCl (W. Bollen and K. Byers Kemper, unreported results, 1967–1969). A comparison of growth with full strength trypticase soy agar (containing 5% NaCl) and that with 0.1% trypticase soy agar (Table 10) shows that microorganisms varied in their halotolerance (ability to grow in 'high' concentrations of salt). The bacteria at site 574-576 grew better in 0.1% trypticase soy broth, but those at site 537-539a were unaffected. One isolant from site 540-543 grew well in 10% added salt.

In addition to the bacteria found at all sites, usually

in the surface as well as in the subsurface soil, a few other groups of microorganisms were present but were not as abundant as the bacteria. Algae were recovered from only site 537-539a and not at all depths. Although subsurface algae could conceivably live heterotrophically, they could be freeze dried remnants from an earlier time when the surface was subjected to flooding and the deposition of steam-laid materials, or they could conceivably have sifted down from the surface through coarse grained materials, such as those encountered previously at depths of 30–60 cm in Wheeler valley after jackhammer dislodgment of the ice-cemented permafrost [Cameron et al., 1970c]. A few yeasts were recovered from site 574-576. Yeasts are the only microorganisms in the antarctic dry valleys that do not seem to follow the ecologic sequence found for the various groups of microorganisms in the dry valleys [Cameron, 1971]. Protozoa, which can occur in the presence of a significant abundance of algae in the dry valleys [Cameron, 1971; Cameron et al., 1970c], were present only in surface sample 537 and were not in all replicate algal cultures of this soil. No specific media were used in attempts to isolate or to cultivate protozoa.

Some groups of microorganisms were determined by the culture methods not to be present in our samples. These microorganisms included molds, streptomycetes, coliforms, and thermophiles. Molds generally have not been grown from dry valley soils [Boyd, 1967] unless algae are also present or there is some other significant source of decomposing organic matter such as guano. Streptomycetes were not found in any of our samples but were found in other parts of the valley around glacial melt water areas. A single species of streptomycete (*Streptomyces longisporoflavus*) is present throughout nearby Wheeler valley in both surface and subsurface soils [Bollen and Nishikawa, 1969], but it is not a common species [Waksman, 1967]. A different species (*Streptomyces exfoliatus*) was recovered from one soil sample in McKelvey valley [Bollen and Nishikawa, 1969]. Although coliforms and thermophiles were not present in any of our samples, they have been reported for antarctic soils subject to human and animal contamination [Boyd and Boyd, 1963a, b]. Coliforms apparently survive for only a relatively short time in antarctic soils [Boyd and Boyd, 1963b], but thermophiles may survive for longer periods in both arctic and antarctic soils [Boyd and Boyd, 1962, 1963a]. The presence of any coliforms or thermophiles in Victoria valley or other dry valley soils probably indicates contamination from previous investigators (Figures 25 and 26), transitory animal invaders, skua droppings (skuas also came to our camp), or personal contamination from sampling procedures. Although antarctic soils may not be readily contaminated with low levels of nonindigenous bacteria, tests have shown that certain species of introduced bacteria can survive in soil under antarctic conditions, that the limiting factors are physical in nature, and that viability will vary among habitats [Boyd et al., 1966].

Kinds of microorganisms. The bacteria most frequently found in our samples (Tables 9 and 10) were members of the diphtheroid (nocardoid) group, which may include the following genera: *Arthrobacter, Brevibacterium, Cellulomonas, Corynebacterium, Erysipelothrix, Jensenia, Kurthia, Listeria, Microbacterium, Mycobacterium, Nocardia* [Davis and Newton, 1969], and others such as *Achromobacter, Mycococcus,* and *Proactinomyces.* The taxonomy and the relationships of these microorganisms are still not well defined, but in general these microorganisms represent a pleomorphic group of aerobic branched or nonbranched cocci rods that are gram positive (or variable), catalase positive, sometimes acid fast and pigment forming, and generally not spore forming. They occur at the surface as well as in the subsurface of the soil, and they may be the only microorganisms occurring as a single population at the drier and less favorable sites. Of the 22 isolants cultured from our samples, only eight were sufficiently studied to be identified by species. The identifications are as follows (asterisk, dagger, and double dagger indicate that the identification was made by K. Byers Kemper, T. d'Arc, or R. M. Johnson, respectively):

Soil Sample	Organism
537	*Bacillus cereus**
	*Mycococcus ruber**
537	*Nitrobacter* sp.†
538	*B. cereus**
539	*Mycococcus albus**
539	soil diphtheroid (rods) *
540	*B. cereus**
543	*B. cereus**
634	*Arthrobacter citreus*‡
634	*A. tumescens*‡

Bacillus samples were identified from two of our sites and were a single species (*B. cereus*). *Bacillus* species are not common in the dry valleys, although *B. megaterium* has been reported from a pond environment in Wright valley [Meyer et al., 1962]. A few *Micrococcus* species have been identified from the dry valley soils (R. Johnson, personal communication,

1969), but none was present in our samples. Some of the colored colonies of micrococci on agar plates may be confused with (or even found similar to) colored colonies of *Corynebacterium* species. *Nitrobacter* was found in only one soil (at 1:10 dilution), as chemical tests [Alexander and Clark, 1965], further isolation, cultivation, and microscopic examination indicated. Because NO_2^- may accumulate in some antarctic soils, the prevalence of NO_2^- oxidizers may be greater in the dry valleys than previously recognized. Tests of a few McKelvey and Victoria valley soils showed that they have microorganisms active in an ammonifying capacity, but tests of samples 537 and 538 showed essentially no nitrifying power, possibly because the *p*H was too high; the tests also showed a very low sulfur-oxidizing power [Bollen, 1968].

The only fungi in our samples were two species of yeasts. They were isolated from dune site 574-575 and grew at both 2° and 20°C. *Cryptococcus albidus* was identified from sample 574, and *Rhodotorula mucilaginosa* from sample 575 (M. di Menna, personal communication, 1968). From soils from the McMurdo Sound area and Wright valley *Cryptococcus* species were most frequently isolated [di Menna, 1960]. *Rhodotorula* and *Cryptococcus* species were found at many locations near Mirnyy [Meyer et al., 1967] and were also the predominant yeasts isolated from a number of other dry valley samples (M. di Menna, personal communication, 1968). As a group, yeasts can grow in relatively dry substrates at high osmotic pressures and are cold adapted. Both properties are advantages in the cold and drought-dominated antarctic ecosystems [di Menna, 1966]. As di Menna [1966] indicated, the incidence of obligately psychrophilic yeasts appears random and does not increase with latitude, although yeasts were also present at the farthest south area of exposed rock and soil at Mt. Howe, 87°21′S, 149° 18′W (R. E. Cameron, unreported results, 1968). A single yeast species may constitute the only population in an otherwise sterile antarctic soil, but it does not necessarily occur in large abundance, i.e., possibly in a concentration of not more than 10 microorganisms per gram of soil.

Algae have been present in Victoria valley for some time, and one sample 20 meters above the present level of Lake Vida has been dated as 9700 ± 350 years old [Calkin, 1964]. Algal material is scattered around the lake shore (Figures 2 and 24). The algae in our samples were cultured from only site 537-539a and included coccoid green and coccoid and filamentous blue-green algae but not at all depths. The algae present in the surface sample (537) were cosmopolitan species: *Chlorococcum humicola, Anacystis marina, Anacystis montana,* and *Schizothrix calcicola.* The oscillatorioid forms are the most common [Drouet, 1962], especially the small filamentous blue-green *Schizothrix calcicola,* which has a number of ecophenes [Drouet, 1963]. This species occurs in other desert soils receiving a sufficient quantity and quality of available water; it can exist as a single population (e.g., in samples 539 and 539a) or in association with other algae (e.g., in the nitrogen-fixing strains of *Nostoc commune*), but, as Holm-Hansen [1963b] has indicated, no *Nostoc* has so far been observed in Victoria valley. In the Antarctic, only *S. calcicola* has been observed to tolerate a wide salinity gradient [Zaneveld, 1969]. This same alga has shown a greater viability after freezing and freeze–thaws than other algae in soils of the McMurdo Sound region [Holm-Hansen, 1963c], but all unicellular green algae tested from the Antarctic also showed a high rate of viability after freezing [Holm-Hansen, 1963a].

There may be more coccoid green species present than there appears, depending on the taxonomic criteria, especially the physiology, accepted for their delineation [Bold and Parker, 1962]. In a glacial stream in Victoria valley additional coccoid and filamentous green algae have been cultured, and an *Anabaena* was observed in a dried specimen [Holm-Hansen, 1964, 1967] that may have been a *Nostoc* species. Material collected in January 1971 from the lake shore several hundred meters west of site 537-539a showed *Chlorococcum humicola, Stichococcus bacillaris, Porphyrosiphon notarisii, Schizothrix calcicola,* and a small *Navicula* species.

The spore-forming algae are rare or absent in antarctic soils, and this finding is consistent with observations in hotter deserts where it is also an advantage to reproduce quickly during the intermittent periods of available moisture and to survive in the vegetative state rather than form spores. Streptomycetes may also exist in soils in either the vegetative or the spore form [Lloyd, 1969], and, as we indicated previously, spore-forming bacteria are not prevalent. Molds may or may not form spores in antarctic soils. Protozoa, sometimes in association with algae, will encyst under adverse conditions. An amoeboid flagellate, *Dimastagoamoeba*(?), was observed in sample 537. *Bodo minimus* grew in the culture of lake shore water. Details of antarctic soil algal ecology, especially with reference to the blue-green algae, have been given previously [Cameron, 1972].

CONCLUDING REMARKS

The soil microbial ecosystem in the antarctic dry valleys is relatively simple in comparison with that of hotter deserts and other areas where it has been complexed by the presence of higher plants, animals, and man [Benoit and Hall, 1970; Cameron et al., 1970b]. The antarctic cold desert provides a unique opportunity for studying soil microbial ecology at its least complex level in terms of gradients in the physical and chemical environment of the soil and microclimate. These gradients, which range from uninhabitable to relatively unfavorable to increasingly favorable environments, demonstrate a new maxim whereby 'ecology recapitulates phylogeny' (first stated by D. Hilchey). This maxim is certainly shown by investigations in the dry valleys, where the solclime is the most important factor and uninterrupted temperature and moisture levels above freezing allow increasingly complex groups of microorganisms to develop on the community level as 'climax' communities. In addition, a combination of other environmental factors is also imposed on the biota and coupled with them. These factors include visible and ultraviolet light intensity and duration, wind intensity and duration and the resultant desiccation effects, freeze–thaw cycles, evaporation rate and intensity, salinity, low organic matter content, and other edaphic properties. Not all the ecologic factors that establish or limit the microbial ecosystem are known or understood, and, because the simplest ecosystems are usually the least stable, caution should be taken not to disturb them. Additional studies to determine the optimum and minimum conditions for the growth of endemic microorganisms, as well as a thorough physiological and taxonomic study of antarctic microorganisms, are needed, including a comparison with soil microorganisms in North Polar deserts and hotter arid regions.

Investigations of soils, microorganisms, and microclimate serve a useful purpose not only in the study of harsh terrestrial soil microbial ecology and the limiting environmental factors for life in this extreme southern continent but also in the application of the resultant information to a better understanding of arid areas in temperate and tropical deserts. In addition, these studies have provided a useful model for the search for possible extraterrestrial life and the establishment of manned bases in an extraterrestrial environment. The kinds, the distribution, the abundance, and the adaptive mechanisms of terrestrial microorganisms in the Antarctic (or the measurement of the physical and chemical environmental factors alone) can assist in determining the possibility of life in an extraterrestrial environment. If microorganisms cannot survive (grow and reproduce) in the Antarctic, which is harsh by terrestrial standards, the possibility of life in a much harsher extraterrestrial environment, such as Mars, is an imposing question.

APPENDIX: CHARACTERISTICS OF SOIL SAMPLING SITES, VICTORIA VALLEY, ANTARCTICA

A. Soil samples 537–539a
  1. Sample depths
    a. Sample 537: 0–2 cm.
    b. Sample 538: 2–15 cm.
    c. Sample 539: 15–25 cm.
    d. Sample 539a: 25–30 cm.
  2. Elevation: 395 meters.
  3. Topography of area: low terrace with sand wedge polygons in outwash fan between alluvial fan and lake.
  4. Location, position, and slope of site: about 50 meters north of northwest end of Lake Vida, near previous field party camp sites and activity; collection site on low terrace and within center of a polygon about 10 meters east of our camp.
  5. Microrelief of surface at site: uneven and pebbly pavement; sand wedge polygons with medium pressure ridges.
  6. Parent material, soil structure, and development: modified glacial till, lacustrine, alluvial, and outwash deposits; azonal structure except for faint alluvial stratification; ice-cemented permafrost at 25 cm.
  7. Microclimate at time of sample collection (January 2 and January 3, 1967): details of microclimate shown in Tables 12–18 and Figures 27–40.

B. Soil samples 540–543
  1. Sample depths
    a. Sample 540: 0–2 cm.
    b. Sample 541: 2–15 cm.
    c. Sample 542: 15–25 cm.
    d. Sample 543: 25–30 cm.
  2. Elevation: 450 meters.
  3. Topography of area: rolling, bouldery glacial debris lobe intersected by drainage channels; outwash fan sloping toward valley floor.
  4. Location, position, and slope of site: about 1500 meters southwest of Lake Vida; collection site on northwest side of large-faceted ventifact; 2°–3° north-facing slope.
  5. Microrelief of surface at site: uneven and pebbly pavement with partially exposed boulders.

6. Parent material, soil structure, and development: glacial till and debris of Bull drift; salt-encrusted stones; xerous, frigic soil with some macro salt crystals; increasingly hard with depth; no ice-cemented permafrost at less than 30 cm.

7. Microclimate at time of sample collection (January 1, 1967, 2030 hours, local time):

a. Air temperature: 1 meter, +1°C; 30 cm, +1.5°C; 15 cm, +2°C.

b. Soil temperature: surface sun, +7°C; shade, +5°C; 5 cm, +4°C; 15 cm, +1.5°C; 30 cm, −1.5°C.

c. Wind: 20–25 mph from the northeast.

d. Light intensity: perpendicular to sun, 1200 ft-c; soil surface, 400 ft-c.

e. Barometric pressure: 27.98 in. Hg.

f. Clouds: mixed, variable, broken cirrocumulus.

C. Soil samples 574-576

1. Sample depths

a. Sample 574: 0–2 cm.

b. Sample 575: 2–15 cm.

c. Sample 576: about 15–25 cm.

2. Elevation: 430 meters.

3. Topography of area: intermittent barchan dunes and sand mantles in glacial morainic outwash.

4. Location, position, and slope of site: about 200 meters northeast of Lake Vida, Victoria valley; collection site on broad expanse of dune sand; about 2° southeast-facing slope.

5. Microrelief of surface at site: faintly rippled dune sand.

6. Parent material, soil structure, and development: aeolian deposits of sand mantles and dune sand; azonal structure; ice-cemented permafrost at about 30 cm.

7. Microclimate at time of sample collection (January 21, 1967, 1630 hours, local time):

a. Air temperature: 1 meter, +6°C (27.5% RH).

b. Soil temperature: surface sun, +15.5°C; shade, +11.5°C; 2 cm, +9.5°C; 15 cm, +1°C; 30 cm, −2.5°C.

c. Wind: 16–18 mph from the east.

d. Light intensity: perpendicular to sun, 8800 ft-c; soil surface, 3000 ft-c.

e. Barometric pressure: 28.00 in. Hg.

f. Clouds: some scattered cirrocumulus.

D. Soil sample 626

1. Sample depth: 0–5 cm.

2. Elevation: 430 meters.

3. Topography of area: gently undulating hummocks and morainic ridges overlooking rocky drainage basin.

4. Location, position, and slope of site: east and about equidistant between Victoria Upper Glacier and west end of Lake Vida; fairly level site on southeast-facing slope.

5. Microrelief of surface at site: rough, uneven, pebbly, and bouldery; weak sand wedge.

6. Parent material, soil structure, and development: recessional and glacial till and debris; xerous, frigic soil developed from dolerite and sandstone.

7. Microclimate at time of sample collection (January 12, 1968, 1100 hours, local time):

a. Air temperature: 1 meter, −2°C.

b. Soil temperature: surface, +2°C; 2 cm, +4°C.

c. Wind: 4–6 mph from the east.

d. Light intensity: perpendicular to sun, 1900 ft-c.

e. Barometric pressure: 27.75 in. Hg.

f. Clouds: about 95% stratocumulus.

E. Soil sample 634

1. Sample depth: 0–5 cm.

2. Elevation: 590 meters.

3. Topography of area: intermittent barchan dunes in glacial morainic outwash.

4. Location, position, and slope of site: about 500 meters northwest of Lake Vida, lower Victoria valley; on gently sloping crest of dune below Packard Glacier.

5. Microrelief of surface at site: slightly rippled drifting dune undergoing saltation.

6. Parent material, soil structure, and development: aeolian deposits of dune sand; azonal structure.

7. Microclimate at time of sample collection (January 15, 1968, 1300 hours, local time):

a. Air temperature: 1 meter, +4°C (37% RH).

b. Soil temperature: surface sun, +11°C; shade, +9°C; 2 cm, +8°C.

c. Wind: 16 mph from the northeast.

d. Light intensity: perpendicular to sun, 3600 ft-c; soil surface, 3200 ft-c.

e. Barometric pressure: 27.60 in. Hg.

f. Clouds: about 20% high cirrus and hazy.

Acknowledgments. During the past 5 years of research on antarctic soils and microorganisms, a number of individuals have provided assistance in various ways. Dr. G. Llano has supported the study of soil microbial ecology in the dry valleys. Dr. N. Horowitz has recognized and suggested the value of the Antarctic as a Mars model for extraterrestrial life detection experiments. Dr. D. Le Croissette of the Jet Propulsion Laboratory and Dr. R. Patterson of Virginia Polytechnic Institute and State University have further encouraged these investigations. In the Antarctic, associate investigators have included R. Benoit, G. Blank, H. Conrow, C. David, C. J. Hall, Jr., J. King, G. Lacy, H. Lowman III, J. Marsh, and F. Morelli. Mrs. D. Gensel has performed most of the microbiological analyses. Other soil and microbiological analyses have been provided by S. Babcock, W. Bollen, K. Byers Kemper, G. Chapel, A. Cherry, T. d'Arc, M. di Menna, M. Frech, R. Haak, J. Hubbard, H. Johnson, R. Johnson, E. Merek, L. Taylor, and J. Sprague. R. Gorny assisted with the instrumentation. Others who have contributed considerably to this work are B. Dowling and M. E. Frias. A debt of gratitude is also owed to a number of personnel from the National Science Foundation Office of Polar Programs, U.S. Navy task force 43 and VX-E6 helicopter group, members of the Jet Propulsion Laboratory bioscience section, and members of various educational institutions, including Prof. J. C. F. Tedrow for reviewing the manuscript. Above all, the study in Victoria valley could not have been undertaken without the assistance of Prof. R. E. Benoit and his graduate student, C. J. Hall, Jr., Virginia Polytechnic Institute and State University, who unselfishly cooperated in a mutual camp and the collection of data after depletion of the Jet Propulsion Laboratory team. This study was supported by National Science Foundation contract NSF-C-585, including logistic support and facilities for the investigations in Antarctica arranged by the Office of Polar Programs, National Science Foundation. This paper presents the results of one phase of research carried out by the Jet Propulsion Laboratory, California Institute of Technology, with additional support provided under contract NAS7-100, sponsored by the National Aeronautics and Space Administration.

REFERENCES

Agarwal, A. S., B. R. Singh, and Y. Kanehiro
1971 Soil nitrogen and carbon mineralization as affected by drying–rewetting cycles. Proc. Soil Sci. Soc. Am., *35:* 96–100.

Ahmadjian, V.
1970 Adaptations of antarctic terrestrial plants. *In* M. W. Holdgate (Ed.), Antarctic ecology. *2:* 801–811. Academic, London.

Alexander, M.
1961 Introduction to soil microbiology. 472 pp. John Wiley, New York.
1965 Nitrification. *In* W. V. Bartholomew and F. E. Clark (Eds.), Soil nitrogen, pp. 307–343. American Society of Agronomy, Madison, Wis.

Alexander, M., and F. E. Clark
1965 Nitrifying bacteria. *In* C. A. Black (Ed.), Methods of soil analysis, pp. 1477–1483. American Society of Agronomy, Madison, Wis.

Allen, A. D.
1962 Formations of the Beacon group in the Victoria valley region. N. Z. Jl Geol. Geophys., *5:* 278–294.

Allen, A. D., and G. W. Gibson
1962 Geological investigations in southern Victoria Land, Antarctica. 6. Outline of the geology of the Victoria valley region. N. Z. Jl Geol. Geophys., *5:* 234–242.

Allen, S. E., and H. M. Grimshaw
1960 Influence of temperature storage on the extractable nutrient ions of soil. J. Sci. Fd Agric., *13:* 525–529.

Anderson, D., E. S. Gaffney, and P. F. Low
1967 Frost phenomena on Mars. Science, *155:* 319–322.

Andreason, R. C.
1952 Soil oxygen: Evaluation by means of redox potentials. Florists' Rev., *110:* 28–29.

Baas Becking, L. G. M., I. R. Kaplan, and D. Moore
1960 Limits of the natural environment in terms of pH and oxidation–reduction potentials. J. Geol., *68:* 243–284.

Bab'eva, I. P., and V. I. Golubev
1969 Psychrophilic yeasts in antarctic oases. Microbiology, *38:* 436–440.

Bailey, L. D., and E. G. Beauchamp
1971 Nitrate reduction and redox potentials measured with permanently and temporarily placed platinum electrodes in saturated soils. Can. J. Soil Sci., *51:* 51–58.

Barghoorn, E. S., and R. L. Nichols
1961 Sulfate-reducing bacteria and pyritic sediments in Antarctica. Science, *134:* 190–191.

Barksdale, L.
1970 *Corynebacterium diphtheriae* and its relatives. Bact. Rev., *34:* 378–422.

Barshad, I.
1964 Chemistry of soil development. *In* F. E. Bear (Ed.), Chemistry of the soil. 2nd ed., 1–70. Reinhold, New York.

Bauman, A. J., R. E. Cameron, G. Kritchevsky, and G. Rouser
1967 Detection of phthalate esters as contaminants of lipid extracts from soil samples stored in standard soil bags. Lipids, *2:* 85–86.

Bauman, A. J., E. M. Bollin, G. P. Shulman, and R. E. Cameron
1970 Isolation and characterization of coal in antarctic dry-valley soils. Antarct. J. U.S., *5:* 161–162.

Becker, E. W.
1970 Studies on low-temperature photosynthesis of algae. Antarct. J. U.S., *5:* 121–122.

Benoit, R. E., and R. E. Cameron
1967 Microbial ecology of some dry valley soils of Antarctica (abstract). Bact. Proc., *31:* 3.

Benoit, R. E., and C. L. Hall, Jr.
1970 The microbiology of some dry valley soils of Victoria Land, Antarctica. *In* M. W. Holdgate (Ed.), Antarctic ecology. *2:* 697–701. Academic, London.

Berg, T. E., and R. F. Black
1966 Preliminary measurements of growth of nonsorted polygons, Victoria Land, Antarctica. *In* J. C. F. Tedrow (Ed.), Antarctic soils and soil forming processes, Antarctic Res. Ser., *8:* 109–124. AGU, Washington, D. C.

Berkowitz, N., S. K. Chakrabartty, F. D. Cook, and J. I. Fujikawa
1970 On the agrobiological activity of oxidatively ammoniated coal. Soil Sci., *110:* 211–217.

Berry, F. A., Jr., et al.
1945 Synoptic meteorology and weather forecasting. *In* F. A. Berry, Jr., E. Bollay, and N. R. Beers (Eds.), Handbook of meteorology, pp. 603–879. McGraw-Hill, New York.

Black, R. F., and T. E. Berg
1963a Hydrothermal regime of patterned ground, Victoria Land, Antarctica. Publ. 61: 121–127. Comm. of Snow and Ice, Int. Ass. of Sci. Hydrol., Berkeley, Calif.
1963b Dating with patterned ground, Victoria Land, Antarctica (abstract). Trans. Am. Geophys. Un., *44*(1): 48.
1964 Glacier fluctuations recorded by patterned ground, Victoria Land. *In* R. J. Adie (Ed.), Antarctic geology, pp. 107–122. North-Holland, Amsterdam.
1965 Patterned ground in Antarctica. Publ. 1287: 121–127. Purdue Univ., Lafayette, Ind.

Bold, H. C., and B. C. Parker
1962 Some supplementary attributes in the classification of *Chlorococcum* species. Arch. Mikrobiol., *42:* 267–288.

Bollen, W. B.
1968 Ammonifying, nitrifying, and sulfur oxidizing capacity of antarctic soils. Progress report of microorganism study, contract 950783: 7 pp. Jet Propul. Lab., Calif. Inst. of Technol., Pasadena.

Bollen, W. B., and S. Nishikawa
1969 Systematic description and key to streptomyces isolants from Chile, Arizona, and Antarctica desert soils. Progress report of microorganism study, contract 950783: 213 pp. Jet Propul. Lab., Calif. Inst. of Technol., Pasadena.

Bouyoucos, G. J., and M. M. McCool
1924 The aeration of soils as influenced by air-barometric pressure changes. Soil Sci., *18:* 53–63.

Boyd, W. L.
1967 Ecology and physiology of soil microorganisms in polar regions. JARE Scient. Rep., spec. issue 1: 265–275.

Boyd, W. L., and J. W. Boyd
1962 Viability of thermophiles and coliform bacteria in arctic soils and water. Can. J. Microbiol., *8:* 189–192.
1963a Soil microorganisms of the McMurdo Sound area, Antarctica. Appl. Microbiol., *11:* 116–121.
1963b Viability of coliform bacteria in antarctic soil. J. Bact., *85:* 1121–1123.

Boyd, W. L., J. T. Staley, and J. W. Boyd
1966 Ecology of soil microorganisms of Antarctica. *In* J. C. F. Tedrow (Ed.), Antarctic soils and soil forming processes, Antarctic Res. Ser., *8:* 129–159. AGU, Washington, D. C.

Buckman, H. O., and N. C. Brady
1960 The nature and properties of soils. 6th ed., 567 pp. Macmillan, New York.

Bull, C.
1966 Climatological observations in ice-free areas of southern Victoria Land, Antarctica. *In* M. J. Rubin (Ed.), Studies in antarctic meteorology, Antarctic Res. Ser., *9:* 177–194. AGU, Washington, D. C.

Bull, C., B. C. McKelvey, and P. N. Webb
1962a Quaternary glaciations in southern Victoria Land, Antarctica. J. Glaciol., *4:* 63–78.
1962b Glacial benches in south Victoria Land. J. Glaciol., *4:* 131–134.

Cairns, R. R., and S. E. Moschopedis
1971 Coal, sulfomethylated coal, and sulfonated coal as fertilizers for solonetz soil. Can. J. Soil Sci., *51:* 59–63.

Calkin, P. E.
1963 Geomorphology and glacial geology of the Victoria valley system, southern Victoria Land, Antarctica. Ph.D. thesis. 293 pp., 1 map. Ohio State Univ., Columbus.
1964 Geomorphology and glacial geology of the Victoria valley system, southern Victoria Land, Antarctica. Inst. Polar Stud. Rep. 10: 66 pp., 42 figs., 2 pls. Ohio State Univ., Columbus.

Calkin, P. E., and C. Bull
1967 Lake Vida, Victoria valley, Antarctica. J. Glaciol., *6:* 833–836.

Calkin, P. E., and A. Cailleux
1962 A quantitative study of cavernous weathering (taffonis) and its application to glacial chronology in Victoria valley, Antarctica. Z. Geomorph., *6:* 317–324.

Cameron, R. E.
1962 Soil studies—Desert microflora. Space Programs Sum. 37-15, *4:* 91–98. Jet Propul. Lab., Calif. Inst. of Technol., Pasadena.
1963 The role of soil science in space exploration. Space Sci. Rev., *2:* 297–312.
1966 Soil sampling parameters for extraterrestrial life detection. J. Ariz. Acad. Sci., *4:* 3–27.
1967 Soil studies—Desert microflora. 14. Soil properties and abundance of microflora from a soil profile in McKelvey valley, Antarctica. Space Programs Sum. 37-44, *4:* 224–236, 239–240. Jet Propul. Lab., Calif. Inst. of Technol., Pasadena.
1969a (Ed.). Soil investigations of antarctic dry valleys. 16-mm color motion picture. Jet Propulsion Laboratory, California Institute of Technology, Pasadena.
1969b Abundance of microflora in soils of desert regions. Tech. Rep. 32-1378: 16 pp. Jet Propul. Lab., Calif. Inst. of Technol., Pasadena.
1969c Cold desert problems and characteristics relevant to other arid lands. *In* W. McGinnies and B. Goldman (Eds.), Arid lands in perspective, pp. 167–205. University of Arizona Press, Tucson.
1970 Soil microbial ecology of Valley of 10,000 Smokes, Alaska. J. Ariz. Acad. Sci., *6:* 11–40.
1971 Antarctic soil microbial and ecological investigations. *In* H. D. Porter (Ed.), Research in Antarctica, pp. 137–189. American Association for the Advancement of Science, Washington, D. C.
1972 (in press). Ecology of blue-green algae in antarctic soils. *In* T. V. Desikachary (Ed.), First international symposium on taxonomy and biology of blue-green algae. Centre for Advanced Studies in Botany, University of Madras, Madras, India.

Cameron, R. E., and R. E. Benoit
1970 Microbial and ecological investigations of recent cinder cones, Deception Island, Antarctica—A preliminary report. Ecology, *51:* 802–809.

Cameron, R. E., and G. B. Blank
1963 Soil organic matter. Tech. Rep. 33-443: 14 pp. Jet Propul. Lab., Calif. Inst. of Technol., Pasadena.
1966 Desert algae: Soil crusts and diaphanous substrata as algal habitats. Tech. Rep. 32-971: 41 pp. Jet Propul. Lab., Calif. Inst. of Technol., Pasadena.
1967 Desert soil algae survival at extremely low temperatures. Cryog. Technol., *3:* 151–156.

Cameron, R. E., and H. P. Conrow
1968 Antarctic simulator for soil storage and processing. Antarct. J. U.S., *3:* 219–221.
1969a Soil moisture, relative humidity, and microbial abundance in dry valleys of southern Victoria Land. Antarct. J. U.S., *4:* 23–28.
1969b Antarctic dry valley soil microbial incubation and gas composition. Antarct. J. U.S., *4:* 28–33.

Cameron, R. E., and J. R. Devaney
1970 Antarctic soil algal crusts: Scanning electron and optical microscope study. Trans. Am. Microsc. Soc., *89:* 264–273.

Cameron, R. E., and E. L. Merek
1971 Growth of bacteria in soils from antarctic dry valleys. Tech. Rep. 32-1522: 11 pp. Jet Propul. Lab., Calif. Inst. of Technol., Pasadena.

Cameron, R. E., G. B. Blank, and D. R. Gensel
1966a Desert soil collection at the JPL soil science laboratory. Tech. Rep. 32-977: 153 pp. Jet Propul. Lab., Calif. Inst. of Technol., Pasadena.
1966b Sampling and handling of desert soils. Tech. Rep. 32-908: 37 pp. Jet Propul. Lab., Calif. Inst. of Technol., Pasadena.

Cameron, R. E., J. King, and C. N. David
1968 Soil microbial and ecological studies in southern Victoria Land. Antarct. J. U.S., *3:* 121–123.

Cameron, R. E., G. B. Blank, and N. H. Horowitz
1969a Bacterial growth in agar subjected to freezing and thawing. Cryog. Technol., *5:* 253–255.

Cameron, R. E., C. N. David, and J. King
1969b Soil toxicity in antarctic dry valleys. Antarct. J. U.S., *3:* 164–166.

Cameron, R. E., J. S. Hubbard, and A. B. Miller
1969c Microflora of the dry valleys of Antarctica (abstract). Bact. Proc., *33:* 15.

Cameron, R. E., D. D. Lawson, and C. Pazaree
1969d Thermoluminescence characterization of antarctic dry valley soils. Agron. Abstr., *S-5:* 106.

Cameron, R. E., G. B. Blank, and N. H. Horowitz
1970a Bacterial growth in agar subjected to freezing and thawing. Cryog. Technol., *6:* 16–18.

Cameron, R. E., J. King, and C. N. David
1970b Microbiology, ecology and microclimatology of soil sites in dry valleys of southern Victoria Land, Antarctica. *In* M. W. Holdgate (Ed.), Antarctic ecology. *2:* 702–716. Academic, London.
1970c Soil microbial ecology of Wheeler valley, Antarctica. Soil Sci., *109:* 110–120.

Cameron, R. E., F. A. Morelli, and H. P. Conrow
1970d Survival of microorganisms in desert soil exposed to five years of continuous very high vacuum. Tech. Rep. 32-1454: 11 pp. Jet Propul. Lab., Calif. Inst. of Technol., Pasadena.

Campbell, I. B., and G. G. C. Claridge
1967 Site and soil differences in the Brown Hills region of the Darwin Glacier, Antarctica. N. Z. Jl Sci., *10:* 563–577.
1969 A classification of frigic soils—The zonal soils of the antarctic continent. Soil Sci., *107:* 75–85.

Cantzlaar, G. L.
1964 Your guide to the weather, pp. 77–79. Barnes and Noble, New York.

Chapman, H. D., and P. F. Pratt
1961 Methods of analysis for soils, plants, and waters. 309 pp. University of California Press, Berkeley.

Chaudhry, I. A., and A. H. Cornfield
1971 Low-temperature storage for preventing changes in mineralizable nitrogen and sulphur during storage of air-dry soils. Geoderma, *5:* 165–168.

Chen, A. W.-C., and D. M. Griffin
1966 Soil physical factors and the ecology of fungi. 5. Further studies in relatively dry soils. Trans. Br. Mycol. Soc., *49:* 419–425.

Chen, C.-P.
1963 The basic characteristics of Ku-erh-pan-t'ung-ku-t'e desert of Dzungaria. Transl. TT: 66-35365. U.S. Dep. of Commer., Washington, D. C.

Christian, J. H. B., and M. Ingram
1959 The freezing points of bacterial cells in relation to halophilism. J. Gen. Microbiol., *20:* 27–31.

Claridge, G. G. C.
1965 The clay mineralogy and chemistry of some soils from the Ross Dependency, Antarctica. N. Z. Jl Geol. Geophys., *8:* 186–220.

Claridge, G. G. C., and I. B. Campbell
1968 Origin of nitrates. Nature, *217:* 428–430.

Clark, F. E.
1967 Bacteria in soil. *In* A. Burges and F. Raw (Eds.), Soil biology, pp. 15–49. Academic, London.

Clark, R. H.
1965 The oases in the ice. *In* T. Hatherton (Ed.), Antarctica, pp. 321–330. Praeger, New York.

Conaway, J., and C. H. M. van Bavel
1967 Radiometric surface temperature measurements and fluctuations in sky radiant emittance in the 600 to 1300 cm^{-1} waveband. Agron. J., *59:* 389–390.

Curl, H., Jr., and E. Sutton
1967 Physiological ecology of cryophilic algae. Semiannu. Progr. Rep. NASA ER-87938: 5 pp. Nat. Aeronaut. and Space Admin., Corvallis, Oreg.

Daniels, F.
1968 Early studies of thermoluminescence in geology. *In* D. J. McDougall (Ed.), Thermoluminescence of geological materials, pp. 3–11. Academic, New York.

Daubenmire, R. F.
1959 Plants and environment. 2nd ed., 422 pp. John Wiley, New York.

Davies, W. E.
1961a Geology of northern Greenland. Polarforschung, *5:* 93–104.
1961b Surface features of permafrost in arid areas. *In* G. O. Raasch (Ed.), Geology of the Arctic. *2:* 981–987.

Davies, W. E., and D. B. Krinsley
1961 Evaluation of arctic ice-free land sites Kronprins Christian Land and Peary Land, North Greenland 1960. Air Force Surv. Geophys., 135: 51 pp.

Davies, W. E., D. B. Krinsley, and A. H. Nicol
1963 Geology of the North Star Bugt area, northwest Greenland. Meddr Grønland, *162:* 1–68.

Davis, G. H. G., and K. G. Newton
1969 Numerical taxonomy of some named coryneform bacteria. J. Gen. Microbiol., *56:* 195–214.

de Vries, D. A.
1958 Note on the heat exchange between soil and air under the influence of an initial temperature difference. *In* Arid zone research. 11. Climatology and microclimatology, pp. 114–122. Unesco, Paris.

di Menna, M. E.
1960 Yeasts from Antarctica. J. Gen. Microbiol., *23:* 295–300.
1966 Yeasts in antarctic soils. Antonie van Leeuwenhoek, *32:* 29–39.

Dodge, C. W.
1965 Lichens. *In* P. Van Oye, J. Van Mieghem, and J. Schell (Eds.), Biogeography and ecology in Antarctica, pp. 194–200. Junk, The Hague.

Dodge, C. W., and G. E. Baker
1938 Lichens and lichen parasites. Ann. Mo. Bot. Gdn, *25*(2) : 515–718, figs. 1–431, pls. 38–65.

Dort, W., Jr.
1967 Internal structure of sandy glacier, southern Victoria Land, Antarctica. J. Glaciol., *6:* 529–540.
1970 Climatic causes of alpine glacier fluctuation, southern Victoria Land. *In* International symposium on antarctic glaciological exploration, pp. 358–362. Hanover, N. H.

Drouet, F.
1962 The Oscillatoriaceae and their distribution in Antarctica. Polar Rec., *11:* 320–321.
1963 Ecophenes of *Schizothrix calcicola*. Proc. Acad. Nat. Sci. Philad., *115:* 261–281.

Durrell, L. W.
1956 Microgreenhouses. Green Thumb, *16:* 17.

Fairbridge, R. W.
1967 Relative humidity. *In* R. W. Fairbridge (Ed.), The encyclopedia of atmospheric sciences and astrogeology, Encycl. Earth Sci. Ser., *2:* 828–829. Reinhold, New York.

Fang, Y.
1963 Basic characteristics, formation, and development of the lake basin topography of Tsinhai Hu. Transl. TT: 66-35365. U.S. Dep. of Commer., Washington, D. C.

Farrell, J., and A. H. Rose
1967 Temperature effects on microorganisms. A. Rev. Microbiol., *21:* 101–120.
1968 Cold shock in a mesophilic and a psychrophilic pseudomonad. J. Gen. Microbiol., *50:* 429–439.

Fiskell, J. G. A.
1970 Cation exchange capacity and component variations of soils of southeastern U.S.A. Proc. Soil Sci. Soc. Am., *35:* 723–727.

Flowers, E. C., and N. F. Helfert
1966 Laboratory and field investigations of Eppley radiation sensors. Mon. Weath. Rev. U.S. Dep. Agric., *94:* 259–264.

Fogg, G. E.
1962 Nitrogen fixation. *In* R. A. Lewin (Ed.), Physiology and biochemistry of algae, pp. 161–170. Academic, New York.

Fogg, G. E., and J. H. Belcher
1961 Physiological studies on a planktonic ‘μ-alga.’ Verh. Int. Verein. Theor. Angew. Limnol., *14:* 893–896.

Fogg, G. E., and A. J. Horne
1970 The physiology of antarctic freshwater algae. *In* M. W. Holdgate (Ed.), Antarctic ecology. *2:* 632–637. Academic, London.

Fristrup, B.
1952 High arctic deserts. *In* Proceedings of the international geological congress. Sect. 7: pp. 91–99. Algiers.
1953 Wind erosion within the arctic deserts. Geogr. Tidsskr., *52:* 51–65.

Fritschen, L. J., and P. R Nixon
1967 Microclimate before and after irrigation. *In* R. H. Shaw (Ed.), Ground level climatology, pp. 351–356. American Association for the Advancement of Science, Washington, D. C.

Gannutz, T. P.
1969 Effects of environmental extremes on lichens. Bull. Soc. Bot. Fr., 169–179.
1970 Photosynthesis and respiration of plants in the antarctic peninsular areas. Antarct. J. U.S., *5:* 49–52.

Gardner, W. E.
1970 The X-ray diffraction analysis and mineral identification of ten antarctic soil samples. Rep. 076011: 4 pp., 2 figs., 2 charts. Sloan Res. Ind., Santa Barbara, Calif.

Gates, D. M.
1968 Energy exchange between organisms and environment. Aust. J. Sci., *31:* 67–74.

Gerasimov, I. P., and R. P. Zimina
1968 Recent natural landscapes and ancient glaciation of the Pamir. *In* H. E. Wright, Jr., and W. H. Osburn (Eds.), Arctic and alpine environments, pp. 267–269. Indiana University Press, Bloomington.

Gibson, G. W.
1962 Geological investigations in southern Victoria Land, Antarctica. 8. Evaporite salts in the Victoria valley region. N. Z. Jl Geol. Geophys., *5:* 361–374.

Goos, R. D., E. E. Davis, and W. Butterfield
1967 Effect of warming rates on the viability of frozen fungus spores. Mycologia, *59:* 58–66.

Goto, S., J. Sugiyama, and H. Iizuka
1969 A taxonomic study of antarctic yeasts. Mycologia, *61:* 748–774.

Greene, C. R., and W. D. Sellers (Eds.)
1964 Arizona climate. 503 pp. University of Arizona Press, Tucson.

Greene, S. W.
1965 Plants of the land. *In* R. Priestley, R. J. Adie, and G. D. Robin (Eds.), Antarctic research, pp. 240–253. Butterworths, London.

Greene, S. W., and R. E. Longton
1970 The effects of climate on antarctic plants. *In* M. W. Holdgate (Ed.), Antarctic ecology. *2:* 786–800. Academic, London.

Greene, S. W., et al.
1967 Terrestrial life of Antarctica. Antarctic Map Folio Ser., folio 5, pp. 1–14. Amer. Geogr. Soc., New York.

Gromov, B. V.
1957 The microflora of rock layers and primitive soils of some northern districts of the USSR. Microbiology, *26:* 57–63.

Gunn, B. M., and G. Warren
1962 Geology of Victoria Land between the Mawson and Mulock glaciers, Antarctica. Bull. Geol. Surv. N. Z., *71:* 157 pp., 2 maps.

Haas, A. R. C.
1944 The turmeric determination of water-soluble boron in soils of citrus orchards in California. Soil Sci., *58:* 123–137.

Hagen, C. A., and E. J. Hawrylewicz
1968a Life in extraterrestrial environments. Rep. IITRI-L6023-14: 13 pp. Ill. Inst. of Technol. Res., Chicago.
1968b Life in extraterrestrial environments. Rep. IITRI-L6023-15. Ill. Inst. of Technol. Res., Chicago.
1969 Life in extraterrestrial environments. Rep. IITRI-L6023-16: 13 pp. Ill. Inst. of Technol. Res., Chicago.

Hall, C. L., Jr.
1968 Isolation of psychrophilic Halophiles from the antarctic polar desert. M.S. thesis. Dep. of Biol., Va. Polytech. Inst., Blacksburg.

Hall, C. L., Jr., and R. E. Benoit
1969 Isolation of halophilic bacteria in the antarctic polar desert (abstract). Bact. Proc., *38:* 4.

Hanks, R. J., S. A. Bowers, and L. D. Bark
1961 Influence of soil surface conditions on net radiation, soil temperature, and evaporation. Soil Sci., *91:* 233–237.

Hariharan, P. S.
1961 The use of the Bellani spherical pyranometer for the measurement of total solar radiation. Indian J. Met. Geophys., *12:* 619–622.

Harrington, H. J.
1965 Geology and morphology of Antarctica. *In* P. Van Oye, J. Van Mieghem, and J. Schell (Eds.), Biogeography and ecology in Antarctica, pp. 1–71. Junk, The Hague.

Hawrylewicz, E. J., C. A. Hagen, and R. Ehrlich
1965 Response to a simulated Martian environment. Life Sci. Space Res., *3:* 64–73.

Hinman, W. C.
1970 Effects of freezing and thawing on some chemical properties of three soils. Can. J. Soil Sci., *50:* 179–182.

Holm-Hansen, O.
1963a Viability of blue-green and green algae after freezing. Physiologia Pl., *16:* 530–540.
1963b Algae: Nitrogen fixation by antarctic species. Science, *139:* 1059–1060.
1963c Effect of varying residual moisture content on the viability of lyophilized algae. Nature, *198:* 1014–1015.
1964 Isolation and culture of terrestrial and fresh-water algae of Antarctica. Phycologia, *4:* 44–51.
1967 Recent advances in the physiology of blue-green algae. *In* Environmental requirements of blue-green algae, pp. 87–96. Federal Water Pollution Control Administration, Corvallis, Oreg., and University of Washington, Seattle.

Holt, D. A., and H. W. Youngberg
1971 Influence of sampling frequency on the precision of estimating accumulated solar radiation. Agron. J., *63:* 240–241.

Hordon, R. M.
1967 Evapotranspiration. *In* R. W. Fairbridge (Ed.), The encyclopedia of atmospheric sciences and astrogeology, Encycl. Earth Sci. Ser., *2:* 372–373. Reinhold, New York.

Horowitz, N. H., et al.
1969 Sterile soil from Antarctica: Organic analysis. Science, *164:* 1054–1056.

Horowitz, N. H., R. E. Cameron, and J. S. Hubbard
1972 Microbiology of the dry valleys of Antarctica. Science, *176:* 242–245.

Hubbard, J. S., R. E. Cameron, and A. B. Miller
1968 Soil studies—Desert microflora. 15. Analysis of antarctic dry valley soils by cultural and radiorespirometric methods. Space Progr. Sum. 37-52, *3:* 172–175. Jet Propul. Lab., Calif. Inst. of Technol., Pasadena.

Hubbard, J. S., G. L. Hobby, N. H. Horowitz, P. J. Geiger, and F. A. Morelli
1970 Measurement of $^{14}CO_2$ assimilation of soils: An experiment for the biological exploration of Mars. Appl. Microbiol., *19:* 32–38.

Ichiye, T.
1967 Pressure—Atmospheric. *In* R. W. Fairbridge (Ed.), The encyclopedia of atmospheric sciences and astrogeology, Encycl. Earth Sci. Ser., *2:* 779–785. Reinhold, New York.

Imshenetskii, A. A.
1969 Microbiology in the United States. Microbiology, *38:* 950–956.

Ingram, M.
1957 Microorganisms resisting high concentrations of sugars or salts. *In* R. E. O. Williams and C. C. Spicer (Eds.), Microbial ecology, pp. 90–133. Cambridge University Press, Cambridge, England.

Jackson, M. L.
1958 Soil chemical analyses. 498 pp. Prentice-Hall, Englewood Cliffs, N. J.
1964 Chemical composition of soils. *In* F. E. Bear (Ed.), Chemistry of the soil. 2nd ed., pp. 71–141. Reinhold, New York.

Jain, B. L., and S. N. Saxena
1970 Distribution of soluble salts and boron in soils in relation to irrigation water. J. Indian Soc. Soil Sci., *18:* 175–182.

Janetschek, H.
1967 Arthropod ecology of south Victoria Land. *In* J. L. Gressitt (Ed.), Entomology of Antarctica, Antarctic Res. Ser., *10:* 205–293. AGU, Washington, D. C.
1970 Environments and ecology of terrestrial arthropods in the high Antarctic. *In* M. W. Holdgate (Ed.), Antarctic ecology. *2:* 871–885. Academic, London.

Jensen, H. I.
1916 Report on antarctic soils. 4. *In* British antarctic expedition, 1907–1909, pp. 89–92.

Jensen, H. L.
1951 Notes on the microbiology of soil from northern Greenland. Meddr Grønland, *142:* 23–29.

Johnson, R. M.
1970 Physiology of desert bacteria (abstract). Bact. Proc., *43:* 41.

Johnson, R. W.
1966 Planning and development of lunar bases. Astronautica Acta, *12:* 359–369.

Johnson, R. W., and P. M. Smith
1970 Antarctic research and lunar exploration. Adv. Space Sci. Technol., *10:* 1–44.

Jones, D., and P. H. A. Sneath
1970 Genetic transfer and bacterial taxonomy. *34:* 40–81.

Jones, R. W., and R. A. Hedlin
1970 Nitrite instability in three Manitoba soils. Can. J. Soil Sci., *50:* 339–345.

Kelley, W. P.
1948 Cation exchange in soils. 144 pp. Reinhold, New York.

Krasil'nikov, N. A.
1956 The microflora of high-altitude rocks and its nitrogen-fixing activity (in Russian). Usp. Sovrem. Biol., *41:* 177–192. (Engl. transl., Jet Propul. Lab., Calif. Inst. of Technol., Pasadena, 1967.)

Kriss, A. E.
1940 Microorganisms in the permanently frozen subsoil (in Russian with English summary). Mikrobiologia, *9:* 879–887.

Lamb, I. M.
1970 Antarctic terrestrial plants and their ecology. *In* M. W. Holdgate (Ed.), Antarctic ecology. *2:* 733–751.

Larsen, S., and A. E. Widdowson
1968 Chemical composition of soil solution. J. Sci. Fd Agric., *19:* 693–695.

Lawson, D. D., J. Ingham, and R. F. Landel
1970 Thermoluminescence as a forensic laboratory tool. New Technol. Rep., Case 2271: 15 pp. Jet Propul. Lab., Calif. Inst. of Technol., Pasadena.

Lederberg, J., and C. Sagan
1962 Microenvironments for life on Mars. Proc. Natn. Acad. Sci. U.S.A., *48:* 1473–1475.

Lindsey, R.
1966 Similarity to space beckons scientists. Technol. Week, *19:* 34, 37.

Linkletter, G. O.
1970 Weathering and soil formation in the dry valleys of Victoria Land, Antarctica. Antarct. J. U.S., *5:* 104.

Llano, G. A.
1962 The terrestrial life of the Antarctic. Scient. Am., *207:* 212–230.
1965 The flora of Antarctica. *In* T. Hatherton (Ed.), Antarctica, pp. 331–350. Praeger, New York.
1970 A survey of antarctic biology: Life below freezing. Bull. Atom. Scient., *26:* 67–74.

Lloyd, A. B.
1969 Behaviour of Streptomycetes in soil. J. Gen. Microbiol., *56:* 165–170.

Lo, L.-H.
1963 A study of the physiographical formation of western Szechuan and Western Yunnan. Transl. TT: 66-35365. U.S. Dep. of Commer., Washington, D. C.

Loewe, F.
1957 Precipitation and evaporation in the Antarctic. *In* P. van Roey (Ed.), Meteorology of the Antarctic, pp. 71–90. Weather Bureau, Pretoria.
1967a The water budget in Antarctica. JARE Scient. Rep., spec. issue 1: 101–110.
1967b On antarctic pressure variations. Q. Jl R. Met. Soc., *39:* 373–380.

Mack, A. R.
1963 Biological activity and mineralization of nitrogen in three soils as induced by freezing and drying. Can. J. Soil Sci., *43:* 316–324.

Marshall, B. J., W. G. Murrell, and W. J. Scott
1963 The effect of water activity, solutes and temperature on the viability and heat resistance of freeze-dried bacterial spores. J. Gen. Microbiol., *31:* 451–460.

Martin, W. P., T. F. Buehrer, and A. B. Caster
1942 Threshold *p*H value for the nitrification of ammonia in desert soils. Proc. Soil Sci. Soc. Am., *7:* 223–228.

Mather, J. R., and C. W. Thornwaite
1958 Microclimate investigations at Point Barrow, Alaska. Publs Clim. Johns Hopkins Univ., *9:* 63–239.

Matthews, S. W., and R. W. Madden
1968 Antarctica: Icy testing ground for space. Natn. Geogr. Mag., 134: 568–592.

Mazur, P.
1961 Manifestations of injury in yeast cells exposed to subzero temperatures. 1. Morphological changes in freeze-substituted and in 'frozen-thawed' cells. J. Bact., *82:* 662–672.

McCraw, J. D.
1960 Soils of the Ross Dependency, Antarctica. N. Z. Soc. Soil Sci. Proc., *4:* 30–35.
1967a Some surface features of McMurdo Sound region, Victoria Land, Antarctica. N. Z. Jl Geol. Geophys., *10:* 394–417.
1967b Soils of Taylor dry valley, Victoria Land, Antarctica, with notes on soils from other localities in Victoria Land. N. Z. Jl Geol. Geophys., *10:* 498–539.

McGeorge, W. T.
1940 Interpretation of soil analyses. Ext. Circ. 108: 9 pp. Univ. of Ariz. Coll. of Agr. and U.S. Dep. of Agr. Agr. Ext. Serv., Tucson.

McKelvey, B. C., and P. N. Webb
1962 Geological investigations in southern Victoria Land, Antarctica. 3. Geology of Wright valley. N. Z. Jl Geol. Geophys., *5*(1): 143–162.

McKenzie, L. J., E. P. Whiteside, and A. E. Erickson
1960 Oxidation–reduction studies on the mechanism of B horizon formation in podzols. Proc. Soil Sci. Soc. Am., *24:* 300–305.

McLaren, D., and J. Skujins
1964 Biochemical activities of terrestrial microorganisms in simulated planetary environments. Final Rep. 4: 11 pp. Space Sci. Lab., Univ. of Calif., Berkeley.

Metson, A. J.
1961 Methods of chemical analysis for soil survey samples. Soil Bur. Bull. 12: 208 pp. N. Z. Dep. of Sci. and Ind. Res., Wellington.

Meyer, G. H., M. B. Morrow, O. Wyss, T. E. Berg, and J. L. Littlepage
1962 Antarctica: The microbiology of an unfrozen saline pond. Science, *138:* 1103–1104.

Meyer, G. H., O. Wyss, and M. B. Morrow
1963 Incidence of soil microorganisms in Antarctica. Bull. Ecol. Soc. Am., *44:* 38–39.

Meyer, G. H., M. B. Morrow, and O. Wyss
1967 Bacteria, fungi, and other biota in the vicinity of Mirnyy Observatory. Antarct. J. U.S., *2:* 248–251.

Mikhaylov, I. S.
1962 The soils of polar wastes and the role of B. N. Gorodkov in their study. Izv. Vses. Geogr. Obschch. Voronezh. Otd., *94:* 520–523. (Engl. transl., Jet Propul. Lab., Calif. Inst. of Technol., Pasadena, 1967.)

Mishustin, E. N., and V. A. Mirzoeva
1964 The microflora of arctic soils. Problemy̅ Sev., no. 8: 170–199. (Engl. transl., Nat. Res. Counc. of Can., Ottawa, 1965.)

Mitchell, R. J.
1964 Trace elements in soils. *In* F. E. Bear (Ed.), Chemistry of the soil. 2nd ed., pp. 320–368. Reinhold, New York.

Mitchell, R. L.
1957 Spectrochemical methods in soil investigations. Soil Sci., *83:* 1–13.

Moghe, V. B., and B. M. Mathur
1966 Status of boron in some arid-soils of western Rajasthan. Soil Pl. Fd, Tokyo, *12:* 11–14.

Munro, D. C., and D. C. MacKay
1964 Effect of incubation method and storage conditions on nitrate production of incubated soil samples. Proc. Soil Sci. Soc. Am., *28:* 778–781.

Murzayve, E. M.
1967 Nature of Sinkiang and formation of the deserts of central Asia. Transl. 67-30944: 617 pp. U.S. Dep. of Commer., Washington, D. C.

Nichols, R. L.
1964 Present status of antarctic glacial geology. *In* R. J. Adie (Ed.), Antarctic geology, pp. 123–137. North-Holland, Amsterdam.
1966 Geomorphology of Antarctica. *In* J. C. F. Tedrow (Ed.), Antarctic soils and soil forming processes, Antarctic Res. Ser., *8:* 1–46. AGU, Washington, D. C.

Nielsen, K. F., and E. C. Humphries
1966 Effects of root temperature on plant growth. Soils Fertil., *29:* 1–7.

Nishita, H., and M. Hamilton
1970 Spurious thermoluminescence of soils. Soil Sci., *110:* 371–378.

Novogrudsky, D. M.
1946 Microbiological processes in semidesert soils. 1. Soil microorganisms and hygroscopic soil moisture (in Russian with English summary). Mikrobiologia, *15:* 177–186.

Olson, J. M.
1970 The evolution of photosynthesis. Science, *168:* 438–446.

Péwé, T. L.
1960 Multiple glaciation in the McMurdo Sound region, Antarctica—A progress report. J. Geol., *68:* 498–514.

Platt, R. B., and J. F. Griffiths
1964 Environmental measurement and interpretation. 235 pp. Reinhold, New York.

Polynov, B. B.
1953 The geological role of organisms. Vop. Geogr., *33:* 45–64. (Engl. transl., Israel Program for Sci. Transl., Jerusalem, 1965.)

Pratt, P. F.
1961 Effect of *p*H on the cation-exchange capacity of surface soils. Proc. Soil Sci. Soc. Am., *25:* 96–98.

Quispel, H.
1946 Measurement of the oxidation–reduction potentials of normal and inundated soils. Soil Sci., *63:* 265–275.

Ragotzkie, R. A., and G. E. Likens
1964 The heat balance of two antarctic lakes. Limnol. Oceanogr., *9:* 412–425.

Rastorfer, J. R.
1970 Effects of light intensity and temperature on photosynthesis and respiration of two east antarctic mosses, *Bryum argenteum* and *Bryum antarcticum.* Bryologist, *73:* 544–556.

Reeve, E., A. L. Prince, and F. E. Bear
1948 The boron needs of New Jersey soils. Bull. New Jers. Agric. Exp. Stn, 709: 4–20.

Rich, C. I.
1962 Removal of excess salt in cation-exchange capacity determinations. Soil Sci., *93:* 87–94.

Richards, L. A. (Ed.)
1954 Diagnosis and improvement of saline and alkali soils. Agr. Handb. 60: 160 pp. U.S. Department of Agriculture, Washington, D. C.

Richards, S. J., R. M. Hagan, and T. M. McCalla
1952 Soil temperature and plant growth. *In* B. T. Shaw (Ed.), Soil physical conditions and plant growth. *2:* 303–480. Academic, New York.

Riederer-Henderson, M.-A., and P. W. Wilson
1970 Nitrogen-fixation by sulphate-reducing bacteria. J. Gen. Microbiol., *61:* 27–31.

Ronca, L. B.
1964 Minimum length of frigid conditions in Antarctica as determined by thermoluminescence. Am. J. Sci., *262:* 767–781.

Rubin, M. J.
1965 Antarctic climatology. *In* J. Van Mieghem, P. Van Oye, and J. Schell (Eds.), Biogeography and ecology in Antarctica, pp. 72–96. Junk, The Hague.
1970 Antarctic meteorology. Bull. Atom. Scient., *26:* 48–54.

Rubin, M. J., and W. S. Weyant
1965 Antarctic meteorology. *In* T. Hatherton (Ed.), Antarctica, pp. 375–401. Praeger, New York.

Rudolph, E. D.
1965 Antarctic lichens and vascular plants: Their significance. Bioscience, *15:* 285–287.
1966 Terrestrial vegetation of Antarctica: Past and present studies. *In* J. C. F. Tedrow (Ed.), Antarctic soils and soil forming processes, Antarctic Res. Ser., *8:* 109–124. AGU, Washington, D. C.
1967 Climate. *In* V. C. Bushnell (Ed.), Antarctic Map Folio Ser., folio 5, p. 5. Amer. Geogr. Soc., New York.
1970 Local dissemination of plant propagules in Antarctica. *In* M. W. Holdgate (Ed.), Antarctic ecology. *2:* 812–817. Academic, London.

Rusin, N. P.
1964 Meteorological and radiation regime of Antarctica (in Russian). Gidrometeorologicheskoe Izdatelstvo, Leningrad, USSR, 1961. (Engl. transl., Israel Program for Sci. Transl., Jerusalem, 1964.)
1965 Some results of and prospects for the study of the meteorology of the air layer near the ground in Antarctica (in Russian). Inf. Byull. Sov. Antarkt. Eksped., *6*(2): 161–162. (Engl. transl., Scripta Technica, Inc., Washington, D. C., 1966.)

Sato, M., and H. Takahashi
1968 Cold shock of germinating *Bacillus subtilus* spores. Agric. & Biol. Chem. (Jap.), *32:* 1270–1274.
1969 Cold shock of bacteria. 2. Magnesium-mediated recovery from cold shock and existence of two critical temperature zones in various bacteria. J. Gen. Appl. Microbiol., Tokyo, *15:* 217–229.

Schofield, E., and E. D. Rudolph
1969 Factors influencing the distribution of antarctic terrestrial plants. Antarct. J. U.S., *4:* 112–113.

Scott, W. J.
1957 Water relations of food spoilage microorganisms. Adv. Fd Res., *7:* 83–127.
1961 Available water and microbial growth. *In* Proceedings of the low temperature microbiology symposium, 1961, pp. 89–105. Campbell Soup Co., Camden, N. J.

Seatz, L. F., and H. B. Peterson
1964 Acid, alkaline, saline, and sodic soils. *In* F. E. Bear (Ed.), Chemistry of the soil. 2nd ed., pp. 292–319. Reinhold, New York.

Sergeyeva, G. G., and S. I. Sokolov
1966 Wind distribution characteristics over Antarctica (in Russian). Inf. Byull. Sov. Antarkt. Eksped., *6*(2): 164–166. (Engl. transl., Scripta Technica, Inc., Washington, D. C., 1967.)

Shaw, N.
1970 Bacterial glycolipids. Bact. Rev., *34:* 365–377.

Shaw, P. J. R.
1960 Local winds in the Mawson area. *In* Antarctic meteorology, pp. 3–8. Pergamon, New York.

Shirley, H. L.
1935 Light as an ecological factor and its measurement. Bot. Rev., *1:* 355–381.
1945 Light as an ecological factor and its measurement. 2. Bot. Rev., *11:* 497–532.

Shneour, E. A.
1966 Oxidation of graphitic carbon in certain soils. Science, *151:* 991–992.

Sinclair, N. A., and J. L. Stokes
1965 Obligately psychrophilic yeasts from the polar regions. Can. J. Microbiol., *11:* 259–269.

Siple, P. A.
1938 Ecology and geographic distribution. Ann. Mo. Bot. Gdn, *25*(2): 467–515, 37 pls.

Slater, P. J.
1969 Soil moisture characteristics and their agricultural significance. Natural vegetation research report for 1968, pp. 16–28. Wellesbourne, Warwick, England.

Slater, P. J., and J. B. Williams
1969 The influence of texture on the moisture characteristics of soil. 10. Relationships between particle-size composition and moisture contents at the upper and lower limits of available water. J. Soil Sci., *20:* 126–131.

Smith, H. V., G. E. Draper, and W. H. Fuller
1964 The quality of Arizona irrigation waters. Agr. Exp. Sta. Rep. 223: 96 pp. Univ. of Ariz., Tucson.

Smith, R. L.
1966 Ecology and field biology. Chap. 16, pp. 289–300. Harper and Row, New York.

Soil Survey Staff
1951 Soil survey manual. Agr. Handb. 18: 503 pp. U.S. Department of Agriculture, Washington, D. C.
1960 Soil classification. 7th approximation, 265 pp. U.S. Department of Agriculture, Washington, D. C.

Solopov, A. V.
1967 Oases in Antarctica (in Russian). Meteorol. Rep. 14. Acad. of Sci. of the USSR, Moscow. (Engl. transl., Israel Program for Sci. Transl., Jerusalem, 1969.)

Stevenson, F. J.
1959 Carbon–nitrogen relationships in soil. Soil Sci., *88:* 201–208.

Stevenson, I. L.
1964 Biochemistry of soil. *In* F. E. Bear (Ed.), Chemistry of the soil. 2nd ed., pp. 242–291. Reinhold, New York.

Stewart, D.
1964 Antarctic mineralogy. *In* R. J. Adie (Ed.), Antarctic geology, pp. 395–401. North-Holland, Amsterdam.

Stone, R. O.
1967 A desert glossary. Earth Sci. Rev., *3:* 211–268.

Straka, R. P., and J. L. Stokes
1960 Psychrophilic bacteria from Antarctica. J. Bact., *80:* 622–625.

Stuhlinger, E.
1969 Antarctic research, A prelude to space research. Antarct. J. U.S., *4:* 5–11.

Sukapure, R. S., M. P. Lechavelier, and H. Prauser
1970 Motile nocardoid Actinomycetales. Appl. Microbiol., *19:* 527–533.

Sushkina, N. N.
1960 Characteristics of the microflora of arctic soils. Soviet Soil Sci., 4: 392–400.

Sutton, C. D.
1969 Effect of low soil temperature on phosphate nutrition of plants—A review. J. Sci. Fd Agric., *20:* 1–3.

Sutton, O. G.
1953 Micrometeorology. 333 pp. McGraw-Hill, New York.

Swan, L. W.
1961 The ecology of the high Himalayas. Scient. Am., *205:* 68–78.
1968 Alpine and aeolian regions of the world. *In* H. E. Wright, Jr., and W. H. Osburn (Eds.), Arctic and alpine environments, pp. 29–54. Indiana University Press, Bloomington.

Swinbank, W. C.
1967 Evaporation. *In* R. W. Fairbridge (Ed.), The encyclopedia of atmospheric sciences and astrogeology, Encycl. Earth Sci. Ser., *2:* 368–372. Reinhold, New York.

Syers, J. K., A. S. Campbell, and T. W. Walker
1970 Contribution of organic carbon and clay to cation exchange capacity in a chronosequence of sandy soils. Pl. Soil, *33:* 104–112.

Tauber, G. M.
1960 Characteristics of antarctic katabatic winds. *In* Antarctic meteorology, pp. 52–64. Pergamon, New York.

Tedrow, J. C. F.
1966a Properties of sand and silt fractions in New Jersey soils. Soil Sci., *101:* 24–30.
1966b Polar desert soils. Proc. Soil Sci. Soc. Am., *30:* 381–387.
1968a Soil investigations in Inglefield Land, Greenland. Final report to U.S. Army Research Office, Durham, contract DA-ARO-D-31-124-G820: 126 pp. Arctic Inst. of N. Amer., Montreal.
1968b Pedogenic gradients of the polar regions. J. Soil Sci., *19:* 197–204.

Tedrow, J. C. F., and J. Brown
1961 Soils of the northern Brooks Range, Alaska: Weakening of the soil-forming potential at high arctic altitudes. Soil Sci., *93:* 254–261.

Tedrow, J. C. F., and F. C. Ugolini
1966 Antarctic soils. *In* J. C. F. Tedrow (Ed.), Antarctic soils and soil forming processes, Antarctic Res. Ser., *8:* 109–124. AGU, Washington, D. C.

Tedrow, J. C. F., F. C. Ugolini, and H. Janetschek
1963 An antarctic saline lake. N. Z. Jl Sci., *6:* 150–156.

Thomas, G. W., and D. E. Kissel
1970 Nitrate volatilization from soils. Proc. Soil Sci. Soc. Am., *34:* 828–830.

Thompson, D. C., and W. J. P. MacDonald
1962 Radiation measurements at Scott Base. N. Z. Jl Geol. Geophys., *5:* 874–909.

Torii, T., and J. Ossaka
1965 Antarcticite: A new mineral, calcium chloride hexahydrate, discovered in Antarctica. Science, *149:* 975–977.

Ugolini, F. C.
1963 Pedological investigations in the lower Wright valley, Antarctica. Proceedings of the Permafrost International Conference. NAS-NRC Publ. 1287: 55–61. Purdue Univ., Lafayette, Indiana.
1967 Soils. *In* V. C. Bushnell (Ed.), Antarctic Map Folio Ser., folio 5, pp. 3–5. Amer. Geogr. Soc., New York.
1970 Antarctic soils and their ecology. *In* M. W. Holdgate (Ed.), Antarctic ecology. *2:* 673–692. Academic, London.

Ugolini, F. C., and J. Bockheim
1970 Biological weathering in Antarctica. Antarct. J. U.S., *5:* 122–123.

Ugolini, F. C., and C. Bull
1965 Soil development and glacial events in Antarctica. Quaternaria, *7:* 251–269.

Ugolini, F. C., and C. C. Grier
1969 Biological weathering in Antarctica. Antarct. J. U.S., *4:* 110–111.

Ugolini, F. C., and M. J. Perdue
1968 Biological weathering in Antarctica. Antarct. J. U.S., *3:* 166.

Van Baalen, C., D. S. Hoare, and E. Brandt
1971 Heterotrophic growth of blue-green algae in dim light. J. Bact., *105:* 685–689.

van Bavel, C. H. M.
1967 Surface energy balance of bare soil as influenced by wetting and drying. Tech. Rep. ECOM 2-67P-2: 72 pp. U.S. Army Electron. Comm., Atmos. Sci. Lab., Fort Huachuca, Ariz.

van Bavel, C. H. M., L. J. Fritschen, and R. J. Reginato
1963 Surface energy balance in arid lands agriculture, 1960–61. Prod. Res. Rep. 76: 46 pp. U.S. Dep. of Agr. and U.S. Army, Washington, D. C.

Van Wijk, W. R., and D. A. de Vries
1963 Periodic temperature variations in a homogeneous soil. *In* W. R. Wijk (Ed.), Physics of plant environment, pp. 102–143. North-Holland, Amsterdam.

Vinogradov, A. P.
1959 The geochemistry of rare and dispersed chemical elements in soils. 2nd ed., 209 pp. Consultants Bureau, New York.

Vlodavets, V. V.
1962 The effect of air humidity on the viability of microorganisms in an aerosol (in Russian). Mikrobiologiya, *31:* 350–356. (Engl. transl., C. T. Ostertag, Jr., Transl. 816, U.S. Army Biol. Lab., Fort Detrick, Md., 1963.)

Wade, F. A., and J. N. de Wys
1968 Permafrost features on the Martian surface. Icarus, *9:* 175–185.

Waksman, S. A.
1967 The Actinomycetes. A summary of current knowledge. 280 pp. Ronald, New York.

Warren, G.
1965 Geology of Antarctica. *In* T. Hatherton (Ed.), Antarctica, pp. 279–320. Praeger, New York.

Webb, P. N., and B. C. McKelvey
1959 Geological investigations in south Victoria Land, Antarctica. N. Z. Jl Geol. Geophys., *2:* 120–136.

Webb, S. J.
1963 The effect of relative humidity and light on air-dried organisms. J. Appl. Bact., *26:* 307–313.

Webb, S. J., D. V. Cormack, and H. G. Morrison
1964 Relative humidity, inositol and the effect of radiations on air-dried microorganisms. Nature, *201:* 1103–1105.

Weyant, W. S.
1966 The antarctic climate. *In* J. C. F. Tedrow (Ed.), Antarctic soils and soil forming processes, Antarctic Res. Ser., *8:* 47–59. AGU, Washington, D. C.

Whetstone, R. R., W. O. Robinson, and H. G. Byers
1942 Boron distribution in soils and related data. Tech. Bull. 797: 32 pp. U.S. Dep. of Agr., Washington, D. C.

Whitman, W. C., and G. Wolters
1967 Microclimatic gradients in mixed grass prairie. *In* R. H. Shaw (Ed.), Ground level climatology. Publ. 86: 165–185. American Association for the Advancement of Science, Washington, D. C.

Wiklander, L.
1964 Cation and anion exchange phenomena. *In* F. E. Bear (Ed.), Chemistry of the soil. 2nd ed., pp. 163–205. Reinhold, New York.

Wilson, A. T.
1964 Origin of ice ages: An ice shelf theory for Pleistocene glaciation. Nature, *201:* 147–149.
1970 The McMurdo dry valleys. *In* M. W. Holdgate (Ed.), Antarctic ecology. *1:* 21–30. Academic, London.

Wilson, A. T., and D. A. House
1965 Chemical composition of South Polar snow. J. Geophys. Res., *70:* 5515–5518.

Wilson, C.
1968 Climatology of the cold regions of the southern hemisphere. Cold Reg. Sci. Eng. Monogr. 1-A3c: 77 pp. U.S. Army Mater. Command Cold Reg. Res. and Eng. Lab., Hanover, N. H.

Winston, P. W., and D. H. Bates
1960 Saturated solutions for the control of humidity in biological research. Ecology, *41:* 232–237.

World Meteorological Organization
1965 Measurement of radiation and sunshine. *In* Guide to meteorological practices. 2nd ed., *8:* 1–36. Geneva.

Young, R. S., P. Deal, J. Bell, and J. Allen
1963 Effects of diurnal freezing–thawing on survival and growth of selected bacteria. Nature, *199:* 1078–1079.

Young, R. S., P. Deal, and O. Whitefield
1967 Response of soil bacteria to high temperatures and diurnal freezing and thawing. Nature, *216:* 355–356.

Yuan, T. L., N. Gammon, Jr., and R. G. Leighty
1967 Relative concentrations of organic and clay fractions to cation-exchange capacity of sandy soils from several soil groups. Soil Sci., *104:* 123–128.

Zaneveld, J. S.
1969 Cyanophytan mat communities in some meltwater lakes at Ross Island, Antarctica. Proc. K. Ned. Akad. Wet., ser. C, *72:* 299–305.

Zeller, A. N.
1968 The influence of microclimate upon the thermoluminescence of rock. *In* D. J. McDougall (Ed.), Thermoluminescence of geological materials, pp. 507–518. Academic, New York.

Zeller, E. J., and W. C. Pearn
1960 Determination of past antarctic climate by thermoluminescence of rocks. IGY Bull., 33: 12–15.

CILIATED PROTOZOA OF THE ANTARCTIC PENINSULA

Jesse C. Thompson, Jr.

Roanoke College, Salem, Virginia 24153

Abstract. From December 17, 1968, to February 5, 1969, extensive collections of ciliated protozoa were made within a 40-mile radius of Palmer Station on Anvers Island, Palmer Archipelago, near the Antarctic Peninsula. Limited collections were made from Deception Island and King George Island in the South Shetland Islands. Samples were taken from melt water pools, fresh-water lakes, tidal pools, and marine waters. The ciliates encountered were cultured, fixed, and stained by the Chatton-Lwoff silver nitrate impregnation technique. Twenty-five genera representing seven orders are recorded. Detailed descriptions of 24 of these ciliates and limited descriptions of seven additional forms are given. Four species of these antarctic ciliates are described as new to science (*Cyclogramma membranella, Sathrophilus antarcticus, Uropedalium antarcticum,* and *Parauronema antarcticum*). Seven genera are new records for Antarctica (*Bursaria, Strombidium, Uropedalium, Parauronema, Diophrys, Urostyla,* and *Pleurotricha*).

From December 17, 1968, to February 5, 1969, extensive collections of ciliated protozoa were made within a 40-mile radius of Palmer Station, Anvers Island, Palmer Archipelago, near the Antarctic Peninsula, and limited collections were made from Deception Island and King George Island in the South Shetland Islands.

The literature presently contains only one detailed description of a ciliated protozoan from Antarctica [Thompson, 1965], although several genera and species are recorded with or without partial descriptions from stations on the continent or adjacent islands [Armitage and House, 1962; Bierle, 1969; Dillon, 1967; Dillon et al., 1968; Flint and Stout, 1960; Hada, 1964, 1966; Murray, 1910; Sudzuki, 1964; Sudzuki and Shimoizumi, 1967; Thomas, 1965].

Specific taxonomic assignments are made when the author believes that the genus is represented by adequately described species. In other cases taxonomic assignments are made only at the generic level when the genus is represented by inadequately described species. Unfortunately many ciliated protozoan genera are in need of revision, and improper specific assignments only compound the confusion that presently exists. Owing to the many and varied genera encountered in this investigation, such generic revisions are impossible at this time. The purpose of this paper is to provide morphologic data on the ciliates encountered for future workers in this field. The scheme of classification used in this paper is based on the taxa as given by Corliss [1961].

MATERIALS AND METHODS

Water samples from melt water pools, fresh-water lakes, tidal pools, and marine waters were collected and cultured in the laboratory. If water samples were devoid of algae or other organic materials, small pieces of dehydrated shrimp or fish muscle were used as a nutrient supplement. Cultures were allowed to stand for a period of time (often several days) and were then examined for ciliated protozoa.

Ciliates were studied with phase microscopy, and certain forms were photographed with 16-mm phase cinematography. Animals were then fixed and permanently stained by the Chatton-Lwoff silver nitrate impregnation technique. Note the importance of this technique in revealing the morphologic data so vital to the taxonomy of this group of organisms. Certain structures such as basal granules or kinetosomes of the somatic and buccal ciliature, cytoproct, contractile vacuole pores (CVP), and complex fibrillar networks selectively accumulate the silver nitrate, and, when this compound is reduced, the preceding structures are readily revealed. Unless it is otherwise noted in the text, 25 silver-impregnated animals were selected randomly for the collection of morphologic data.

SYSTEMATIC DISCUSSION OF CILIATED PROTOZOA

The following classification of the class Ciliata, subclass Holotricha, is used:

Order GYMNOSTOMATIDA

Family ENCHELYIDAE

***Urotricha* sp.**

Fig. 1

Collection site. This ciliate was collected from a fresh-water rock pool on top of the hill behind 'Old' Palmer Station.

Morphology. This ciliate is ovoid in shape and circular in cross section. The cytostome is located in a slightly flattened area at the anterior end. A short conical cytopharynx is ringed by what appear to be short trichites. The cytostome is greatly expansible because the ciliates were seen feeding on large protozoan cysts. The ciliary meridians number 22–25 and most often terminate short of the posterior end. The posterior end is free of cilia except for a single long caudal cilium, but this area contains scattered argentophilic granules and fibrils. Three argentophilic groups of granules are located just posterior to the cytostome and probably possess cilia. It could not be determined whether they were membranous in nature. One or two CVP are present in the cilia-free end just posterior to the ciliary meridians. A large spherical macronucleus is located near the center of the body and most often in the anterior half. Fusiform trichocysts are often present and are usually more concentrated in the clear posterior end. In silver-impregnated specimens the average size is 35.0 × 20.0 μ. The ranges in length and width are 40.5–29.4 and 25.0–15.7 μ, respectively.

Discussion. Urotricha has been reported in the older literature [Kahl, 1930–1935] to contain a ring of heavier cilia surrounding the cytostome. This ring was not clearly distinguished in the present investigation. However, the somatic ciliature, as indicated by the argentophilic kinetosomes in silver preparation, are closely opposed in rings around the cytostome and may explain the heavier cilia described in the older literature. Hada [1966] reported *Urotricha farcta* from Showa, McMurdo, and Mirnyy stations but gave no descriptions.

***Lacrymaria* sp.**

Fig. 2

Collection site. This species was collected in a rock tidal pool in the Argentine Islands.

Morphology. The animal is flask shaped and has a long necklike region in the anterior end one-third to one-half of the length of the body. A nipplelike protrusion is found at the anterior end. The animal constricts somewhat after fixation, most of the constriction occurring in the necklike region. Approximately 25–30 ciliary meridians are present. Silver-impregnated specimens show the meridians to be spiraled except in the necklike region, where they are straight. About two-thirds of the nipplelike protrusion is ciliated. Several small CVP appear at the bluntly pointed to pointed posterior end. The macronucleus, which is located in the posterior half of the body, is elongate and often appears somewhat bilobed. The macronucleus is approximately 35 μ long. The average size of silver-impregnated specimens is 207.4 × 63.2 μ. The range in length is 265.1–188.4 μ, and the range in width is 76.7–49.9 μ.

Discussion. No attempt was made to determine the species of *Lacrymaria* described above. It is hoped that the morphologic data included here will be useful in future studies of this genus. Bierle [1969] reported *Lacrymaria* sp. from Coast Lake at Cape Royds, Ross Island, in Victoria Land.

Family NASSULIDAE

***Cyclogramma membranella* n. sp.**

Fig. 3

Collection site. This ciliate was collected from a fresh-water rock pool on top of a hill behind Old Palmer Station. The rock pools were teeming with protozoa. This form was also collected from a melt water pool on Litchfield Island near Palmer Station.

Morphology. The animal is elongated and more or less circular in cross section. The anterior end is bent slightly to the left. The anterior ventral surface is slightly flattened, and the cytostomal region is slightly depressed. The small cytostome is located near the anterior end and is surrounded by trichites. Three membranellelike structures lie to the left of the cytostome and in silver preparation consist of two or three rows of basal granules. The anteriormost membranelle is also the farthest left, and its basal granular field averages 2.4 μ in length. The second membranelle lies just to the left and anterior to the cytostome, and its basal granular field averages 3.2 μ in length. The third membranelle is posterior and slightly to the left of the cytostome, and its basal granular field averages 2.5 μ in length. An undulating membranelike structure arches to the left over the cytostomal area, and the basal granular field averages 6.8 μ in length. In silver preparations it appears to consist of about two rows

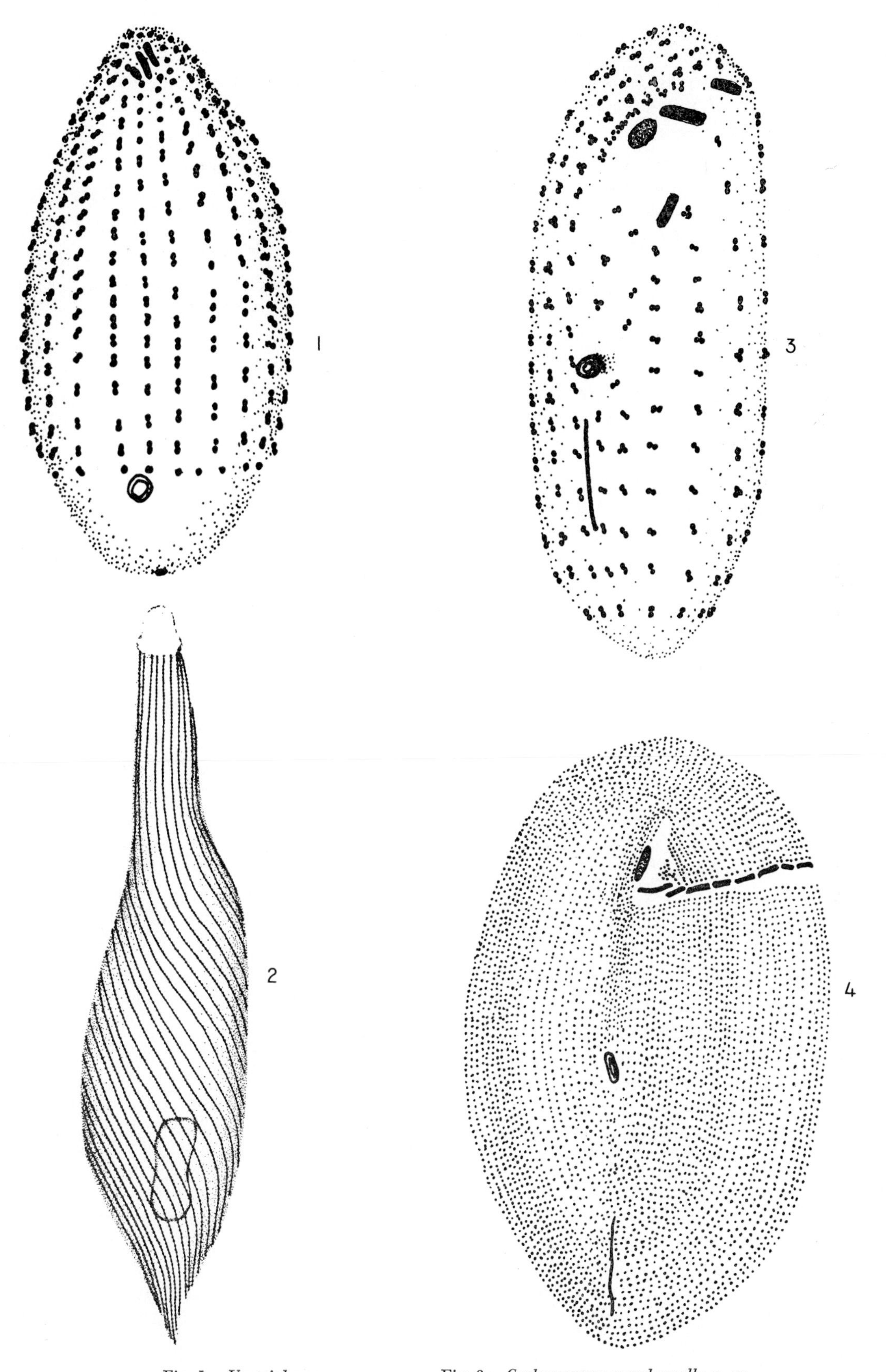

Fig. 1. *Urotricha* sp.
Fig. 2. *Lacrymaria* sp.
Fig. 3. *Cyclogramma membranella* n. sp.
Fig. 4. *Nassula* sp.

of granules. It was not possible to determine the nature of the membrane of this structure. The ciliate possesses about 25 ciliary meridians, of which five to seven are postcytostomal in position. The meridians on the right ventral and right dorsal surfaces arch to the left. A single CVP is located on the ventral surface near the center of the body but usually lies in the posterior half. A tubelike canal leads from the CVP into the cytoplasm. This canal appears to contain fiberlike structures. In silver preparation the cytoproct appears slitlike and is located just posterior to the CVP. Trichocysts are present. No caudal cilium was observed. In silver-impregnated specimens the average size is 37.8 × 16.5 μ. The range in length is 44.3–27.9 μ, and the range in width is 20.5–13.7 μ.

Discussion. In a recent review of the genus *Cyclogramma*, Fauré-Fremiet [1967] recognizes the following six species: *C. trichocystis*, *C. protectissima*, *C. sorex*, *C. rubens*, *C. tricirrata*, and *C. blochmanni* (which he describes as a new species).

Cyclogramma membranella n. sp. is the smallest member of this genus. The maximum length given for *C. membranella* is about 5 μ less than the minimum length given for any of the other species, and the minimum length is about 22 μ less than the minimum length given for any of the other forms.

Cyclogramma membranella n. sp. possesses seven fewer ciliary meridians than the minimum number described for any other species and eleven fewer than the maximum number given for any other form.

The position and the arrangement of the three membranellelike structures of *C. membranella* are unique for the genus. In no other species described is membranelle 2 (M_2) the largest, are membranelle 1 (M_1) and M_2 located anterior to the cytostome in a position approximately perpendicular to the long axis of the body, and is membranelle 3 (M_3) located posterior to the cytostome in a position approximately parallel to the long axis of the body. The third membranelle of *C. membranella* n. sp. is more posteriorly located in relation to the cytostome than the third membranelle of any other form and is the only species to show an argentophile-free area posterior to the cytostome.

Cyclogramma membranella n. sp. is characterized as the smallest species and has fewer meridians and an argentophile-free zone posterior to the cytostome. The position and the arrangement of the three membranelles are unique.

Bierle [1969] reports a form, *Cyclogramma* sp., from Coast Lake at Cape Royds, Ross Island, in Victoria Land, but insufficient morphologic data prevent taxonomic consideration.

Nassula sp.

Fig. 4

Collection site. This fresh-water ciliate was collected from a melt water pool on Humble Island and also from a fresh-water rock pool on the hill behind Old Palmer Station.

Morphology. The animal is ellipsoid to ovoid in shape and more or less circular in cross section. It appears slightly flattened on the ventral surface and has a shallow depression in the area of the cytostome. The cytostome is located about 25 μ from the anterior end. It is surrounded by a shallow vestibulum whose wall appears striated in silver-impregnated animals and may be membranous in nature.

About 20 large trichites arch posteriorly from the cytostome and pass deeply into the cytoplasm. The cytopharyngeal basket is somewhat expanded in the anterior region. From 12 to 16 membranelles appear at the anterior end and form a band that begins just posterior to the vestibulum. This adoral zone of membranelles extends laterally toward the left side of the ciliate and continues posteriorly around the body to the left dorsal surface. In silver-impregnated animals the bases of the membranelles were composed of four to five rows of basal granules. A cilia-free zone lies adjacent to the adoral zone of membranelles as well as in the area surrounding the vestibulum. The number of ciliary meridians was difficult to ascertain but was estimated to range from 95 to 120. A single CVP is located near the center of the ventral surface on a line just to the right of the vestibulum. It opens from a short tube deeper in the cytoplasm. This tube is often seen greatly enlarged. In silver preparations the cytoproct is a long wavy line posterior to the CVP and near the posterior end of the animal. No caudal cilium was observed. In silver-impregnated specimens the average size is 123.3 × 83.1 μ. The range in length is 148.0–102.6 μ, and the range in width is 109.0–66.2 μ.

Discussion. The genus *Nassula* was recorded by Sudzuki and Shimoizumi [1967] and Armitage and House [1962] from Skua Lake and Coast Lake on Ross Island and in the area of Commonwealth Glacier near McMurdo Sound. It was also reported by Dillon [1967] and Dillon et al. [1968] from melt water ponds near McMurdo Station. Bierle [1969] reports

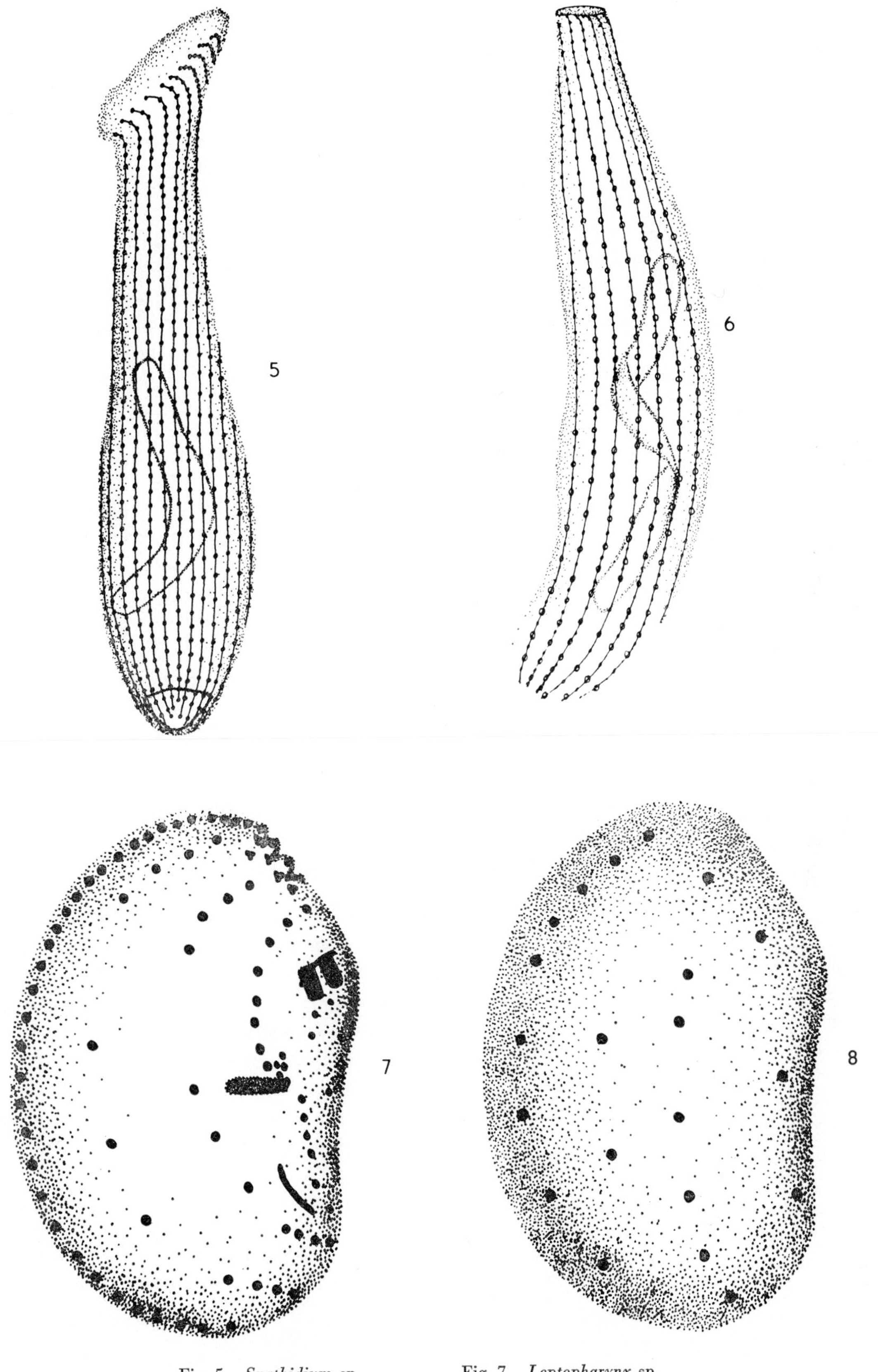

Fig. 5. *Spathidium* sp.
Fig. 6. *Spathidium* sp.

Fig. 7. *Leptopharynx* sp.
Fig. 8. *Leptopharynx* sp.

Nassula aurea and *N. ornata* from Coast Lake, Victoria Land.

Family SPATHIDIIDAE

***Spathidium* sp.**

Fig. 5

Collection site. This fresh-water ciliate was collected from a melt water pool on Torgersen Island and also from Humble Island near Palmer Station.

Morphology. The animal is elongated, there being a gradual tapering in the anterior portion of the body. The tapered necklike region flares out to form a somewhat truncate anterior end that is set at about a 45° angle to the long axis of the body. The necklike region is somewhat flattened, but the posterior region is more or less rounded in cross section. The cytostome seems to extend along the entire anterior end. The number of ciliary meridians was extremely difficult to determine but was estimated to range from 26 to 40. The meridians are terminated short of the anterior end. The granules shown in Figure 5 are used to indicate the position of meridians and do not indicate the exact size or the number of basal granules of each meridian. The ciliary basal granules are numerous in each meridian and closely opposed. Several CVP are located near the rounded posterior end. A large spherical CVP that is often somewhat flattened in the anterior region is located at the posterior end of the animal. An elongated macronucleus that is often curved is located in the posterior half of the animal. In silver-impregnated specimens the average size is 152.5 × 39.5 μ. The range in length is 187.4–121.7 μ, and the range in width is 50.0–27.6 μ.

Discussion. Spathidium spathula was reported by Dillon et al. [1968] from a melt water pool near Palmer Station. However, only a photograph with no description was included. The genus was reported by Hada [1966] from Mirnyy Station, by Sudzuki [1964] from Langhovde Hills, and by Sudzuki and Shimoizumi [1967] in samples collected from Langhovde Hills, Showa and McMurdo stations, and Ongul and East Ongul islands. However, Sudzuki and Shimoizumi reported no specific station for the ciliate. Bierle [1969] reported *Spathidium spathula* and *Spathidium* sp. from Coast Lake, Victoria Land.

***Spathidium* sp.**

Fig. 6

Collection site. This fresh-water species was collected from a melt water pool on Bonaparte Point, Anvers Island, near Palmer Station.

Morphology. The animal is elongated and somewhat slender and has a slight taper in the anterior end, although in well-fed cultures the ciliates increase appreciably in size. However, if this increase occurs, the anterior portion of the body still remains slender. The animal is generally rounded in cross section, and this symmetry continues throughout the anterior end. The anterior end is truncated and set more or less perpendicular to the long axis of the body. The cytostome is located at the anterior end in a small funnel-shaped depression from which a short cytopharynx appears to extend into the cytoplasm. The number of ciliary meridians is estimated to range from 14 to 17 and to extend to the truncated anterior end. Trichocystlike pores are scattered on or near the meridians. Several CVP are present near the rounded posterior end. The macronucleus is very elongated and often appears loosely coiled near the middle of the body. In silver-impregnated specimens the average size is 74.4 × 18.1 μ, the range in length is 86.8–58.3 μ, and the range in width is 15.1–21.7 μ.

Discussion. The genus *Spathidium* contains >70 described species. In view of the lack of descriptive detail in the older literature, specific determination is practically impossible. Note that the presence of the small funnel-shaped depression at the anterior end is apparently new for this genus.

Family CHLAMYDODONTIDAE

***Chilodonella* sp.**

Collection site. This genus was collected in a tidal pool on one of the Argentine Islands.

Morphology. This form was identified with phase microscopic studies of the living culture, and only the size was determined. The average length and width were 112 and 48 μ, respectively.

Discussion. Bierle [1969] reported *Chilodonella uncinata* from Coast Lake, Victoria Land.

Order TRICHOSTOMATIDA

Family MICROTHORACIDAE

***Leptopharynx* sp.**

Figs. 7 and 8

Collection site. This fresh-water species was collected from a melt water pool on Litchfield Island.

Morphology. The animal is laterally compressed and very thin. The anterior end is somewhat rounded and notched. The dorsal surface is convex, and the ventral surface is more or less concave, especially in the anterior half, which contains the buccal apparatus. It is difficult to determine the ciliation in the living form, but in silver-impregnated specimens four anterior–posterior directed rows of argentophilic structures are found on both the right and the left sides. A series of grooves corresponding to the rows of argentophilic structures is present on both the right and the left sides of the animal. Most of these argentophilic structures appear compound and may represent both trichocyst pores and ciliary basal granules. Row 1, which is just to the right of the buccal apparatus, appears to end near the CVP, but a row of smaller granules just to the left of the CVP continues posteriorly. The number and the position of the argentophilic structures are shown in Figure 7. A row of smaller granules presumed to be ciliary basal granules lies just to the left of the buccal apparatus. A few argentophilic granules appear just posterior to the buccal apparatus and also just anterior to the CVP. Several clusters of small granules appear near the notched anterior end on the right side of the animal. The left side contains few argentophilic structures (Figure 8). The buccal apparatus is located in a shallow concavity on the right ventral surface and consists of a short undulating membranelike structure, two membranelles, a cytostome, and a cytopharyngeal basket. The undulating membranelike structure is very short and lies just to the right of the two membranelles. In silver-impregnated specimens the area is occupied by a short series of closely opposed granules presumed to be ciliary basal granules. In silver-impregnated animals the bases of the two membranelles are rectangular in shape and consist of approximately three rows of granules each. The cytostome lies deepest in the ventral concavity near the membranelles, and a small cytopharyngeal basket passes into the cytoplasm. A single CVP opens on the right ventral surface, and a tubelike canal leads from it into the cytoplasm. In silver-impregnated specimens the cytoproct shows up as a short dark line posterior to the CVP. A spherical macronucleus lies near the center of the body. The average size of silver-impregnated specimens is $27.7 \times 18.3\ \mu$. The range in length is $30.0–23.8\ \mu$, and the range in width is $20.5–15.4\ \mu$.

Discussion. The type species of this genus, *Leptopharynx costatus,* was described by Mermod [1914], but Kahl [1930–1935] considered this species synonymous with *Trichopelma sphagnetorum,* which was described by Levander [1900]. However, Corliss [1960] pointed out that *Trichopelma* was a preoccupied name and should have been replaced with *Leptopharynx.*

Six species of *Leptopharynx* (> *Trichopelma*) have been described in the literature. Only two species, *Leptopharynx costatus* as redescribed by Prelle [1961] and a new species *Trichopelma* (< *Leptopharynx*) *agilis* Savoie [1957], have been described by means of modern techniques. Prelle was the first worker to point out the presence of the two buccal membranelles, which had been overlooked in the work of Savoie.

In view of the small size of the antarctic form and the differences in the number of argentophilic complexes, it is most probable that this organism represents a new species. However, a complete restudy of the genus should be made before a new species is proposed.

Flint and Stout [1960] recorded *Trichopelma sphagnetorum* from McMurdo Station, and Sudzuki and Shimoizumi [1967] listed *Trichopelma* in samples collected from Langhovde Hills, Showa and McMurdo stations, and Ongul and East Ongul islands.

Order HYMENOSTOMATIDA

Family TETRAHYMENIDAE

Sathrophilus antarcticus n. sp.

Fig. 9

Collection site. This fresh-water species was collected on Bonaparte Point near Palmer Station in a melt water pool.

Morphology. The animal is elongated and has a somewhat pointed anterior end in slimmer forms and a more bluntly pointed anterior end in larger forms. The posterior end is rounded. The ciliary meridians number 10–11, 11 being the usual number. The first meridian just to the right of the cytoproct is postoral and ends just to the right of the base of the undulating membrane (UM). All other meridians terminate near the anterior and posterior ends. The buccal cavity occupies the anterior one-third of the ventral surface of the animal and extends almost to the anterior end of the animal. It is somewhat shallow in the anterior end but deepens in the area of the cytostome. The right lateral border of the buccal cavity is occupied by the UM. In silver-impregnated specimens the basal granules of this membrane begin just opposite the posterior end of M_2, continue posteriorly, and curve

Fig. 9. *Sathrophilus antarcticus* n. sp.

slightly around the right posterior border of the buccal cavity.

The number of basal granules in the UM is characteristically nine. The first seven granules are closely opposed and border the buccal cavity, but the last two granules are farther apart and slightly posterior to the buccal cavity. Occasionally the last two granules are not present. The line of basal granules in the UM measures 5.3 μ in length. The buccal cavity contains three ciliary membranelles, whose linear bases contain about three rows of granules. Membranelle 1 is the longest of the three, and its linear axis is oriented almost parallel to the long axis of the body. It is the anteriormost membranelle, and in silver-impregnated specimens the base measures 3.0 μ in length. Membranelle 2 is located just anterior to the anterior end of the UM, and the silver-impregnated base measures 2.5 μ in length. Membranelle 3 is the most posterior of the three, and its silver-impregnated base measures 2.0 μ along its longest axis. Silver-impregnated lines are present in the buccal cavity and are more numerous in the posterior end. The cytostome opens from the left dorsal wall of the buccal cavity, and the cytopharynx curves sharply into the posterior cytoplasm. The cytopharynx is often almost parallel to the long axis of the body. A single CVP opening from a short tubular canal is located near meridians 2–4 and the posterior end.

In silver-impregnated specimens the cytoproct appears as a long granular line posterior to the buccal cavity. Several argentophilic granules often appear near the anterior end of the cytoproct and posterior to the buccal cavity. A single long caudal cilium is present. The average size of silver-impregnated animals is 24.7 $\times$ 10.5 μ. The range in length is 28.2–23.1 μ, and the range in width is 12.0–9.4 μ.

Discussion. The type species of this genus, *Saprophilus agitatus,* was described by Stokes [1887], but Corliss [1960] found the genus name preoccupied by a coleopteran and proposed the generic name *Sathrophilus.* Kahl [1926] described four new species: *S. mobilis, S. putrinus, S. ovatus,* and *S. oviformis.* Later Kahl [1930–1935] described two additional species: *S. chlorophagus* and *S. muscorum.* Stout [1956] and Thompson and Cone [1963] redescribed *S. muscorum.*

Sathrophilus muscorum is easily distinguished from among the seven species listed above by its smaller size, larger buccal cavity, fewer meridians, and tube-like canal leading into the cytoplasm from a subterminal CVP. *Sathrophilus antarcticus* n. sp. is similar to *S. muscorum,* but there are important differences. There are fewer ciliary meridians in *S. antarcticus* (10–11) than in *S. muscorum* (15–17). The buccal cavity is located closer to the anterior end in *S. antarcticus,* and the length of the UM, as indicated by the basal granules, is shorter (5.3 μ compared to 8.8 μ in *S. muscorum*). The basal granules of the UM are larger in size and fewer in number in *S. antarcticus* (nine granules compared to approximately 25 in *S. muscorum*). Membranelle 1 is located closer to the anterior end of the body in *S. antarcticus,* and the basal granules of M_1 are oriented more perpendicular to the long axis of the body. Membranelle 2 in *S. antarcticus* does not possess the posteriorly directed bar of granules shown by Thompson and Cone [1963] for *S. muscorum.* The cytopharynx in *S. antarcticus* as seen in silver-impregnated animals is more parallel to the long axis of the body.

Because of the differences cited above, the present author believes that a new species, *Sathrophilus antarcticus,* should be proposed for this ciliate.

Family Uronematidae

***Uropedalium* sp.**

Fig. 10

Collection site. This species was collected in a tidal pool located on one of the Wauwermans Islands.

Morphology. The animal is elongated and has somewhat rounded anterior and posterior ends. In slimmer forms a polar cap is noticeable at the anterior end. The ciliary meridians number 15. These bipolar meridians are more or less symmetrical except that the posterior half of M_1 curves noticeably around the posterior portion of the buccal cavity. The remaining meridians on the right ventral surface curve slightly around the posterior end of the buccal cavity. Meridian 1 occurs just to the right of the buccal cavity, and the terminal granules of this meridian continue farther posteriorly than the granules of other meridians on the right ventral surface. The first meridian often appears devoid of basal granules in the area of the UM. A prominent space occurs between the last two posterior basal granules of all meridians except meridian 2, which is associated with the CVP. In meridian 2 a space occurs between the second and the third posterior basal granules. The space separating the last two basal granules is more prominent on the right ventral surface. The terminal basal granule of meridian 2 is located farther from the posterior end of the animal than the terminal granules of the other meridians. The terminal granules of the successive meridians on the right side become progressively closer to the posterior end until approximately at meridian 7 or 8 the terminal granules become equidistant from the posterior pole. Argentophilic lines connect the basal granules in each meridian. Anteriorly these lines join the anterior circular line. Posteriorly the lines except that from meridian 15 join the posterior circular line, which almost encircles the polar basal body complex. The posterior circular line terminates at meridian 14 and leaves an opening between meridian 14 and the cytoproct. An argentophilic line from meridian 15 joins the polar basal body complex. The buccal cavity extends along two-thirds or more of the ventral surface. There is no noticeable depression in the anterior portion, but it deepens in the posterior portion, especially in the area of the cytostome. The depression in the right posterior portion of the buccal cavity is further enhanced by an indentation in the surface of the animal just to the right of and posterior to the cytostomal area. The cytostome is located approximately in the posterior third of the animal and appears to lie close to the posterior half of the UM. The UM is located in the posterior half of the animal, and in silver-impregnated animals the base averages 6.1 μ in length. The anterior end of the UM is slightly anterior to the posterior end of M_2, and the posterior end terminates just posterior to the cytostomal area. Membranelle 3 is the most posterior of the three membranelles, and in silver-impregnated specimens its longest axis is perpendicular to the long axis of the body. The argentophilic base is presumed to consist of several basal granules. Membranelle 2 appears to be bipartite in silver-impregnated animals. The posterior portion is a small compact mass similar in composition to M_3 but slightly smaller, and the anterior portion consists of either smaller compact masses and single basal granules connected by an argentophilic line or merely single basal granules connected by an argentophilic line. The number of smaller compact masses and basal granules in M_2 is not consistent but generally ranges from five to eight. The length of M_2 exceeds the length of the UM. Membranelle 1 is anterior to and often slightly to the left of M_2. In silver-impregnated animals it usually consists of three or four isolated basal granules connected by an argentophilic line. Its length is approximately half that of M_2. The single CVP is associated with the posterior end of meridian 2. A single basal granule usually lies just anterior and posterior to the CVP. A long space occurs between the anterior granule associated with the CVP and the next anterior granule in meridian 2. The cytoproct is located posterior to the buccal cavity and very near the posterior end. In silver-impregnated animals the cytoproct appears as a heavy granular line. An argentophilic line continues posteriorly from the cytoproct and becomes continuous with the posterior circular line. Trichocystlike structures are occasionally observed, especially at the anterior end of the animal. A long caudal cilium is present. A rather large spherical macronucleus appears near the middle of the body. The average size of silver-impregnated animals is 24.7 $\times$ 16.8 μ, the range in length is 27.0–20.7 μ, and the range in width is 19.8 $\times$ 13.0 μ.

Discussion. No published report of the genus *Uropedalium* has appeared since Kahl [1928] described the type species *Uropedalium pyriforme*. Kahl [1930–1935] reported as *Uropedalium opisthostomum* a ciliate that Lepsi [1926] had described as *Uronema opisthostomum*. Both descriptions almost totally lack morphologic details, but they do call attention to the posteriorly located 'mouth.' Because of this characteristic and the general similarity of these animals to

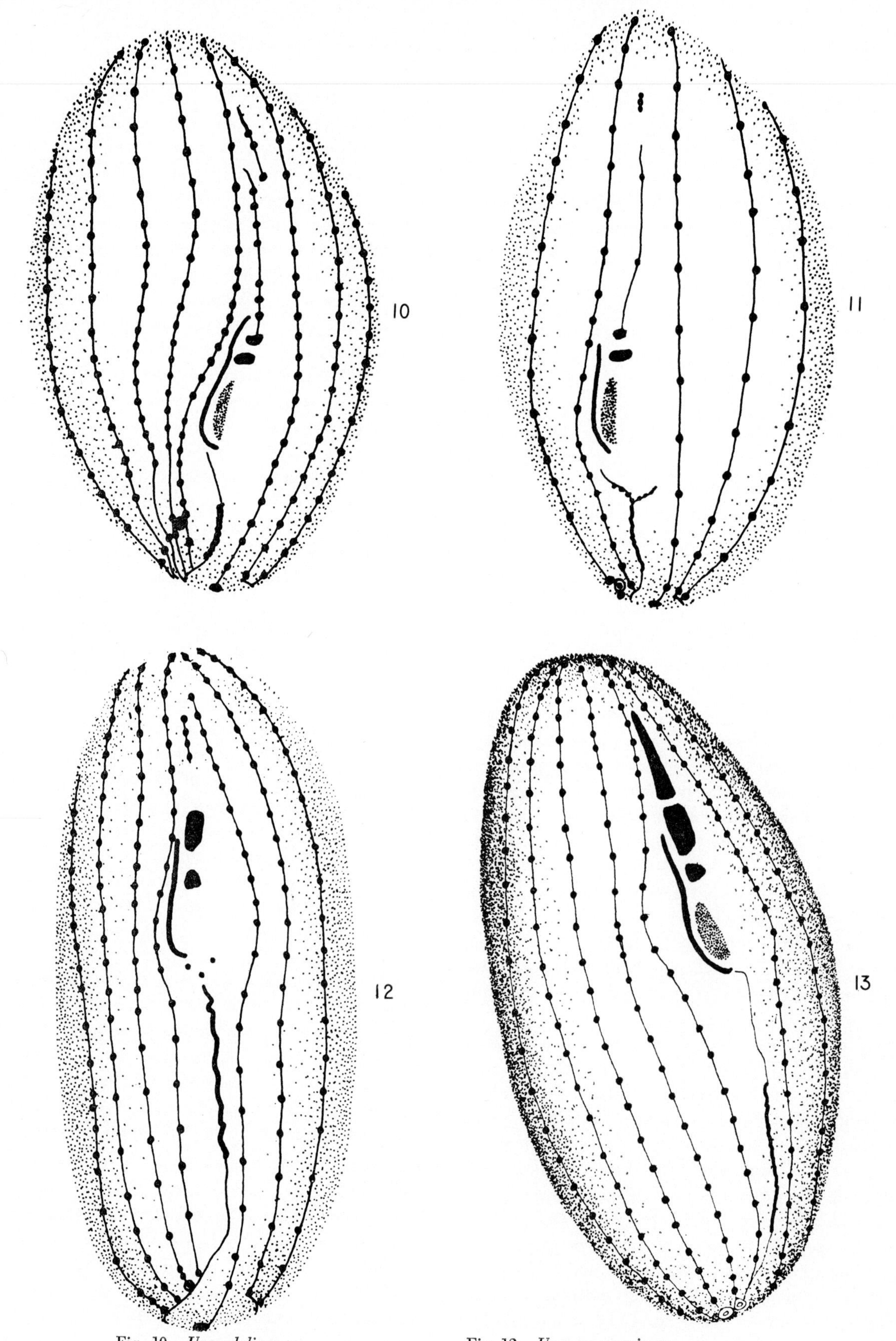

Fig. 10. *Uropedalium* sp.
Fig. 11. *Uropedalium antarcticum* n. sp.
Fig. 12. *Uronema marinum.*
Fig. 13. *Parauronema antarcticum* n. sp.

the drawings given for *U. pyriforme*, the present author believes that these two ciliates are congeneric with *Uropedalium* sp. as described above. A new species described as *Uronema parva* by Czapik [1968] is undoubtedly a member of the genus *Uropedalium*. The photographs of the silver-impregnated infraciliature, particularly those of the buccal apparatus, shown in Plate I of Czapik are strikingly similar to the description of the infraciliature of *Uropedalium* sp. Arthur C. Borror, University of New Hampshire, and Frederick R. Evans, University of Utah, have also studied forms similar to the ciliate described above (personal communication, 1970). Because of the similarity of the several forms studied by Borror and Evans and the ciliate described by Czapik to *Uropedalium* sp. described from Antarctica, it seems expedient to postpone a species designation until all workers have reviewed these data.

Uropedalium antarcticum n. sp.

Fig. 11

Collection site. This species was collected in a freshwater pool on Bonaparte Point near Palmer Station.

Morphology. The animal is elongated, and the anterior end tapers to form a blunt polar cap that is more obvious in slimmer forms. The posterior end shows less tapering and is more or less rounded at the caudal pole. The body is somewhat indented on the right ventral surface near the cytostomal area. The ciliary meridians number 12. The bipolar meridians are usually sinuously curved, the curve being greatest in the posterior portion of the meridians on the right ventral surface. Meridian 1 lies to the right of the buccal cavity, and the basal granules of this meridian are usually more closely opposed in the portion posterior to the UM than the basal granules of other meridians. The basal granules beside the UM are often absent in this meridian. The terminal basal granules of all the meridians seem to be about equidistant from the caudal pole. This position is in marked contrast to the position of the terminal granules of *Uropedalium* sp. Argentophilic lines join the basal granules in each meridian. The lines also pass anteriorly to join the anterior circular line and pass posteriorly, except the line from meridian 12, to join the posterior circular line. The posterior circular line almost encircles the polar basal body complex and joins all meridians except meridian 12. A line from meridian 12 joins the polar basal body complex. The buccal cavity extends along approximately two-thirds of the ventral surface of the body. There is no noticeable depression in the anterior portion, but the depression deepens in the area of the cytostome. The cytostome is located in the posterior third of the animal just to the right of the UM. The UM is located in the posterior half of the animal, and in silver-impregnated animals the base averages 5.5 μ in length. The base of the UM is slightly indented in the vicinity of M_3 and usually terminates just anterior to M_3. Occasionally granules pass anteriorly as far as the posterior end of M_2, but these granules are not closely opposed. Membranelle 3 is the most posterior of the three membranelles, and in silver-impregnated specimens its longest axis is perpendicular to the long axis of the body. Membranelle 2 lies anterior to M_3 and is less consistent in form. It often appears as a single argentophilic mass. At other times the single larger argentophilic mass will have smaller granules located anteriorly to it. The granules are attached to a long argentophilic line. The number of granules is usually two or three, but as many as seven have been counted. Membranelle 1 is located close to the anterior end and consists of from one to four argentophilic granules usually closely opposed. It was not possible to determine whether cilia were attached to the isolated basal granules of the membranelles. A single CVP is located near the caudal pole and is associated with meridian 2. One basal granule is usually just anterior to the CVP, and one is usually posterior to the CVP. The cytoproct is posterior to the buccal cavity near the posterior end of the animal. In silver-impregnated animals it appears as a granular line that is often bifurcated near the posterior end of the buccal cavity. Trichocystlike pores are associated with the ciliary meridians. A long caudal cilium is present. A large macronucleus is located near the center of the body most often in the anterior half. The average size of silver-impregnated specimens is 28.9–15.8 μ. The range in length is 34.0–24.8 μ, and the range in width is 18.6–13.0 μ.

Discussion. Morphologic similarities between *Uropedalium* sp. and *Uropedalium antarcticum* indicate their congeneric relationship. The posteriorly located cytostomal area, the depression on the right posterior surface of the body, and in particular the similarities in the buccal apparatus and body ciliation attest to this relationship. However, in the opinion of the present author, specific differences do exist. *Uropedalium antarcticum* is more tapered at the anterior and posterior ends, and the anterior polar cap is more prominent. There are fewer ciliary meridians in *U. antarcticum* (12) than in *U.* sp. (15). In *U. antarcticum* there are fewer basal granules for each ciliary meridian, and they are not so closely opposed. The location of the terminal basal granules of the meridians on the right ventral surface is noticeably different in *U. antarcticum*. In *U. antarcticum* they are approximately equidistant from the caudal pole, whereas in *U.* sp. from meridian 7 or 8 to meridian 3 they become progressively farther removed. The UM is shorter in *U. antarcticum* and less definite in the anterior region. Membranelle 2 of *U. antarcticum* contains fewer basal granules in the extended portion, and the basal granules of M_1 are more closely opposed. *Uropedalium antarcticum* was collected from a fresh-water habitat, whereas *Uropedalium* sp. was isolated from a marine tidal pool.

Because of the lack of morphologic data for *U. pyriforme* and *U. opisthostomum*, it is most difficult to compare these forms with *U. antarcticum*. Note that unnecessary confusion results when attempts are made to resurrect many poorly described forms from the older literature. It is most probable that *U. pyriforme*, *U. opisthostomum*, and *Uronema* (< *Uropedalium*) *parva* will be more specifically similar to *Uropedalium* sp. All these ciliates are marine forms.

Uronema marinum

Fig. 12

Collection site. This marine species was collected at the west end of Port Foster on Deception Island in the South Shetland Islands.

Morphology. The animal is elongated with a rounded posterior end. The anterior end is somewhat rounded but possesses a polar cap directed to the right side of the animal and also toward the posterior end of the body. The bipolar ciliary meridians number 15–17 and are more or less equally spaced except where the first and last meridians curve around the posterior region of the buccal cavity. The anteriormost basal granule of each meridian except that of the last lies near the anterior circular line that encloses a cilia-free area at the anterior pole. The anteriormost basal granule of the last meridian lies just anterior to M_1. The posteriormost basal granules of each meridian lie near the posterior circular line, which almost encircles the polar basal body complex. The posterior circular line is open between the next to last meridian and the first meridian. The basal granules are usually more closely opposed in the anterior half of the animal. In silver-impregnated specimens argentophilic lines join the basal granules of each meridian. Anteriorly these lines join the anterior circular line, and posteriorly they join the posterior circular line except the line from the last meridian, which joins the polar basal body complex. The buccal cavity extends along most of the anterior ventral surface of the body and contains a tetrahymenal buccal apparatus. The buccal cavity is more depressed in the cytostomal region. A UM extends along the right posterior border of the buccal cavity, and in silver-impregnated specimens its base averages 7.4 μ in length. The base of the UM is relatively straight in the anterior portion, but the posterior portion curves around the base of the buccal cavity. The anterior end of the argentophilic base lies near the middle of M_2, and the UM is slightly indented near M_3. The posterior edge of the UM is an average of 18 μ from the anterior end of the animal. The argentophilic base of M_1 consists of a single row of granules that appears to vary from four to six granules and averages 3.0 μ in length. The anterior end of M_1 is an average of 4.8 μ from the anterior end of the animal. The argentophilic base of M_2 averages 2.8 $\times$ 1.4 μ and is irregularly rectangular in shape. The argentophilic base of M_3 is also irregularly rectangular in shape and averages 1.5 $\times$ 1.3 μ. Only one CVP is usually present near the posterior end of meridian 2. Occasionally two CVP are present and are associated near the posterior ends of meridians 2 and 3. In silver-impregnated specimens the cytoproct appears as a heavy wavy line posterior to the buccal cavity. An argentophilic line continues from the posterior end of the cytoproct and joins the argentophilic line at the base of meridian 1. An argentophilic line continues anteriorly from the cytoproct and ends near the base of the buccal cavity. Several argentophilic granules are located anterior to the cytoproct and posterior to the buccal cavity. Trichocystlike pores are often observed in association with the ciliary meridians. A caudal cilium is present and is presumed to originate from the polar basal body complex. The average size of silver-impregnated specimens is 36.3 $\times$ 15.7 μ. The range in length is 44.0–31.1 μ, and the range in width is 18.8–12.8 μ.

Discussion. Species of the genus *Uronema* were the most commonly collected marine ciliates and appeared in 34 cultures collected at 10 sites (Arthur Harbor, a small island in Neumayer Channel, Port Lockroy, Wauwermans Islands, a small island in Hamburg Bay, Lemaire Island, Joubin Islands, Bonaparte Point, Deception Island, and Greenwich Island). The *Uronema marinum* described above is very similar to the *U. marinum* redescribed by Thompson [1964] and would appear to be conspecific with this form. However, a

TABLE 1. Comparison of *Uronema* Cultures

Culture	Average Size, μ	Contractile Vacuole Pores	Number of Meridians
A	33.1 × 17.0	1 near meridian 2	15–16
B	39.5 × 20.0	1 near meridian 2	15–16
C	40.6 × 20.7	1–2 near meridians 2 and 3	18–20
D	36.3 × 15.7	1 near meridian 2	15–17
E	39.3 × 22.1	1 near meridian 2	15–18
F	37.4 × 19.4	1–2 near meridians 2 and 3	15–17
G	43.3 × 22.2	1–2 near meridians 2 and 3	18–22
H	37.5 × 20.9	1 near meridian 2	15–18
I	37.8 × 19.6	1 near meridian 2	16–18
J	45.5 × 21.6	1–2 near meridians 2 and 3	16–21

comparative study of 10 of the cultures of *Uronema* reveals a range in body size, location of CVP, and number of meridians (Table 1). From the available data it would seem advisable to consider all these forms to be members of the *Uronema marinum* complex. Such complexes already exist for three species in the genus *Tetrahymena* [Corliss, 1970]. Bierle [1969] reported *Uronema* sp. from Coast Lake, Victoria Land.

Parauronema antarcticum n. sp.

Fig. 13

Collection site. This marine species was collected from a rock tidal pool on Greenwich Island, South Shetland Islands, which is located approximately 135 miles north-northeast of Palmer Station.

Morphology. The animal is elongated and has rounded anterior and posterior ends. It is rounded in cross section. The bipolar ciliary meridians number 18–19 and are more or less evenly distributed except where the first and last meridians curve around the posterior portion of the buccal cavity. The anteriormost basal granule of each meridian lies near the anterior circular line. The anteriormost granule of the last meridian often lies slightly posterior to the anterior granules of the other ventral meridians. The posteriormost basal granules of the ciliary meridian lie close to the posterior end of the animal. The posteriormost basal granule of meridian 1 usually lies farther from the caudal pole than the basal granules of the other meridians. The posterior terminal basal granule of the last meridian lies very close to the polar basal body complex. In silver-impregnated specimens argentophilic lines join the basal granules of each meridian. Anteriorly these argentophilic lines join the anterior circular line, and posteriorly the lines from all meridians except the last join the posterior circular line. The line from the last meridian joins the polar basal body complex. The posterior circular line almost encircles the polar basal body complex but is open between the next to the last meridian and the first meridian. The buccal cavity extends along the anterior ventral surface of the body and contains the tetrahymenal buccal apparatus. The buccal cavity is deepest in the area of the cytostome. A UM extends along the right side of the posterior portion of the buccal cavity. In silver-impregnated specimens the base of the UM averages 9.0 μ in length. Anteriorly it terminates near the middle of M_2, but it occasionally extends to near the anterior edge. The argentophilic base of the UM is relatively straight in the anterior portion, but it curves along the right side of the buccal cavity just posterior to M_3. The posterior edge of the UM averages 19.7 μ from the anterior end of the animal. In silver-impregnated specimens the triangular base of M_1 consists of numerous closely opposed basal granules and averages 5.5 μ in length and 1.4 μ at its greatest width. The apex of the triangular base is anteriorly directed and is an average of 3.1 μ from the anterior end of the animal. The argentophilic base of M_2 is irregularly rectangular in shape, although the posterior edge is often angular, and the base averages 3.3 × 1.9 μ. The argentophilic base of M_3 is irregularly rectangular in shape and averages 1.5 × 1.4 μ. The CVP number two to three. If two are present, they are associated at the terminal ends of ciliary meridians 2 and 3 near the posterior end of the animal and, if three are present, they are associated at those portions of meridians 2–4. The cytoproct is located between the first and the last meridian posterior to the buccal cavity. In silver-impregnated specimens the cytoproct appears as a heavy wavy line. An argentophilic line continues posteriorly from the cytoproct and joins the posterior circular line near meridian 1. A caudal cilium is present. A large spherical macronucleus is present near the middle of the body but usually in the anterior half. The average size of silver-impregnated specimens is 38.2 × 17.8 μ. The range in length is 45.0–33.0 μ, and the range in width is 21.7–14.8 μ.

Discussion. The genus *Parauronema* was erected and placed in the family *Uronematidae* by Thompson [1967]; it contains a single species, *P. virginianum.* A review of the generic characteristics of *Parauro-*

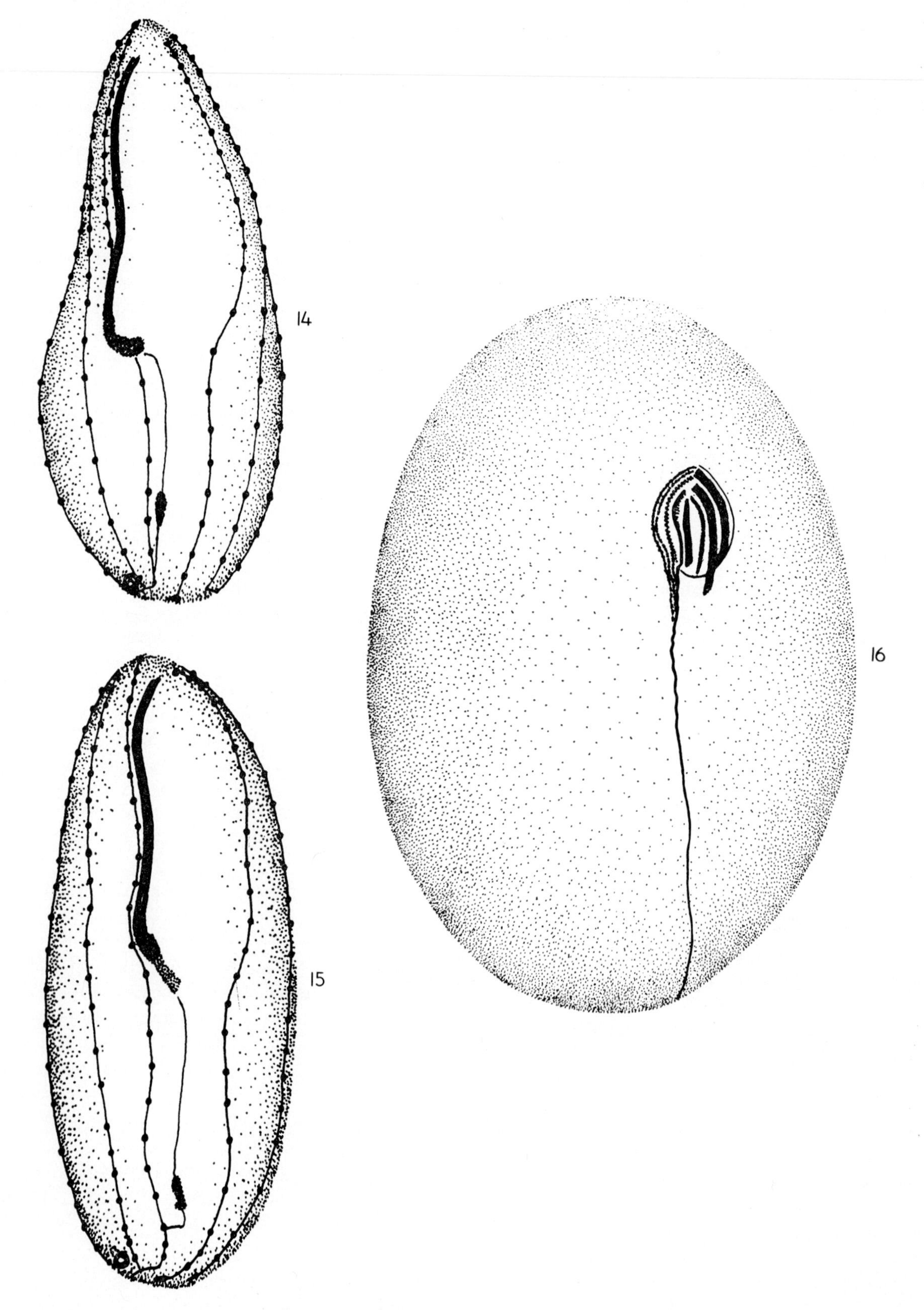

Fig. 14. *Pseudocohnilembus* sp. (strain 1).
Fig. 15. *Pseudocohnilembus* sp. (strain 2).
Fig. 16. *Frontonia* sp.

nema leaves no doubt as to the generic assignment of the ciliate described above. The somatic and buccal infraciliatures of *P. virginianum* and *P. antarcticum* are quite similar. There are certain differences between these two organisms that the present author believes important enough to warrant a new species name for the antarctic form. *Parauronema antarcticum* is a larger species (the size range for *P. virginianum* being 35.8–23.4 μ), and its anterior end is not bluntly pointed. *Paraurenema antarcticum* possesses a larger number of ciliary meridians (10–11 being given for *P. virginianum*). The number of CVP is two to three in *P. antarcticum* and only one in *P. virginianum*. Another difference is the greater length of M_1 in *P. antarcticum*. It is approximately twice the length of M_1 in *P. virginianum*. Also the length of M_1 in *P. antarcticum* is nearly twice the length of M_2, but the lengths of M_1 and M_2 in *P. virginianum* are approximately equal.

Family Pseudocohnilembidae

Pseudocohnilembus sp. (Strain 1)

Fig. 14

Collection site. This species was collected at the wave line of a small island in the Neumayer Channel. Melt water from the ice-covered island must have altered the salinity of the habitat.

Morphology. The animal is elongated and has a rounded posterior end. The anterior half of the body is noticeably tapered and curved slightly to the right. The anterior end is somewhat pointed. A polar cap is present and is directed posteriorly. The bipolar ciliary meridians number nine to 12 and are more or less evenly spaced except where the first and last meridians curve around the buccal cavity. The basal granules most often occur in pairs and are connected in silver-impregnated specimens by argentophilic lines. The anteriormost basal granules of all meridians lie near the anterior circular line although the basal granules of the dorsal meridians are slightly posterior. The posteriormost basal granules of all meridians lie very close to the posterior end of the animal and are closely opposed to the posterior circular line. The posterior circular line in silver-impregnated animals nearly encircles a small area at the posterior end containing the polar basal body complex. The posterior circular line is open between the next to last meridian and the first meridian, which lies to the right of the buccal cavity. The buccal cavity occupies most of the anterior half of the ventral surface of the animal and contains two long membranes along the right side. The buccal cavity appears shallow except in the area of the cytostome, where it is depressed. In silver-impregnated specimens the farthest right granular base of the outer membrane is longer and posteriorly appears to curve down into the depressed cytostomal area. The inner membrane appears to terminate just anterior to the cytostome. The argentophilic base of the inner membrane averages 17.0 μ in length. The cytopharynx lies just under the right posterior surface of the buccal cavity and in silver-impregnated specimens appears as a tubelike structure that is often bulbous at the posterior end. Argentophilic lines occur in the buccal cavity. The CVP number one to two and are associated with the posterior terminal ends of meridians 3 and 4. The CVP are closely opposed to the posterior circular line. In silver-impregnated specimens the cytoproct appears to consist of several granular clumps located posterior to the buccal cavity but usually closer to the posterior end of the body. An argentophilic line continues anteriorly from the cytoproct and joins the argentophilic lines in the posterior end of the buccal cavity. An argentophilic line continues posteriorly from the cytoproct and joins another argentophilic line between the last two or three basal granules of meridian 1. A long caudal cilium is present and is presumed to be associated with the polar basal body complex. The average size of silver-impregnated specimens is 40.8 × 15.7 μ. The range in length is 43.7–35.9 μ, and the range in width is 18.5–14.1 μ.

Discussion. Four species of the genus *Pseudocohnilembus* have been described. Evans and Thompson [1964] erected the genus and described three new species: *P. persalinus, P. hargisi,* and *P. longisetus.* Thompson [1966] described a new species, *P. marinus.* A description of *P. longisetus* from Antarctica was published by Thompson [1965]. Table 2 compares the important characteristics of *Pseudocohnilembus* sp. (strain 1) with those of all other described species. Although the antarctic form differs somewhat from all the other species, the present author is reluctant to describe *Pseudocohnilembus* sp. (strain 1) as a new species. Note that speciation based primarily on morphologic data is extremely difficult in many genera. However, the seeming tolerance of the antarctic forms to fresh water is noted as a first for the genus, since all the other forms were marine species. The present author may have been hasty in describing the first antarctic ciliate [Thompson, 1965] as *Pseudocohnilembus longisetus.* This ciliate was collected from unidentified plant material at an elevation of about 100 feet

TABLE 2. Comparison of Species of *Pseudocohnilembus*

	P. persalinus	*P. hargisi*	*P. longisetus*, Florida Strain	*P. longisetus*, Antarctic Strain	*P. marinus*	*P.* sp., Strain 1
Shape of anterior end	somewhat pointed	blunt	somewhat blunt	tapered and somewhat blunt	tapered and somewhat blunt	tapered and somewhat pointed
Average size, μ	30.2 × 14.0	44.3 × 18.3	26.6×11.5	28.5 × 14.3	32.2 × 12.8	40.8 × 15.7
Meridians, no.	8–9	13–15	10–11	9–11	10	9–12
Length of inner buccal membrane, μ	?	21.8	12.8	15.4	12.0 and lateral bar	17.0
Contractile vacuole pores	1 near meridian 3	1–3 near meridians 3 and 5	1 near meridian 3	1–3 near meridians 2 and 5	1 near meridian 4	1–2 near meridians 3 and 4

on Nelly Island and survived several weeks in a frozen state. Unique characteristics may exist in the antarctic forms that permit them to tolerate fresh water and also to form resistant stages in dried or frozen habitats. This interesting question can be answered only after future study of the antarctic forms. Bierle [1969] reported *Pseudocohnilembus longisetus* from Coast Lake, Victoria Land.

Pseudocohnilembus sp. (Strain 2)

Fig. 15

Collection site. This species was collected at the wave line of Arthur Harbor near the dock at Palmer Station.

Morphology. This species is very similar to *Pseudocohnilembus* sp. (strain 1) but has some slight differences. The animal possesses a blunter anterior end, and the meridians number 11–12. The CVP are identical to those of *Pseudocohnilembus* sp. (strain 1). The argentophilic base of the inner buccal membrane averages 18.6 μ in length. The average size of silver-impregnated specimens is 45.3 × 22.5 μ. The range in length is 50.9–38.4 μ, and the range in width is 27.2–18.0 μ.

Discussion. Although this species is blunter at the anterior end and slightly larger than *P.* sp. (strain 1), the two forms are similar in all other taxonomic characteristics and would appear to be conspecific.

Pseudocohnilembus sp. (Strain 3)

Collection site. This species was collected from a rock pool about 12 feet above the wave line at Palmer Station. The salinity of the culture was not determined, but the location of the rock pool would indicate that it was probably melt water.

Morphology. This form is also quite similar to *Pseudocohnilembus* sp. (strain 1). It possesses a somewhat pointed anterior end, and the meridians number 11–12. Only one CVP is present near meridian 3. The argentophilic base of the inner membrane averages 18.0 μ in length. The average size of silver-impregnated specimens is 41.3 × 18.0 μ. The range in length is 47.5–35.6 μ, and the range in width is 20.5–15.1 μ.

Discussion. This form is probably conspecific with *Pseudocohnilembus* sp. (strain 1).

Family Frontoniidae

Frontonia sp.

Fig. 16

Collection site. This marine species was collected from a rock tidal pool on Lemaire Island.

Morphology. Only two specimens of this animal were discovered, but a brief description will be included in this report. The animals are ovoid in shape and show the typical buccal infraciliature of the genus *Frontonia.* Only three polykineties are present to the right of the cytostome. The base of the buccal cavity is located about 90 μ from the posterior end of the animal, and the anterior end of the buccal cavity about 30 μ from the anterior end of the animal. Approximately 150–175 somatic meridians are present. One specimen measures 137.5 × 82.2 μ, and the other 144.8 × 96.5 μ.

Discussion. Because of the lack of specimens, no attempt is made to compare this form with other species of *Frontonia.* The genus *Frontonia* was identified by Sudzuki and Shimoizumi [1967] in collections made from Langhovde Hills, Showa and McMurdo stations, and Ongul and East Ongul islands, but no specific site was designated for this ciliate. Bierle [1969] reported *Frontonia* sp. from Coast Lake, Victoria Land.

Family PLEURONEMATIDAE

Pleuronema sp.

Fig. 17

Collection site. This marine species was collected from a rock tidal pool on one of the Joubin Islands. It was also collected on Lemaire Island in a rock tidal pool.

Morphology. The animal is elongated and somewhat flattened on the ventral surface. The anterior and posterior ends are convex in shape, especially the posterior end. The anterior end often appears somewhat blunt. The posterior pole of the animal seems to be more ventrally located. Except for the flattened ventral surface the animal appears to be rounded in cross section.

The number of ciliary meridians range approximately from 53 to 62. Four to seven meridians on the left side of the buccal cavity terminate either anterior to the cytostome or along the left side of the cytostomal area but do not extend posteriorly as far as the base of the UM. Numerous cilia longer than the normal somatic cilia extend from the posterior end of the animal. In silver-impregnated animals numerous trichocystlike pores are associated with both the somatic and the buccal ciliature. The buccal cavity extends along more than half of the length of the ventral surface and is depressed, especially in the cytostomal area. The greatest width of the buccal cavity, which is just posterior to the cytostome, averages 21.0 μ. The UM extends along the right side of the buccal cavity, and in silver-impregnated specimens its base measures 57.5 μ in length. The UM curves around the left posterior border of the buccal cavity and ends just anterior to the cytostome. This large membrane is quite obvious in the living animal, since it is extended during the feeding process. The base of the UM is an average of 19.6 μ from the posterior end of the animal. Three membranellelike structures are present in the buccal cavity and are designated membranelles 1–3. Membranelle 1 lies in the left anterior region of the buccal cavity, and its argentophilic base averages 10.6 μ in length. It appears to consist of two or more lateral rows of basal granules. Membranelle 2 lies to the left of the UM and extends from near the anterior end of the UM to the anterior cytostomal region. Its argentophilic base averages 34.4 μ in length and is more or less straight except that the posterior end possesses a hooklike process. Membranelle 2 appears to consist of more than one row of granules in the anterior portion and in the region of the hooklike process. Membranelle 3 lies just to the right of the cytostome and in silver-impregnated specimens consists of two parts, a V-shaped structure lying farthest to the right and a bifurcated structure lying near the cytostome. The cytostome is just posterior to M_2 and in silver-impregnated animals appears oval in shape. Numerous argentophilic lines originate near the inner edge of the posterior region of the UM and continue to the edge of the cytostome. The cytoproct is located posterior to the buccal cavity and in silver-impregnated animals appears as a heavy dark line. The CVP number one to three (usually one) and are located dorsally. Numerous dorsal and lateral meridians terminate in the region of the CVP. A single macronucleus is present, usually in the anterior half of the animal. The average size of silver-impregnated animals is 95.9 $\times$ 72.8 μ. The range in length is 127.0–71.2 μ, and the range in width is 94.0–51.6 μ.

Discussion. From recent studies of the genus *Pleuronema* by Borror [1963] and Dragesco [1968] the species described above appears to be *Pleuronema coronatum*. However, note that *Pleuronema* sp. is larger in size and possesses more ciliary meridians than *P. coronatum*, as described by Borror and Dragesco. The genus *Pleuronema* was listed by Dillon et al. [1968] from Cape Royds on Ross Island, but no description of the ciliate was given. *Pleuronema* sp. was reported by Bierle [1969] from Coast Lake.

Cyclidium sp.

Fig. 18

Collection site. This marine species was collected along the shore line of Arthur Harbor, Anvers Island, near Old Palmer Station.

Morphology. The animal is elongated, and the anterior end tapers to a blunt polar cap. The posterior end is less tapered and also blunt. The animal is more or less rounded in cross section, but the cytostomal area is slightly flattened. The bipolar meridians number 11 and are more or less evenly distributed except where meridians 1 and 11 curve around the buccal apparatus. The anterior basal granules of all meridians lie very close to the anterior circular line. The terminal basal granules of meridians on the left ventral surface usually do not extend as far posteriorly as the terminal basal granules of other meridians, and

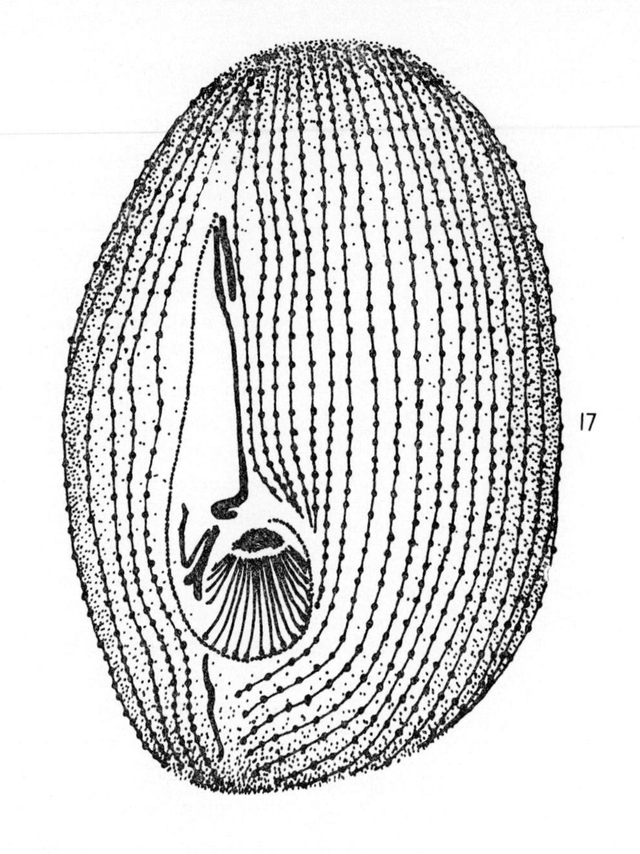

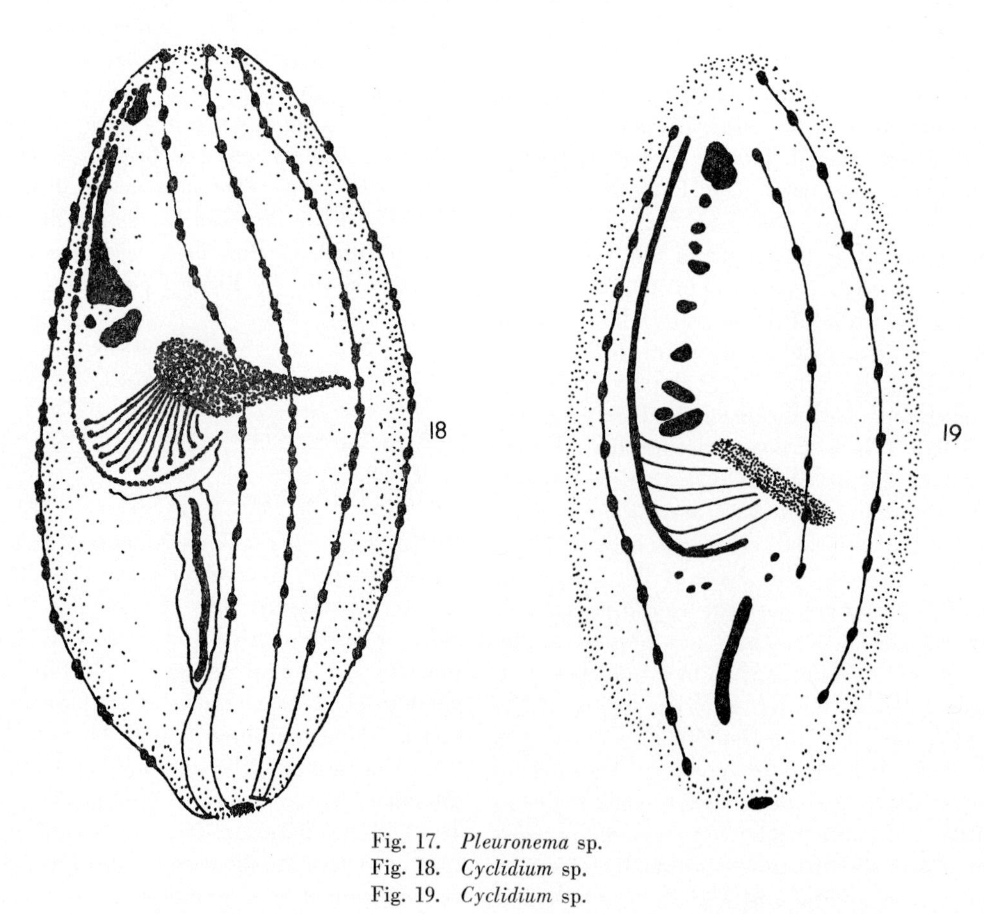

Fig. 17. *Pleuronema* sp.
Fig. 18. *Cyclidium* sp.
Fig. 19. *Cyclidium* sp.

the terminal basal granule of meridian 11 is usually farther from the caudal pole. The basal granules of meridians are more numerous and closely opposed in the anterior half of the animal. The basal granules often appear doubled. Argentophilic lines connect the terminal basal granules of each meridian except that of meridian 11 to the posterior circular line. The line from meridian 11 joins the polar basal body complex. The posterior circular line almost encircles the polar basal body complex but is open between meridians 10 and 1 and usually between meridians 3 and 4. The area enclosed by the posterior circular line is very small. The buccal cavity extends along approximately half of the anterior ventral surface of the animal and contains a UM along the right border and three membranelles. The distance from the base of the UM to the anterior end averages 13.5 μ. The length of the UM as measured from silver-impregnated specimens averages 12.2 μ. The anterior end of the UM terminates just posterior to the anterior end of meridian 1. The ciliary basal granules of the UM are small and closely opposed. Argentophilic lines extend from the posterior curve of the UM to the cytostomal area. Membranelle 1 is just posterior to the anterior end of meridian 11. In silver-impregnated specimens its granular mass can be clearly distinguished. Membranelle 2 is the longest membranelle, and its silver-impregnated base averages 4.3 μ in length. The arrangement of the basal granules of M_2 is triangular in shape, and the granules are closely opposed. Membranelle 3 is the most posterior membranelle and is just anterior and to the right of the cytostome. A single argentophilic granule can often be observed next to the UM near the posterior end of M_2. In silver-impregnated specimens the cytopharynx appears long and approximately triangular in shape. The long axis of the cytopharynx is usually perpendicular to the long axis of the body. A single CVP is associated with the posterior end of meridian 2. In silver-impregnated specimens the cytoproct appears as a heavy line completely enclosed by argentophilic lines. An argentophilic line from this complex usually passes posteriorly to join the posterior circular line near the posterior end of meridian 1. A caudal cilium is present. A spherical macronucleus is located near the middle of the body. Trichocystlike pores are associated with the ciliary meridians. The average size of silver-impregnated specimens is 23.1 $\times$ 14.0 μ. The range in length is 26.4–20.4 μ, and the range in width is 16.7–12.4 μ.

Discussion. Approximately 50 species of *Cyclidium* have been described; however, most of these descriptions are based on superficial observations of living specimens, and there is little hope of future recognition of these forms. Only two marine species of *Cyclidium* have been described by means of modern techniques. Borror [1963] described a new species, *Cyclidium marinum*, but he did not show the details of the buccal and somatic infraciliature. Dragesco [1963] decribed a new species, *Cyclidium plouneouri. Cyclidium* sp. differs from *C. marinum* in body size and number of ciliary meridians. Since Borror gave no details of the buccal or somatic infraciliature, no comparisons can be made by using these data. *Cyclidium* sp. differs from *C. plouneouri* in body size, ciliary meridians, and details of the somatic and buccal infraciliature. The shape and the position of the infraciliature of the three buccal membranelles differ greatly, and *Cyclidium plouneouri* shows an additional curvature in the infraciliature of the posterior portion of the UM that is not present in the antarctic form. Owing to the present taxonomic confusion in this genus, the present author believes it would be more sensible to postpone a species designation until a later time. Antarctic forms of *Cyclidium* have been listed by Sudzuki [1964], Sudzuki and Shimoizumi [1967], and Bierle [1969].

Cyclidium sp.

Fig. 19

Collection site. This species was collected from a fresh-water pool on Bonaparte Point.

Morphology. The animal is elongated, and the anterior end tapers to a blunt polar cap. The posterior end is tapered but has a rounded caudal pole. The animal is more or less rounded in cross section, but the cytostomal area is slightly flattened. The bipolar ciliary meridians number 10–11 and are more or less evenly distributed except where the first and the last meridians curve around the buccal cavity. The basal granules are more numerous and closely opposed in the anterior end of the animal. The anteriormost basal granule in meridian 1 is located just anterior to the UM. The posterior terminal basal granule is located posterior to the cytoproct near the posterior end of the animal. The second meridian is short, the last basal granule being near the posterior end of the UM. The first basal granule of the last meridian is located near M_1, and the posterior basal granule lies just posterior to the base of the UM. The buccal cavity extends along more than half of the ventral surface of the body. It is shallow but deepens in the cytostomal area. The UM runs along the rght border of the buccal cavity, and in silver-impregnated specimens its base averages 13.0 μ in length. The distance from the posterior end of the UM to the anterior end of the animal averages 14.5 μ. The ciliary basal granules of the UM are small and closely opposed. Membranelle 1 lies just posterior to the anterior end of the UM, and in silver-impregnated specimens its granular base appears to consist of two groups of granules. The anterior group is larger, and the smaller group is located posterior and often slightly to the left of the larger group. Membranelle 2 is the longest of the three membranelles, and its argentophilic base consists of from six to seven separate groups of granules. The posterior groups of granules are wider, and the anterior groups narrower. In silver-impregnated specimens, M_3 is the widest of the membranelles and appears to consist of a single group of granules. A single granule is often visible between M_3 and the base of M_2 near the UM. In silver-impregnated specimens the cytopharynx is tubular in shape and directed posteriorly at an angle to the long axis of the body. The cytoproct in silver-impregnated specimens appears as a heavy line just posterior to the base of the buccal cavity. Several argentophilic granules are usually present anterior to the cytoproct and posterior to the buccal cavity. The details of the argentophilic lines associated with the cytoproct, the polar basal body complex, and the ciliary meridians were not clearly visible in the slide preparations, but these lines are assumed to be similar to the argentophilic lines in the species of *Cyclidium* described earlier. The single CVP is located at the posterior end of meridian 2 and lies to the right of the posterior end of the UM. A caudal cilium is present. No trichocystlike pores are visible in this species. The average size of silver-impregnated specimens is 22.1 $\times$ 11.6 μ. The range in length is 23.6–19.9 μ, and the range in width is 13.4–9.8 μ.

Discussion. Only two fresh-water species of *Cyclidium* have been described in which the somatic and buccal infraciliatures were portrayed. *Cyclidium citrullus* was described by Czapik [1963], and *Cyclidium glaucoma* was redescribed by Berger and Thompson [1960]. It seems unlikely that *Cyclidium* sp. is conspecific with either of these two forms. *Cyclidium citrullus* is a larger species (about 30 μ in length) and possesses 16 meridians. It does not show the shortened second ciliary meridian with the CVP near the posterior end of the UM shown in *Cyclidium* sp. The three buccal membranelles, as shown in silver-impregnated animals, are different in the number and the position of the granules. *Cyclidium glaucoma* is similar in size and number of meridians to *Cyclidium* sp., but it possesses a long second meridian, the CVP being near the posterior end of the body. The last meridian in *C. glaucoma* is not shortened like that in *Cyclidium* sp. The argentophilic base of M_2 of *Cyclidium glaucoma* is wider and does not show the distinct groups of granules seen in *Cyclidium* sp. For reasons stated earlier, species designation should be reserved until further study is made of this genus.

Order PERITRICHIDA

Family VORTICELLIDAE

Vorticella sp.

Fig. 20

Collection site. This genus was collected from the melt water pond on Humble Island near Palmer Station.

Morphology. This form was identified by phase microscopic studies of living cultures, and only limited morphologic data were obtained. The animals were observed to attach to debris by means of a short stalk and, when they were disturbed, to contract into a ball-like mass. After a short period they extend and continue feeding. When they are not attached, they move about in the culture at great speed. The posterior girdle of cilia is present at all times. In free swimming forms the posterior end is somewhat tapered and has a short tip, but the stalk is not present. A large coiled or bent macronucleus is present. The average size of silver-impregnated specimens is 79.0 $\times$ 42.7 μ. The range in length is 95.9–65.4 μ, and the range in width is 51.3 $\times$ 33.8 μ. Fixed animals show various degrees of contraction.

Discussion. The genus *Vorticella* was reported from Cape Royds by Murray [1910]. Hada [1964, 1966] reported *Vorticella microstoma* and *V. pusilla* from Showa. Dillon et al. [1968] reported *Vorticella* from Cape Royds on Ross Island. Sudzuki and Shimoizumi [1967] reported *Vorticella* in collections from Langhovde Hills, Showa and McMurdo stations, and Ongul and East Ongul islands, but no specific site was given. Armitage and House [1962] reported *Vorticella* from the McMurdo Sound area, and Bierle [1969] reported *Vorticella* sp. from Coast Lake.

The following classification of the subclass Spirotricha is used:

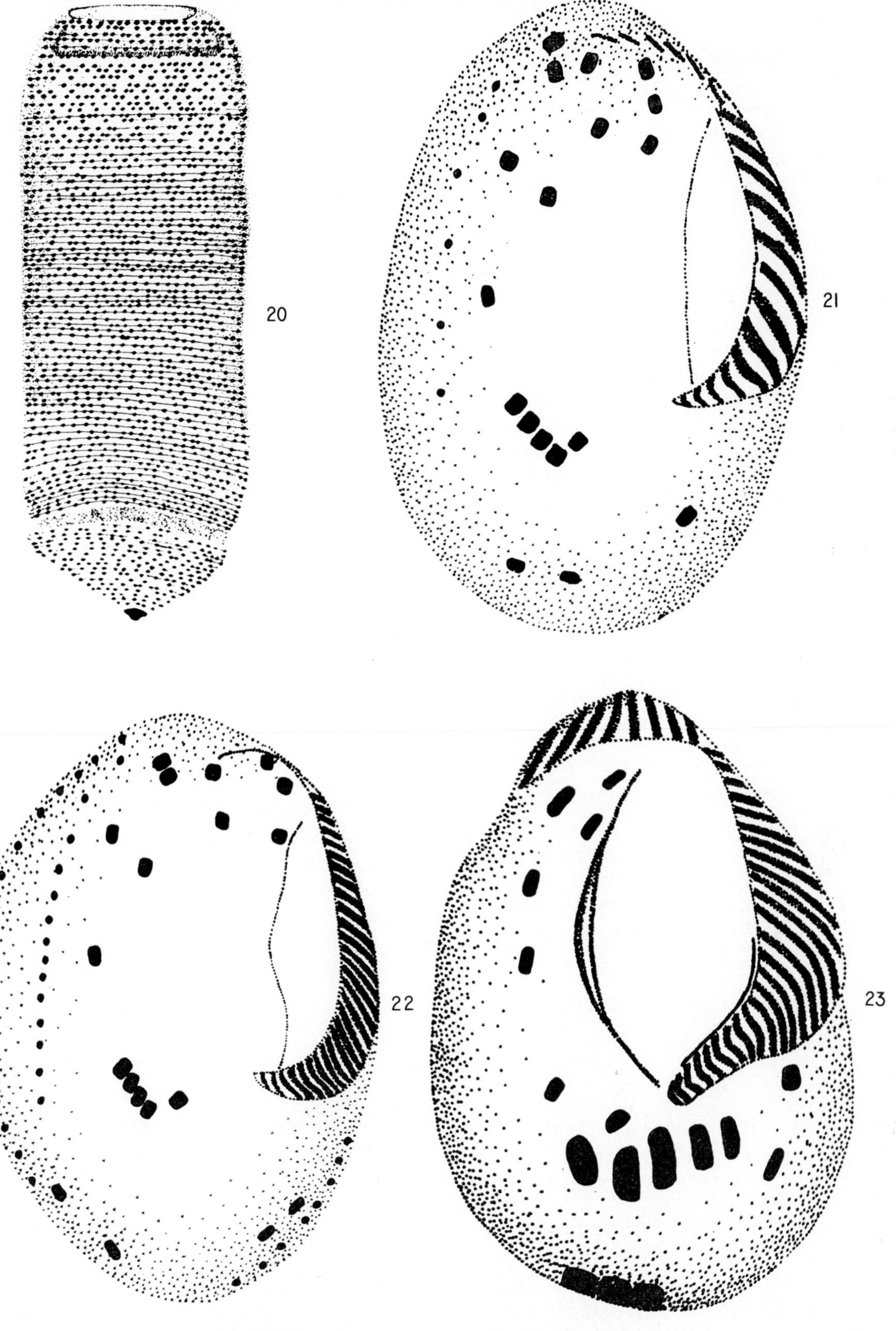

Fig. 20. *Vorticella* sp.
Fig. 21. *Euplotes* sp.
Fig. 22. *Euplotes* sp.
Fig. 23. *Diophrys* sp.

Order HETEROTRICHIDA

Family Bursariidae

Bursaria sp.

Collection site. This fresh-water form was collected in a melt water pond near Old Palmer Station.

Morphology. The animal is elongated and broad. The anterior end is somewhat truncate, whereas the posterior end is bluntly tapered toward the left side. The anterior dorsal surface is somewhat straight, but the posterior dorsal surface is convex. The ventral surface is somewhat flattened, and the lateral surfaces are convex. Most of the anterior half of the animal is occupied by a large buccal cavity divided into two portions by a longitudinal fold. The ventral portion of the buccal cavity contains a wide slitlike opening that continues from the anterior end to near the middle of the body. The buccal membranelles originate on the left lateral anterior wall of the buccal cavity and spiral posteriorly across the longitudinal fold to the right side of the animal and then back toward the left side. The buccal membranelles continue to near the posterior end. Somatic cilia are uniform and very dense. The bandlike macronucleus is much longer than the body and is curved and bent within the cytoplasm. The average size of silver-impregnated specimens is 293.6 × 179.4 μ. The range in length is 342.9–241.3 μ, and the range in width is 203.2–146.7 μ.

Discussion. The description of *Bursaria* sp. agrees generally with the description given for *Bursaria truncatella*. However, *Bursaria* sp. is a much smaller form. This species is probably new, but a more complete study should be made for confirmation.

Order OLIGOTRICHIDA

Family Halteriidae

Halteria sp.

Collection site. This genus was collected from melt water pools on Bonaparte Point and Litchfield Island.

Morphology. This form was identified by phase microscopic studies of living cultures, and no morphologic data were obtained.

Discussion. The genus *Halteria* was listed by Sudzuki and Shimoizumi [1967] and Armitage and House [1962] from antarctic collections. *Halteria grandinella* was reported by Bierle [1969] from Coast Lake.

Strombidium sp.

Collection site. This form was collected from tidal pools on one of the Argentine Islands and also on the Roca Islands.

Morphology. This form was identified by phase microscopic studies of the living culture, and no morphologic data were obtained.

Order HYPOTRICHIDA

Family Euplotidae

Euplotes sp.

Fig. 21

Collection site. This ciliate was collected in a tidal pool near the sewage outlet at Palmer Station.

Morphology. The body is somewhat elliptic in shape, the dorsal surface being convex and the ventral surface being flattened. The cirri consist of seven frontals, three ventrals, five anals, and three caudals. About seven rows of dorsal bristles were present, and the bristles numbered about eight to nine in the dorsalmost rows. The adoral zone of membranelles averaged 25.6 μ in length. The body in silver-impregnated specimens averaged 42.7 × 27.7 μ in size, the range in length being 52.6–37.2 μ and the range in width being 32.4–24.5 μ.

Discussion. According to Dr. Arthur C. Borror, University of New Hampshire, this form is apparently a new species, but this classification should be confirmed by further study.

The genus *Euplotes* has been reported by Dillon et al. [1968] and by Armitage and House [1962] from Antarctica. Bierle [1969] reported *Euplotes patella*, *E. aediculatus*, and *Euplotes* sp. from Coast Lake.

Euplotes sp.

Fig. 22

Collection site. This ciliate was collected in a rock pool in the Wauwermans Islands. The water was probably fresh, but no salinity tests were made.

Morphology. The body is somewhat elliptic in shape, the right side showing more curvature. The dorsal surface of the body is convex, and the ventral surface flattened. The cirri consist of seven frontals, three ventrals, five anals, and four caudals. About 12 rows of dorsal bristles were present, and the number of bristles in the dorsalmost rows varied from 11 to 16. The adoral zone of membranelles averaged 45.9 μ in length. The body in silver-impregnated specimens averaged 79.0 × 59.2 μ in size, the range in length being 83.3–71.4 μ and the range in width being 64.8–54.1 μ.

Discussion. According to Dr. Arthur C. Borror this form is apparently a new species, but this classification should be confirmed by further study.

Diophrys sp.

Fig. 23

Collection site. This ciliate was collected in a tidal pool on a small island in Hamburg Bay.

Morphology. The body is somewhat elliptic in shape, the anterior end being more pointed. The dorsal surface of the body is convex, whereas the ventral surface is flattened. Two large sausage-shaped macronuclei are present. One macronucleus is curved and located in the anterior half of the body, whereas the other macronucleus is straighter and located in the posterior half of the body. The long axis of the posterior macronucleus is more or less parallel to the long axis of the body. The positions of the two macronuclei appear to be constant. The cirri consist of five frontals, two ventrals anterior to the anal cirri, five anals, two postoral laterals, and three large posterior cirri. Four or five rows of dorsal bristles are present and curve slightly to the right side of the body. There are few bristles in each row. The maximum number of bristles is 12, and the minimum number is six. The adoral zone of membranelles averages 36.4 μ in length. The body in silver-impregnated specimens averages 59.1 × 41.6 μ in size, the range in length being 66.2–53.1 μ and the range in width being 46.6–37.2 μ.

Discussion. This ciliate is similar to *Diophrys oligothrix,* a new species described by Borror [1965]. However, there are important differences. The antarctic form is a much smaller species both in the range of body length and width and in the length of the adoral zone of membranelles. The antarctic form often shows an additional row of dorsal bristles, and the number of bristles in each row in *Diophrys oligothrix* is greater. The maximum number of bristles in a row of *D. oligothrix* is 24, and the minimum number is nine. No mention was made in the description of *D. oligothrix* of the constant position of the two macronuclei. The antarctic form is probably a new species, but further studies should be made for confirmation.

Family Oxytrichidae

Uroleptus sp.

Fig. 24

Collection site. This ciliate was collected in a freshwater pool on Bonaparte Point near Palmer Station.

Morphology. The body is elongated and has a rounded anterior end and a pointed posterior end. In living specimens the posterior end is often very pointed and drawn out. The body is convex dorsally and flattened ventrally. Two oval to elongate macronuclei are present near the central portion of the body. The number of frontal cirri is difficult to discern and is either three or five, depending on interpretation. Two cirri that may be considered frontals are located on the right side of the buccal cavity near the middle row of ventral cirri. It is difficult to determine whether they are ventral or frontal cirri. The posteriormost frontal cirrus is located near the anterior end of the UM, and in silver-impregnated animals the argentophilic base is often obscured by the UM. Three rows of ventral cirri are present. The middle row of ventral cirri is usually doubled throughout most of its length. The ventral cirri extend to the pointed posterior end. Five to seven rows of lateral and dorsal bristles are present. The adoral zone of membranelles averages 45.5 μ in length. The body in silver-impregnated specimens averages 141.1–47.8 μ in size. The range in body length is 181.8–116.6 μ, and the range in body width is 66.5–34.2 μ.

Discussion. Hada [1964] listed *Uroleptus gibbus* from Showa, and Hada [1966] recorded *Uroleptus gibbus* and *U. musculus* from McMurdo Station. Bierle [1969] reported *Uroleptus* sp. from Coast Lake. Dr. Arthur C. Borror believed the *Uroleptus* sp. described above to be *Uroleptus caudatus.* The present author does not believe that *Uroleptus* sp. is either *U. gibbus* or *U. musculus. Uroleptus gibbus* possesses four ventral rows of cirri and does not possess the pointed posterior end. *Uroleptus musculus* is a plump form, the length of the peristome being one-third that of the body. *Uroleptus* sp. may well be the species *U. caudatus,* but the strong frontal cirri are not shown in this form. The taxonomic position of *Uroleptus* sp. must be decided after a review of the genus.

Oxytricha sp.

Fig. 25

Collection site. This ciliate was collected from a rock pool at the highest elevation behind Old Palmer Station.

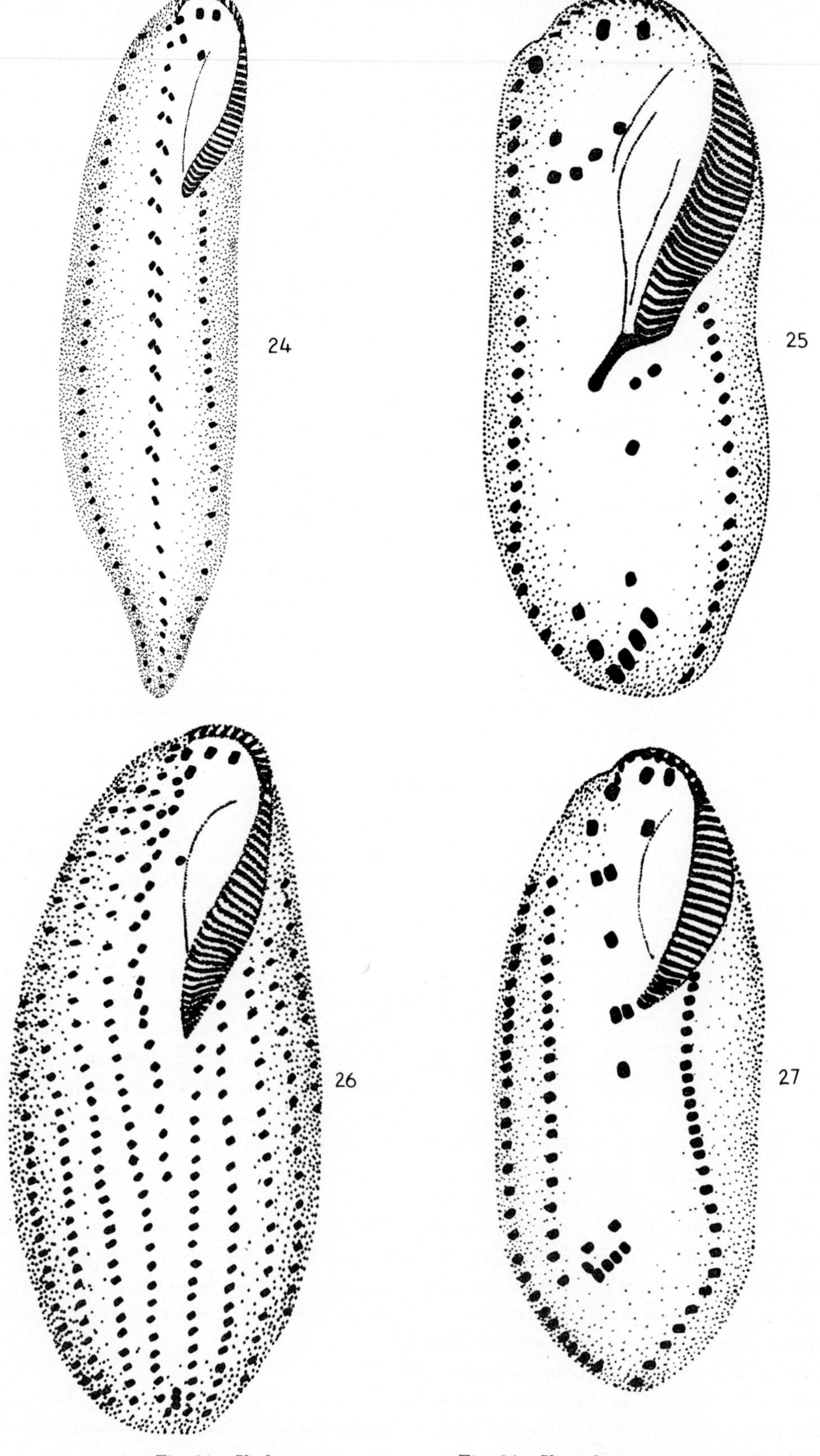

Fig. 24. *Uroleptus* sp.
Fig. 25. *Oxytricha* sp.
Fig. 26. *Urostyla* sp.
Fig. 27. *Pleurotricha* sp.

Morphology. The body is elongated and has rounded anterior and posterior ends. The dorsal surface is slightly convex, and the ventral surface is flattened. Three or four oval to elongated macronuclei are located in the posterior two-thirds of the body. Eight frontal cirri are present. These cirri consist of two isolated groups. The anterior group is made up of three strong cirri, and the posterior group of five weaker ones. Five ventral cirri are also located in two isolated groups. One group of three cirri is found just posterior to the buccal cavity, and another group of two just anterior to the five strong anal cirri. Two rows of marginal cirri are present. Four dorsal rows of bristles are present. The adoral zone of membranelles averages 50.8 μ in length. The body in silver-impregnated specimens averages 108.3 $\times$ 42.6 μ in size. The range in length is 120.3–90.7 μ, and the range in width is 49.7–38.7 μ.

Discussion. According to Dr. Arthur C. Borror *Oxytricha* sp. is probably a form of *Oxytricha cavicola.* However, there are noticeable differences. *Oxytricha cavicola* is a much larger species, size variations being listed as 180–250 μ. Only four anal cirri are shown for *O. cavicola,* and no mention is made of some forms possessing three macronuclei. *Oxytricha cavicola* also shows an adoral zone of membranelles much shorter in relation to body length. Whether the antarctic form is a new species of *Oxytricha* will require a restudy of this genus.

The genus *Oxytricha* was recorded by Dillon et al. [1968] from Cape Royds on Ross Island. Bierle [1969] reported *Oxytricha fallax, O. lundibunda,* and *Oxytricha* sp. from Coast Lake.

Urostyla sp.

Fig. 26

Collection site. This ciliate was collected from a tidal pool on Lemaire Island.

Morphology. The body is somewhat elliptic in shape, the anterior end being more tapered. The posterior end is rounded. The dorsal surface of the body is convex, whereas the ventral surface is flattened. The macronucleus appears to be fragmented into several score of oval to elongated pieces. The ventral cirri are arranged in 11–13 rows. The first row of ventral cirri to the right of the buccal cavity is shortened and ends posterior to the midbody region. The exact number of frontal cirri is difficult to discern. The frontal cirri are scattered among the first two rows of ventral cirri near the anterior end of the animal. Except for the first three or four frontal cirri the argentophilic bases of the frontal cirri are not noticeably larger than the bases of the ventral cirri. Three to five anal cirri are oriented in an oblique row posterior to the short row of ventrals near the posterior end of the animal. About seven rows of dorsal bristles are present. The adoral zone membranelles average 47.1 μ in length. The body in silver-impregnated specimens averages 124.6 $\times$ 60.8 μ in size. The range in length is 150.0–103.1 μ, and the range in width is 74.5–52.2 μ.

Discussion. The first short row of ventral cirri, the fragmented macronuclei, and the small size of this species of *Urostyla* appear to be a unique combination of characters. However, if the shortened first row of ventral cirri described above is actually a posterior prolongation of the frontal cirri, this form is related to *Urostyla cristata,* which was described by Jerka-Dziadosz [1964] to possess such frontal cirri. *Urostyla cristata* is a much larger form that varies in length from 300 to 450 μ, and the posterior prolongation of the frontal cirri continues all the way to the anal cirri, which number eight to 12. According to Dr. Arthur C. Borror the antarctic form of *Urostyla* is a new species.

Pleurotricha sp.

Fig. 27

Collection site. This ciliate was collected from one of the Joubin Islands in a rock tidal pool with fresh-water dilution.

Morphology. The body is elongated and has a rounded posterior end. The anterior end is somewhat rounded, and the organism is usually more tapered in the anterior half. The body is flattened ventrally and convex dorsally. Two elongated macronuclei are present. The frontoventral cirri number 13 and appear in three isolated groups. The anteriormost group contains eight cirri. The middle group, which is located near the posterior end of the buccal cavity, contains three cirri, and the posteriormost group, which is located just anterior to the transverse cirri, contains two cirri. Five transverse cirri are present. Two rows of marginal cirri are found on the right ventral surface of the body, the inside row always being shorter at the posterior end and usually being shorter at the anterior end. The single left marginal row of cirri begins near the posterior end of the buccal cavity and terminates near the posterior end of the body. About five rows of dorsal bristles are present. The adoral zone membranelles average 45.9 μ in length. The body in silver-impregnated specimens averages 111.3 × 47.8 μ in size. The range in length is 124.8–100.5 μ, and the range in width is 54.9–41.1 μ.

Discussion. According to Dr. Arthur C. Borror this form is a variety of *Pleurotricha lanceolata.* Because a variety of forms have been described in the literature as *Pleurotricha lanceolata,* the present author believes that specific assignment should be postponed until further studies have been made on the antarctic form.

DISCUSSION

Taxonomic studies on the ciliated protozoa of Antarctica have been almost totally neglected. A preliminary account of ciliated protozoa from Cape Royds by Murray [1910] lists the genus *Vorticella* and a *Nassula*-type form. A few other drawings of ciliates are included but are not given generic names. In studies on the microbiology of some soils in the McMurdo Sound region Flint and Stout [1960] list three forms: *Colpoda cucullus, Homalogastra setosa,* and *Trichopelma sphagnetorum.* Armitage and House [1962] made limnologic studies on melt water ponds and lakes in the McMurdo Sound area and listed the ciliated genera *Nassula, Euplotes, Vorticella, Intrastylum,* ? *Opisthonecta, Halteria,* and *Tracheophyllum.* During an investigation of fresh-water protozoa from Showa Station, Hada [1964] recorded the following ciliate species: *Coleps hirtus* (corrected to *Glaucoma scintillans* in the reprint), *Prorodon teres, Colpoda cucullus, Epistylis* sp., *Vorticella microstoma, Vorticella pusilla, Strombilidium gyrans, Uroleptus gibbus,* and *Holosticha intermedia.* His paper includes very short descriptions of each species and eight drawings. The drawings appear to have been made from living organisms and are of little use in identification. Sudzuki [1964] lists 10 genera of ciliates from moss–water communities at Langhovde Hills. The study was made from frozen moss collections after their return to the zoological institute of the Tokyo University of Education. The 10 genera include *Colpoda* sp., ? *Cyclidium, Dileptus* sp., ? *Balantidioides,* ? *Keronopsis, Opisthotricha* sp., *Paradileptus, Parauroleptus* sp., *Pyxidium,* and *Spathidium* sp. Short descriptions are given of *Opisthotricha, Paradileptus,* and *Pyxidium.* There are drawings of six genera, but they appear to have been made from living animals and contain little if any taxonomic data.

Pseudocohnilembus longisetus was redescribed by Thompson [1965] from unidentified plant material from an exposed rock surface at an elevation of about 100 feet on Nelly Island near Wilkes Station. This description is the only one in the earlier literature resulting from the use of modern techniques. A single ciliate, *Stylonychia* sp., has been reported from melt water pools on the Clark Peninsula by Thomas [1965], but he gives no data on this form. Hada [1966] investigated fresh-water protozoa in mosses growing on sandy beaches and rocks and from algae in small streams and pools collected from Showa, McMurdo, and Mirnyy stations. He lists the following ciliated protozoa: *Spathidium lieberkuhni, Didinium balbianii* var. *nanum, Prorodon teres, Urotricha farcta, Colpoda cucullus, Glaucoma scintillans, Epistylis* sp., *Vorticella microstoma, Vorticella pusilla, Strombilidium gyrans, Uroleptus gibbus, Uroleptus musculus, Holosticha intermedia,* and *Holosticha vernalis.* In an ecologic investigation Dillon [1967] recorded the genus *Nassula* from a melt water pond near McMurdo Station. In a study of the stability of the faunistic composition of antarctic microorganisms of Langhovde Hills, Showa and McMurdo stations, and the Ongul Island area, Sudzuki and Shimoizumi [1967] list as representative of the antarctic microfauna the following ciliate genera: *Colpoda, Homalogastra, Cyclidium, Trichopelma, Spathidium, Dileptus, Paradileptus,* ? *Opisthotricha, Holotricha, Parauroleptus, Pyxidium,* ? *Vorticella, Blepharisma, Nassula, Frontonia, Halteria,* and *Strombilidium.* In an ecologic study of a fresh-water lake on Cape Royds Dillon et al. [1968] list the ciliate genera *Nassula, Pleuronema, Vorticella, Oxytricha, Trachelophyllum, Euplotes,* and *Paramecium.* Also included in this paper is a photograph of a ciliate listed as *Spathidium spathula* collected from a melt water pond near Palmer Station.

Bierle [1969] lists 48 species of ciliated protozoa (appendix) in an ecologic study of Coast Lake, a small fresh-water lake in Victoria Land. Although this work represents the most extensive list of ciliated protozoa from Antarctica, it is primarily ecologic in nature. A drawing or photograph represents each species, but the morphologic data are extremely limited and insufficient for taxonomic review.

In the present limited investigation of the ciliated protozoa of the Antarctic Peninsula, 25 genera representing seven orders are recorded. Detailed descriptions of 24 of these forms and limited descriptions of seven additional forms are given. Four species of these antarctic ciliates are described as new to science (*Cyclogramma membranella, Sathrophilus antarcticus, Uropedalium antarcticum,* and *Parauronema antarcticum*), and seven genera (*Bursaria, Strombidium, Uro-*

pedalium, *Parauronema*, *Diophrys*, *Urostyla*, and *Pleurotricha*) are new records for Antarctica. The present paper contains the only study of marine ciliates from Antarctica.

There is little doubt that many of the forms to which specific names have not been assigned will be determined as new species in future analyses. The present author is reluctant to erect new species in genera that are in need of revision. It is obvious to workers in this field that hasty specific assignments only add to the taxonomic confusion that presently exists. Furthermore there is the added danger of confusing the possible unique ciliate fauna of Antarctica with forms described from other areas of the world.

Although this investigation was primarily taxonomic in nature, it has brought to light some other interesting points that bear further investigation. Most of the genera of hypotrichs studied here were noted for the small size of the organisms, whereas in general these forms in other environments are much larger. This observation was also noted for the small heterotrich *Bursaria*. Only further investigations will determine if body size is correlated with factors such as longer encystments, shorter periods of metabolic activity, lower environmental temperatures, and fewer available nutrients. Two genera, *Pseudocohnilembus* and *Uropedalium*, were collected from melt water pools, but all the former species of these genera are marine forms. Also many forms collected from tidal pools and assumed to be marine in nature may have been living in water of very low salinity because of melting. Only future studies will determine whether this finding is simply a factor of low temperature or involves the osmoregulatory mechanism.

The present author hopes that this investigation of the ciliated protozoa of the Antarctic Peninsula will stimulate further interest in the protozoa of Antarctica.

APPENDIX: CILIATED PROTOZOA FROM COAST LAKE, VICTORIA LAND

Order Gymnostomatida: *Enchelyodon* sp., *Enchelys* sp., *Lacrymaria* sp., *Spasmostoma viride*, *Trachelophyllum* sp. 1, *Trachelophyllum* sp. 2, *Dileptus anser*, *Acineria* sp., *Amphileptus claparedei*, *Didinium balbianii*, *Didinium nasatum*, *Enchelydium virens*, *Homalozoon vermiculare*, *Spathidioides* sp., *Spathidium spathula*, *Spathidium* sp., *Chilodonella uncinata*, *Chilodontopsis* sp., *Cyclogramma* sp., *Nassula aurea*, *Nassula ornata*, *Paranassula* sp.

Order Trichostomatida: *Colpoda cucullus.*

Order Suctorida: *Podophrya* sp.

Order Hymenostomatida: *Philaster armata*, *Uronema* sp., *Pseudocohnilembus longisetus*, *Homalogastra setosa*, *Paramecium trichium*, *Frontonia* sp., *Cyclidium* sp., *Pleuronema* sp.

Order Peritrichida: *Intranstylum* sp., *Vorticella* sp., *Vaginicola* sp., *Epistylis* sp., *Pyxidiella* sp.

Order Heterotrichida: *Blepharisma steini.*

Order Oligotrichida: *Halteria grandinella.*

Order Hypotrichida: *Euplotes patella*, *Euplotes aediculatus*, *Euplotes* sp., *Oxytricha fallax*, *Oxytricha lundibunda*, *Oxytricha* sp., *Stylonychia mytilis*, *Opisthotricha* sp., *Uroleptus* sp.

Acknowledgments. The identification of the ciliates in the order Hypotrichida was made by Dr. Arthur C. Borror, University of New Hampshire. The author acknowledges the complete support of this investigation by the Office of Polar Programs, National Science Foundation. Further acknowledgment is given to John M. Croom, research associate, who made invaluable contributions to this work. The taxonomic survey was greatly enhanced by helicopter support from the USCGC *Edisto* and support by the RV *Hero.* The research was conducted in the new laboratory facilities at Palmer Station. Slides of the silver-impregnated protozoa described in this study will be deposited in the U.S. National Museum, Washington, D. C.

REFERENCES

Armitage, K. B., and H. B. House
1962 A limnological reconnaissance in the area of McMurdo Sound, Antarctica. Limnol. Oceanogr., 7(1): 36–41, fig. 1, tables 1–4.

Berger, J., and J. C. Thompson, Jr.
1960 A redescription of *Cyclidium glaucoma* O.F.M., 1786 (Ciliata: Hymenostomatida), with particular attention to the buccal apparatus. J. Protozool., 7(3): 256–262.

Bierle, D. A.
1969 The ecology of an antarctic fresh-water lake with emphasis on the ciliate protozoa. Ph.D. thesis. Univ. of S. Dak., Vermillion.

Borror, A. C.
1963 Morphology and ecology of the benthic ciliated protozoa of Alligator Harbor, Florida. Arch. Protistenk., *106:* 465–534.
1965 New and little-known tidal marsh ciliates. Trans. Am. Microsc. Soc., *84*(4): 550–565.

Corliss, J. O.
1960 The problem of homonyms among generic names of ciliated protozoa, with proposal of several new names. J. Protozool., 7(3): 269–278.
1961 The ciliated protozoa: Characterization, classification, and guide to the literature. 310 pp. Pergamon, New York.

1970 The comparative systematics of species comprising the hymenostome ciliate genus *Tetrahymena*. J. Protozool., *17*(2) : 198–209.

Czapik, A.
1963 La morphogenèse du cilié *Cyclidium citrullus* Cohn (Hymenostomatida, Pleuronematina). Acta Protozool., *1*(2) : 5–11.
1968 La morphologie de *Uronema elegans* Maupas et de *Uronema parva* sp. n. Acta Protozool., *5*(11) : 225–229.

Dillon, R. D.
1967 The ecology of free-living and parasitic protozoa of Antarctica. Antarct. J. U.S., *2*(4) : 104.

Dillon, R. D., D. Bierle, and L. Schroeder
1968 Ecology of antarctic protozoa. Antarct. J. U.S., *3*(4) : 123–124, fig. 1.

Dragesco, J.
1963 Compléments à la connaissance des ciliés mésopsammiques de Roscoff. 1. Holotriches. Cah. Biol. Mar., *4:* 91–119.
1968 Les genres *Pleuronema* Dujardin, *Schisocalyptra* nov. gen. et Histiobalantium Stokes (Ciliés Holotriches Hymènostomes). Protistologica, *4*(1) : 85–106.

Evans, F., and J. C. Thompson, Jr.
1964 Pseudocohnilembidae n. fam., A hymenostome ciliate family containing one genus, *Pseudocohnilembus* n. g., with three new species. J. Protozool., *11*(3) : 344–352.

Fauré-Fremiet, A.
1967 Le genre *Cyclogramma* Perty, 1852. J. Protozool., *14*(3) : 456–464, figs. 1–15.

Flint, E. A., and J. D. Stout
1960 Microbiology of some soils from Antarctica. Nature, *188*(4752) : 767–768.

Hada, Y.
1964 The fresh-water fauna of the protozoa in the region of the Showa Station in Antarctica. Bull. Suzugamine Women's Coll. Natur. Sci., *11:* 5–21, figs. 1–23.
1966 The fresh-water fauna of the protozoa in Antarctica. JARE Scient. Rep., spec. issue 1: 209–215.

Jerka-Dziadosz, M.
1964 *Urostyla cristata* sp. n. (Urostylidae, Hypotrichida) ; The morphology and morphogenesis. Acta Protozool., *2*(11) : 123–128, 2 pls.

Kahl, A.
1926 Neue und wenig bekannte Formen der holotrichen und heterotrichen Ciliaten. Arch. Protistenk., *55:* 197–438.
1928 Die Infusorien (Ciliata) der Oldesloer Salzwasserstellen. Arch. Hydrobiol., *19:* 50–123, 189–246.
1930–1935 Urtiere oder Protozoa. 1. Wimpertiere oder Ciliata (Infusoria), eine Bearbeitung der freilebenden und ectocummensalen Infusorien der Erde, unter Ausschluss der marinen Tintinnidae. *In* E. Dahl (Ed.), Die Tierwelt Deutschlands. Teil. 18, 21, 25, and 30, pp. 1–886. G. Fisher, Jena.

Lepsi, J.
1926 Zur Kenntnis einiger Holotrichen. Arch. Protistenk., *53:* 378–406.

Levander, K. M.
1900 Zur kenntnis des Lebens in den Stehenden Kleingewässern auf den Skäzeninseln. Acta Soc. Fauna Flora Fenn., *18*(6) : 1–107.

Mermod, G.
1914 Recherches sur la faune infusioriennes des tourbières et des eaux voisines de Sainte-Croix (Jura vaudois). Revue Suisse Zool., *22:* 31–114.

Murray, J.
1910 Microscopic life at Cape Royds. Sci. Rep. Br. Antarct. Surv., *1:* 17–22.

Prelle, A.
1961 Contribution à l'étude de *Leptopharynx costatus* (Mermod). Bull. Biol. Fr. Belg., *95:* 731–752.

Savoie, A.
1957 Le cilié *Trichopelma agilis* n. sp. J. Protozool., *4*(4) : 276–280.

Stokes, A. C.
1887 Notices of new fresh-water infusoria. Proc. Am. Phil. Soc., *24:* 244–255.

Stout, J. D.
1956 *Saprophilus muscorum* Kahl, a tetrahymenal ciliate. J. Protozool., *3*(1) : 28–30.

Sudzuki, M.
1964 On the microfauna of the antarctic region. 1. Moss–water community at Langhovde. JARE Scient. Rep., ser. E, 19: 1–43, 6 pls., tables 1–9.

Sudzuki, M., and J. Shimoizumi
1967 On the fresh-water microfauna of the antarctic region. 2. Stability of faunistic composition of antarctic microorganisms. JARE Scient. Rep., spec. issue 1: 216–235, figs. 1–3, tables 1–6.

Thomas, C. W.
1965 On populations in antarctic meltwater pools. Pacif. Sci., *19:* 515–521.

Thompson, J. C., Jr.
1964 A redescription of *Uronema marinum* and a proposed new family Uronematidae. Va. J. Sci., *15:* 80–87.
1965 *Pseudocohnilembus longisetus,* A hymenostome ciliate from Antarctica. Va. J. Sci., *16*(2) : 165–169.
1966 *Pseudocohnilembus marinus* n. sp., A hymenostome ciliate from the Virginia coast. J. Protozool., *13*(3), 463–465.
1967 *Parauronema virginianum* n. g., n. sp., A marine hymenostome ciliate. J. Protozool., *14*(4) : 731–734.

Thompson, J. C., Jr., and V. Cone
1963 A redescription of the hymenostome ciliate *Sathrophilus* (*Saprophilus*) *muscorum* (Kahl, 1931) Corliss, 1960, with particular attention to the buccal apparatus. Va. J. Sci., *14*(1) : 16–22.

LIFE CYCLE STUDIES OF SOME ANTARCTIC MITES AND DESCRIPTION OF A NEW SPECIES, *PROTEREUNETES PAULINAE* (ACARI: EUPODIDAE)

Elmer E. Gless

Montana College of Mineral Science and Technology, Butte, Montana 59701

Abstract. Information acquired at Hallett Station, Antarctica, during three austral summer seasons (1965–1966, 1966–1967, and 1967–1968) relating to the biology of four species of antarctic mites is presented. *Stereotydeus belli* (Trouessart, 1902) was reared in vitro from tritonymph to adult, after which one of its offspring was reared through all the immature stages to the adult stage. Throughout the second and third seasons (1966–1967 and 1967–1968) mites were kept in dishes containing an artificial medium on which algae and moss were grown to supply natural food. A new species of mite of the family Eupodidae and the genus *Protereunetes* was found and reared in vitro from the adult stage through all the immature stages but not again to adult. Another species, *Eupodes wisei* Womersley and Strandtmann, 1963, was also reared from adult through all the immature stages but not again to adult. A fourth species, *Coccorhagidia gressitti* Womersley and Strandtmann, 1963, which was found to be predaceous, was reared from tritonymph to adult and from adult to deutonymph but not to tritonymph. The time spent in the various stages of development and the techniques of rearing are presented. The morphological changes characterizing each stage have been determined and are also presented. All stages of the new species of mite, *Protereunetes paulinae*, are described.

This account of life cycle studies and description of a new species of the order Prostigmata are based on work conducted during three austral summer seasons at Hallett Station, Antarctica. The periods were approximately October through February of 1965–1966, 1966–1967, and 1967–1968.

The first entomological collections in Antarctica were made by members of the *Belgica* expedition of 1897–1899. The *Southern Cross* expedition of 1898–1900 was the first to overwinter on the continent proper and the second to make entomological collections. Its senior zoologist, Dr. Nikolai Hanson, who died at the end of that winter, had collected extensively at Cape Adare, just 75 miles from Hallett Station. However, it was not until 1958, at the end of the International Geophysical Year, that Madison E. Pryor was sent by the Polar Committee of the U.S. National Academy of Sciences to Hallett Station to investigate arthropod ecology [Gressitt, 1967].

Pryor subsequently spent two seasons at Hallett Station (1958–1959 and 1959–1960). His work was concerned with basic environmental features related to soil arthropods. Pryor [1962] reported four species of Collembola, three of which had been previously described and one of which was believed to be new. He also reported two prostigmatid mite species. One, *Penthaleus belli* Trouessart, 1903 [sic], which has been reported to be of the family Eupodidae, has since been redescribed by Womersley and Strandtmann [1963] as *Stereotydeus belli* (Trouessart, 1902) and correctly placed in the family Penthalodidae. The second, *Stereotydeus (Tectopenthalodes) villosus* (Trouessart), 1903, has since been redescribed by Womersley and Strandtmann [1963]. It was correctly placed in the family Penthalodidae by Pryor, but it has never been reported by other workers. Pryor may have misidentified it, since *Stereotydeus villosus* has been reported from only the South Shetland Islands and in the region of the Antarctic Peninsula, approximately 2700 miles away.

Halozetes antarctica (Michael, 1903) and *Pertorgunia belgicae* (Michael, 1903) are two oribatid mites also reported by Pryor [1962]. They have never been reported from Hallett Station since that time. *Halozetes antarctica* is now known as *Alaskozetes antarctica* (Michael, 1903) by Wallwork's [1962] redescription.

A thorough examination of the Hallett Station study area by the writer has never revealed specimens of the two oribatid mites mentioned above. One mite, *Maudheimia petronia* Wallwork, 1962, is prevalent in the study area and has been found in all ecological habitats described for the two species named by Pryor except on the surface of melt pools. Therefore, as J. L. Gressitt suggested (data obtainable from E. E. Gless records, Department of Biological Sciences, Montana College of Mineral Science and Technology, Butte, Montana), it is assumed that the mites collected by Pryor are merely different stages of *M. petronia*.

After Pryor's two seasons at Hallett Station several researchers made entomological collections. During the summer of 1960–1961 C. Bailey, E. B. Fitzgerald, and B. Reid, who constituted a New Zealand ornithological team, submitted specimens to the Bernice P. Bishop Museum. J. C. L. M. Mather, who was employed by the Bishop Museum to operate insect nets aboard a ship in antarctic waters during the 1962–1963 summer, collected briefly at Hallett. And in 1964 J. L. Gressitt, K. A. J. Wise, and J. Shoup, a Bishop Museum entomology team, ran several transects, made several 24-hour microclimatological observations, and collected mites and Collembola extensively [Gressitt, 1967].

Until this writing the free-living mites described from Hallett Station and vicinity were as follows:

PROSTIGMATA
- Penthalodidae
  - *Stereotydeus belli* (Trouessart, 1902)
  - *S. punctatus* Strandtmann, 1967
  - *S. delicatus* Strandtmann, 1967
- Eupodidae
  - *Eupodes wisei* Womersley and Strandtmann, 1963
- Rhagidiidae
  - *Coccorhagidia gressitti* Womersley and Strandtmann, 1963
- Tydeidae
  - *Tydeus wadei* Strandtmann, 1967
  - *Tydeus setsukoae* Strandtmann, 1967
- Pachygnathidae
  - *Nanorchestes antarcticus* Strandtmann, 1963

CRYPTOSTIGMATA (ORIBATEI)
- Oribatulidae
  - *Maudheimia petronia* Wallwork, 1962

METHODS

At the beginning of the first season (1965–1966), when the initial collections of mites were made in the field for species evaluation, some specimens were retained alive in snap cap vials. The vials were interchangeable on an aspirator mounting. *Stereotydeus belli* was the only species that survived removal from its natural habitat after collection. The other, more delicate species could not tolerate the change and succumbed before they could be transported the short distance to the laboratory to be placed on an artificial medium. The first attempts to keep mites alive in vitro were simply to drop the mites from the snap cap vial onto wet filter paper in a petri dish. The petri dish was then placed in a refrigerated incubator at 5°–10°C. Although the temperature was satisfactory, the humidity was not. Some of the mites ran about on and under the edges of the filter paper and were readily observed to be negatively phototrophic. Others became entrapped by adhesion to the water droplets on the sides and the covers of the dishes.

The next attempt was to take rocks bearing mites to the laboratory and to place additional mites on the rocks in finger bowls to which water had been added. The mites moved about freely on the rocks but eventually became trapped in the water moat. Also water in the dishes evaporated very quickly in the frequently low humidity of the laboratory and the refrigerated incubator.

An additional attempt to keep mites alive in vitro was to sift quantities of the coarse soil found along the talus of the skua rookery. A U.S. standard no. 20 mesh screen (0.84-mm opening) was used to collect the finer particles. Sufficient sifted soil was accumulated to place a layer approximately 3–4 mm deep in the bottom of the dishes. Petri dishes 5 cm in diameter were used in that trial and in all trials thereafter. Water was added until the soil particles were thoroughly wet, and any excess water was withdrawn with ink-blotting paper. Mites were placed in this preparation with no long-term success. Condensation was always excessive in the closed petri dishes, and mold growth soon overtook the soil surface and made it uninhabitable. Neither time nor facilities permitted identification of the mold. Some samples were sterilized in an autoclave at 15 psi for 10 min prior to wetting and placing mites. Although mold contamination was greatly reduced, the success was little or no better.

It was apparent that controlled low humidity was desirable. Nighttime, low sunlight intensities, and temperatures were conducive to high humidity rates, which were indicated by heavy condensation in the petri dishes even though they were enclosed within the refrigerated incubators. To compensate for the fluctuation, the dishes were placed in glass chambers of the type used for chemical desiccation. The humidity within the chambers was regulated by means of a saturated solution of calcium chloride [Solomon, 1951]. In all subsequent experiments the temperature in the incubator was regulated to a range of 5°–10°C except during observations, when the mite containers became somewhat warmer. The observation periods were subsequently limited to 2 min in an effort to prevent temperature increases, which were considered detrimental.

After the improvement in the control of relative humidity and the limitation on the observation periods, the mites were noted to be healthier and more active. Most *S. belli* could be kept alive for 2–3 weeks. From regular observation it became second nature to recognize their actions and habits. In time the creatures appeared to be starving. Food offerings of *Prasiola crispa* alga and *Bryum argentium* moss were ignored. The mites spent a great deal of time on the *B. argentium;* however, no feeding was observed. An increase in the water content of the soil in conjunction with the placement of ? *Oscillatoria* sp. alga was not tolerated by the mites.

A medium in the petri dish nearly matching the microhabitat observed under rocks in the skua rookery was desired. Hence mixtures of charcoal and plaster of Paris were tried in accordance with the experience of Evans et al. [1961].

The wetting agent used was an alga-rearing solution of the type described by Bold [1957]. Since this solution had already been prepared to isolate cultures of filamentous algae, it was incorporated into the solid medium to support *Ulothrix* and ? *Oscillatoria* sp. in and on the moss and the surrounding areas.

The combination of 50 grams of charcoal, 50 grams of plaster of Paris, and 150 ml of alga-rearing solution seemed to yield an excellent surface for growing algae and moss. The air bubbles that had migrated to the surface formed small air cavities that, when they were dry, became small and usable hiding places for mites. These cavities were similar to those found on the many volcanic rocks in the natural habitat of the mites [Gless, 1967].

Late in the 1965–1966 season little or no progress had yet been made with *S. belli.* The larger *C. gressitti* were repeatedly taken to the laboratory, and failures were always registered until the petri dishes containing the artificial medium were taken to the field and *C. gressitti* brushed into them. Information about the biology of *C. gressitti* learned from this early attempt has been presented by Gless [1967].

On our arrival at Hallett Station in October 1966 some of the previously prepared culture dishes were located and examined for possible continued use. The cultures had been placed in the supply cabinet before the station was closed several months earlier. The addition of water showed that the algae and the moss were still viable; however, there were no living mites.

Shrinkage was a problem with the charcoal–plaster of Paris medium. The medium shrank as much as 3 mm from the edge of the petri dish. The gap was sufficient for the wandering mites to enter and to disappear. The addition of dry fine, particle-sized clay did not prevent the development of the gap. A second attempt was made by sifting the soil with a U.S. standard no. 60 sieve (0.250-mm) and placing the fine particles in a thin layer around the edge of the charcoal–plaster of Paris medium in the petri dish. The sifted soil effectively plugged up the shrinkage gaps around the edges. However, the soil had to be sterilized prior to its addition to the medium because mold was a continuing problem. Since the mites are negatively phototropic, they attempted to hide in the *B. argentium* and under the *P. crispa.* In later studies the lights above the dishes in the refrigerated incubator were shut off, and the lights below the dishes were left on. That arrangement worked out very satisfactorily, since the vegetative growth did not suffer and the mites seemed to fare better. Sterile distilled water was used to maintain a proper moisture balance, since the use of an alga-rearing solution was believed to cause the nutrients in the culture to become too concentrated. After repeated observations the amount of water needed for the desired correct moisture balance could be deduced. The factors involved were the water condensate on the inside of the dish, the reflective sparkle from the charcoal–plaster of Paris surface, the intensity of the green color of the algae and moss, and above all the actions of the mites.

Another continuing problem was the placement of healthy active stages of mites on the culture medium. Precooled culture dishes were taken to the field locations of large mite populations. When a rock of suitable size bearing numbers of mites was held over the dish and tapped lightly with another rock, many clinging specimens were dislodged and fell onto the medium.

In the laboratory the culture dishes containing mites were continuously cooled by placement in an ice water bath while observations with a dissection microscope were made. The observation periods could then be extended without detrimental increases in

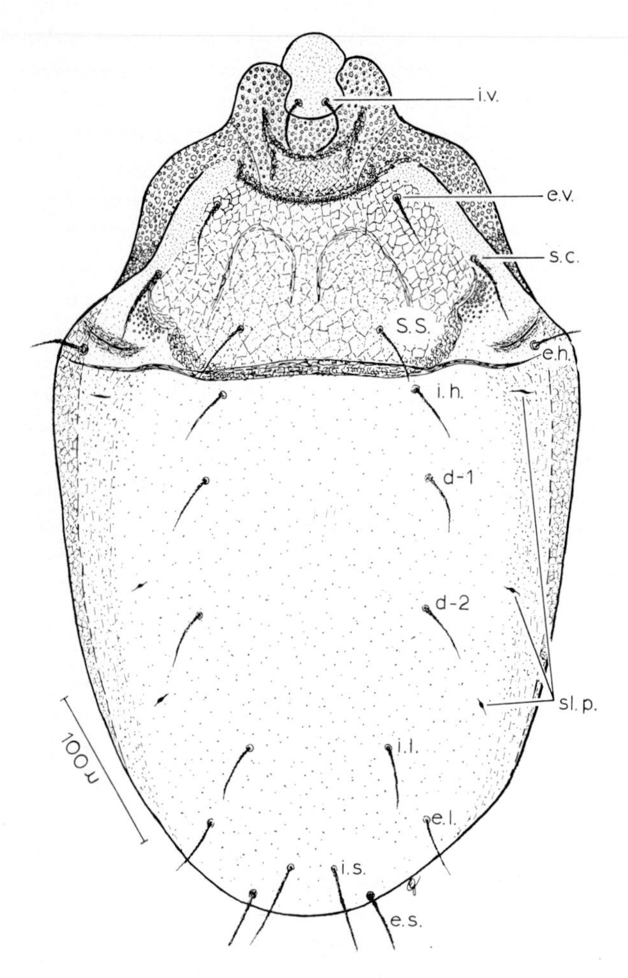

Fig. 1. Idiosoma, *Stereotydeus belli* (Trouessart, 1902), family Penthalodidae. Adult dorsal chaetotaxy of families Penthalodidae, Eupodidae, and Rhagidiidae: i.v., internal vertical; e.v., external vertical; s.c., scapular; S.S., sensory seta (trichobothria); e.h., external humeral; i.h., internal humeral; d-1 and d-2, dorsal 1 and dorsal 2; i.l., internal lumbar; e.l., external lumbar; i.s., internal sacral; e.s., external sacral; and sl.p., slit pores.

temperature. For further selection an Irwin loop (Welch Manufacturing Co., stainless steel and rust resistant in chloral hydrate clearing solutions) was dipped in sterile distilled water. The loop bearing the small droplet of water was touched to the dorsum of a selected mite, which was then transferred to another culture dish. However, the mite could not free itself from the adhesive forces of the water droplet without the aid of the manipulator. The Irwin loop had to be oriented so as to drain the water off and into the medium and thus to free the captive mite.

The methods described were not successful in all cases. *Eupodes wisei* could seldom be found by tapping rocks over the culture dishes. That species was captured by means of an aspirator prepared with a snap cap vial. When several *E. wisei* were collected in such a manner, the vial was immediately emptied into the precooled culture dish, the mites thus not being subjected to the hazards of transportation to the laboratory in a detrimental environment. *Coccorhagidia gressitti* were later collected in the same manner. *Protereunetes paulinae* were never found during this investigation by any method other than flotation, a method patterned after that of Winkler [1912]. Although this species is delicate and fragile, many mites survived the rough treatment of the soil collection method. By means of Irwin loops usable mites could be collected from the surface of the flotation and placed in suitable culture dishes. Mites of other species were also occasionally added to individual cultures from flotation collections.

Mold was a problem throughout the entire program. It was encountered in every season and in every life cycle study. Daily observations were necessary. Sparse growth could be removed with a fine pointed probe. Heavy growth was not easily controlled and was generally observed to be indicative of excess water in the culture dish. Attempts at removal usually failed. Methods of water removal by adding sterile soil and pieces of blotting paper and by opening the dish to air were usually ineffective. The addition of dry bits of *P. crispa* sometimes helped; however, the pieces were more often detrimental to the mites and a hindrance to observations.

By repeated observations and adjustments of water content in seemingly countless culture dishes, representatives of the various intermediate stages were eventually collected.

Specimens to be studied were placed in 30-mm stender dishes containing 80% ethanol until they could be prepared for microscopic observation. Mites selected for close observation were later transferred to a clearing solution [Evans et al., 1961, p. 79] by a fine tipped dropping pipette or an Irwin loop. After several trials the clearing agent was replaced by one prepared without 2.5% HCl, i.e., one with 40 grams of chloral hydrate to 25 ml of water. The more delicate and fragile immature stages seemed to be too severely damaged by the acid to be useful. The experience of trial and error dictated how long to leave the specimens in the clearing solution. Dates of collection and illustrations of morphological differences for most developmental stages are presented below.

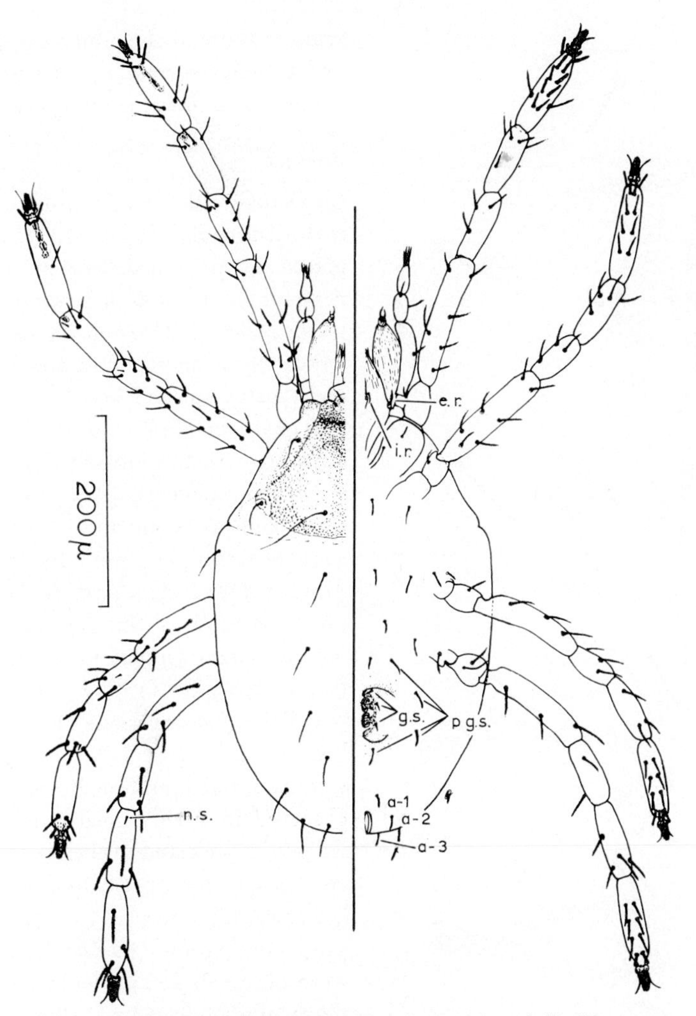

Fig. 2. Dorsal and ventral aspects of *Stereotydeus belli* (Trouessart, 1902), tritonymph: i.r., internal rostral seta; e.r., external rostral seta; g.s., genital seta; pg.s., paragenital setae; a-1, a-2, a-3, anal setae; and n.s., nude seta.

BIOLOGY AND SYSTEMATICS

The families Penthalodidae, Eupodidae, and Rhagidiidae, of the terrestrial prostigmatid mites living in north Victoria Land, Antarctica, are similar with respect to reproduction, general body shape, setation, and numbers and feeding habits of their immature stages. The dorsal chaetotaxy of *S. belli* (Figure 1) is representative of all three families [Strandtmann, 1967]. Other consistent setal numbers are afforded by the coxae ventrally and the tarsi both dorsally and ventrally. Leg setation is less fully developed in the immature stages but with further study could possibly be incorporated as a lesser taxonomic reference.

The external genital setae and the internal genital knobs are probably the most important characters for differentiating immature stages. The tritonymph of *S. belli* (Figure 2) does not have completely developed genitalia, body sclerotization, leg setation, or femoral divisions; however, its structures are sufficiently developed to serve as a guide to the identification of all species included in this paper.

Family Penthalodidae, Sig Thor, 1933

Stereotydeus belli (Trouessart, 1902)

Figs. 1–3

Penthaleus belli Trouessart, 1902. Coll. Nat. Hist. '*Southern Cross*,' p. 225.

Chromotydeus belli: Thor and Willman, 1941. Tierreich, *71:* 66.

Stereotydeus belli: Womersley and Strandtmann, 1963. Pacif. Insects, *5:* 458.

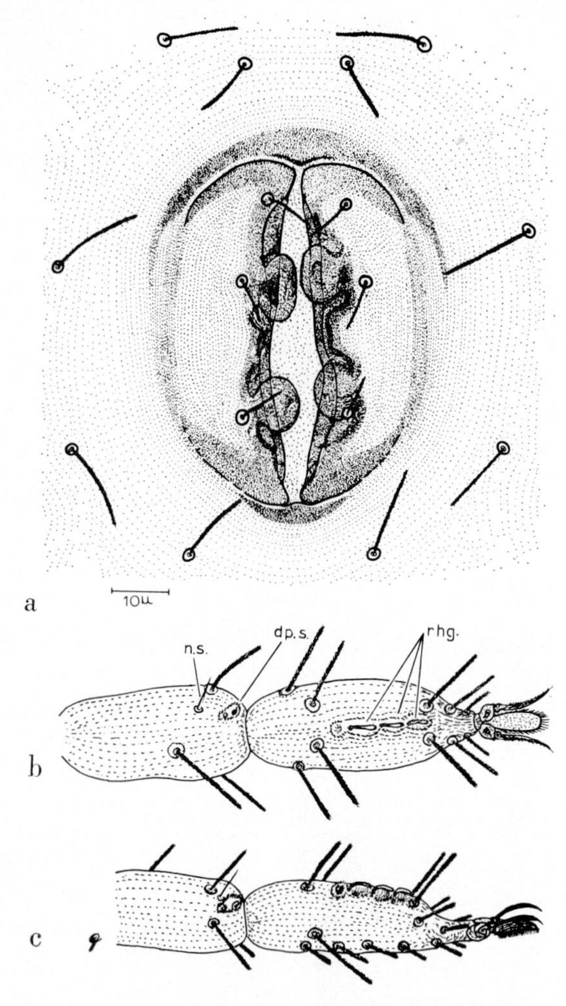

Fig. 3. *Stereotydeus belli* (Trouessart, 1902), tritonymph. (*a*) Genital field. (*b*) Tarsus 1, dorsal: n.s., nude seta; dp.s., depressed solenidion; and rhg., rhagidiforms. (*c*) Tarsus I, lateral.

At the beginning of the 1965–1966 season large groups of mites were found at the base of the talus slope directly east of the station. Some specimens were collected and identified as *S. belli.* Initial attempts to keep specimens alive in culture were unsuccessful.

After one season of experience with culture media (at the beginning of the 1966–1967 season), more mites of the same species were collected and taken alive to the laboratory. The living mites were sorted with a dissection microscope according to size and general appearance. Adults could be clearly identified and separated from immatures. The size next smaller than the adult was assumed to be the tritonymph. When the mites of this size were cleared and examined, their morphological features were nearly the same as those of the adults except for reduced sclerotization, reduced numbers of external genital setae, and absence of internal genital setae [Strandtmann, 1967].

Biology

On October 20, 1966, approximately 75 culture dishes, each dish containing 10–15 *S. belli* tritonymphs, were placed in an atmosphere of 32% relative humidity. A small clump of *B. argentium* with filamentous blue-green algae (? *Oscillatoria* sp.) growing around and among the gametophytes and the patches of *P. crispa* and *Nostoc* sp. had been in culture on the surface of the charcoal–plaster of Paris medium for about 6 days prior to the introduction of the mites.

On November 10, 1966, two males and one female were collected from the culture dishes. Two exuviae were observed but were too fragile to be saved for demonstration. The next day many more adults were observed, and at the end of the fourth day all remaining tritonymphs were removed to separate culture dishes. No feeding had been witnessed.

Routine observations for maintenance of water balance and removal of mold mycelia and dead mites were continued, and on November 26, 1966, eggs were observed in various places in several cultures. They were rose pink and averaged 140 μ in length. Neither copulation nor spermatophores had been observed.

Larvae were observed feeding on a filamentous blue-green alga species (? *Oscillatoria* sp.). Protonymphs were observed feeding only once. Deutonymphs were never observed feeding, although it was obvious from their enlarged and dark colored abdomens that food was being acquired. Tritonymphs were once seen eating *P. crispa;* however, the filamentous blue-green alga may be preferred, since bits of it disappeared, although the tritonymphs were never actually observed feeding on it. In an effort to supply better food, samples of a golden brown diatom (*Navicula* sp.) were placed in several cultures. The diatom was taken from large masses growing along the freshets of melt water in the skua rookery. Although the mites were never observed feeding on that particular alga, cleared specimens often contained diatoms.

By mid-February it became doubtful that any tritonymph would molt to adult before the end of the season. Most immatures found in nature were deutonymphs. The living mites in culture became reduced to approximately eight in six cultures. The season was ending, and the time to close the station was near. The laboratory was to be closed for the winter

TABLE 1. *Stereotydeus belli*, Days Spent in Each Stage of Development

Date Placed in Culture or Molting Observed	Stage	Days
Oct. 20, 1966	tritonymph	
Nov. 10, 1966	adult	22
Nov. 26, 1966	egg	16
Dec. 11, 1966	larva	15
Dec. 26, 1966	protonymph	15
Jan. 9, 1967	deutonymph	15
Jan. 30, 1967	tritonymph	21
Feb. 22, 1967	adult	23

on February 24. A final observation of the cultures was made on February 22 to make adjustments. One mite was recumbent, and, since it would be many days before the cultures could be observed again, that specimen was prepared. It was a tritonymph molting to adult and thus constituted the completion of the F-1 generation. The adult genitalia can be seen within the tritonymphal skin (slide IH79-67, Gless collection). Other morphological characters conform to the description given by Womersley and Strandtmann [1963]. On our arrival in the United States 17 days later the mites in the cultures were dead. The mites were tritonymphs.

The days spent in each stage of development are indicated in Table 1. Adults constituted 10% of the field samples taken 2 days prior to closing the station, and 90% of the immatures present were in the tritonymphal stage. Deutonymphs constituted most of the remaining samples. Some protonymphs were collected, but no counts were made. No larvae were present.

Tritonymphs had been collected in large numbers during the previous October; however, only one mite of the eight tritonymphs in the in vitro studies transformed to adult. Hence the tritonymphal (earlier) stage is assumed to be the condition in which most *S. belli* spend the winter. It is quite possible that more maturing takes place during early winter, since there are many warm days during March and April before it becomes extremely cold. The number of mites in this study is in reality too small for any comparison with in vivo samples that may have been taken at the beginning of the next season.

TABLE 2. *Stereotydeus belli*, Number of Specimens Examined and Individual Lengths

Stage	Number Examined		Length, μ
	in vitro	Field-Collected	
Larva	5	8	130–150
Protonymph	5	32	235–265
Deutonymph	6	36	250–290
Tritonymph	13	63	455–480
Adult	1	18	530–575

TABLE 3. *Stereotydeus belli*, Coxal Setal Numbers

Stage	Region I	Region II	Region III	Region IV
Larva	2	1	2	. . .
Protonymph	2	1	2	0
Deutonymph	2(3)	1	3(4)	2
Tritonymph	3	1	3(4)	3(4)
Adult	3	1	4	3

The numbers in parentheses indicate that the number can be present on one or both legs at the same time.

Morphology

Dorsal chaetotaxy is incomplete in the larva and complete in all the nymphal stages. There is a gradual transition from the smooth anterior of the larva to the pronounced trilobed and sclerotized condition of the adult propodosoma. The epivertex with trichobothria is well developed in all stages. The dorsal slit pores are present but difficult to see in the nymphal stages. They could not be demonstrated in the larval stage. Comparisons of the body size ranges measured from the tip of the epivertex to the posterior of the hysterosoma are shown in Table 2. A summary of the coxal setal formulas is shown in Table 3. Coxal pits in the region of epimera III do not appear until the tritonymphal stage. The trochanteric setae of all stages of development are shown in Table 4.

Probably the most important morphological characters for differentiating immature stages of *S. belli* are the presence or the absence of external genital setae, internal genital knobs, and paragenital setae. Their numbers are compared in Table 5. Immature stages do not have internal genital setae. Reproductive structures, such as a sperm sac in adult males, have never been observed.

TABLE 4. *Stereotydeus belli*, Number of Trochanteric Setae

Stage	Region I	Region II	Region III	Region IV
Larva	0	0	0	. . .
Protonymph	0	0	1	0
Deutonymph	1	1	1	0
Tritonymph	1	1	1	1(2)
Adult	1	1	1	2

The number in parentheses indicates that the number can be present on one or both legs at the same time.

TABLE 5. *Stereotydeus belli,* Comparison of Genital Structures

Stage	Pairs of Setae on Flaps	Pairs of Paragenital Setae	Pairs of Internal Genital Knobs
Larva	...	0	0
Protonymph	1	0	1
Deutonymph	2	2	2
Tritonymph	3	5	2
Adult	6	8–10	2

Rhagidial solenidia on tarsi I and II increase in numbers with developmental stages, whereas depressed solenidia remain the same. Their distribution is shown in Table 6. Positions of nude solenidia are compared in Table 7.

Family Eupodidae, C. L. Koch, 1842

Eupodes wisei Womersley and Strandtmann, 1963. Pacif. Insects, *5:* 541.

Fig. 4

The coarse talus from the cliff of Hallett Peninsula east of the station has fanned out into part of the study area. The lower extreme or terminal area of the fan is the annual nesting site for numbers of South Polar skuas. At the close of the 1965–1966 season *E. wisei* was found in small numbers on the undersides of rocks in that portion of the study area. The most frequent collections were in areas immediately below the vacant nest sites of previous years. Later these mites were found under the edges of much larger rocks at a depth suitable to their environmental requirements.

Tritonymphs of *E. wisei* have three pairs of external genital setae, as those of *S. belli* do. The enlarged femora IV clearly distinguish them from all other species found in the study area.

Early attempts to take *E. wisei* to the laboratory failed because the mites could not withstand the transit.

TABLE 6. *Stereotydeus belli,* Distribution of Rhagidial and Depressed Solenidia

Stage	Tarsi I	Tibiae I	Tarsi II	Tibiae II
Larva	1	1	1	1
Protonymph	1	1	1	1
Deutonymph	2	1	2	1
Tritonymph	3	1	3	1
Adult	3	1	3	1

The *E. wisei* were much more difficult to work with than the *S. belli,* since only two or three of them at a time could be located on the under surface of any given rock. Since the mites are saltatorial, as their enlarged femora IV indicate, a snap cap vial containing a 1-cm square of moist blotting paper was more useful than a culture dish in collecting and retaining large numbers. When sufficient numbers were collected, they were transferred to precooled culture dishes prepared with the growth of algae and moss as was done for *S. belli.* Sorting for uninjured tritonymphs was done later in the laboratory.

Attempts at collecting *E. wisei* early in the season failed. In mid-November approximately 38 tritonymphs were placed in culture. Twelve teneral adults observed on November 27 and 28 were dead on December 2. Adults were collected in larger numbers from the field to replenish the cultures.

The first ivory colored eggs were observed on December 19. They averaged 135 μ in length. As with *S. belli* neither copulation nor spermatophores were observed.

Biology

Although *E. wisei* was effectively retained through several molts in vitro, little is known concerning its feeding habits. The protonymph was the only stage observed to feed. The mites were apparently subsisting on plant life in the cultures; however, the inadequacy of the food supply was suggested by the rapid dying

TABLE 7. *Stereotydeus belli,* Comparison of Positions of Nude Solenidia

Stage	Tibia I	Tibiae II	Tibiae III	Tibiae IV	Genua I	Genua II	Genua III	Genua IV
Larva	A	A	0	...	0	M	0	...
Protonymph	M	M	0	0	M	M	M	0
Deutonymph	A	M/A	A	B	M	M/A	B	0
Tritonymph	A	A	A/M	B	A/M	M/A	M	0
Adult	A	A	B/M	M	A	A	M	0

A, apical; B, basal; M, medial; M/A, more mites found with solenidion medial than found with solenidion apical; A/M, more mites found with solenidion apical than found with solenidion medial; B/M, more mites found with solenidion basal than found with solenidion medial.

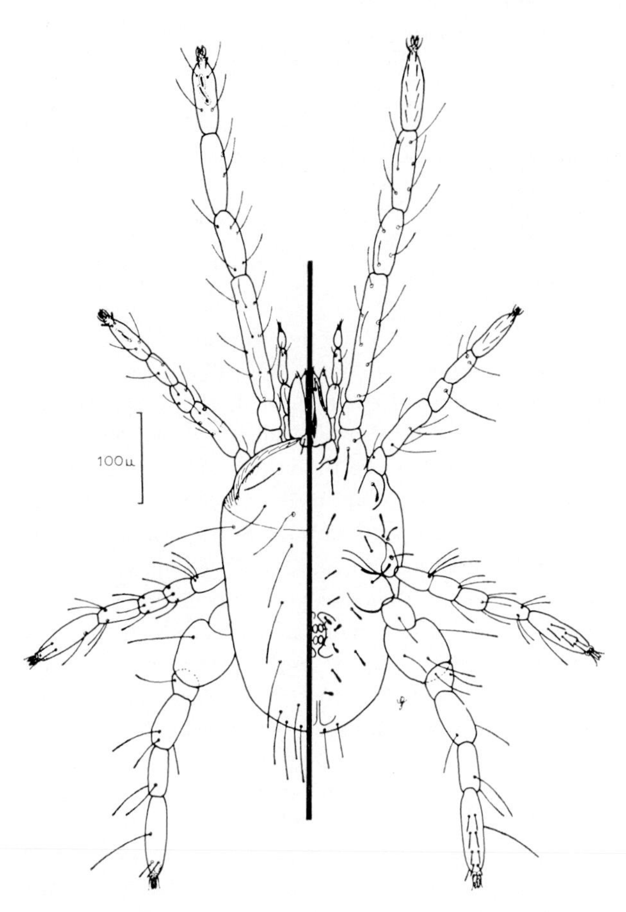

Fig. 4. Dorsal and ventral aspects of *Eupodes wisei* Womersley and Strandtmann, tritonymph.

off of the original in vitro adults, the later deutonymphs, and the tritonymphs of the 1967–1968 season. The diatom in the culture was identified as *Navicula muticopsis.* It is a heavy xanthin producer and

TABLE 8. *Eupodes wisei,* Days Spent in Each Stage of Development

Date Placed in Culture or Molting Observed	Stage	Days
Nov. 27–28, 1966	tritonymph	5–6
Dec. 2, 1966	adult	dead
Dec. 2, 1966	adult (field-collected)	17
Dec. 19, 1966	egg	14
Jan. 2–12, 1967	larva	10–18
Jan. 12, 1967	protonymph	14
Jan. 26–Feb. 10, 1967	deutonymph	14–28
Feb. 14, 1967	tritonymph	dead
Dec. 29, 1967	tritonymph	dead
Jan. 11, 1968	tritonymph	dead
Jan. 25, 1968	tritonymph	dead
Feb. 18, 1968	adult	dead (?25)

TABLE 9. *Eupodes wisei,* Number of Specimens Examined and Individual Lengths

Stage	Numbers Examined: in vitro	Numbers Examined: Field-Collected	Length, μ
Larva	4	9	170–195
Protonymph	2	13	200–225
Deutonymph	2	27	215–240
Tritonymph	6	41	300–325
Adult	6	107	350–375

TABLE 10. *Eupodes wisei,* Number of Trochanteric Setae

Stage	Region I	Region II	Region III	Region IV
Larva	0	0	0	. . .
Protonymph	0	0	1	0
Deutonymph	1	0	1	0
Tritonymph	1	1	1	1
Adult	1	1	1	1

TABLE 11. *Eupodes wisei,* Number of Coxal Setae

Stage	Region I	Region II	Region III	Region IV
Larva	2	1	2	. . .
Protonymph	3	1	2	0
Deutonymph	3	1	4	2
Tritonymph	3	1	4	2
Adult	3	1	4	3

is suspected of contributing to the generally darker and more reddish appearance of *E. wisei* specimens in vitro. All stages of *E. wisei* taken from their natural habitat had ivory to orange colored legs, and the dorsal longitudinal stripe of the hysterosoma was a light yellow. In the laboratory cultures it was a deep brownish red. *Navicula* sp. is seldom found in the native microhabitat of *E. wisei.* The natural food for these mites is believed to be an alga peculiar to their particular microhabitat. However, none has been isolated for demonstration. A schedule of days and dates in culture for each stage appears in Table 8.

Morphology

Dorsal chaetotaxy is complete in all stages. The division between the propodosoma and the hysterosoma can be seen in all the immature stages as well as in the adult. Epirostral shoulders are evident in all immatures; however, they are not evident in the adult. The numbers of specimens examined and their lengths are given in Table 9. The trochanteric setal numbers are given in Table 10.

TABLE 12. *Eupodes wisei,* Comparison of Genital Structures

Stage	Pairs of Setae on Flaps	Pairs of Paragenital Setae	Pairs of Internal Genital Knobs
Larva	. . .	0	0
Protonymph	1	0	1
Deutonymph	2	2	2
Tritonymph	3	4	2
Adult	6	5	2

TABLE 13. *Eupodes wisei,* Comparison of Tarsi I and II Rhagidial Solenidia

Stage	Tarsus I	Tarsus II
Larva	1	1
Protonymph	1	1
Deutonymph	2	1
Tritonymph	2	2
Adult	2	2

The coxal setae, which are pronouncedly claviform in the early developmental stages, are less so in the tritonymphal and adult stages. The coxal setal numbers are compared in Table 11.

As was true in *S. belli,* the most important morphological characters for differentiating immatures are the presence or the absence of external genital setae, internal genital knobs, and paragenital setae. Additionally the clavate coxal setae and the enlarged femora of the nymphs serve well for identifying *E. wisei* immatures. Genital structures are compared in Table 12.

Immature stages do not have internal papillae or internal genital setae. Additional reproductive structures, such as the sperm sac in adult males, are not present in immatures. Rhagidial solenidia on tarsi I and II are compared in Table 13. No other solenidia are found on other leg segments. In the tritonymphal and adult stages there are usually one or more long filiform and slightly pilose setae on each leg segment.

Remarks on Adult Morphology

Strandtmann described the ventral setae of adult *E. wisei* in Womersley and Strandtmann [1963] on the basis of the epimeral regions. Accordingly the epimeral setal counts were presented as: I, 2; II, 1; III, 3; IV, 2. Later Strandtmann decided to delete the use of epimera and to describe the ventral setae on the basis of the coxal fields. This method includes the medial setae, which other authors have mentioned separately.

Thus the coxal setal formula for *E. wisei,* as described and illustrated by Womersley and Strandtmann, would be 3, 2, 5, 3. The writer has examined >100 adult specimens from the type area and has never found such an adult setal formula. The ventral view of the adult was one that had been found among Womersley's drawings by Strandtmann [Womersley and Strandtmann, 1963]. It had been made some time prior to his death and was subsequently published in coauthorship with Strandtmann. Strandtmann (unpublished data, 1968) states that the extra setae (i.e., one lateral seta and one medial seta of coxa III and one seta of trochanter III) had erroneously been left on the drawing. Additionally the body length reported by Womersley and Strandtmann [1963] is somewhat longer than that observed by the writer.

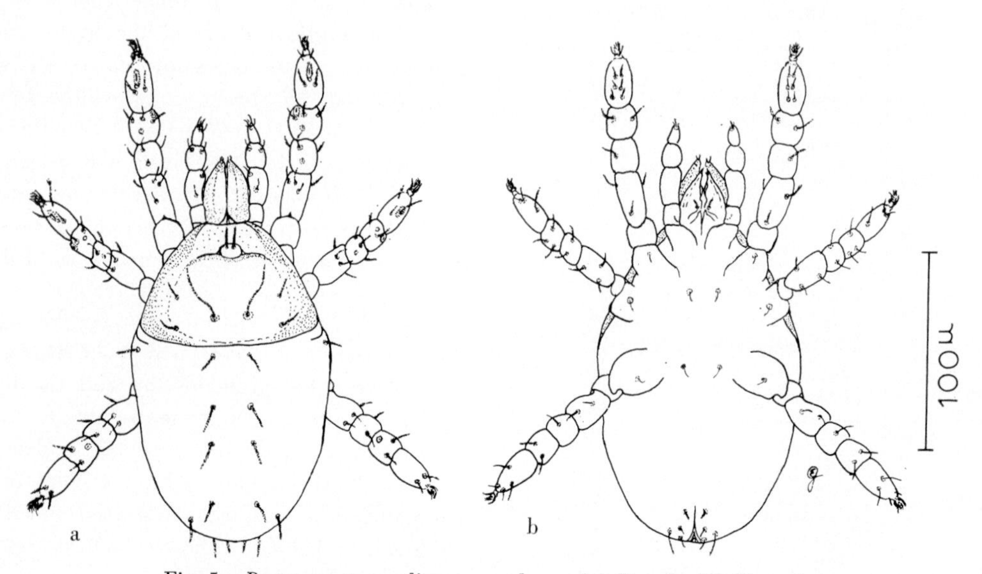

Fig. 5. *Protereunetes paulinae* n. sp., larva. (*a*) Dorsal. (*b*) Ventral.

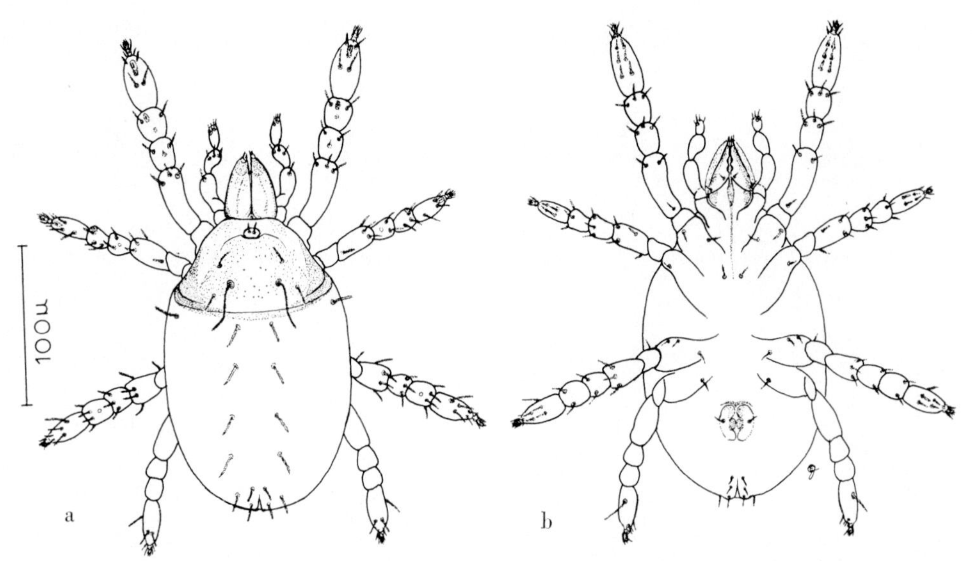

Fig. 6. *Protereunetes paulinae* n. sp., protonymph. (*a*) Dorsal. (*b*) Ventral.

Protereunetes paulinae sp. n.

Figs. 5–11

Aspirator collections on November 20, 1966, taken north-northeast of research site A were immediately placed in a chloral hydrate clearing solution. These collections contained mites of several species in various stages of development, and, when they were examined closely, two specimens were found to be of the genus *Protereunetes*. No previous record can be found of that genus from north Victoria Land or Hallett Peninsula.

On December 25, 1966, numbers of adults of the new species were found in soil flotations from about

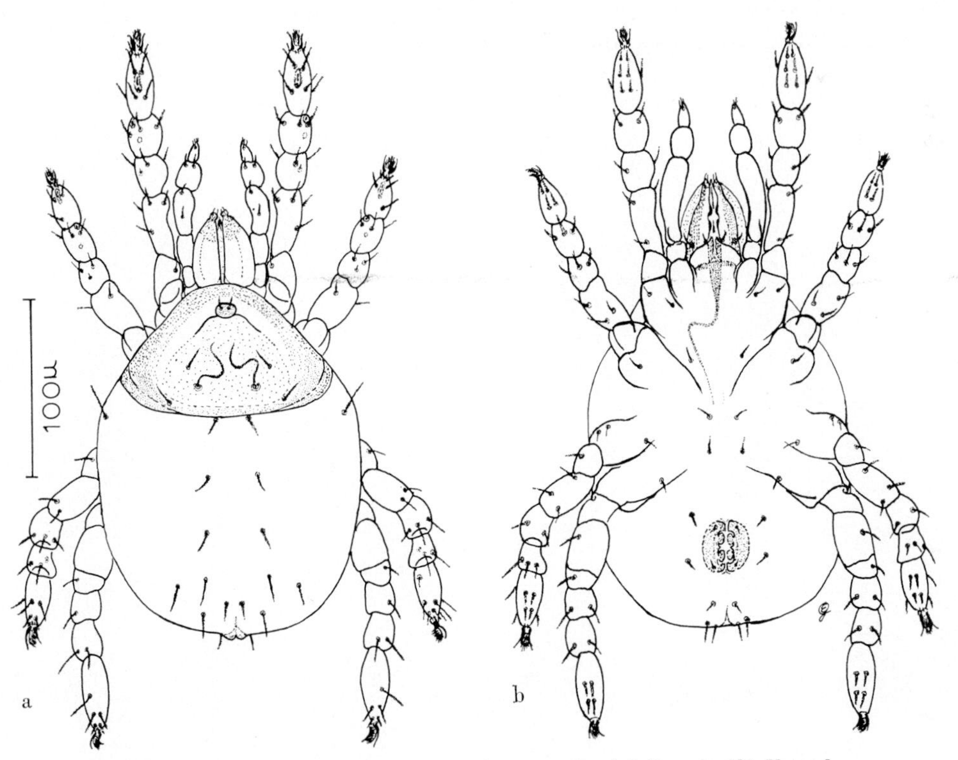

Fig. 7. *Protereunetes paulinae* n. sp., deutonymph. (*a*) Dorsal. (*b*) Ventral.

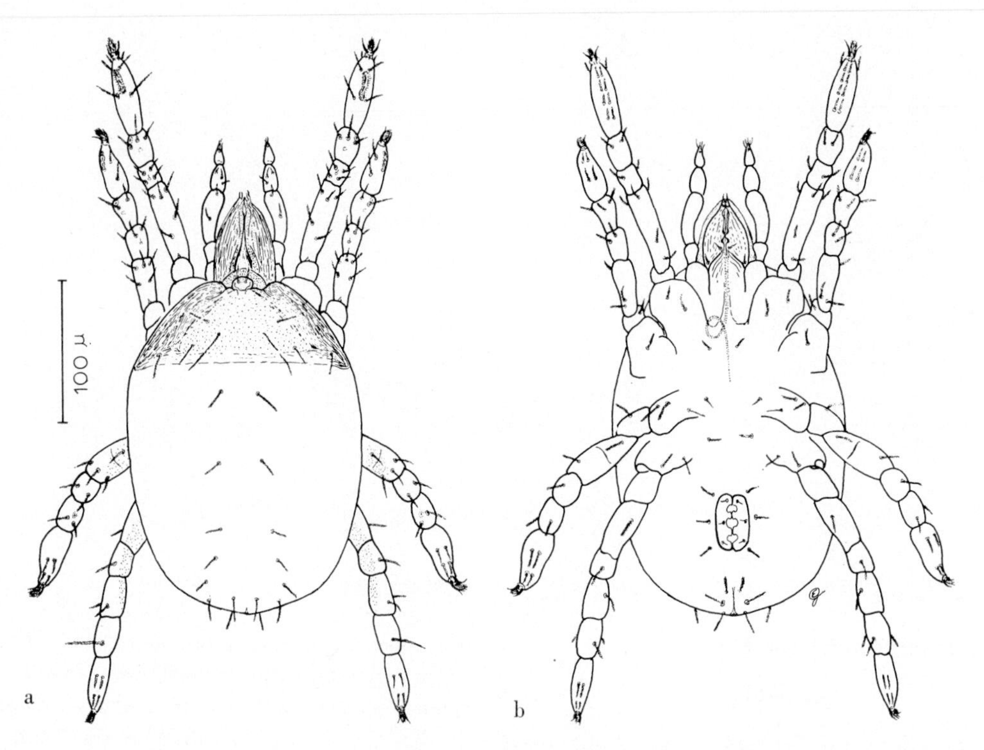

Fig. 8. *Protereunetes paulinae* n. sp., tritonymph. (*a*) Dorsal. (*b*) Ventral.

70 meters southeast of the station at an elevation of approximately 15 meters. The vegetation of the microhabitat consisted of *B. argentium* clumps, of which portions were dead and decaying and on which *Nostoc* sp., ? *Oscillatoria,* and an abundance of the golden diatom of the genus *Navicula* were growing. *Prasiola crispa* was found in small amounts throughout the area.

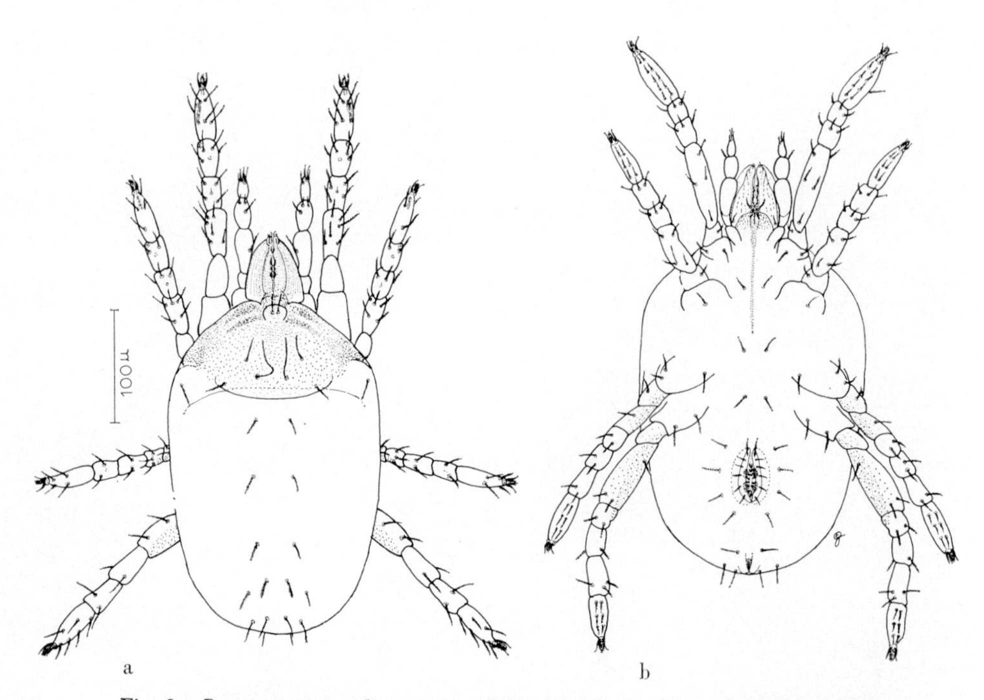

Fig. 9. *Protereunetes paulinae* n. sp., adult. (*a*) Dorsal male. (*b*) Ventral female.

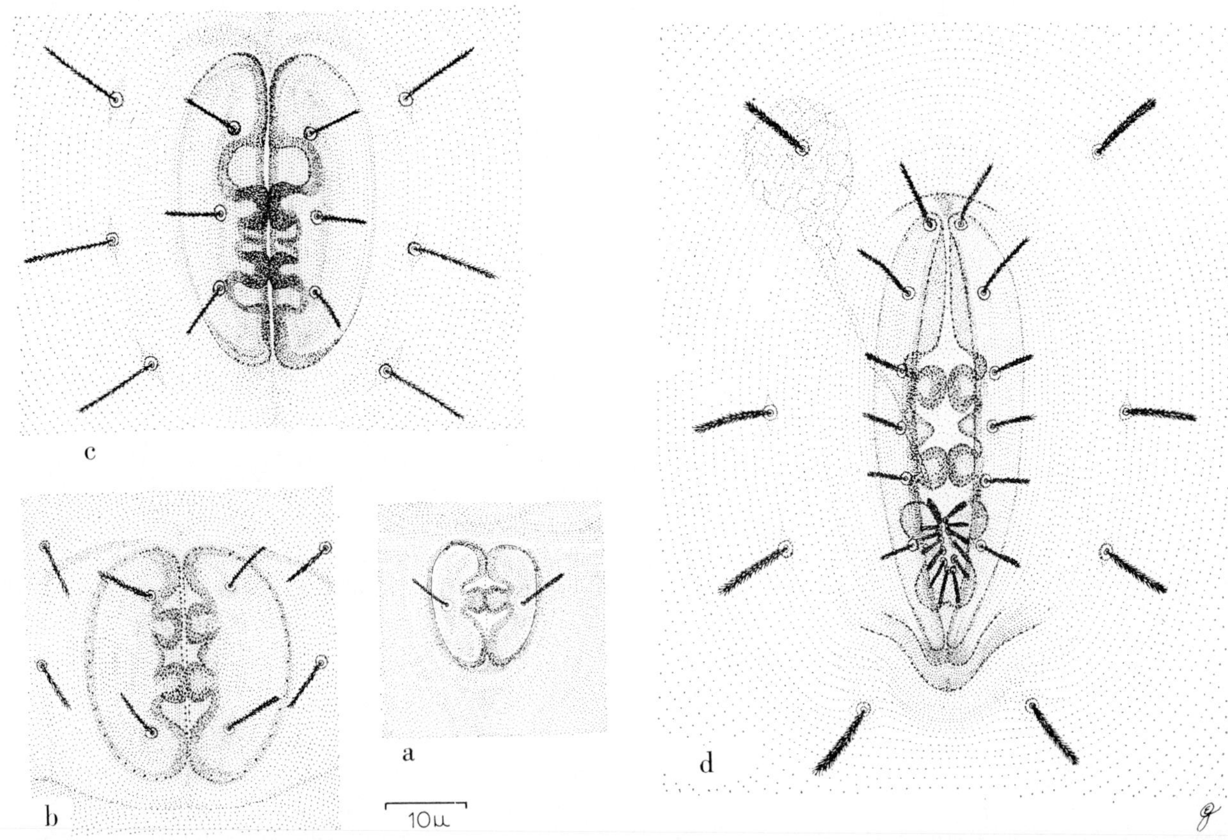

Fig. 10. *Protereunetes paulinae* n. sp., genital fields. (*a*) Protonymph. (*b*) Deutonymph. (*c*) Tritonymph. (*d*) Adult male.

Although all other prostigmatid mites found in the area are brightly colored (i.e., they have legs of red, orange, or yellow and stripes of similar colors on their striking black bodies), this species has a dark- to light-green body with a white longitudinal stripe on the hysterosoma. Occasionally the major color of the idiosoma will be deep maroon to orange. The body setae are white and can be seen with the dissection microscope. The legs are nearly always white or cream.

Additional soil and vegetation samples were placed in controlled humidity chambers and observed regularly. Adults, as determined from cleared specimens, were collected from the later samples and placed on prepared mite-free vegetation that had been collected from the same area.

Biology

All attempts to rear *P. paulinae* to the second filial generation in both the 1966–1967 and the 1967–1968 seasons were failures. The mites were reared in vitro from adult through egg, larva, protonymph, deutonymph, and tritonymph, but not again to adult. Larvae were observed to move very slowly in the culture dishes and were believed to feed on the filamentous ? *Oscillatoria* growing on the wet decaying surface of *B. argentium*. Protonymphs and deutonymphs actively feed on the same alga. A schedule of days and dates in culture for each stage appears in Table 14.

Since tritonymphs were not found in culture or in vivo until late February, the tritonymphal stage is suspected to be the one in which this species spends the winter months.

Larval Morphology

Dorsal. Epivertex pronounced with one pair short pilose setae; in some specimens appears to be positioned one-fourth length of propodosoma above and posterior to bases of chelicerae. Propodosoma soft in region bounded by sensory, external, and scapular setae and lightly punctate outside that area. Trichobothria 2–3 times longer than remaining dorsal setae. Division between propodosoma and hysterosoma distinct. One pair sacral setae absent; one pair ventral setae present.

Ventral. Rostral setae: internals basal, externals

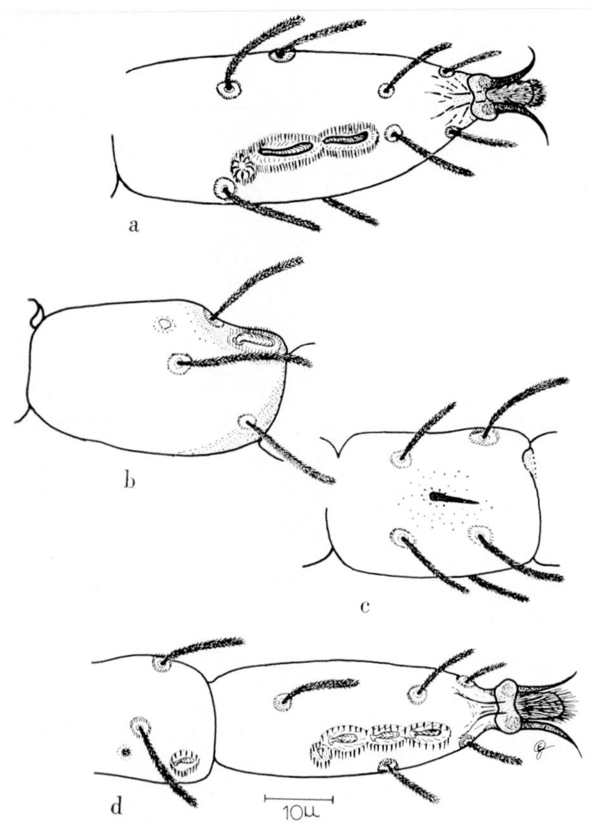

Fig. 11. *Protereunetes paulinae* n. sp., adult. (*a*) Tarsus I dorsal. (*b*) Tibia I dorsal. (*c*) Genu I dorsal. (*d*) Tarsus II and part of tibia II dorsal.

absent. Coxal setal formula: 2, 1, 2. Trochanteric setae: none. Genital structures: none. Paragenital setae: none. Anal pore terminal with three pairs of setae.

Appendages. Chelicerae well developed and lightly punctate. Pedipalps two-thirds length legs I and nude ventrally except for apex. Legs, dorsal: tarsi I and II with one rhagidiform with a short pilose seta at its base. Tibiae I and II with small rhagidiform anterior dorsal and apical. Tibia I with very delicate globose and medial solenidion. Tibia II with smooth area where delicate globose solenidion may or may not be. Tibia III with delicate nude solenidion basal. Genu I with medial nude solenidion. Genu II with smooth area where solenidion may or may not be. Legs, ventral: tarsi I with three pairs setae. Tarsi II with one pair setae and one seta probably of an incompletely developed second pair. Tarsi III with two pairs setae.

Protonymphal Morphology

Dorsal. Same as larva except for slight sclerotization of propodosoma in region of propodosomal trichobothria and complete dorsal chaetotaxy.

Ventral. Coxal setal formula: 3, 1, 3, 1. Trochanteric setal formula: 0, 0, 1, 0. Internal and external rostral setae basal. Genital flaps indistinct. One pair external genital setae and one pair internal genital knobs. No paragenital setae. Anal pore terminal with three pairs setae; tarsi II and III with two pairs setae and tarsi IV nude.

TABLE 14. *Protereunetes paulinae*, Days Spent in Each Stage of Development

Date Placed in Culture or Molting Observed	Stage	Days
Dec. 25, 1966	Adults	
?	(Eggs)	26
Jan. 20, 1967	Larva	13
Feb. 1, 1967	Protonymph	11
Feb. 12, 1967	Deutonymph	10
Feb. 22, 1967	Tritonymph	?dead

Deutonymphal Morphology

Dorsal. Chaetotaxy complete. Sclerotization of propodosoma slightly increased from protonymphal condition. Hysterosoma apparently broad (two-thirds length) in only culture specimen examined. Field-collected mites with same setation and genital structures did not indicate equivalent comparative width. Such a condition believed due to newly molted state combined with mounting medium. Division between propodosoma and hysterosoma evident.

Ventral. Coxal setal formula: 3, 1, 4, 2(3). Trochanteric setal formula: 0, 0, 1, 0. Rostral setae basal. Genital flaps distinct with two pairs external genital setae and two pairs internal genital knobs. Two pairs paragenital setae. Anal pore terminal with three pairs setae.

Appendages. Chelicerae and pedipalps same as protonymph. Legs, dorsal: tarsus I with one rhagidiform with a stellate seta at its base. Remaining dorsal solenidia of legs same as protonymph. Legs, ventral: tarsus I with three pairs setae; remaining tarsi with two pairs.

Tritonymphal Morphology

Dorsal. Epirostral shoulder region more pronounced than that in preceding stages and lightly sclerotized. Sclerotization on region of sensory setae increased. Division between propodosoma and hysterosoma evident. Hysterosoma broad (three-fourths length);

TABLE 15. *Protereunetes paulinae,* Numbers of Setae for Each Leg Segment

	Tarsi	Tibiae	Genua	Femora	Tro-chanters
Region I	20	6	6	11	1
Region II	11	5	4	8	1
Region III	11	4	3	7	1
Region IV	11	5	3	6	0

however, does not appear as ballooned as that in the deutonymph. Field-collected mites with same morphology with hysterosomal width of same general appearance.

Ventral. Coxal setal formula: 3, 1, 4, 3. Trochanteric setal formula: 0, 1, 1, 0. Genital flaps distinct with three pairs external genital setae and two pairs of internal genital knobs. Three pairs of paragenital setae present. Anal pore terminal with three pairs para-anal setae.

Appendages. Chelicerae and pedipalps same as those of protonymphs and deutonymphs. Legs, dorsal: tarsus I with two rhagidiforms and short nude seta. Tarsus II with three rhagidiforms subtended with stellate solenidion. Remaining dorsal solenidia of legs same as those of protonymphs and deutonymphs. Femur III with slight indication of division. Femur IV divided. Femora III and IV and genu IV with posterior granulations. Legs, ventral: tarsus I with four pairs setae, remaining tarsi with two pairs setae.

Adult Morphology

Dorsal. Propodosomal trichobothria finely pilose and about one-third longer than external verticals, two-thirds as long as the internal verticals, and about twice as long as scapulars and internal and external humeral setae. Scapulars and humerals essentially same length as remaining dorsal setae of hysterosoma. Division between propodosoma and hysterosoma evident. Body more elongate than that of deutonymph or tritonymph.

TABLE 16. *Protereunetes paulinae,* Number of Specimens Examined and Individual Lengths

	Number Examined		
Stage	in vitro	Field-Collected	Length, μ
Larva	6	8	140–165
Protonymph	4	27	165–185
Deutonymph	1	28	190–205
Tritonymph	2	53	240–260
Adult	0	180	270–295

TABLE 17. *Protereunetes paulinae,* Numbers of Coxal Setae

Stage	Region I	Region II	Region III	Region IV
Larva	2	1	2	. . .
Protonymph	3	1	3	1
Deutonymph	3	1	4	2(3)
Tritonymph	3	1	4	3
Adult	3	1	4	3

The number in parentheses indicates that the number can be present on one or both legs at the same time.

Ventral. Coxal setal formula: 3, 1, 4, 3. Trochanteric setal formula: 1, 1, 1, 0. Genitalia in both sexes similar. Genital flaps elongate and easily distinguished, each with six external genital setae. Anteriormost pair genital setae slightly longer than remaining six pairs internal genital setae and can be seen medial or posterior to two pairs genital knobs. Pouch-like structure evident in some specimens and considered a sperm sac, since never before observed in study preparations when eggs present. Four paragenital setae. Anal pore posterior ventral. Three pairs para-anal setae with a-1 much shorter than other two.

Appendages. Chelicerae well developed with small terminal chelae. Internal and external rostral setae basal. Pedipalps one-half as long as legs I; terminal segment with three or four setae apical, next to terminal segment and one posterior ventral seta apically. Remaining segments of pedipalps nude ventrally. Legs, dorsal: solendia as in tritonymph. Femora III with slight ventral division forming basifemoral and telofemoral regions. Femora IV completely divided to basifemoral and telofemoral regions. Both femora III and IV with posterior granulations that extend to include trochanter. Legs, ventral: tarsi I with six pairs setae; remaining tarsi with 3 pairs. Numbers of setae for each leg segment recorded in Table 15.

Morphological Summary

The body sizes of the developmental stages are compared in Table 16.

TABLE 18. *Protereunetes paulinae,* Numbers of Trochanteric Setae

Stage	Region I	Region II	Region III	Region IV
Larva	0	0	0	. . .
Protonymph	0	0	1	0
Deutonymph	0	0	1	0
Tritonymph	0	1	1	0
Adult	1	1	1	0

TABLE 19. *Protereunetes paulinae*, Comparison of Genital Structures

Stage	Pairs of Setae on Genital Flaps	Pairs of Para-genital Setae	Pairs of Internal Genital Knobs
Larva	...	0	0
Protonymph	1	0	1
Deutonymph	2	2	2
Tritonymph	3	3	2
Adult	6	4	2

Dorsal. One pair sacral setae ventral, and one absent in larval stage. All succeeding stages with complete chaetotaxy. Epirostral region anterior and ventral to the epivertex rounded and easily seen in larva, protonymph, and deutonymph. Propodosomal shoulders pronounced and appear to encroach on epirostrum and epivertex.

Ventral. Chelicerae well developed and very lightly sclerotized. Lightly chitinized ? esophageal tube can be seen at bases of submental rostrum and chelicerae in all nymphal stages. Structure clearly visible, sometimes looped, and extends to area of coxa III. Coxal setal numbers compared in Table 17. Trochanteric setal numbers compared in Table 18.

As was true in *S. belli* and *E. wisei,* the most important morphological characters for differentiating immatures are the presence or the absence of external genital setae, internal genital knobs, paragenital setae, and rostral setae. The rostral setae of *P. paulinae* are basal in all stages: larvae, one pair; all nymphs and adults, two pairs.

The genital structures increase in complexity with each molt. They are compared in Table 19. Immatures do not have internal genital setae, whereas adults have six pairs, which are usually medial or posterior to the genital knobs. The male sperm sac seen in adults has never been observed in immatures.

Rhagidiforms are compared in Table 20. The two anterior rhagidial solenidia of adult tarsus II are sometimes oriented in a parallel fashion so that they

TABLE 20. *Protereunetes paulinae*, Comparison of Rhagidial Solenidia on Tarsi I and II

Stage	Tarsus I	Tarsus II
Larva	1	1
Protonymph	1	1
Deutonymph	1	2
Tritonymph	2	3
Adult	2	3

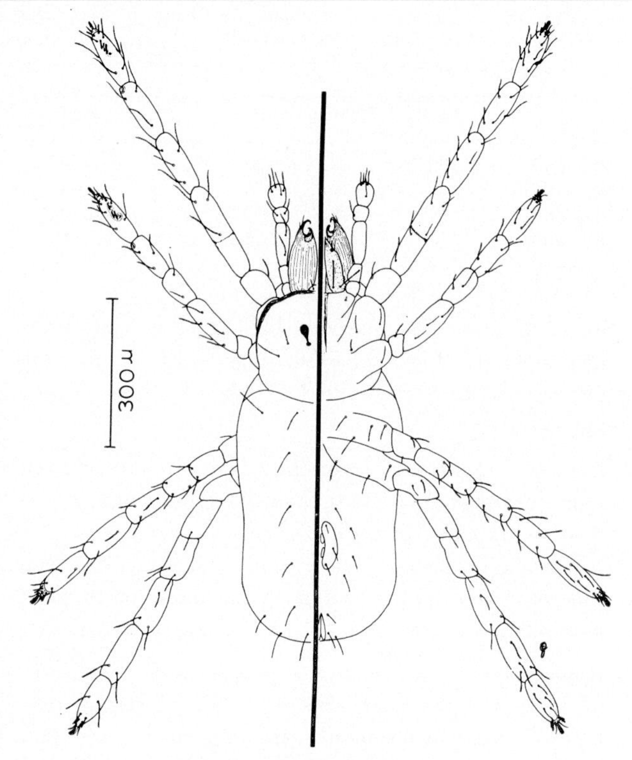

Fig. 12. *Coccorhagidia gressitti* Womersley and Strandtmann, tritonymph, dorsoventral aspect.

form a triangular pattern. The accompanying setae are short and pilose in the larval and protonymphal stages. The deutonymph, the tritonymph, and the adult have stellate setae accompanying the rhagidiform(s) of tarsus I and a nude seta with the rhagidiform(s) of tarsus II.

Holotype. ♀ (Bishop 7986), Cape Hallett, 72°20′S, 170°10′E, December 25, 1966, E. Gless, collector.

Paratypes. 1 ♀, December 25, 1966; 1 ♀, January 15, 1967; 5 ♀ and 1 ♂, January 22, 1967; 2 ♀,

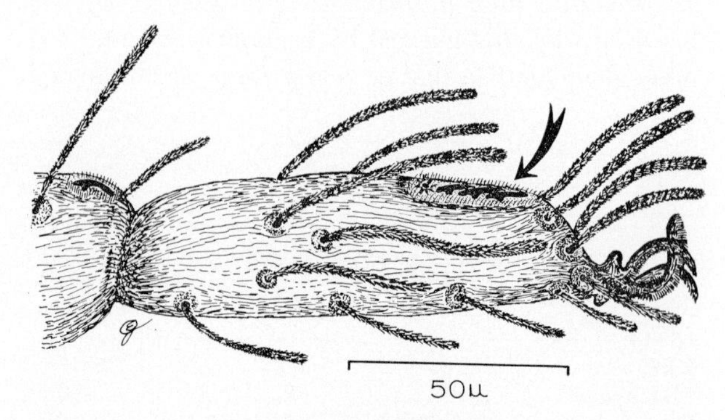

Fig. 13. *Coccorhagidia gressitti* Womersley and Strandtmann, tritonymph, tarsus and tibia I; lateral indicates rhagidiforms (arrow) and related stellate solenidia.

January 26, 1967; 1 ♂, February 15, 1967; 2 ♀, January 4, 1968; all in loose soil in north shadow of rock measuring 1 meter × 1 meter × 2 meters about 1000 meters southeast of Hallett Station on a talus slope.

Remarks on P. paulinae

This species is the third of the genus *Protereunetes* to be reported from Antarctica. It differs from *P. minutus* and *P. maudae* Strandtmann 1967 [Strandtmann, 1967] in the adult over-all body length, the presence of solenidia on both genua I and II, and the absence of a seta on trochanter IV. Other differences are found in general body shape and leg setation. The name is derived from the name of the writer's wife to fulfill a promise from early graduate school days.

Family Rhagidiidae, Oudemans, 1922

Coccorhagidia gressitti Womersley and Strandtmann, 1963. Pacif. Insects, *5:* 467.

Figs. 12–14

Coccorhagidia gressitti have been found in all parts of the study area from 2 meters above the shore line of Willett Cove to the top of Hallett Peninsula (about 310 meters). Suggested predaceous feeding habits reported by Gressitt and Shoup [1967] were subsequently confirmed by Gless [1967].

During early studies, late immature stages of adult *S. belli* were used as food for *C. gressitti* adults.

Fig. 14. *Coccorhagidia gressitti* Womersley and Strandtmann. Photomicrograph of adult holding and eating a *Stereotydeus belli* (Trouessart) tritonymph, 20×.

TABLE 21. *Coccorhagidia gressitti*, Days Spent in Each Stage of Development

Date Placed in Culture or Molting Observed	Stage	Days
Dec. 24, 1965	adult	1
Dec. 25, 1965	egg	29
Jan. 22, 1966	larva	±53
Dec. 4, 1966	tritonymph and adult	14
Dec. 4, 1966	adult	18–32
Dec. 22, 1966	egg	24
Jan. 14, 1967	larva	21
Feb. 4, 1967	protonymph	12
Feb. 16, 1967	deutonymph	

When the larvae hatched in vitro during the initial study, they died without molting. Since vegetation in the culture was the only food offered, it is probable that they starved to death.

It is impossible to differentiate between living tritonymphs and adults when one uses a dissection microscope. Specimens were collected and placed in culture on November 20, 21, and 22, 1966. Twenty-five cultures, each culture containing three or four mites, were prepared as previously described. On December 4, 1966, several shed skins were found on the surface of the charcoal–plaster of Paris medium. Only one was suitable for study. The portion of the exoskeleton revealing the genital region could be definitively examined. External genital setae numbered three on one flap and four on the other, and paragenital setae, five on one side and six on the other. The additional setae were confusing, since they did not follow the pattern of the three species previously described.

On December 22, 1966, many pale yellow eggs were found in the cultures. They averaged 165 μ in length. As many adults as could easily be transferred were removed to the other culture dishes. All living mites were removed by the tenth day after the first eggs were observed. Several attempts to transfer eggs to consolidation cultures were made but were discontinued as too harmful.

TABLE 22. *Coccorhagidia gressitti*, Number of Specimens Examined and Individual Lengths

	Numbers Examined		
Stage	in vitro	Field-Collected	Length, μ
Larva	2	6	335–380
Protonymph	3	19	440–495
Deutonymph	2	23	590–645
Tritonymph	3	72	670–745
Adult	0	84	995–1225

TABLE 23. *Coccorhagidia gressitti*, Comparison of Genital Structures

Stage	Pairs of Setae on Flaps	Pairs of Para-genital Setae	Pairs of Internal Genital Knobs
Larva	...	...	...
Protonymph	1	0	1
Deutonymph	2	0	2
Tritonymph	3(4 or 5)	4(5 or 6)	2
Adult	7(8)	5	2

Numbers in parentheses indicate that the number can be present on one or both legs at the same time.

Biology

Larvae of *C. gressitti* were observed to feed on larvae of other mites. It was not until Collembola eggs and nymphs were offered to the deutonymphs that it was learned that they could be used as food for immature *C. gressitti*. Hence it is assumed that, since Collembola eggs are consumed by deutonymphs, they are also consumed by other stages of *C. gressitti*. Tritonymphs and adults readily eat tritonymphs of *S. belli*. In the laboratory, *C. gressitti* tritonymphs and adults were never observed to eat Collembola in any stage of collembolan development.

A schedule of dates and days in culture is given in Table 21. Nymphs were reported to have been collected by Gressitt and Shoup [1967] in November. The writer collected tritonymphs in late October, and deutonymphs were observed to molt in the laboratory in late February. Consequently it is assumed that the deutonymphs and the tritonymphs represent the main overwintering forms of *C. gressitti;* however, adults are occasionally collected in early October.

Morphology

Body sizes of the developmental stages are given in Table 22.

As was true for the other prostigmatid mites reported in this paper, the genital structures are of major importance for differentiating immatures, and except for the tritonymph the genital structures follow the same pattern. The structures are compared in Table 23. Immatures do not have internal papillae or genital setae, whereas adults have 10 pairs of papillae, a single seta arising from each papilla.

Acknowledgments. During the 1965–1966 and 1966–1967 seasons the writer was employed by the Bishop Museum to study 'a mite' at Hallett Station. During a third season (1967–1968) the writer also worked at Hallett Station under grant GA1259 from Iowa State University. The information presented here was accumulated during those three seasons and was used as a portion of a thesis submitted in partial fulfillment of the requirements for the Ph.D. under the direction of Dr. Ellis A. Hicks. The study presents partial results of work on grant GA219 to the Bernice P. Bishop Museum from the Office of Antarctic Programs, National Science Foundation.

REFERENCES

Bold, H. C.
1957 Morphology of plants. 360 pp. Harper and Row, New York.

Evans, G. O., J. G. Sheals, and D. MacFarlane
1961 The terrestrial Acari of the British Isles. 1. An introduction to their morphology, biology and classification. Bartholomew, Dorking, England.

Gless, E. E.
1967 Notes on the biology of *Coccorhagidia gressitti* Womersley and Strandtmann. *In* J. L. Gressitt (Ed.), Entomology of Antarctica, Antarctic Res. Ser., *10:* 321–323. AGU, Washington, D. C.

Gressitt, J. L.
1967 Introduction. *In* J. L. Gressitt (Ed.), Entomology of Antarctica, Antarctic Res. Ser., *10:* 1–23. AGU, Washington, D. C.

Gressitt, J. L., and J. Shoup
1967 Ecological notes on free-living mites in north Victoria Land. *In* J. L. Gressitt (Ed.), Entomology of Antarctica, Antarctic Res. Ser., *10:* 307–320. AGU, Washington, D. C.

Pryor, M. E.
1962 Some environmental features of Hallett Station, Antarctica, with special reference to soil arthropods. Pacif. Insects, *4:* 681–728.

Solomon, M. E.
1951 Control of humidity with potassium hydroxide, sulfuric acid, or other solutions. Bull. Ent. Res., *42:* 543–554.

Strandtmann, R. W.
1967 Terrestrial Prostigmata (trombidiform mites). *In* J. L. Gressitt (Ed.), Entomology of Antarctica, Antarctic Res. Ser., *10:* 51–103. AGU, Washington, D. C.

Wallwork, J. A.
1962 Notes on the genus *Pertorgunia* Dalenius, 1958 from Antarctica and Macquarie Island. (Acari: Oribatei). Pacif. Insects, *4:* 881–885.

Winkler, A.
1912 Eine neue Sammeltechnik für Sabterrankäfer (Schwemm-Methode). Koleopt. Rdsch., *1:* 119–124.

Womersley, H., and R. W. Strandtmann
1963 On some free-living prostigmatic mites of Antarctica. Pacif. Insects, *5:* 451–472.

SUBANTARCTIC RAIN FOREST OF MAGELLANIC CHILE: DISTRIBUTION, COMPOSITION, AND AGE AND GROWTH RATE STUDIES OF COMMON FOREST TREES

Steven B. Young

Department of Botany and Institute of Polar Studies, Ohio State University, Columbus 43210

Abstract. The southernmost forests in the world grow in Magellanic Chile, where forest species extend their range to within about 500 miles of Antarctica. The characteristic vegetation of this region is a dense evergreen rain forest consisting mainly of the evergreen beech *Nothofagus betuloides*. Low-grown alpine vegetation is also widespread, particularly on exposed areas and at higher altitudes. However, there is no true timber line, and large expanses of treeless moorland are confined to the more inland areas of the Magellanic region. The distribution of the Magellanic rain forest is correlated with an oceanic microthermal temperature regime and high, constant precipitation. Thus true rain forest vegetation does not extend significantly east of the Andes. In southern Patagonia there is a belt of transitional forest between the rain forest and the summergreen steppe forest. Age and growth rate studies conducted on most of the important tree species of the rain forest indicate that forest growth is slow and that many of the trees reach great age. Some specimens of *Nothofagus betuloides* are probably 1500–2000 years old, although trees of this age were not sampled because of their great size. *Pilgerodendron uvifera* and *Podocarpus nubigena* may reach ages of ⩽1000 years. Even some of the small understory trees are probably several hundred years old. In contrast *Drimys winteri* and *Pseudopanax laetevirens* grow comparatively rapidly and probably do not reach great age. The variation in growth rates among the species studied indicates that growth rate is determined by genetic as well as environmental factors. Because of their great age, some Magellanic forest trees should be useful in indicating the extent of previous glaciation in the region. To do so would probably require the sampling of large specimens by means of power equipment. Regeneration of forest growth in the Magellanic region would obviously proceed slowly. The rain forest is in little immediate danger of exploitation for economic reasons, but the slow regeneration of the forest should be considered in any future plans for economic development.

The Magellanic rain forest of southern Chile is the southernmost forest formation in the world. The major elements of the Magellanic forest reach almost to latitude 56°S in the Islas Hermite [Skottsberg, 1916], only about 500 miles from the northern tip of the Antarctic Peninsula. The Magellanic flora extends its range farther south than any other complex vascular flora. Only two species of vascular plants, the grass *Deschampsia antarctica* Desv. and the caryophyllaceous *Colobanthus quitensis* (Kunth) Bartl., both of which occur on the antarctic continent, range farther south than elements of the Magellanic flora do.

The heart land of the true Magellanic rain forest is in the rugged fiord lands and islands west of the Andes and south of the Golfo de Penas (latitude 48°S). In this wild wet country the dark-green evergreen forest stretches as an unbroken thicket along thousands of miles of sinuous indented coast line. The heavy underbrush and the dense growth of epiphytes make the forest among the most impenetrable in the world. Lianas and shrubs with brilliant red or orange flowers give the rain forest a tropical aspect that is heightened by the hummingbirds that visit these flowers. Another characteristic that the Magellanic rain forest shares with tropical areas is the lack of climatic seasonality. Both winter and summer are cool, windy, and wet. Winter brings a lowering of the snow lines in the uplands, but there is no freeze-up along the shores. In the Magellanic forest some plant is in flower at all times of the year.

A point of dissimilarity between the Magellanic forest and the rain forests of the tropical highlands is the paucity of species in the Magellanic forest. Although tropical forests are often made up of scores of species, a single species of evergreen beech, *Nothofagus betuloides* (Mirb.) Blume., dominates the Magellanic forest, often nearly to the exclusion of other species.

In the uplands and on the southern and western borders of the southern Chilean Archipelago the evergreen forest becomes patchy and disjunct. It is largely replaced by shrubbier vegetation interspersed with extensive areas of bare ice-scoured bedrock.

The natural history of southernmost South America has long been the subject of intensive scientific investigation. Since the days of Charles Darwin and J. D. Hooker a series of expeditions has worked the area and produced a voluminous literature. However, much of this work has centered in Tierra del Fuego and in the pampas and mesic forests east of the Andes. These areas are suitable for herding and some agriculture, and access to them has been comparatively easy. The rugged drowned coast line of the Magellanic forest region is much more difficult of access, and rough seas and heavy weather usually compound the logistic problems. Therefore the rain forest, particularly the part north of the Strait of Magellan, has largely been neglected by scientific expeditions. Since the turn of the century most of the botanical collections from the area have been made by three major expeditions. The great Swedish botanist Carl Skottsberg, leader of the Swedish expedition of 1907–1909 to Patagonia and Tierra del Fuego, collected in the area during the austral autumn of 1908. Most of Skottsberg's collections were from the inner reaches of the archipelago, but he also visited Grupo Evangelistas [Skottsberg, 1910, 1916].

During the course of the Royal Society expedition to southern Chile in 1958–1959 the New Zealand botanist E. J. Godley made extensive collections at Puerto Eden on Isla Wellington. A small but significant collection was also made at the limestone formations at Isla Guarello. Extensive collections were made at Isla de Chiloé and on Isla Navarino outside the main area of the Magellanic rain forest [Godley, 1963, 1965]. A survey of the ecology and soils of the area was also made [Holdgate, 1961].

In the austral spring of 1969 the RV *Hero*, a National Science Foundation research vessel, made three cruises through the channels of the southern Chilean Archipelago. The first cruise (cruise 69-4, September to October 1969) was devoted mainly to cryptogamic and phanerogamic botany, and large collections were made throughout the archipelago [Imshaug, 1970]. The second cruise (cruise 69-5, October to November 1969) was devoted primarily to marine biology [Kaesler, 1970]. The third cruise (cruise 69-6, November to December 1969) supported shore-based geologic and geophysical work [Halpern, 1970]. During the three cruises approximately 1000 collections of vascular plants were made, as well as about 35 wood samples for the age and growth studies discussed later in this paper.

GEOGRAPHY AND CLIMATOLOGY

Geography. The terrain on which the Magellanic rain forest is developed is a heavily glaciated fiord land similar in topography to the coasts of Norway and southeastern Alaska. In southern Chile the western margin of the Andes dips below sea level and forms an archipelago of rugged mountainous islands. These islands are separated from each other and from the mainland by a complex system of channels. The coasts of the islands and of the mainland are highly convoluted, fiords and blind channels penetrating far into the interior. Inside the coastal fringe the mainland rises sharply to the high summits and ice fields of the main spine of the Andes Mountains.

Several major fiord systems reach deep into the mountains and often terminate in active valley glaciers. South of latitude 52°S fiords pierce the mountain chain and extend into the steppe, broad shallow bays such as Seno de Otway being formed.

The rugged topography of the Magellanic region has been accentuated by repeated glaciations during the Pleistocene [Auer, 1960]. The total extent of Pleistocene glaciation in southern South America has not yet been fully determined [Flint, 1957; Mercer, 1965], but it seems certain that the habitat of the Magellanic rain forest was totally glaciated. Most of the outermost islands of the archipelago are rounded in profile (Figure 1). They appear to have been completely scoured by a piedmont ice sheet, and their surface is mainly exposed bedrock, soil being developed mainly in pockets and ravines. Farther inland the topography is more alpine in aspect. Sharp peaks, oversteepened slopes, talus slopes, and deposits of unconsolidated material are common (Figure 2).

Climatology. The climate of coastal Magellanic Chile shows a number of unique features strongly correlated with the distribution of vegetal types in the area. Unfortunately there is no body of long-term climatic data available from stations in the central portions of the rain forest region. Grupo Evangelistas (Figures 3 and 4) can be assumed to have a climatic regime comparable to that of much of the Magellanic rain forest region, although these tiny exposed islands do not support typical rain forest vegetation [Skottsberg, 1916]. However, the location of the Grupo Evangelistas station in an area of comparatively low relief probably tends to keep precipitation at a relatively

Fig. 1. Topography and vegetation of the exposed outer islands of the southern Chilean Archipelago (Puerto del Morro, western end of Canal Trinidad, 50°03′S, 75°02′W). The rounded ice-scoured summits of the hills and the large expanses of exposed bedrock indicate that glaciation was complete. Dark areas are patches of forest, which reach nearly to the summits even in this exposed situation. The main forest species is *Nothofagus betuloides; Pilgerodendron uvifera* and *Drimys winteri* are also important locally. Even these dwarfed forests support dense growths of epiphytes such as *Hymenophyllum* spp. Lighter patches are boggy areas, where the rush, *Marsippospermum grandiflorum,* is the dominant species. *Sphagnum* is rare or absent in these situations. Mammalian life was never observed on these outer islands, and only a few land birds were seen. The R/V *Hero* is in the foreground.

low level. At Isla Guarello (50°23′S, 75°20′W) in an area of high rugged mountains, the mean annual precipitation reaches the extraordinary level of 7.5 meters [Almeyda and Saez, 1958]. The highest levels of precipitation normally occur some distance inland from the outher coastal fringes of the archipelago [Almeyda and Saez, 1958].

Cabo Raper is situated in the area of contact between the Magellanic forest and the more northerly Valdivian forest. Temperatures at Cabo Raper are probably slightly higher than those in more central portions of the Magellanic forest.

The two most salient features of the climate of the Magellanic rain forest region are the cold microthermal temperature regime and the constant excessively high precipitation and humidity. These features set the Magellanic region off climatically from the rest of southern South America, with the partial exception of the Valdivian forest region. This contrast is clearly illustrated by the climatic profiles of selected stations in southern South America (Figure 3).

A transect from west to east in southern South America shows a dramatic climatic trend. The major feature of this trend is the rapid diminution of precipitation as one crosses the rain barrier of the Andes. At Punta Arenas only a few miles outside the eastern boundary of the rain forest the precipitation is less than one-sixth that at Grupo Evangelistas (Figure 3). At Puerto Natales the mean annual precipitation is only 308 mm [Almeyda and Saez, 1958], about one-twentieth that at Isla Guarello. Coupled with increasing dryness toward the east is an increase in the continentality of the temperature regime. The mean annual temperatures of the inland stations are comparable to those of the coastal rain forest region, but the difference between the mean summer temperatures

Fig. 2. Topography and vegetation of the more protected inner reaches of the southern Chilean Archipelago (Puerto Charrúa, Isla Wellington, 50°01'S, 74°42'W). Deep fiords and steep sided valleys are indicative of intense valley glaciation. The waterfall is about 150 meters high. The dense forest is developed mainly on unconsolidated talus derived from the steep slopes. The dominant forest tree species is *Nothofagus betuloides.* Under these conditions *Nothofagus* trees may approach 30 meters in height and $\leqslant 2$ meters in trunk diameter. Other large trees are *Drimys winteri, Pseudopanax laetevirens,* and *Podocarpus nubigena. Maytenus magellanica* and *Tepaulia stipularis* are important elements of the understory. *Pilgerodendron uvifera* is restricted mainly to bogs and openings in the forest. The floor of the dense forest is often covered to a depth of 2–3 meters by a layer of undecomposed tree trunks and branches. Travel here is difficult and hazardous.

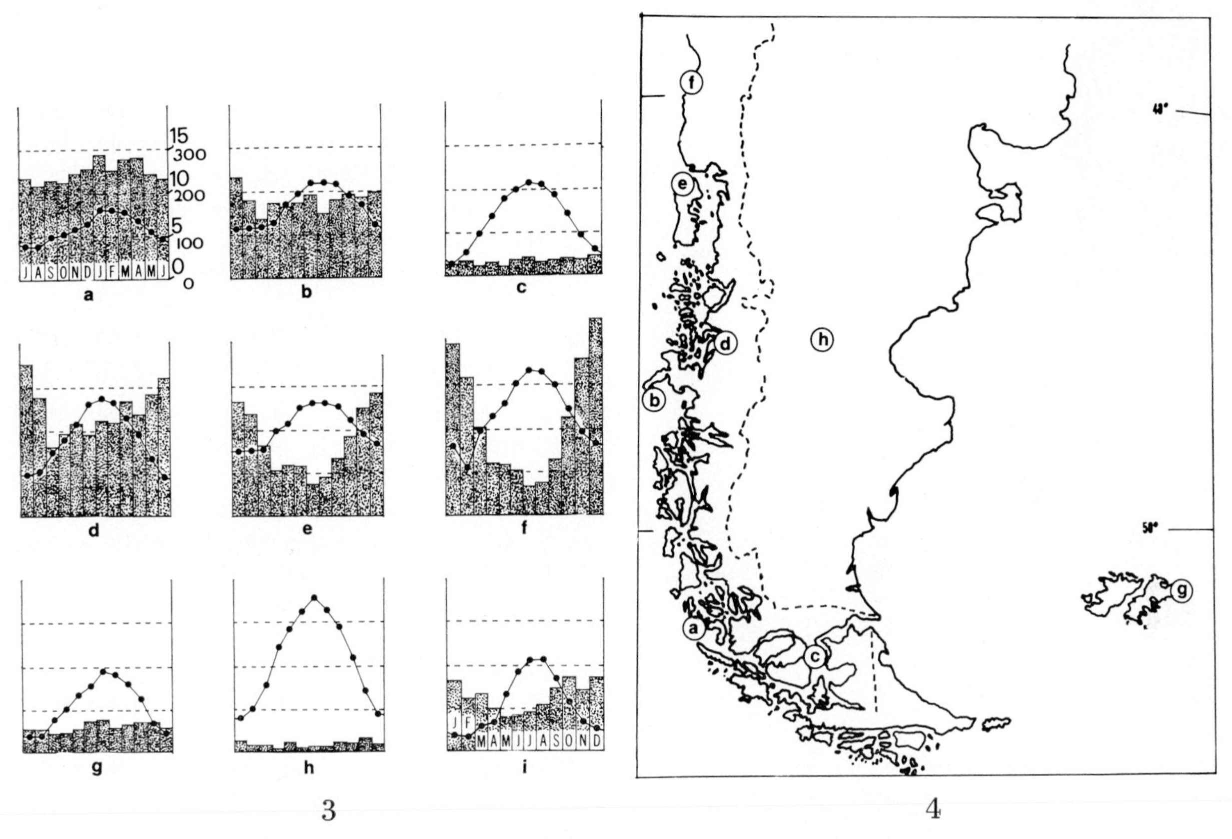

Figs. 3–4. Selected stations in southern South America whose climatic data are presented and a station in Iceland (Figure 3 only) given for comparison with a highly maritime situation in the cold temperate northern hemisphere (note that the sequence of months for the Icelandic station has been altered to conform with the austral seasons). (*a*) Grupo Evangelistas (52°25′S, 74°55′W), typical of the most exposed portions of the Magellanic forest region. (*b*) Cabo Raper (46°55′S, 75°40′W), near the transitional zone between the Magellanic and the Valdivian forests. (*c*) Punta Arenas (53°25′S, 70°40′W), in the zone of summergreen forest bordering the Patagonian steppe. (*d*) Puerto Aisen (46°25′S, 72°55′W), in the Valdivian forest region. (*e*) Ancud (42°00′S, 74°02′W), in the Valdivian forest region. (*f*) Valdivia (39°50′S, 73°40′W), in the Valdivian forest region. (*g*) Stanley (51°45′S, 57°55′W), in the Falkland Islands, with a climate comparable to that of many of the subantarctic islands. (*h*) Sarmiento (45°40′S, 69°12′W), in the heart of the Patagonian Steppe. (*i*) Vestmannaeyjar (Iceland). Note that seasonality in terms of both temperature and precipitation increases rapidly to the north and the east of the Magellanic rain forest region (compiled mainly from Almeyda and Saez [1958], Skottsberg [1916], van Rooy [1957], and Walter [1960]). Fig. 3, annual temperatures (points) in degrees Celsius and precipitation (bars) in millimeters. Fig. 4, station locations.

increases considerably from west to east. The difference between the mean January and the mean July temperatures at Grupo Evangelistas is only 4.1°C, whereas it is 8.9°C at Punta Arenas and 10.4°C at Puerto Natales [Almeyda and Saez, 1958].

There is a less dramatic climatic trend along a transect from the coastal Magellanic region northward into the Valdivian region. There the trend is toward an increase in seasonality for both precipitation and temperature. At Ancud (Figures 3 and 4) the total precipitation is of the same order of magnitude as that at Grupo Evangelistas, but rainfall during the summer month of January is only one-third that during July.

At Puerto Aisén the difference between the January and the July mean temperatures is 8.9°C. These differences become even more marked farther north, as at Valdivia. Therefore north of the Magellanic forest region there is a definite comparatively warm and dry summer. Within the Magellanic rain forest seasonal differences in temperature are minimal, and seasonal differences in precipitation insignificant. The rain forest is continually saturated with water during the entire year.

No other area has a climatic regime closely com-

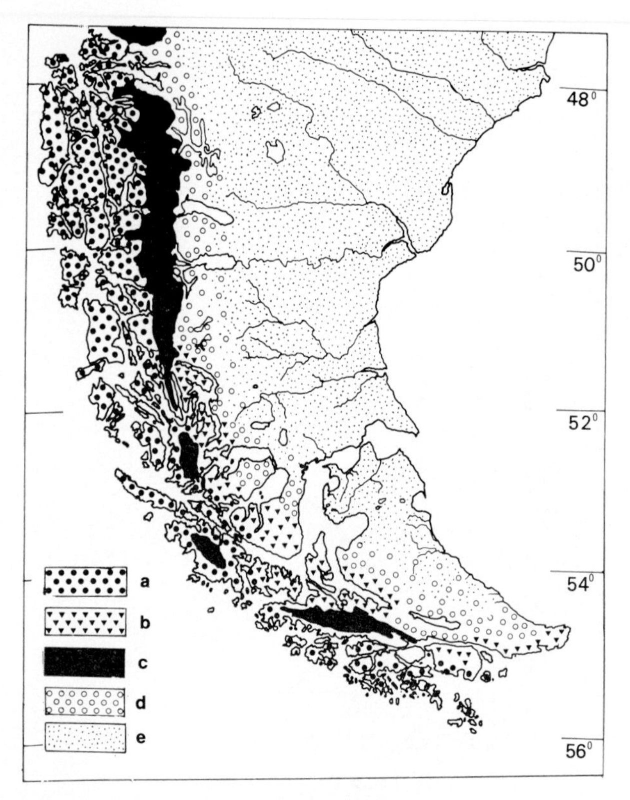

Fig. 5. General distribution of the vegetation formations in southern South America: *a*, evergreen rain forest; *b*, evergreen transitional (mesic) forest; *c*, ice fields; *d*, summergreen forest; and *e*, steppe and semidesert. The dominant species in the evergreen rain forest is *Nothofagus betuloides. Drimys winteri* is common and may be dominant in some better drained areas. *Pilgerodendron uvifera* is locally abundant along the forest borders. *Podocarpus nubigena* is locally important but apparently does not occur south of latitude 51°S. *Maytenus magellanica, Tepaulia stipularis,* and *Pseudopanax laetevirens* are the main understory species. In this formation, undergrowth is abundant, and epiphytes are of great importance. *Nothofagus betuloides* is also dominant in most of the transitional forest, although *N. pumilio* may be locally dominant. In this formation there is little undergrowth except along the forest borders, and epiphytes are rare or absent. An exception is *Myzodendron* spp. ('mistletoe'), which is parasitic on *Nothofagus. Pseudopanax* is locally important as an understory species. The summergreen forest consists mainly of *Nothofagus pumilio*, although stands of *N. betuloides* commonly occur on the higher elevations. Epiphytes, except for *Myzodendron*, seldom occur. The undergrowth has largely been destroyed by intensive sheep grazing. Local variation due to elevation is not shown on the map. *Nothofagus antarctica* may be locally dominant at high elevations in the rain forest, and small areas of rain forest may be found at high elevations on some of the eastern slopes of the mountains.

parable to that of coastal Magellanic Chile, although some of the subantarctic islands show comparable temperature data [van Rooy, 1957]. None of these islands has an exceptionally heavy rainfall, and none has a true forest cover. It is interesting that some stations in Iceland (Figure 3) and in southwestern Alaska have a climatic regime not greatly dissimilar to that of the eastern edge of the Magellanic rain forest. These areas are outside the range of northern hemisphere timber trees. Experimental plantings of Magellanic trees such as *Nothofagus betuloides, N. antarctica,* and *Pilgerodendron uvifera* in Iceland or in the Aleutian Islands might be of considerable interest.

DISTRIBUTION AND COMPOSITION OF THE MAGELLANIC RAIN FOREST

The general distribution of the Magellanic rain forest is shown in Figure 5. In the vicinity of the south shore of the Golfo de Penas there is a transition zone in which the Magellanic forest comes in contact with the Valdivian forest of the more northerly regions. North of this zone the Magellanic forest is confined to highlands and exposed situations. Isolated areas of Magellanic vegetation occur as far north as the highlands of Isla de Chiloé and on the Cordillera Pelada (about latitude 40°S) [Skottsberg, 1916; Holdgate, 1961].

Much of the eastern boundary of the Magellanic rain forest region is formed by the western slope of the main range of the Andes. The eastern slope of the mountains is far too dry to support a rain forest. A scrubby summergreen forest consisting mainly of *Nothofagus pumilio* (Poepp. et Endl.) Krasser occurs there.

South of latitude 52°S the mountains become disjunct and are separated by straits, channels, and lowlying valleys. There the eastern boundary of the rain forest is associated with climatic rather than topographic features, and there is a broad zone of transition from the rain forest to the summergreen forest. The transitional forest is superficially similar to the rain forest, since the same tree species, *Nothofagus betuloides,* is dominant in both. However, the two formations differ greatly in other aspects. The eastern border of the rain forest is marked by the disappearance of a number of characteristic species. Among these species are *Philesia magellanica* Gmel., the dominant undergrowth shrub of the rain forest, and *Tepaulia stipularis* (Hook. ex Arn.) Griseb., an important shrub or small tree. Several important fern species, including the epiphyte *Hymenophyllum pectinatum* Cav., the ground-dwelling *Blechnum magellanicum* (Desv.) Mett., and *Gleichenia quadripartita* (Poir.) Moore, abruptly terminate their ranges at the

Fig. 6. Transitional forest of *Nothofagus betuloides* south of Puerto Natales (approximately 52°00′S, 72°05′W). The tall straight trees and the lack of undergrowth and epiphytes are characteristic of this forest formation. Professor R. M. Schuster of the University of Massachusetts gives the scale. Although *Nothofagus* grows more rapidly in this zone than in the rain forest, we estimate the large tree to be approximately 500 years old. Timbering and grazing operations threaten most remaining stands of virgin timber in the transitional forest zone.

eastern edge of the rain forest. At the eastern edge of the transitional forest most of the rain forest species have disappeared or are confined to *Sphagnum* bogs. The forest consists of a nearly pure stand of *Nothofagus betuloides,* little undergrowth (Figure 6), and no epiphytic vascular plants. At its eastern edge the transitional forest interdigitates with the summergreen forest, the two seldom forming mixed stands. The line of demarcation between the two forest types is no longer clear, since there have been extensive timber cutting and forest destruction for sheep grazing. The summergreen forest (Figure 7) forms a narrow belt along the edge of the steppe and occurs as isolated copses in valleys for some distance into the steppe. The undergrowth there consists mainly of drought resistant grassland species. Fairly extensive patches of summergreen *Nothofagus pumilio* forest occur in some parts of the transitional forest, particularly in low-lying areas along the Strait of Magellan.

Sphagnum bogs, 'Magellanic moorland,' and the question of a coastal timber line. There are two distinct low-vegetation formations in coastal southern Chile to which the term 'moorland' could be applied. One occurs extensively on gently rolling or flat situations and is characterized by a dense growth of *Sphagnum.* There a layer of peat has been built up to a depth of ⩾2 meters (Figure 8). The deep peat tends to mask the topography on which it is developed, and the aspect is of smooth countryside, little bedrock being exposed. The trees are small and infrequent, probably because of the instability of the substrate. Most of the herbaceous plants common in the coastal regions also occur in sphagnum moorland. Cushion plants such as *Astelia pumila* Banks et Sol., *Azorella* species, and *Donatia fascicularis* are common, but they do not usually cover extensive areas, since the *Sphagnum* tends to overgrow them. *Hymenophyllum secundum* Hook. et Grev. and *Nanodea muscosa* Gaertn. are among the few plants able to maintain themselves within the thick sphagnum mats.

Sphagnum moorland often covers extensive areas. However, it is almost entirely confined to the transitional forest zone. To the west of the boundary between the rain forest and the transitional forest, *Sphagnum* occurs occasionally or is absent. It never covers extensive areas. Small raised sphagnum bogs extend eastward to the edge of the summergreen forest and occasionally even into it.

The second type of low vegetation of the Magellanic region contains no *Sphagnum.* The most important species are the cushion plants mentioned above and some taller forms such as *Rostkovia magellanica*

Fig. 7. Summergreen scrub forest of *Nothofagus pumilio* near the border of the Patagonian steppe (approximately 52°30′S, 71°25′W). The dense appearance of the leafless canopy is caused by the heavy growth of 'mistletoe' (*Myzodendron* spp.) and lichen (*Usnea*). This forest is of little or no commercial importance. Large areas are being destroyed mainly by burning to increase grazing land.

(Hook.) Desv. and *Marsippospermum grandiflorum* (L.f.) Hook. These species do not form peat as fast as *Sphagnum* does, and the peat layer associated with them is usually <1 meter thick. The topography on which this vegetal formation is developed is usually rugged and heavily scoured by glacial action. As a result this formation, which will be referred to henceforth as alpine moorland, is seldom developed over extensive areas. It occurs most commonly in pockets in the bedrock where drainage is impeded or on patches of thin soil over bedrock. Most of the larger areas of alpine moorland contain a fairly heavy cover of trees, although these trees may be dwarfed and spreading (Figures 9 and 10), as are the 'elfin forests' of other alpine areas. There is seldom a clear line of demarcation between the alpine moorland and the rain forest as there is between sphagnum moorland and forest (Figure 9). The rain forest is less prevalent in exposed situations and at higher altitudes, where it is replaced by alpine moorland, bare rock, or permanent ice and snow.

Earlier workers [e.g., Skottsberg, 1916; Holdgate, 1961] have not distinguished clearly between the two types of Magellanic moorland, and they have implied that many of the western and southern parts of the southern Chilean Archipelago support extensive formations of rolling treeless vegetation. However, as we have seen, the vegetal formation most appropriately called moorland is essentially confined to the innermost part of the archipelago. Skottsberg [1916] implied that there is a definite coastal timber line in the western and southern portions of the archipelago. Skottsberg apparently considered it to be a line beyond which trees do not grow to timber size rather than a floristic boundary beyond which tree species do not grow. Patches of typical rain forest actually occur in ravines and other sheltered areas even in the outermost coastal areas. Specimens of *Nothofagus betuloides* having trunk diameters of >1 meter were seen within 1 mile of the outer coast of Isla Madre de Dios (about latitude 50°20′S) and on the western end of Isla Desolación (about latitude 52°50′S). Extensive forests apparently containing large trees were seen from the ship on the exposed outer coasts of these islands.

High-elevation forests. On some of the more rugged islands one can often see fairly extensive patches of deciduous forest at elevations of ⩽1000 meters. These forests are often in the most precipitous areas, and getting to them is a major undertaking. The few high-elevation forests that I was able to reach were small and consisted of shrubby growth up to about 2 meters in height, mainly *Nothofagus antarctica* (Forst.) Oerst. However, other high-altitude forests appeared from a distance to contain deciduous trees of considerable size. Some high forests may contain *N. pumilio*, a species that is rare or absent in lowland coastal areas in the rain forest region. It is remarkable that large forest trees can live at high altitudes in these latitudes, and a detailed investigation of the high-forest flora might be rewarding.

Character of the rain forest. The character of the Magellanic rain forest differs radically from that of most subarctic forests in similar latitudes in the northern hemisphere. The northern forests consist mainly of conifers, whereas the rain forest consists mainly of the evergreen beech *Nothofagus betuloides.* In the

Fig. 8. Section through sphagnum moorland near Bahía San Nicolás, Península de Brunswick (53°51′S, 71°07′W). A heavy build-up of peat (approximately 3 meters in this view) blankets large areas in the transitional forest zone. As a result extensive rolling moorlands with only scattered tree growth are common. This phenomenon is not seen in the rain forest region, where *Sphagnum* is rare or absent.

Fig. 9. Dwarf rain forest and 'moorland' on the exposed western end of Isla Desolación (Bahía Tuesday, 52°51′S, 74°31′W). This forest is low and interspersed with herbaceous vegetation but maintains the same species composition as a normal rain forest. The main tree species are *Drimys winteri* (large leaves), *Nothofagus betuloides* (small leaves), and *N. antarctica* (leafless).

Fig. 10. High moorland near Bahía Tuesday (elevation of about 500 meters). *Nothofagus antarctica* (pictured) is the main tree species in these situations. Even on the exposed outer coasts this species may reach an altitude of 1000 meters. It is often reduced to a twisted creeping shrub. The dominant herbaceous species in the photograph is *Marsippospermum grandiflorum.* Small shallow lakes such as the one shown often support a dense growth of *Isoetes savatieri* Franch., a remarkable plant apparently endemic to the southern part of the Magellanic forest region.

northern forest all growth must take place within a period of some 3 months; the forest is frozen and dormant during the rest of the year. In the rain forest, as we have seen, there is no period of complete dormancy. In the northern forest the undergrowth consists mainly of tundra species, and epiphytes are rare. There is little floristic difference between the northern forests and the adjacent tundra areas. In the Magellanic forest many species, particularly the epiphytes, are closely associated with the forest and do not occur in treeless areas.

A striking feature of the Magellanic rain forest is the almost complete lack of mammalian life. Of the comparatively few species of mammals native to southern South America [Darlington, 1965], only the otter ranges widely in the rain forest. Land birds are comparatively rare in both numbers of species and numbers of individuals. Flying insects are seldom abundant; the hordes of mosquitoes found in the subarctic forests have no counterpart in the rain forest.

In spite of the lush growth of the rain forest, it is floristically depauperate. As yet no final tabulation of the number of vascular species found in the rain forest region has been made, but the total flora probably includes <250 species. This number is strikingly low for a temperate continental area. No more than 60–70 of these species are of broad distribution and major ecologic importance. The only group that seems to have undergone a major radiation in the vicinity of the Magellanic coastal region is the Hymenophyllaceae [Diem and de Lichtenstein, 1959]. Approximately 15 species of this group are represented in the rain forest flora, mostly as epiphytes.

The Magellanic forest differs from the Valdivian forest to the north mainly in the loss of the Valdivian species. Most Magellanic species also occur in the Valdivian region, but the majority of the Valdivian species reach the southern limit of their range in the vicinity of the north shore of the Golfo de Penas. A few species of essentially Valdivian distribution extend their range into the northern portion of the Magellanic forest, e.g., *Hymenophyllum cruentum* Cav., *Podocarpus nubigena* Lindl., *Lomatia ferruginea* (Cav.) R. Br., *Weinmannia trichosperma* Cav., and *Campsidium valdivianum* (Phil.) Skottsb.

Three arborescent species make up essentially the

entire overstory of the Magellanic rain forest. *Nothofagus betuloides* is by far the most important of these species and forms extensive pure stands. *Pilgerodendron uvifera* (D. Don.) Florin and *Drimys winteri* Forst. are important locally, especially along the forest borders. *Nothofagus antarctica* occurs mainly as a small tree or a shrub in areas of alpine vegetation. Important understory trees include *Maytenus magellanica* (Lam.) Hook. f., *Tepaulia stipularis* (Hook. ex Arn.) Griseb., *Pseudopanax laetevirens* (Gay) Seem., and *Desfontainea spinosa* Ruiz et Pav. The role of each of the preceding species is treated in detail in the next section.

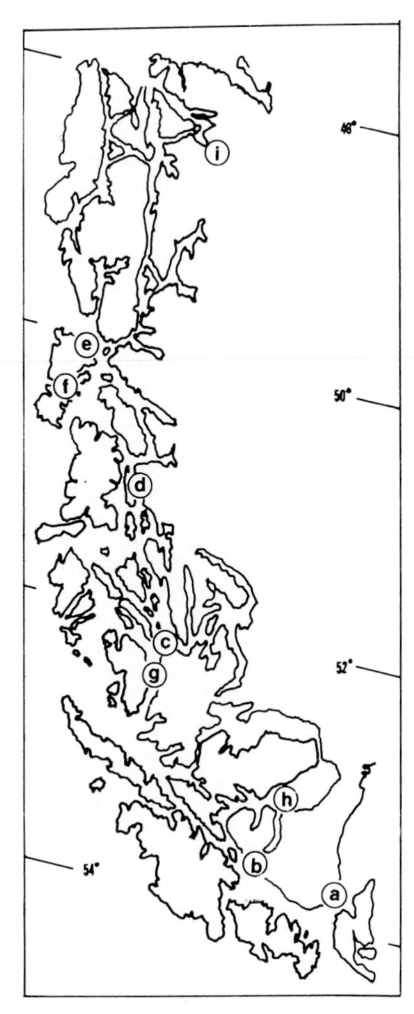

Fig. 11. Collection locations of the wood samples discussed in the text: *a*, Bahía San Nicolás; *b*, Bahía Borja; *c*, Bahía Isthmus; *d*, Puerto Bueno; *e*, Estero Henderson; *f*, Isla Madre de Dios; *g*, Islas Otter; and *h*, Isla Vivian.

Fig. 12. Two specimens of *Pilgerodendron uvifera* on the edge of a raised bog in typical sphagnum moorland in the transitional forest zone. (Bahía san Nicolás, Península de Brunswick, 53°51′S, 71°07′W). The smaller tree (Young 1 in Table 1) was 111 years old, and the larger (Young 2) was 97 years old. Johannes Bielecki, engineer of R/V *Hero*, gives the scale.

Important undergrowth species are *Philesia magellanica*, *Berberis ilicifolia* Forst., *Escallonia serrata* Sm., *Pernettya pumila* (L. f.) Hook., and *Gaultheria serphyllifolia* (Lam.) Skottsb.

AGE AND GROWTH RATE STUDIES OF COMMON FOREST TREES

During RV *Hero* cruise 69-5 a group of wood samples was collected from a series of stations in the Magellanic rain forest and in some transitional forest areas. Most of the important arborescent species were sampled at least once. The samples were collected with a hand bow saw, the sampling of the largest trees thus being precluded. No specimens of more than about 25 cm in diameter could be sampled. The trees were severed as close to ground level as possible, and transverse sections of the trunk were cut to a convenient thickness.

On their arrival at the laboratory the sections were sanded smooth and finished with damar. In a few

Fig. 13. *Pilgerodendron uvifera*, supporting dense growth of mosses and hepatics as epiphytes. This growth sometimes nearly covers the foliage of the slow growing trees (Fiordo Témpano, 48°43'S, 74°05'W).

cases in which the growth rings were extremely narrow, it was found that they could be observed best on the surface of a fresh razor cut. The rings were counted with the aid of a stereoscopic dissecting microscope. The locations of the stations from which wood samples were collected are shown in Figure 11.

Pilgerodendron uvifera (D. Don.) Florin. Pilgerodendron is the most important arborescent conifer in the Magellanic region. It occurs most commonly along the interface between the dense forest and the alpine moorland. Scattered individuals, often of large size, are found within the forest. Individuals and small groves of *Pilgerondendron* are common on alpine moorland. In the transitional forest region *Pilgerodendron* is confined to sphagnum moorlands (Figure 12). There it seldom reaches a height of >5 meters. In the Magellanic forest region *Pilgerodendron* is a lowland species. Few specimens were seen at an elevation of >100 meters.

In woodlands *Pilgerodendron* commonly reaches a height of 10–15 meters and a trunk diameter at breast height of 50 cm. The largest individuals noted were >20 meters tall and had trunks as much as 1 meter in diameter. In exposed areas *Pilgerodendron* may be shrubby, and specimens were found that were almost entirely overgrown with mosses and hepatics (Figure 13).

Pilgerodendron wood is soft, fine grained, and aromatic. The broad sapwood is yellow, and the heartwood reddish brown. The growth rings are distinct and evenly spaced. The bark is thin and reddish brown and consists of thin exfoliating plates. The results of the age and growth rate studies of *Pilgerodendron* are given in Table 1.

In the samples of larger specimens of *Pilgerodendron* growth appears to have begun slowly, continued

TABLE 1. Ages and Growth Rates of Selected Specimens of *Pilgerodendron uvifera*

Sample	Location	Habitat	Height, meters	Diameter, mm	Age, years	Mean Annual Increase in Diameter, mm
Young 1	Bahía San Nicolás	sphagnum moor	1.6	47	111	0.42
Young 2	Bahía San Nicolás	sphagnum moor	2.0	58 (61 × 56)	97	0.60
Young 5	Bahía Borja	forest edge	3.1	145	247	0.59
Young 6	Bahía Borja	forest edge	3.9	218	308	0.71
Young 9	Bahía Isthmus	forest edge	2.3	82 (76 × 88)	65	1.26
Young 12	Puerto Bueno	forest edge	3.8	133 (136 × 130)	225	0.58
Young 13	Puerto Bueno	alpine moor	1.1	83 (88 × 78)	69	1.20
Mercer*	Fiordo Témpano	alpine moor		190 (150 × 230)	263	0.72
Mercer	Fiordo Témpano	alpine moor		170 (150 × 190)	256	0.43
Mercer	Fiordo Témpano	alpine moor		140	186	0.75
Mercer	Fiordo Témpano	alpine moor		80	108	0.74
Mercer	Fiordo Témpano	alpine moor		90	100	0.90

The mean annual increase in diameter for all samples was 0.74 mm.
* Collected by Dr. John Mercer in January 1968 near the snout of a sea level glacier.

TABLE 2. Ages and Growth Rates of Selected Specimens of *Drimys winteri*

Sample	Location	Habitat	Height, meters	Diameter, mm	Age, years	Mean Annual Increase in Diameter, mm
Young 11	Bahía Isthmus	forest edge	8.0	145	74	1.95
Young 17	Puerto Bueno	rain forest	10.0	212	90	2.36
Young 21	Estero Henderson	rain forest	6.5	94 (85 × 103)	40*	2.45
Young 31	Islas Otter	forest edge	5.0	119	57*	2.09

The annual increase in diameter for all samples was 2.21 mm.
* Approximate.

slowly for 20–100 years, speeded up considerably for a period of years, and finally slowed down again. The apparent temporary increase in the growth rate is explained by the development of a thickened portion at the base of the trunks of vigorous young trees.

An increment in trunk diameter of 0.70–0.80 mm/yr appears to be normal for *Pilgerodendron* over a period of time. The only samples that showed an average increment of >1.0 mm/yr were from two young trees in which the thickening at the base was already marked. One of these samples was from a tree only about 1 meter tall. Although the trunk diameter was comparatively large, the actual increment of wood per year was not great.

The slow growth of *Pilgerodendron* is apparently not correlated with competition from other trees. A small seedling about 50 cm tall and 15 mm in diameter at the base that was growing in open sphagnum moorland was at least 21 years old. In spite of its small size this specimen bore numerous mature cones.

On the assumption that the figures given here are representative of the growth rates of larger trees, average sized specimens of *Pilgerodendron* having trunk diameters of about 50 cm would be at least 500 years old, and the largest trees would be at least 1000 years old. There is no evidence that dwarfed specimens from exposed areas are of exceptional age.

Along the coast of the western portion of the Strait of Magellan there are often signs of fairly intensive cutting of *Pilgerodendron*. The cutting is done in a more or less haphazard fashion, usually with axes. It is not known what the wood is used for. Since *Pilgerodendron* is seldom abundant and grows at such a slow rate, stands of this species can obviously withstand only the most limited pressure from lumbering. There is virtually no indication of use of *Pilgerodendron* wood from the rain forest north of the Strait of Magellan. A few Indian boats and huts made of *Pilgerodendron* were seen, but this use can hardly affect the stands.

Drimys winteri Forst. *Drimys winteri* is an important forest tree in both the Valdivian and the Magellanic forests. It ranges from the warm temperate coastal regions of central Chile south to the Islas Hermite. Individual specimens of *Drimys* are scattered through the dense stands of *Nothofagus betuloides*. *Drimys* is particularly abundant along the forest–moorland border, and shrubby specimens occur commonly on the better drained areas of alpine moorland (Figure 11). Although it is found throughout the rain forest, *Drimys* seems to grow best in areas of good drainage or reduced rainfall. The species forms nearly pure stands on some low sandy islands a few miles north of the Strait of Magellan; pure stands are also found in mesic situations in the western portion of the transitional forest. *Drimys* is normally confined to the lower elevations in the Magellanic rain forest region. Its range does not extend to the eastern border of the transitional forest. In the forest *Drimys* commonly reaches a height of 20–30 meters or more. The largest trees may have trunks of ⩽1 meter in diameter, but trunk diameters of 50–75 cm are more common.

Drimys wood is soft, white or pale yellow, and highly aromatic. In healthy trees of the sizes sampled there is no heartwood. The bark is smooth, light gray, and usually <1 cm thick. The growth rings in the wood are evenly spaced and complete but rather indistinct. Several attempts at sampling were abortive because of heart rot. The results of the age and growth rate studies on *Drimys* are given in Table 2.

Drimys appears to be the fastest growing and shortest-lived of the large trees in the Magellanic rain forest. Although the average increment in trunk diameter in the specimens studied was only slightly more than 2.2 mm/yr, many specimens showed growth rates of ⩽5.0 mm/yr for periods of several years. The

TABLE 3. Ages and Growth Rates of Selected Specimens of *Nothofagus betuloides*

Sample	Location	Habitat	Height, meters	Diameter, mm	Age, years	Mean Annual Increase in Diameter, mm
Young 7	Bahía Borja	sphagnum moor	2.0	85	72	1.18
Young 8	Bahía Borja	mesic forest	10.0	211	78	2.70
Young 10	Bahía Isthmus	rain forest	3.8	115 (108 × 121)	62	1.85
Young 15	Puerto Bueno	forest edge	5.1	165	104	1.59
Young 22	Estero Henderson	rain forest	7.0*	125 (118 × 132)	138	0.90
Young 27	Isla Madre de Dios	rain forest	12.0*	185	125	1.47
Young 28	Islas Otter	rain forest	10.0	205*	187	1.10
Young 29	Islas Otter	forest edge	4.9	104 (100 × 108)	143	0.73
Young 30	Islas Otter	forest edge	7.0	157 (145 × 170)	142	1.10

The mean annual increase in diameter for all samples was 1.42 mm.
* Approximate.

largest specimens of *Drimys* are probably 200–400 years old.

There was no indication that *Drimys* timber is ever harvested in the rain forest region.

Nothofagus betuloides (Mirb.) Blume. This species is by far the most important element in both the rain forest and the transitional forest. In both forests it forms extensive pure stands. North of the Golfo de Penas *N. betuloides* is of scattered occurrence and little importance. In the rain forest, specimens of *N. betuloides* are normally gnarled and spreading and have a thick crown of large horizontal branches. The diameter of the trunk above the swell of the base often exceeds 1 meter. Trunks of twice this diameter are common in some places.

Extensive stands of *N. betuloides* occur at elevations of ≤500 meters in exposed areas like the western end of Isla Desolación. In other areas evergreen forests were seen at elevations of ≥1000 meters. In the transitional forest, *N. betuloides* grows tall and straight. Specimens ≥50 meters tall and having trunk diameters of 2 meters were seen near the eastern border of the transitional forest (Figure 6). The species extends its range into the pampas region on hilltops and in south-facing hillsides and ravines.

The wood of *N. betuloides* is close grained and hard. The sapwood is nearly white and varies in thickness. The heartwood is dark brown and often discolored by blackish streaks. The growth rings are usually fairly clear but are often discontinuous or highly contorted. Accurate counts are difficult to make on transverse sections and would be nearly impossible on cores obtained by boring. I believe that the counts given here are accurate to ±5% of the total age. Many samples were rejected because of heart rot. These samples were mostly from small specimens growing in moorland. Growth in these specimens was usually much slower than that in the specimens included in Table 3.

The growth rates in *N. betuloides* appear to be less uniform than those in other species studied, but all specimens from the rain forest area increased their trunk diameters at an average rate of <2.00 mm/yr. If these figures are indicative of the average growth rates of large specimens in the rain forest, old trees may be 1000–1500 years old or more. Growth is apparently much faster in the transitional forest. The single specimen from the western border of the transitional forest (Young 8) grew at nearly twice the mean rate.

In the transitional forest region, *N. betuloides* is used extensively for lumber, fence posts, and so forth. It is the only important timber tree of the region. At the present rate of cutting, both for timber and for increasing grazing land, it is doubtful that in a few years many virgin stands of *N. betuloides* will be left in the eastern portion of the transitional forest. There is no indication of timber cutting in the rain forest, and the gnarled trees would be of little value.

Understory trees and minor elements. Maytenus magellanica (Lam.) Hook. f. is normally a large shrub or small tree; exceptional specimens of *Maytenus* growing in the dense rain forest may reach a height of 20 meters and a trunk diameter of ≥50 cm. This species is of shrubby growth on alpine moorland. *Maytenus* is common in the rain forest but rare in the transitional forest. The wood is hard and pale brown, and there is no clear distinction between sapwood and heartwood. The figures given in Table 4 indicate that *May-*

TABLE 4. Ages and Growth Rates of Understory Species and Minor Elements of the Magellanic Forest

Sample	Location	Habitat	Height, meters	mm Diameter,	Age, years	Mean Annual Increase in Diameter, mm
Maytenus magellanica						
Young 14	Puerto Bueno	forest edge	3.5	100	101	1.00
Young 20	Estero Henderson	rain forest	6.5	103 (93 × 113)	190*	0.53
Mean for all samples						0.77
Tepaulia stipularis, Young 23	Estero Henderson	rain forest	5.0	95*	136*	0.70
Pseudopanax laetevirens						
Young 26	Isla Madre de Dios	rain forest	10.0*	180	68*	2.65
Young 33	Isla Vivian	dry forest	6.5	106	38*	2.79
Mean for all samples						2.72
Podocarpus nubigena						
Young 18	Estero Henderson	rain forest	8.5*	125	164	0.76
Young 19	Estero Henderson	rain forest	9.0*	120 (113 × 127)	196	0.61
Mean for all samples						0.69

* Approximate.

tenus grows slowly, particularly in the rain forest. The largest specimens are certainly several hundred years old.

Tepaulia stipularis (Hook. et Arn.) Griseb. is normally a shrub; it is common along shore lines and forest borders in the Valdivian and Magellanic rain forests but apparently does not range far south of the Strait of Magellan [Skottsberg, 1916]. It barely reaches the western border of the transitional forest. *Tepaulia* occasionally reaches tree stature in the rain forest. The wood is dark red, hard, and heavy. Even when it is thoroughly dry, it will not float in fresh water. There is no clear demarcation between heartwood and sapwood. Accurate ring counts are difficult to obtain. All specimens appeared to have a mean annual growth rate of considerably less than 1.0 mm/yr. Old specimens would therefore be several hundred years old.

Pseudopanax laetevirens (Gay) Seem. is common throughout both the rain forest and the transitional forest. It is normally a small loosely spreading tree, but large specimens may reach a height of >20 meters and a trunk diameter of ⩾50 cm. The wood is white with prominent rays; it is light and fairly hard. Older specimens show a prominent greenish-brown heartwood. Growth rings are indistinct, and the ages given in Table 4 are approximate. The trees grow rapidly, and specimens of >100 years old are probably rare.

Desfontainia spinosa Ruiz et Pav. is a loosely spreading shrub or small tree common in the understory of dense forest. It ranges through the Valdivian forest and through the Magellanic forest at least as far south as the southern shores of the Strait of Magellan. The largest specimens seen were <10 meters tall and had stems of ⩽10 cm in diameter. The wood is white and has narrow growth rings that could not be counted accurately. A specimen having a trunk diameter of 75 mm appeared to be >100 years old.

Podocarpus nubigena Lindl. is a Valdivian species ranging into the northern part of the Magellanic forest. Specimens were seen as far south as Puerto Bueno (latitude 51°S). In the northern part of the Magellanic region, *Podocarpus* is a fairly common medium sized forest tree. Specimens ⩽20 meters tall having trunk diameters of ⩽75 cm were seen. Specimens growing in open areas have a densely pyramidal 'Christmas tree' shape. The wood is yellowish brown and has distinct growth rings, and there was no heartwood in the specimens sampled. The bark is thin and exfoliates in flat plates. The specimens sampled showed a slow growth rate; thus large trees growing in dense forest may reach an age of 500–1000 years.

No useful samples of either *Nothofagus antarctica* or *N. pumilio* were obtained.

DISCUSSION

Most forest trees of the Magellanic rain forest clearly grow at an exceedingly slow rate. There can be little doubt that the rain forest contains many specimens of *Nothofagus betuloides* well over 1000 years old, and there are probably specimens ⩾2000 years old. That the slow growth rate is a function of genetics as well as of the cold wet environment is indicated by the comparatively rapid growth rate of some forest elements, notably *Drimys winteri* and *Pseudopanax laetevirens.*

The great age of some Magellanic trees has implications for the study of glacial fluctuations in southern Chile. Studies of the ages of forest trees growing near the glacial borders and on old trim lines should provide information on the glacial regime of the area for a period of ⩾1000 years. It is unfortunate that the growth rings in *Nothofagus betuloides* are so indistinct and irregular. To accumulate the data that trees of this species are capable of providing, it would be necessary to fell large trees by means of power equipment. Although *Pilgerodendron* is neither as abundant nor apparently as long-lived as *N. betuloides,* it would lend itself to age studies by means of borings.

The regeneration of the Magellanic forest in areas destroyed by glacial advance or other causes is clearly a long-term process. Fortunately at least at present there is little or no economic value in timber from the Magellanic rain forest. Nevertheless the regulation of the cutting of *Pilgerodendron* should probably be considered in the future. Intensive harvesting has already taken place over large areas in the eastern part of the transitional forest. Extensive stands of large virgin timber in this region will probably become rare before many more years have passed.

Acknowledgments. The author's work in southern Chile was sponsored by National Science Foundation grant GA-13977. The work is entitled 'Systematics and Phytogeography of Vascular Plants of the Antarctic and Islands in the Southern Ocean,' and was logistically supported through the U.S. Antarctic Research Program. Reduction of data is being carried on at the Institute of Polar Studies, Ohio State University. Among the many people who have been of help during the study I wish particularly to thank Dr. Henry A. Imshaug of the Department of Botany and Plant Pathology, Michigan State University. Dr. Imshaug's knowledge of the Magellanic flora and the literature pertaining to it has always been freely shared, and it has been of great help to me. Dr. John H. Mercer of the Institute of Polar Studies supplied several wood samples from the northern portion of the Magellanic forest region. Contribution 215 of the Institute of Polar Studies, Ohio State University, Columbus, Ohio.

REFERENCES

Almeyda A., E., and F. Saez S.
1958 Recopilación de datos climáticos de Chile y mapas sinopticos respectivos. Min. de Agr., Santiago.

Auer, V.
1960 The quaternary history of Fuego-Patagonia. Proc. R. Soc., ser. B, 152: 507–516.

Darlington, P. J.
1965 Biogeography of the southern end of the world. 236 pp. Harvard University Press, Cambridge, Mass.

Diem, J., and J. S. de Lichtenstein
1959 Las Himenophylláceas del área argentino-chilena del sud. Darwiniana, *11:* 611–760.

Flint, R. F.
1957 Glacial and Pleistocene geology. 550 pp. John Wiley, New York.

Godley, E. J.
1963 The Royal Society expedition to southern Chile: Botanist's itinerary. N. Z. J. Bot., *1:* 316–324.
1965 Contributions to the plant geography of southern Chile. Revta Univ., Santiago, *48:* 31–39.

Halpern, M.
1970 Hero cruise 69-6. Antarct. J. U.S., *5:* 44.

Holdgate, M. W.
1961 Vegetation and soils in the South Chilean Islands. J. Ecol., *49:* 559–580.

Imshaug, H. A.
1970 Hero cruise 69-5. Antarct. J. U.S., *5:* 41–42.

Kaesler, R. L.
1970. Hero cruise 69-5. Antarct. J. U.S., *5:* 43–44.

Mercer, J. H.
1965 Glacier variations in southern Patagonia. Geogrl Rev., *55*(3): 390–413.

Skottsberg, C.
1910 Übersicht über die wichtigsten Pflanzenformationen Südamerikas s. von 41°, ihre geographische Verbreitung und Beziehungen zum Klima. 1. Botanische Ergebnisse der Swedischen Expedition nach Patagonien und dem Feuerland 1907–1909. K. Svenska Vetensk-Akad. Handl., 46: 1–26.
1916 Die Vegetationsverhältnisse längs der Cordillera de los Andes s. von 41°. 5. Botanische Ergebnisse der Swedischen Expedition nach Patagonien und dem Feuerland 1907–1909. K. Svenska Vetensk-Akad. Handl., 56: 1–366.

van Rooy, M. P. (Ed.)
1957 Meteorology of the Antarctic. 240 pp. Weather Bur., Dep. of Transp., Pretoria.

Walter, H.
1960 Klimadiagram Weltatlas. Fischer, Jena.